Sustainable Ecological Restoration and Conservation in the Hindu Kush Himalayan Region

Sustainable Ecological Restoration and Conservation in the Hindu Kush Himalayan Region

Edited by

Zhanhuan Shang

College of Ecology, Lanzhou University, China

Allan Degen

Desert Animal Adaptations and Husbandry, Wyler Department of Dryland Agriculture, Blaustein Institutes for Desert Research, Ben-Gurion University of Negev, Israel

Devendra Gauchan

Alliance of Bioversity International and CIAT, Nepal

Madan Koirala

Central Department of Environmental Science, Tribhuvan University, Nepal

Muhammad Khalid Rafiq

Rangeland Research Institute, National Agricultural Research Center, Pakistan

CABI

CABI is a trading name of CAB International

CABI
Nosworthy Way
Wallingford
Oxfordshire OX10 8DE
UK

Tel: +44 (0)1491 832111
E-mail: info@cabi.org
Website: www.cabi.org

CABI
200 Portland Street
Boston
MA 02114
USA

Tel: +1 (617)682-9015
E-mail: cabi-nao@cabi.org

A catalogue record for this book is available from the British Library, London, UK.

ISBN-13: 9781800622555 (Hardback)
9781800622562 (ePDF)
9781800622579 (ePUB)

DOI: 10.1079/9781800622579.0000

Commissioning Editor: Ward Cooper
Editorial Assistant: Helen Elliott
Production Editor: Shankari Wilford

Typeset by Straive, Pondicherry, India
Printed and bound in the UK by CPI Group (UK) Ltd, Croydon, CR0 4YY

Contents

About the Editors

Zhanhuan Shang is a full professor at College of Ecology, Lanzhou University. He started his research career at Lanzhou University in 2006. He conducted and participated in more than 20 projects funded by NSFC, MOST, MOE, CAS, FAO, GEF/WB, EC, IAEA, ICIMOD and DFG. His research interests include ecological restoration, grassland ecology, biodiversity conservation, carbon management and sustainable development. He has more than 250 publications in international and Chinese journals and books.

Allan Degen joined Ben-Gurion University of the Negev, Beer Sheva, in 1972, where he held the Slome Chair for Desert Animal Production and became an Emeritus Professor in 2014. He does research on livestock production under harsh conditions and has co-operated in projects in Kazakhstan, Uzbekistan, Nepal, Ethiopia, Afghanistan, Kenya, South Africa and China. In Israel, he is studying rural and urban livestock and crop production among Bedouin in the Negev. He has authored/co-authored 25 books or chapters and over 300 peer-reviewed publications. He has received numerous research grants, and his research has been presented at many conferences worldwide.

Devendra Gauchan joined Bioversity International in 2014 as a National Project Manager and is currently serving as Honorary Research Fellow at the Alliance of Bioversity International and CIAT-CGIAR, Rome. He is also an Adjunct Professor at the Institute of Agriculture and Animal Sciences, Tribhuvan University, Kathmandu. Previously, he served as a Senior Scientist and Head of the Socioeconomics & Agricultural Research Policy Division at Nepal Agricultural Research Council, Kathmandu. He has over 25 years of postgraduate experience in agriculture, agrobiodiversity, food security and natural resource management, with over 100 research papers in peer-reviewed journals, proceedings and books.

Madan Koirala joined Tribhuvan University in 1981 and retired as Professor of Environmental Science in 2018. His areas of work are mountain environment, biodiversity and climate change. He has authored 15 books or chapters and over 30 journal articles; reviewed in 10 national and international journals; and was Editor-in-Chief of the *Journal of Environmental Science*, Nepal. As a climate change expert he has participated in UNFCCC COP meetings, representing Nepal. He has been involved in national-level study projects on scope and expansion of science & technology education in Nepal; HUC projects in Nepal, India, Bhutan, China and Pakistan; and EU-funded projects in Spain, UK, Portugal, Cyprus, Uzbekistan, Germany, Norway, Poland, Cyprus, Latvia, Kyrgyzstan, India and Nepal.

Muhammad Khalid Rafiq is Director/Principal Scientific Officer at Rangeland Research Institute (RRI), National Agricultural Research Centre (NARC), Islamabad, Pakistan. His research interests include grassland ecology, biochar production and applications for carbon sequestration. He has also attended conferences at national and international level in the UK, China, Nepal and Pakistan. He has published many research articles in journals, including *Journal of Hazardous Materials* and *Science of the Total Environment*, and in books.

Contributors

Editors

Allan Degen Desert Animal Adaptations and Husbandry, Wyler Department of Dryland Agriculture, Blaustein Institutes for Desert Research, Ben-Gurion University of Negev, Beer Sheva 8410500, Israel. Email: degen@bgu.ac.il

Devendra Gauchan Alliance of Bioversity International and CIAT, Khumaltar, Lalitpur, Nepal.

Madan Koirala Central Department of Environmental Science, Tribhuvan University, Kathmandu, Nepal. Email: madankoirala@gmail.com

Muhammad Khalid Rafiq Rangeland Research Institute, National Agricultural Research Center, Islamabad, Pakistan. Email: khalid_parc@hotmail.com

Zhanhuan Shang College of Ecology, Lanzhou University, Lanzhou, China. Email: shangzhh@lzu.edu.cn

Contributors

Diwakar Adhikari Central Department of Environmental Science, Institute of Science and Technology, Tribhuvan University, Nepal.

Huma Ali College of Ecology, Lanzhou University, Lanzhou, China. Email: huma2020@lzu.edu.cn

Usman Ali Rangeland Research Institute, National Agricultural Research Center, Islamabad, Pakistan.

Rukhsanda Aziz Center for Interdisciplinary Research in Basic Sciences, International Islamic University, Islamabad, Pakistan.

Zhiqiang Dang College of Ecology, Lanzhou University, Lanzhou, China.

Rashila Deshar Central Department of Environmental Science, Tribhuvan University, Kathmandu, Nepal.

Qinghui Fang College of Ecology, Lanzhou University, Lanzhou, China. Email: 120220907251@lzu.edu.cn

Narayan Prasad Gaire Department of Environmental Science, Patan Multiple Campus, Tribhuvan University, Nepal.

Yamuna Ghale Research Lead, Nepal Center for Contemporary Research (NCCR), Civil Homes, Kathmandu, Nepal.

Monika Ghimire Central Department of Environmental Science, Tribhuvan University, Kathmandu, Nepal.
Haonan Guo College of Ecology, Lanzhou University, Lanzhou, China.
Jalal Hayat Khan Rangeland Research Institute, National Agricultural Research Center, Islamabad, Pakistan.
Jiayi He State Key Laboratory of Herbage Improvement and Grassland Agro-ecosystems, College of Ecology, Lanzhou University, Lanzhou, China.
Mei Huang College of Ecology, Lanzhou University, Lanzhou, China. Email: huangm2018@lzu.edu.cn
Awais Iqbal College of Ecology, Lanzhou University, Lanzhou, China.
Xiaoping Jing College of Ecology, Lanzhou University, Lanzhou, China.
Ritika K.C. Department of Environmental Science, Amrit Campus, Tribhuvan University, Kathmandu, Nepal.
Udhab Raj Khadka Central Department of Environmental Science, Tribhuvan University, Kathmandu, Nepal. Email: ukhadka@cdes.edu.np
Srijana Khanal College of Ecology, Lanzhou University, Lanzhou, China. Email: Srijana.khanalgautam@gmail.com
Furbe Lama College of Ecology, Lanzhou University, Lanzhou, China. Email: lama2021@lzu.edu.cn
Dongmei Li College of Ecology, Lanzhou University, Lanzhou, China. Email: lidm21@lzu.edu.cn
Jie Lian Naiman Desertification Research Station, Northwest Institute of Eco-Environment and Resources, Chinese Academy of Sciences (CAS), Lanzhou, China.
Peipei Liu College of Ecology, Lanzhou University, Lanzhou, China. Email: liupp@lzu.edu.cn
Youyan Liu College of Ecology, Lanzhou University, Lanzhou, China. Email: liuyouy@lzu.edu.cn
Binyu Luo College of Ecology, Lanzhou University, Lanzhou, China. Email: luoby18@lzu.edu.cn
Ramesh Raj Pant Central Department of Environmental Science, Institute of Science and Technology, Tribhuvan University, Nepal. Email: rpant@cdes.edu.np
Zhen Peng College of Ecology, Lanzhou University, Lanzhou, China. Email: pengzh20@lzu.edu.cn
Tianyun Qi College of Ecology, Lanzhou University, Lanzhou, China. Email: qity21@lzu.edu.cn
Usha Rai Central Department of Environmental Science, Tribhuvan University, Kathmandu, Nepal.
Ramesh Prasad Sapkota Central Department of Environmental Science, Tribhuvan University, Kathmandu, Nepal. Email: sapkotaramesh@hotmail.com
Wenyin Wang College of Ecology, Lanzhou University, Lanzhou, China. Email: wangwy@lzu.edu.cn
Hui Xu School of Economics, Lanzhou University, Lanzhou, China. Email: xhhui@lzu.edu.cn
Dawei Zhang State Key Laboratory of Herbage Improvement and Grassland Agro-ecosystems, College of Ecology, Lanzhou University, Lanzhou, China.
Rui Zhang Naiman Desertification Research Station, Northwest Institute of Eco-Environment and Resources, Chinese Academy of Sciences (CAS), Lanzhou, China. Email: zhangrui@lzb.ac.cn

Preface

In *A Short Walk in the Hindu Kush*, British travel writer Eric Newby's understatement of the region's difficulties and dangers makes it difficult to imagine that the Hindu Kush Himalaya (HKH) is the highest, most harsh and rugged mountain region in the world. It is also the world's largest and poorest mountain area. The HKH region is rich in ecosystems, landscapes, ethnic groups, cultures and traditions and it is also recognized as the world's repository of biodiversity and ecosystem services – the 'pulse' of the world. However, under global change, erratic rainfall and anthropogenic activities, the region is facing difficult challenges. The region is very sensitive to climate change and the degradation of ecosystems, and the melting of glaciers and permafrost are exacerbating serious ecological and environmental problems. All eight countries in the HKH region – namely Afghanistan, Pakistan, Nepal, India, China, Bangladesh, Myanmar and Bhutan – are facing these challenges and environmental difficulties.

The landscape and environment of the HKH region have been severely damaged. The HKH countries and the UN Decade for Ecosystem Restoration (2021–2030) are facing unprecedented challenges. Because of poverty, inaccessability and poor market access in the mountain areas, local people are very dependent on immediate natural resources for their livelihood and survival, which will exert even greater pressure on the conservation and restoration of natural resources in the region. Consequently, there are urgent needs for policies and funding for conservation and restoration of the ecological systems and functions to protect livelihoods and improve food security in the region.

With these issues in mind, we organized this book: (i) to provide a systematic account of the past and present ecological problems of the HKH region; (ii) to describe the threats and challenges faced by the local communities and people; and (iii) to propose solutions to adapt to or mitigate the problems. The book is particularly concerned with the development of transformative adaptations and sustainable management for ecological conservation and restoration. It is guided by three main questions: (i) Are the conservation strategies of existing international organizations and government policies protecting ecosystems and addressing biodiversity? (ii) Are these management measures a one-time solution or are they long-term?; and (iii) Based on the history of ecological conservation and restoration, what is the strategic framework and scenario forecast for the future? We also analyze the HKH region's protection and restoration programmes and new models for climate change and globalization. We provide suggestions for sustainable and adaptive development of the HKH region through discussion of potential future scenarios, and we examine how to improve local livelihoods and food security.

Sustainable Ecological Restoration and Conservation in the Hindu Kush Himalayan Region provides a comprehensive overview of the ecological state of the HKH region, the challenges of global change for the region, and possible future ecological strategic shifts and adaptations. The protection of the HKH region – a global mountain centre with multiple political, cultural, historical and ecological dimensions – helps to safeguard the ecological environment and human well-being around the world. Taking transformative adaptation and sustainable management as a starting point, the book offers a new perspective on the issues and urgent needs of ecological conservation and restoration in the region. The chapters of the book highlight the ecological issues of the region, biodiversity conservation issues and the livelihoods of local people, providing a detailed analysis of the challenges. The links between government policies, ecology and livelihoods are explored from social, economic and policy perspectives in order to propose sustainable strategies for the protection of ecological balance and restoration and to address the livelihood development of mountain peoples.

The book brings together scholars from Biodiversity International, ICIMOD, China, Israel, Nepal and Pakistan to illustrate their accounts of biodiversity, livelihood options and food security in the HKH region. They also shed light on the culture and traditions of the HKH mountain communities, the political systems of the Third Pole countries, and the livelihoods and living conditions of mountain people. We hope that in the context of global climate change, this book will enable readers to better understand the outstanding contribution and efforts of mountain countries and local residents to ecological conservation and restoration in the HKH region and attract more people to participate in ecological conservation and development in this region.

The authors attempted to provide accurate, comprehensive and up-to-date information but, as with all books, there may be some errors due to the limitations of the authors' knowledge, data collection and depth of analysis. We apologize for these. Today, more and more research and survey methods are being applied to this area, so that much detailed information is being reported as precisely as possible. We hope that our book will prove to be a reference point and provide some guidance for future research.

Zhanhuan Shang, Allan Degen, Devendra Gauchan,
Madan Koirala and Muhammad Khalid Rafiq
September 2023

Acknowledgements

Many people deserve thanks for their help in getting this book published. We thank all the authors who contributed chapters and the editors at CAB International for their kind help. We thank Zhen Peng and Dongmei Li who did considerable editorial work during the chapters' organization. Some of the contents of this book were the result of the authors' research, some of which was supported by granting agencies. These grants include the Natural Science Foundation of China (31961143012, U21A20183), the Second Tibetan Plateau Scientific Expedition and Research (STEP) program (Grant no. 2019QZKK0302), the Ministry of Science and Technology Foreign Experts Program Project (G2022175008L), the Fundamental Research Funds for the Central Universities (lzujbky-2021-ct10), the '111' Programme 2.0 (BP0719040) and the US Agency for International Development (USAID) and US-Israel AID-CDR-CAR programmes.

1 Environmental Problems in the History and Development of the Hindu Kush Himalayan Region

Zhanhuan Shang[1,*], Allan Degen[2], Zhiqiang Dang[1], Zhen Peng[1] and Devendra Gauchan[3]

[1]*College of Ecology, Lanzhou University, Lanzhou, China;* [2]*Institutes for Desert Research, Ben-Gurion University of the Negev, Israel;* [3]*Alliance of Bioversity International and CIAT, Lalitpur, Nepal*

Abstract

The Hindu Kush Himalayan (HKH) people have a long history that has shaped their development into today's societies and has contributed to their unique ethnic cultural traditions. However, it is not accidental that the HKH countries face common environmental problems, as they share the same geographical and climatic conditions. In this chapter, we describe the connection between the historical environmental problems during the development history and the current cultural, social and economic statuses of the ethnic peoples in the HKH region. The evolution, driving factors and the prospect of solutions to the environmental problems are discussed.

Key words: environmental problems, economic development, HKH region, climate change, globalization

1.1 Introduction

The Hindu Kush Himalayan (HKH) region, consisting of alpine, subtropical forest and desert ecosystems, is characterized by the world's highest mountain peaks, and contains rich resources that support the livelihoods of more than 240 million mountain-dwellers and 1.65 billion people downstream. Approximately 1.9 billion people live in the ten major river basins of this vast region, and about 3 billion people, just under half the global population, depend on the food produced in these river basins (Biswas *et al.*, 2016). The region has a long history in which the unique natural environment shaped the activities of the local people, their cultures and their traditions.

Even though it is rich in plant biodiversity, water and energy resources and ecosystem services, the HKH region faces multiple challenges, including land use, land cover, climate changes and invasive species infestation, leading to species loss and ecosystem degradation (Wester *et al.*, 2019). Natural disasters and frequent local wars and conflicts have caused great ecological environmental damage, while the damage due to climate change has been substantial and is aggravating the current situation of the local people. The mountains have become a limiting factor for livelihood diversification and are the

*Corresponding author: shangzhh@lzu.edu.cn

DOI: 10.1079/9781800622579.0001

home to more than 40% of the world's poor. Poverty has always accompanied the development of human society in the HKH region, and most people do not have the resources to leave the mountains. The situation has forced the local residents to exploit the natural resources to maintain basic livelihoods, but this approach is not sustainable, as it is at the expense of the depletion of grasslands, forests and soils. Furthermore, poor health and living conditions of the residents have led to high mortality rates. Nonetheless, the population is growing, but there are insufficient livelihood resources to support this increase (Poudel and Shaw, 2015). The local people have strong cultural and religious traditions to protect nature and wildlife, which may also be a factor in their poverty.

It is expected that climate change will have a serious negative impact on the local people in the HKH region, as the current ecological services will worsen, and the interaction between climate change and poverty will make the sustainable development of the region more difficult. Land degradation, forest reduction, wildlife loss, water pollution, pest outbreaks and carbon emissions are becoming more prominent, and are expected to increase with climate warming.

Soil degradation and forest depletion are the most serious environmental problems in the region. Although these problems are greatly influenced by the mountainous terrain and geography, there are also profound historical reasons. This chapter focuses on the ecological and environmental problems in the HKH region, and the strategies and solutions to mitigate them from the perspectives of human activities and environmental history. In addition, scenarios are presented for a sustainable future for the HKH region based on global and regional developments. We hope that through this chapter readers will have an overall understanding of the history and expected future ecological and environmental changes in the HKH region.

1.2 Main Environmental Problems of the Hindu Kush Himalayan Region

The history of human social development is linked with the natural resources, which support the survival and development of people in the HKH region (Sharma *et al.*, 2006). Approximately 39% of the HKH region is grassland, 33% is protected area, 21% is forest, and 5% is farmland. The continuous increase in population is not accompanied by improvements in social development but by social and environmental problems, such as a reduction in the human development index (HDI), gender inequality and adult illiteracy (Poudel and Shaw, 2015).

Asian countries have lost nearly 70% of their forest cover in the last 8000 years and 95% of their primary forests (Malcolm and Hunter, 1999). This has had a large impact on people relying on forests for their traditional livelihood in the HKH region (Fig. 1.1), with forest loss as high as 90% in some areas (ICIMOD, 2012). The causes of such large-scale forest depletion are diverse, including legal and illegal timber harvesting (Fig. 1.2), expansion of government-organized and voluntary agricultural activities, war, fire, infrastructure construction, such as roads and dams, and crop rotation. The situation has not been mitigated in the past 30 years, although this has been a period of rapid development in the global developing countries (Karki *et al.*, 2012).

In the HKH region, as in other forested areas, energy consumption is dominated by deforestation. However, in grassland supporting grazing livestock, dung is an important source of heat. Today, there are many areas that still retain this traditional way of energy use, as fossil energy is expensive and insufficient. Hydropower could be a solution as a source of power, but the construction of such stations in backward areas is problematic. Solar energy is another solution, and is becoming more popular on a small scale. With population growth and a concomitant need for more fuel, the areas of forest are declining. In south and south-east Asian countries alone, the forest area decreased from 323.2 to 283.1 million ha between 1990 and 2005, with an annual loss of 2.58 million ha between 1990 and 2000 and 2.85 million ha between 2000 and 2005 (FAO, https://www.fao.org/forest-resources-assessment/zh/, accessed 3 April 2024). The development of community forestry in south-east Asia, which is aimed at improving the management of forests by local residents, has a long history. However, in reality, the role of community forestry is relatively limited, mainly because there is no

Fig. 1.1. Women bringing fodder leaves for livestock from forests in mid-hills of Nepal. (Photo: Allan Degen)

suitable livelihood to replace this traditional livelihood based on forests. Due to the geographical restrictions of the mountains and cultural isolation, a single resource is difficult to replace.

The most important environmental change in the HKH region is the melting of glaciers and the formation of glacial lakes, especially in the past 30 years (Fig. 1.3) (Li *et al.*, 2022). On the one hand, the glacial lakes provide a large amount of water resources, but on the other hand, they also cause natural disasters, such as the collapse of glacial lakes and flooding. Glacial lakes inundate valleys, grassland, farmland and residential areas, causing the residents to leave and use other areas for crop production. Glacial lake collapses have increased by 26% in the HKH region and by 7% in northern Pakistan since 2001 (Ashraf *et al.*, 2021).

Flood, landslide and fire disasters, particularly on the southern slopes of the Himalayas, have caused great damage in the HKH countries (Rusk *et al.*, 2022). Floods are the most common hazard in mountainous areas and often occur in conjunction with landslides. In some mountainous areas, deforestation causes landslides and flooding. Local people convert forest land to cropland by cutting down trees, and the method of 'shifting cultivation' in the forest can cause these. Some fires are caused by people, as traditional farming and hunting historically burn forests to clear the land.

Fig. 1.2. Wood collected in the Lower Mustang, Nepal. (Photo: Allan Degen)

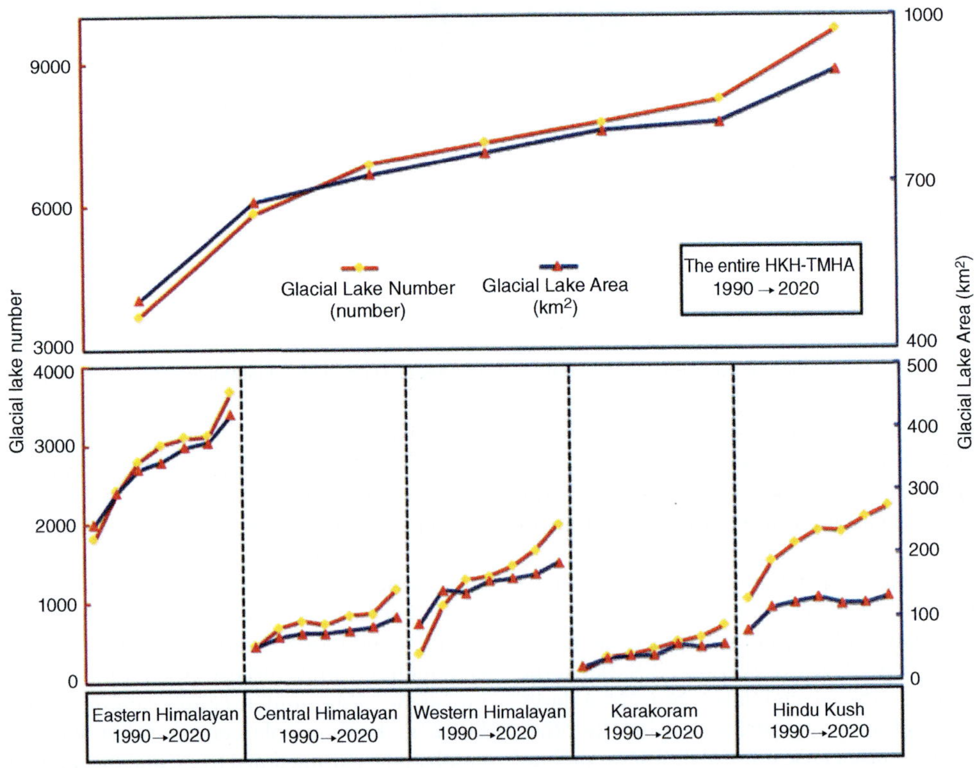

Fig. 1.3. Changes in the number and area of glacial lakes in five mountain ranges in the Hindu Kush Himalayan region from 1990 to 2020. HKH-TMHA: Himalayan, Karakoram, and Hindu Kush. (From Li *et al.*, 2022).

Soil loss is the most common long-term cause of the decline in soil fertility in the HKH region, and has always been an important environmental problem in the development of human society. The main cause of soil loss is excessive or poor farming practices with 60–80% of soil loss due to anthropogenic activities that expose bare soil to the erosive forces of water and wind. In the HKH region, forest agriculture has caused substantial soil loss. The HKH region has suffered a large amount of soil erosion since people inhabited the area and is continuing to do so today (McNeill, 2000). Currently, the use of industrial fertilizers has reached its peak worldwide, enabling food production to maintain a high level in spite of soil loss. However, this has not occurred in the HKH region, indicating that the advantages of the industrial society have not been transferred to the HKH region, which may also be a reason why the residents in the HKH region remain poor. Livestock manure is commonly used as fertilizer in the HKH countries (Fig. 1.4). The problem of poor mountain terrain is likely to deepen the problem of poverty and backwardness in the HKH region and persist in the history of environmental problems. Furthermore, the acquisition of modern tools has made it easier for people to cut down trees, which has exacerbated the problem of soil loss.

Fig. 1.4. Women transporting livestock manure to fertilize the fields in mid-hills Nepal. (Photo: Allan Degen)

1.3 Common and Different Environmental Problems among Countries of the Hindu Kush Himalayan Region

1.3.1 Land-use change

The driving forces of land-use change are mainly due to meteorological, economic and policy reasons, with temperature change as the strongest explanatory power. Analyses of land-use change and its driving forces are essential to formulate land management plans and ecological environment governance and restoration measures in the HKH region. In 2020, the Himalayan region consisted of 35% forest land, 19% grassland and 18% bare land, while bodies of water, wetlands and construction land accounted for less than 1%. From 1992 to 2020, the areas of shrubland, construction land, wetlands, and glaciers and snow increased, while the areas of woodland, grassland, cultivated land, bodies of water and bare land decreased. Close to 19% of land usage types have been altered, with the main changes from woodland to scrubland, scrubland to grassland, woodland to cultivated land, grassland to bare land, and glacial snow to bare land. From the perspective of landscape changes, ecological landscape diversity and fragmentation increased, and connectivity decreased, while the overall ecological environment continued to deteriorate. Deterioration was particularly noticeable at the junction of the north and south slopes of the Himalayan region, areas that exhibited large elevation fluctuations.

From 1992 to 2020, the changes in shrubland, construction land and wetland in the HKH region were substantial, and the type of changes differed among countries. Forests in Pakistan exhibited an increasing trend, with an increase of 8.38%, mainly at the expense of glacier snow on the north slope of the Himalayas. In China, under the influence of land and forest policies and regulations, forests only decreased by 2.01%. India was affected by the conversion of large areas of forests into cultivated land in the Kashmir Valley, decreasing forests by 0.91%. Shrubland in all countries displayed an increasing trend, mainly at the expense of forests and grassland, with the fastest increase in Bhutan.

In China, due to the combination of climate change and human disturbance, grassland has been seriously degraded, which not only reduces biodiversity and natural resources but also reduces the ecosystem services. Scrubland, construction land and wetland have changed greatly, with the change in construction land reaching 246% from 1992 to 2020. However, the area of construction land is very small – 37.3 km^2 in 2020 – accounting for only 0.02% of the land in the HKH region. The grassland had the largest change in area, with a decrease of 2466 km^2. Grassland is the most widely distributed land type in the Qinghai-Tibetan Plateau, at 70%, followed by unused land, such as sandy land and bare land, at 15.8%. Urban land expanded rapidly, with a growth rate of 2.34% from 1992 to 2005 and 69% from 2005 to 2015. From 1992 to 2015, 9.14% of unused land on the Qinghai-Tibetan Plateau was turned into grassland, 3.27% of shrubland and arable land into forest land, and 5.5% of glaciers into water (Peng *et al.*, 2022).

In India, the scrubland, construction land and wetland changed considerably, with the change of construction land reaching 80%, while the area of shrubland increased by 6265 km^2. In Nepal, construction land and wetland changed greatly, with the change in construction land reaching 504% because Kathmandu and other cities were expanding rapidly, whereas the area of forests decreased by 5563 km^2. In Pakistan, the greatest increase was in wetland, at 147%, while the area of glacier snow cover decreased by 3659 km^2. In Bhutan, wetland increased by 957%, while shrubland increased by 1678 km^2. Bhutan had a small wetland area of 0.69 km^2 in 1992, which increased to 7.27 km^2 in 2020, accounting for only 0.02% of the land area. This change was due mainly to the impact of seasonal events such as flooding.

In recent decades, the HKH region has undergone significant changes in land use, cropping systems and market access. Farmers in the region have gradually shifted from subsistence to high-value farming. A number of factors have contributed to this shift, including a growing awareness of the niche of high-value crops such as fruits, vegetables and spices, including ginger, turmeric and cardamom (Adhikari, 2014). Other factors, such as improved road networks, market access for previously isolated communities, growth in remittance inflows, expansion of co-operatives, increase in non-governmental organizations and targeted government activities have also contributed to the shift to high-value crops. More recently, high-value crops have been introduced into crop rotation, especially in low-altitude areas with access to markets, technology and irrigation. In China, high-value cash crops provide farmers with four times as much income as rice. Although Tibet has a vast land area, the resources that can be exploited and utilized are very limited due to the fragile ecological environment. Under the perspective of climate change and globalization, China has begun to pay attention to grassland restoration, and the management of livestock and grassland in pastoral areas.

The value of the top three high-value crops in Nepal, namely lentils, tea and cardamom, exceeds the import value of cereals and dairy products. Tea, coffee, cardamom, ginger and turmeric have become important cash crops in some mountainous areas of Nepal (Khattri, 2012). In Bhutan, traditional crops such as buckwheat, sweet potato, millet, wheat, naked barley and green beans have been replaced by potatoes. Intensively grown potato fields are heavily fertilized and hand-held traditional farm tools have been replaced by small tractors (Wangchuk and Siebert, 2013). In northern Pakistan, cash crops such as potatoes and fruits, including almonds, apricots, grapes and cherries, have been introduced gradually since the 1980s. In addition to cash crops, improved connectivity, such as the construction of the Karakoram Highway, has greatly improved trade and income in the region (Kreutzmann, 2006; Gioli *et al.*, 2014). Despite repeated crop failures, most households in the Indian Himalayas continue to grow potatoes intensively in the hope of generating and diversifying household income, as farm income is crucial. The deterioration of pasture vegetation in Afghanistan is a process that dates back hundreds of years. By 2030, drought could become the norm, leading to land degradation and desertification in Afghanistan. Approximately 80% of Afghans rely on rain-fed agriculture and grazing for their income, both of which are threatened by rising temperatures and erratic rainfall.

1.3.2 Social livelihood

While disasters cause greater economic and infrastructure losses in developed countries, they typically claim more human lives in developing countries. Economic vulnerability is particularly important in building resilience and reducing disaster risk, and is, therefore, particularly important in the HKH region. Five of the eight Himalayan countries, namely Afghanistan, Bangladesh, Bhutan, Myanmar and Nepal, are classified as least-developed countries (UN, 2017). The multi-dimensional poverty index (MPI), a measure of poverty that covers 100 developing countries, assesses poverty at the individual level and complements traditional income-based measures of poverty by identifying the deprivation each person faces in terms of education, health and standard of living (Alkire and Robles, 2017). The MPI for countries in the HKH region in 2022 is presented in Table 1.1.

As measured by the MPI, Afghanistan ranks highest in multi-faceted poverty (0.272), China ranks lowest (0.016), followed by India (0.069). While the MPI and the Human Development Index (HDI) calculate national poverty rates differently, most HKH countries score poorly on both. With the exception of China, countries in the HKH region rank highly in the MPI, indicating a high level of economic vulnerability. The HKH region is already at high risk because of natural disasters, but the impacts of climate change will further exacerbate this situation. Like other mountainous areas, the HKH region has experienced above-average warming, which has adversely affected freshwater, snow, glaciers and permafrost (Yao *et al.*, 2012), and it is widely believed that the region will be one of Earth's hotspots for future climate change. Primary-sector livelihoods, such as agriculture, are becoming increasingly uncertain and risky, and poor households have limited options to cope with inadequate resources. Mountain communities have a long history of adapting to extreme conditions; however, traditional adaptation mechanisms are often inadequate to address recent

Table 1.1. The multi-dimensional poverty index (MPI) for countries in the Hindu Kush Himalayan region in 2022. (UNDP, Bangkok Regional Hub, 2020)

Country	MPI
Afghanistan	0.272
Bangladesh	0.104
Bhutan	0.175
China	0.016
India	0.069
Myanmar	0.176
Nepal	0.074
Pakistan	0.198

socio-economic and environmental changes. Economic development in mountain areas has lagged behind lowlands, foothills and urban areas.

Although the livelihoods of the majority of the people revolve around agriculture, there is insufficient arable land in the HKH region to support the rural families in the mountains. With increasing population pressure, poverty, soil erosion and degradation, and the decline in food security and environmental protection, the management of marginal lands is receiving increasing attention. Diversifying agricultural activities into high-value commercial crops, and processing agricultural and other natural resource-based materials, while protecting soils, forests and other natural resources, are the most logical steps to improve the livelihoods of mountain people. Poverty and food insecurity are key challenges facing the HKH mountain region and a holistic approach is needed to address these challenges to achieve improvement and sustainable development. With expanding urbanization, industrialization and infrastructure, the problem of shrinking agricultural land is further exacerbated.

In recent years, horticultural crops, including fruits and vegetables, have played an important role in mountain livelihoods, although animal husbandry, especially buffalo and goats, provides more income-generating opportunities. In the HKH region, many changes have taken place in the allocation of land resources, production of cereals, food crops and horticultural crops, and livestock structure and composition, all of which are generally influenced by forests and other natural resources, input availability, marketing and the socio-economic infrastructure. The key challenges facing mountain agriculture are cropland security, soil erosion and declining soil fertility, increasing poverty and marginal farming. The HKH region has large areas of marginal land, including wasteland, shrubland and uncultivated forests, which can support agriculture and improve livelihoods in mountain areas.

The livelihoods and social developments of poor people in the HKH region are dependent on biodiversity and ecosystem services. In Nepal, mountain farmers adopted a variety of strategies in an attempt to improve food security and livelihoods. These include expanding cultivation to steep slopes (Fig. 1.5) and increasing male

Fig. 1.5. Terraced agriculture on slopes of mid-hills of Nepal. (Photo: Allan Degen)

migration to plains and urban areas. However, the net results were increased environmental degradation, land marginalization, increased work for women and, ultimately, increased poverty. In Afghanistan, as female participation has increased in supporting the household, water has become an area where the intersection of poverty and gender has made some women's livelihoods particularly vulnerable. Women are unable to pay for petrol for pumps to draw water, and community sanctions prevent women from sharing water. In India, the creation of a one-health programme created new jobs for workers in non-health occupations, thereby improving health standards and strengthening efforts to achieve universal health coverage. There are many other areas that can improve sustainable living standards through job creation and increased income. Such investments not only reduce the burden of disease but also empower women through job creation. As men move out in large numbers to seek work elsewhere, much of the workload and responsibilities for the household fall on women. A study in northern Pakistan suggests that positive gender change processes are more likely to be inter-generational and driven by increased educational opportunities for girls.

As a complementary livelihood option, the development of tourism will not only bring socio-economic benefits to the region but also potentially address broader social and socio-cultural issues. China is the first country to achieve the United Nations Millennium Development Goals (MDGs) and its ecological poverty alleviation programme has mitigated absolute poverty in the region. The Qinghai-Tibetan Plateau is a typical ecologically fragile region, which is caught in a vicious cycle between poverty and ecosystem degradation (Niu *et al.*, 2022). The livelihood capital on the Qinghai-Tibetan Plateau is increasing gradually, but it is still in a state of relative scarcity and spatial heterogeneity. To prevent and reverse the degradation of grasslands on the Qinghai-Tibetan Plateau, the Chinese government has implemented a series of ecological projects and policies, including the Natural Forest Protection Project, the project of returning farmland to forest and grassland, the policy of banning grazing, the Three North shelterbelt system project, the ecological protection and construction project of the Three River Sources in Qinghai Province, and the construction of the shelterbelt system in the middle and upper reaches of the Yangtze River (Nyima and Hopping, 2019; Xiong *et al.*, 2021).

1.3.3 Environmental risks

Although there are many excellent risk assessments, there are relatively few in practice. Characterized as a hotspot for natural disasters, the HKH region is considered a high disaster risk area in addition to having a low human development index. According to the Asian Water Development Outlook (AWDO, 2013), India is the country most prone to hydrometeorological hazards among HKH region countries, followed by China, Bangladesh, Pakistan and Nepal (indices are not available for Afghanistan, Bhutan and Myanmar). The AWDO report considers vulnerability as a function of exposure, and soft and hard coping capacities. Bangladesh is at the highest risk level, followed by Pakistan, Nepal, India and China. Bangladesh, Pakistan and Nepal also have high levels of underlying vulnerability (as measured by indicators such as poverty levels) when compared to China and India.

Countries in the HKH region are classified as highly vulnerable by many global risk assessments (Table 1.2). Climate change is expected to exacerbate existing hazards and lead to the emergence of new ones. The sensitivity component includes the environmental aspects contained in the biophysical characteristics of an area, the social factors closely related to the nature and scope of available livelihood options, and access to resources.

Details of the risk calculation approach and indicators used can be found at www.WorldRiskReport.org (accessed 23 January 2024) (licensed under CC BY 4.0).

The International Centre for Integrated Mountain Development (ICIMOD) is deeply concerned about the deteriorating air quality in Bangladesh, Myanmar, Nepal and parts of India. Recent findings revealed hazardous levels of particulate matter (PM), posing a severe risk to human health. In Kathmandu, Nepal's capital, $PM_{2.5}$ levels exceeded 205 μg/m^3 in April 2023, with PM_{10} levels reaching a staggering 430 μg/m^3. These alarming PM values highlight the urgent need for action to address this critical issue. Air

Table 1.2. Risk index in the HKH countries (2022).

Country	World Risk Index (%)	Exposure (%)	Vulnerability (%)	Susceptibility (%)	Lack of coping capacity (%)	Lack of adaptive capacity (%)	Rank
Afghanistan	4.05	0.25	65.7	55.6	77.4	65.8	98
Bangladesh	27.9	16.6	47.0	36.8	59.2	47.6	9
Bhutan	1.09	0.10	11.9	14.3	2.62	45.2	176
China	28.7	64.6	12.8	15.8	12.1	10.8	8
India	42.3	36.0	49.8	39.5	55.4	56.3	2
Myanmar	35.5	22.4	56.1	53.4	58.9	56.3	6
Nepal	2.62	0.25	27.5	27.2	13.5	56.9	131
Pakistan	26.8	13.1	54.6	41.4	61.0	64.4	10

pollution poses severe threats to human health, contributing to neonatal morbidity, stunting and learning delays in children, and with prolonged exposure linked to respiratory illnesses, heart diseases and other health complications, especially for vulnerable populations. The spike in air pollution in the region is due to forest fires, which are increasing in number and severity due to climate change and residents burning biomass. As well as the impacts on human health, pollutants cause extensive damage to the region's rich biodiversity and accelerate climate impacts. Rising temperatures are causing glaciers to melt at an alarming rate: if global warming exceeds 2°C, 50% of the glaciers in the region will be lost, leading to changes in river flow, which can have serious consequences for freshwater biodiversity, agriculture, drinking water and other human needs. Black carbon or soot emitted from forest fires and burning crop residues after harvest can accelerate the melting of high mountain glaciers, further contributing to their decline.

Delegates from the eight HKH countries, including UNFCCC national focal points and HKH High-Level Task Force members – along with representatives from the COP26 presidency – met in July 2021 to raise a unified voice for the HKH in the 26th Conference of Parties (COP26) to the United Nations Framework Convention on Climate Change (UNFCCC). The meeting also launched the HKH2Glasgow campaign under the title '#HKH2Glasgow Road to COP26'. The campaign worked on amplifying mountain voices to promote ambitious climate action for the HKH region at COP26 and scale up investment in mountain-specific priorities. ICIMOD's Regional Drought Monitoring and Outlook System (RDMOS) for South Asia, unveiled in 2019, produced reliable drought indicators for the region with a specific focus on Afghanistan, Bangladesh, Nepal and Pakistan.

Over the past 30 years, China has made remarkable achievements in restoring degraded ecosystems and improving the environment. For example, in the Tibetan plateau and Hengdan mountain areas of the HKH region, large-scale projects were implemented to protect natural forests and return farmland to forests and grazing land to grassland by introducing a sustainable path that combines ecological management with mountain development. These successful projects and models can provide inspiration and examples for developing countries, especially south Asian countries in the HKH region. Large-scale, policy-backed initiatives can make a big difference in the fight against climate change. Global and regional drivers of change in biodiversity and ecosystem loss are prevalent and increasing in the HKH region. These drivers include land use and land-cover change, pollution, climate change, invasive species, solid waste, habitat degradation and resource over-exploitation, which are affecting biodiversity, ecosystem services and human well-being.

1.3.4 Food and nutrition security

There are two main determinants of nutritional security: food intake (quantity and quality) and health status; and they are influenced by economic, social, biological and public health issues (UNICEF, 1998). Food and nutritional insecurity in the HKH region remains a serious challenge; more than 30% of the population is food-insecure and about 50% is suffering from some form of malnutrition, with women and children the most vulnerable. Insecurity is greatest in remote mountain areas. Food security challenges in mountain regions differ from those in the plains due to inaccessibility, vulnerability, seasonality, limited economic opportunities, poor market access and harsh biophysical conditions. The causes of food and nutritional insecurity in the HKH region are multi-faceted and complex, and are influenced by a range of factors, including high level of poverty, degradation of natural resources, climate change, low level of market development, uncertain food support, and inadequate policy and institutional support. Traditional mountain food systems are currently under threat from rapid socio-economic and environmental changes, including changing dietary habits, a shift to monocultures and commodity crops, loss of water sources, soil degradation and declining market values. Mountain agriculture is relatively less competitive, and young people are increasingly abandoning agricultural livelihoods, leading to a decline in food production and adversely affecting local food systems.

Between a third and a half of all children (< 5 years) in the region are stunted, and the incidence of wasting and underweight is also high, particularly in mountainous areas. Eating habits in the HKH region are changing. This has led to the deterioration of traditional mountain food systems and a decline in agricultural biodiversity. Factors such as high poverty and low dietary energy intake, lack of sanitation, inadequate nutrition knowledge, climate change and environmental degradation also affect food and nutrition security in the HKH region. In Pakistan, nearly 61% of the population in mountainous areas is food insecure, compared to 49% at the national level (Rasul *et al.*, 2016), whereas in Nepal, nearly 60% of the mountain population and 53% of the hill population are food insecure, compared to 48% in the plains. In India, 18% of people in the mountain states are under-nourished, slightly lower than the national rate, while the average caloric intake in urban and rural areas in the mountain states tends to be slightly lower than the national average.

Another clear indicator of nutritional insecurity in the HKH region is the high incidence of stunting, wasting and underweight in children under five years of age. In Nepal, the percentage of stunted children in the eastern, central and western mountains ranges from 45% to nearly 60%, while wasting and underweight in the western mountains are greater than in other subregions (Ministry of Health and Population (Nepal), 2011). The percentages of stunted and wasted children in Pakistan are higher in mountainous than in plain areas, with particularly high rates of stunting (82%), wasting (13%) and underweight (37%) in Balochistan (NIPS/Pakistan and ICF International, 2013). In Afghanistan, the prevalence of stunting is very high in the north-east, north-west and east, ranging from 45% to 53%, with stunting (53%), wasting (18%) and underweight (41%) particularly high in the east (USAID, WHO, UNICEF, JNFPA and Indian Institute of Health Management Research Jaipur (India), 2010). Except for Kachin state, stunting and wasting percentages in the mountains of Myanmar are higher than the national-average areas; while the proportion of underweight children in Chin and Shan states is high (Ministry of National Planning and Economic Development and Ministry of Health, 2011). The incidence of stunting, wasting and underweight among children in the hills of India is slightly lower than in the plains, but is still very high at nearly 43%, 16% and 35%, respectively (NFHS-4). Meghalaya is the only mountain state where these indices are higher than the national average. Meghalaya's transition from subsistence to commercial agriculture, from multi-crop to single-crop, from traditional to modern crops, and from food to non-food crops has had a negative impact on production and consumption patterns and has led to a decline in food diversity, thereby affecting nutritional security.

The number of underweight women aged 15–49 is relatively high in the HKH region, with percentages of 9%, 36%, 18% and 14% in Afghanistan, India, Nepal and Pakistan, respectively, compared to the world average of 10%. In India, as a whole, the prevalence of underweight women is lower in the hill states than in the plains (30% vs. 36%). In Nepal, the proportion of food-insecure households, and stunting and underweight among children (< 5 years) in the mountains is higher, but the incidences of anaemia and underweight in women (15–49 years) are lower than in the plains (Terai) and at the national level. The prevalence of underweight women is higher in mountain than in lowland areas (17% vs. 12%), but the average body mass index (BMI), a measure of body fat based on height and weight of adults, is within the normal range for both groups. In contrast, in Pakistan, the incidence of underweight among women is low, and lower in the mountainous provinces than in the plains (NIPS/Pakistan and ICF International, 2013).

Only 6% of the population in Bhutan lives below the food poverty line (Bangladesh Bureau of Statistics, 2012), while 96% of households have access to improved drinking water sources and 63% have access to improved sanitation facilities. Despite this, the incidence of stunting and underweight remains high, indicating a lack of dietary diversity and micronutrient deficiencies. In Tibet, growth was retarded in 11.4% of children, and body weight was low in 7.9% of children, while the prevalence of malnutrition was high and the diet often consisted of a single item. With increasing levels of food insecurity, mountain people generally face deficiencies in most micronutrients.

1.3.5 Urbanization and water resources

The HKH region is characterized by a unique hydrogeology, in which springs play a crucial role. Springs are a safe source of drinking water for rural and urban communities, and with increasing population and urbanization there is an ever-increasing demand for spring water in the region. It was estimated that there are nearly 4 million springs in the Himalayas and that more than 60% of the population living in the Indian Himalayan Region (IHR) is directly dependent on springs. Springs also contribute to base flows in the streams and rivers. Aquifers feed into springs, which are the main water sources for urban and rural areas. However, in many parts of the HKH, springs that are fed during the monsoon by groundwater or underground aquifers are drying up, threatening a way of life for local communities (Niti Aayog, 2018). Several cities in the HKH region are suffering from water crises and have put severe pressure on existing water resources (Table 1.3).

In Indian Himalayan urban settlements, changes in climatic conditions, population growth, tourism pressures and unplanned development activities were noted. Reduced rainfall is creating a water shortage, which is further exacerbated by tourism. In Nepal, the Kathmandu Valley is the traditional urbanized centre and is markedly different from other mountainous areas. Due to tourism, demographic pressures and poor infrastructure and governance, the gap between demand and supply of water is about 70% (Snyder and Iowa, 2014). Water scarcity, ineffective water-management practices and a growing population are major problems.

Xining city, as the big city of the Qinghai-Tibetan Plateau, faces the challenges of water scarcity and deteriorating water quality. Severe degradation of catchments and pollution of the nearby Huangshui river has been a major problem. The International Bank for Reconstruction and Development (IBRD) financed a

Table 1.3. Demand and supply of water in million litres per day (MLD) and major issues in selected cities of the Hindu Kush Himalayan region. (From Singh and Pandey, 2020, licensed under CC BY 4.0)

City	Supply (MLD)	Demand (MLD)	Major issues
Kabul (Afghanistan)	52.1	–	Limited distribution due to high energy cost. Only 20% of population supplied with water
Quetta (Pakistan)	106	170	Rapid population growth, low rainfall, groundwater depletion, water deficit of 50%
Shimla (India)	54.5	64.7	Lack new sources, high tourist influx, leakages in pipelines; 84.2% demand is met
Mussoorie (India)	7.67	14.4	Poor infrastructure, tanker economy, high tourist influx, recharge zones degraded
Nainital (India)	20.5	18.2	Depletion of lake level which is main source of water, encroachment over recharge zones, increase in impermeable areas
Kathmandu (Nepal)	86	280	Rapid influx of migrants, tourist influx, poor infrastructure, poor governance
Xining (China)	–	–	Water shortage, deterioration of water quality
Thimpu (Bhutan)	–	9.9	Water contamination, increase in impermeable surface causing flooding

US$150 million water environment project in Xining. The project aims to invest in rainwater and sewage-collection pipes, restore the environment along the Beichuan river, and strengthen the city's capacity for integrated water environmental management. Pakistan is one of the most arid and water-deprived countries in the world (Khan, 2009). Due to over-exploitation, the groundwater has dropped to 300 metres. The Water and Sanitation Agency (WASA) can produce only 100 million barrels of water per day through 400 tubewells, still leaving a 50% shortage. The local population relies heavily on tankers operating through informal arrangements and is consuming water dangerously. The water crisis in Afghanistan's capital, Kabul, is causing concern, as the residents are facing water shortages daily. Currently, less than 20% of Afghanistan's population has access to water (Mukherjee *et al.*, 2006). Water energy costs are high and distribution is limited. Due to the lack of a sewage treatment network (only 3%), water-borne diseases are rampant and infant mortality is high (Himmelsbach, 2010). Urban areas in Bhutan are facing drinking water and sanitation problems due to migration of people from rural areas. Limited supplies of treated water, inadequate distribution systems, and flooding due to increased impervious surfaces are the major issues (Bhujel, 2019). The effects of climate change are being felt by the increased incidence of floods and droughts in the region, and these effects are greater in urban than in rural areas due to the disorderly development of cities.

The rapid expansion of urban centres in low-lying cities under the International Health Regulations (IHR) in Afghanistan, Pakistan and India has led to severe groundwater extraction and depletion (Rodell *et al.*, 2009). There has been little development on mountain and hilly irrigation, and there is a lack of an overall policy of mountain water governance. The provision of infrastructure including roads, electricity, drainage, water supply, sanitation, recreation, communications and trade facilities require significant investment and participation from the government.

Most water-improvement/safety projects in the HKH region are concentrated in large city areas, although only 27% of the population in the IHR is urban (Sharma and Singh, 2014).

Most cities in the HKH area rely on spring water to varying degrees; Shimla and Musashi are completely dependent on springs. Directly related to the resilience of urban water in the region is the widely reported drying up of springs, depleting groundwater resources on which many urban water supply systems depend (Pandey, 2018). Some towns and cities also depend on surface water, and their catchment areas are degrading due to urbanization and increased built-up areas. This unregulated growth prevents the recharge of underground aquifers, as former watercourses have materialized, and soil compaction has increased run-off, potentially leading to flooding and soil erosion. The current practice of collecting water from a distant source is the norm, but is expensive and technically challenging.

Most HKH urban centres are unable to meet their water needs year-round, and as a result, many urban centres have seasonal population movements. Current planning fails to provide sufficient water for the large influx of tourists into urban centres, leading to severe water shortages, congestion and pollution. The most common way to address water shortage is to pump groundwater from springs, or by digging and drilling wells. However, given the inherent vulnerability of mountain aquifers, over-reliance on groundwater can lead to harmful consequences in the future, unless it is accompanied by appropriate and well-planned recharge plans. Therefore, there is a need to develop long-term strategies which take into account the fragility of mountain ecosystems and the ecological restoration of forest highlands that feed urban water-supply systems. Without long-term and sustainable urban planning and stakeholder accountability, many urban centres in the HKH region will face difficulties in securing sufficient water resources in the future, which will be exacerbated by climate change.

1.4 Strategy and Tactics for Solving Environmental Problems in the Hindu Kush Himalayan Region

1.4.1 Strategies

Action for disaster risk reduction, mitigation of and adaptation to climate change and good governance promote sustainable production systems to strengthen food and nutrition security and to provide more income-generating opportunities for mountain people, particularly women. These actions can reduce poverty and ensure that women, men and children lead healthy lives. Therefore, it is urgent to protect this global asset at the local, national, regional and global levels, increase investments and strengthen regional co-operation to limit global warming to less than 1.5°C by 2100 (Sharma, 2017).

Given the inter-connectedness of the mountains, HKH countries must enhance regional and international co-operation for climate action. This includes co-operation to facilitate data sharing to address transboundary climate risks; collaboration to support regional and international learning; and technology transfer for innovation, capacity building and green infrastructure. This is extremely important when HKH countries implement adaptation policies, and also equally important when developing post-pandemic recovery plans. Scaled-up investment in terms of financial and technological resources is required to achieve the ambitious targets set by the member countries for climate adaptation and reduction of carbon emissions through actions in key sectors like agriculture, water, energy, industry and transport. While the member countries and local communities have already allocated substantial resources to support these actions, corresponding commitments from the international community will enable mountain people to act at the scale and pace required to adapt to a changing climate and minimize global temperature rise. This is particularly relevant to the mountain areas as they bear the brunt of climate change impacts in the form of rapidly melting snow and glaciers, expansion of glacial lakes, unpredictable rainfall, and increasing frequency and intensity of natural disasters, among other impacts.

ICIMOD and its partners in Afghanistan, Bhutan, China, India, Myanmar, Nepal and Pakistan designed the conceptual framework for the Trans-Himalayan Transect and transboundary landscapes (TBLs, Fig. 1.6; Kotru *et al.*, 2020). It stems from the recognition that addressing threats to biodiversity and ecosystem services is a global issue that requires strong interfaces between science, practice, policy and society. Its aim is to create an effective pathway for regional co-operation, capturing the milestones and objectives set by global multi-lateral

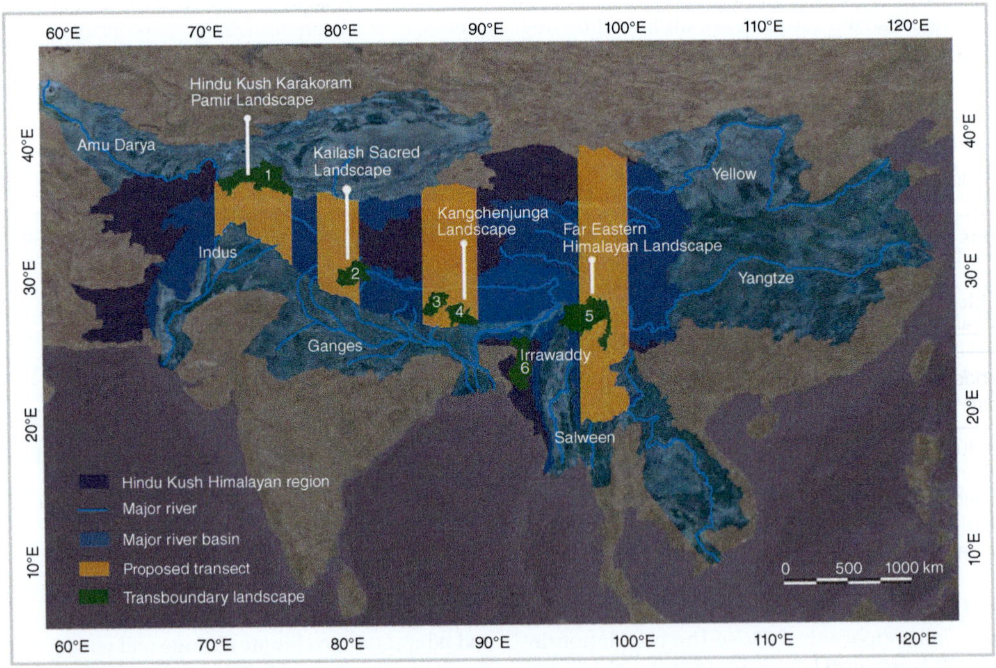

Fig. 1.6. ICIMOD's transboundary landscapes (TBLs) embedded within four north–south trans-Himalayan transects. (From Kotru *et al.*, 2020)

decisions. The framework sets out four basic principles, namely, promoting an integrated and participatory approach to ecosystem management, integrating multi-stakeholder priorities, promoting a common framework among countries of regional acceptance, and jointly contributing to global policies. As noted in the Fourth Report of the Intergovernmental Panel on Climate Change, it is an overall framework to address biodiversity data gaps in the region, while addressing the environmental, socio-economic and political challenges of biodiversity conservation and management beyond a country's administrative and management boundaries (Kotru *et al.*, 2020).

TBLs are designated as intended geographical units to implement the above framework. They are conceptualized as entry points for long-term regional co-operation to promote biodiversity conservation and monitoring, and trigger ecological and socio-economic resilience. Six TBLs were identified to represent the entire HKH region, and were put into use in four landscapes (Table 1.4). The rationale was to choose a landscape in which two or more countries can participate, and in which the countries have some degree of ecosystem, economic and socio-cultural ties. The 'four TBLs' approach to conservation and development includes co-operating with institutions in partner countries to identify challenges and opportunities, building a common understanding of pressing issues that cut across countries, developing pathways based on shared transnational contexts and visions, and promoting interventions that foster cross-border co-operation.

1.4.2 Tactics

1. Promote and strengthen co-operation at all levels in the HKH region and act at national, regional and international levels. The focus is on increasing investment and creating momentum for stronger regional co-operation to sustain the mountain environment and improve livelihoods in the HKH region.

2. Recognize and prioritize the uniqueness of HKH mountain people in national, regional and global decision-making bodies and processes. Through regional co-operation among countries, the HKH countries should contribute to the development of mountain-specific agendas to achieve the United Nations Sustainable Development Goals. There should be more investment in mountain infrastructure to build resilient, equitable and inclusive mountain communities.

3. Take decisive action to enhance ecosystem resilience by halting biodiversity loss and land degradation, and by promoting transboundary co-operation in landscapes and river basins to manage HKH's forests and other ecosystems sustainably. Evidence-based research should be employed to develop innovations to ensure sustainable adaptation to climate change and improve the ability to reduce disaster risk.

4. Accelerate action to achieve HKH's Sustainable Development Goals, in line with the 9 Mountain Priorities. The focus should be on transformative adaptation, poverty mitigation, gender and social equity, and development, especially for the least-developed countries.

Table 1.4. Key geographical features of the four transboundary landscapes (TBLs) in the Hindu Kush Himalaya region. (From Kotru *et al.*, 2020)

Landscape (west to east)	Participating countries	Key landscape feature	Total area (km^2)	Total population
Hindu Kush Karakoram Pamir Landscape (HKPL)	Afghanistan, China, Pakistan, Tajikistan	High alpine and endangered species	67,506	1,000,000
Kailash Sacred Landscape (KSL)	China, India, Nepal	Sacred landscape with high cultural significance	31,000	1,300,000
Kangchenjunga Landscape (KL)	Bhutan, India, Nepal	Corridors and habitat connectivity	25,085	7,248,293
Landscape Initiative for the Far-Eastern Himalaya (HI-LIFE)	China, India, Myanmar	Biodiversity hotspots and high endemism	71,452	213,600

5. Take concerted action to keep global warming to a minimum, with regional, national and local actions on reducing short-lived climate pollutants and achieving carbon-neutral societies in the HKH by 2035.

HKH is an important regional lifeline, but anthropogenic activities and climate change pose direct threats to the region's livelihoods, biodiversity and, ultimately, sustainability. Changes in this high-altitude region are and will continue to have a significant impact, not only regionally but also globally. National, regional and global actions are needed urgently to sustain this global asset, with a focus on increasing investment and strengthening regional co-operation to sustain the mountain environment and improve HKH livelihoods.

How to achieve a win-win situation between economic development and ecological protection, and how to create a mountain society where economy and environment complement each other are still major challenges for the development of the HKH region. To effectively cope with various challenges in mountain protection and raise the public's awareness of ecological protection in mountain areas, we must rely on scientific and technological co-operation, coordinate economic development and ecological protection, develop mountain areas, improve living conditions and the environment of mountain residents, and accelerate the harmonious coexistence between man and nature.

This is a new era full of opportunities and challenges. The HKH region is facing multiple challenges such as climate change, biodiversity loss and environmental pollution. Member countries should provide think-tanks, plan innovative initiatives, explore cutting-edge directions, and actively organize diverse activities such as mountain science forums. In addition, they should focus on green and sustainable development, strengthening regional exchanges and co-operation, and participating in talent exchange, capacity building, joint research and mountain spirit dissemination.

1.5 A Sustainable Future for the Hindu Kush Himalaya Region

The HKH region is a vital global and environmental asset that extends from Afghanistan to Myanmar, an area of more than 4.2 million km^2 (Bajracharya *et al.*, 2015). Ten river basins originate in the HKH and it holds the largest mass of ice cover apart from the North and South Poles. In terms of species conservation, the HKH plays a vital role, containing four global biodiversity hotspots, 330 important bird areas and hundreds of mountain peaks higher than 6000 m. The downstream importance of these peaks cannot be overstated as the water, food and energy sourced in the HKH sustain the lives of 210 million people and benefit nearly 1.3 billion. Three billion more people are provided with food produced in this region. Apart from the material importance, the HKH region also boasts unparalleled cultural richness, and is the home to a vast array of religions, languages and traditional knowledge.

Eight countries share the HKH region, namely Afghanistan, Bangladesh, Bhutan, China, India, Myanmar, Nepal and Pakistan. Regional co-operation among these countries is critical for maintaining sustainable development of this region and the ecosystem services. Protecting and developing the vast mountainous areas of the HKH is not only a local or regional issue but is also a national and global concern. As a major country in the region, China should take the lead in the international community in calling for the protection and development of mountain areas. The landmark Hindu Kush Himalaya Ministerial Summit and Declaration stressed the importance of the Hindu Kush Himalaya Action Initiative, paving the way for cross-border co-operation on biodiversity conservation and a unified voice in the international arena. The international community has helped many mountain people in poverty alleviation and development by creating development opportunities. In addition, the establishment of ecological compensation schemes was established under the framework of the United Nations Convention on Biological Diversity. Co-operation among the relevant countries in the region is essential to protect the HKH region and initiate sustainable development.

The HKH region and its people are on the front line of climate change, and even a 1.5°C increase would have adverse impacts on water availability, ecosystem health and livelihoods. Mitigation of the effects of an increase in air temperature requires awareness, investment and technology from all the HKH countries and the global community. In November 2021, the

Conference of the Parties (COP) to the United Nations Framework Convention on Climate Change (UNFCCC) met in Glasgow, UK, for COP26 to deliver the ambitious plans set out in the Paris Agreement. It was proposed to limit the rise in global temperature to 1.5°C by 2100, strengthen the capacity of nature and people to adapt to climate change, and align financial flows to zero-carbon and climate-resilient pathways. ICIMOD worked with its regional member countries of the HKH region to ensure mountain voices were heard at COP26. To do so, they launched the HKH2Glasgow campaign with the aim to: (i) promote ambitious climate action for the HKH; and (ii) launch a framework to scale up investment in mountain-specific climate priorities (Kaur *et al.*, 2021).

By promoting mountain voices at COP26 and by launching the framework, the HKH2Glasgow campaign contributed to climate-resilient mountain societies, carbon-neutral mountain economies and climate-aligned financial flows in the HKH region.

As part of the science policy uptake process in 2020, beginning with the publishing of the Hindu Kush Himalaya Assessment report (Wester *et al.*, 2020), ICIMOD facilitated a historic Ministerial Mountain Summit with ministers from all eight HKH countries. The ministers signed a declaration agreeing to strengthen regional co-operation in the HKH (Wester *et al.*, 2020); to promote a united voice for the HKH at regional, global and UN platforms; to promote policies focusing on mountain environments and livelihoods; and to assess the feasibility of establishing a regional institutional mechanism.

References

Adhikari, J. (2014) Agriculture adaptation practices in South Asia: case of Nepal. SAWTEE Working Paper No. 01(iii)/14. South Asia Watch on Trade, Economics and Environment (SAWTEE).

Alkire, S. and Robles, G. (2017) *Multidimensional Poverty Index Summer 2017: Brief Methodological Note and Results*. University of Oxford, UK.

Ashraf, A., Bilal, I.M., Mustafa, N., Naz, R. and Ahmad, B. (2021) Prevalent risk of glacial lake outburst food hazard in the Hindu Kush–Karakoram–Himalaya region of Pakistan. *Environmental Earth Sciences* 80, 451. DOI: 10.1007/s12665-021-09740-1

AWDO (2013) Asian Water Development Outlook 2013: Measuring Water Security in Asia and the Pacific. Asian Development Bank, Mandaluyong City, Philippines. Available at: https://www.adb.org/publications/asian-water-development-outlook-2013 (accessed 31 January 2024).

Bajracharya, S.R., Maharjan, S.B., Shrestha, F., Guo, W., Liu, S. *et al.* (2015) The glaciers of the Hindu Kush Himalayas: current status and observed changes from the 1980s to 2010. *International Journal of Water Resources Development* 31(2), 161–173. DOI: 10.1080/07900627.2015.1005731

Bangladesh Bureau of Statistics (2012) Statistical pocketbook of Bangladesh 2019. Available at: http://203.112.218.65/WebTestApplication/userfiles/Image/LatestReports/PB2012.pdf.

Bhujel, G.D. (2019) Water woes: investments, infrastructures, policies and advocacies, Kuensel. Available at: http://www.kuenselonline.com/water-woes-investments-infrastructures-policies-and-advocacies (accessed 3 September 2023).

Biswas, A.K. and Tortajada, C. (2016) Water security, climate change and sustainable development: an introduction. In: Smith, O. (ed.) *Water Resources Development and Management*. Springer, Singapore, pp. 65–82.

Gioli, G., Khan, T. and Scheffran, J. (2014) Climatic and environmental change in the Karakoram: making sense of community perceptions and adaptation strategies. *Regional Environmental Change* 14(3), 1151–1162. DOI: 10.1007/s10113-013-0550-3

Himmelsbach, T. (2010) Groundwater Resources at Risk: Assessment of the Present Situation and Consequences for the Future. Karlsruhe, Germany, pp. 24–25. Available at: chrome-extension://oemmndcbldboiebfnladdacbdfmadadm/https://www.bgr.bund.de/EN/Themen/Wasser/Projekte/abgeschlossen/TZ/Afghanistan/summary_report.pdf?__blob=publicationFile&v=2 (accessed 24 January 2024).

ICIMOD (2012) *Sustainable Mountain Development 1992, 2012, and Beyond-Rio+20. Assessment Report for the Hindu Kush Himalayan Region*. UN Department of Economic and Social Affairs. Available at: https://sdgs.un.org/publications/sustainable-mountain-development-1992-2012-and-beyond-rio20-assessment-report-hindu (accessed 24 January 2024).

Karki, M. *et al.* (2012) *Sustainable Mountain Development in the Hindu Kush – Himalaya: From Rio 1992 to Rio 2012 and Beyond*. International Centre for Integrated Mountain Development, Kathmandu.

Kaur, N., Sharma, D., Rana, P.J.B., Mishra, U., Shrestha, A.B. *et al.* (2021) HKH2Glasgow: promoting ambitious climate action for the Hindu Kush Himalayas. *Mountain Research and Development* 41(4), P1–P3. DOI: 10.1659/MRD-JOURNAL-D-21-00049.1

Khan, F. (2009) *Water Governance and Corruption in Pakistan. Running on Empty: Pakistan's Water Crisis*. Woodrow Wilson International Centre for Scholars, Asia Program, Washington, DC.

Khattri, M.B. (2012) Climate change, millet and ritual relationship with the magars of Argal, Baglung, Nepal. *Dhaulagiri Journal of Sociology and Anthropology* 6, 107–124. DOI: 10.3126/dsaj.v6i0.8481

Kotru, R.K., Shakya, B., Joshi, S., Gurung, J., Ali, G. *et al.* (2020) Biodiversity conservation and management in the Hindu Kush Himalayan region: Are transboundary landscapes a promising solution? *Mountain Research and Development* 40(2), 15–23. DOI: 10.1659/MRD-JOURNAL-D-19-00053.1 (CC by 4.0 Deed)

Kreutzmann, H. (2006) High mountain agriculture and its transformation in a changing socio-economic environment. In: Kreutzmann, H. (ed.) *Karakoram in Transition, Culture, Development and Ecology in the Hunza Valley*. Oxford University Press, Oxford, UK, New York, and Karachi, pp. 329–358.

Li, W., Wang, W., Gao, X., Wang, X. and Wang, R. (2022) Inventory and spatiotemporal patterns of glacial lakes in the HKH-TMHA region from 1990 to 2020. *Remote Sensing* 14(6), 1351. DOI: 10.3390/rs14061351

Malcolm, L. and Hunter, J. (1999) Maintaining biodiversity in forest ecosystems. In: Salim, E. and Ullsten, O. (eds) *Our Forests Our Future: Report of the World Commission on Forests and Sustainable Development*. Cambridge University Press, Cambridge, UK, pp. 147–153.

McNeill, J.R. (2000) *Something New under the Sun: An Environmental History of the Twentieth-century World*. W.W. Norton & Co., New York.

Ministry of Health and Population (Nepal), New ERA and ICF International Inc. (2012) *Nepal Demographic and Health Survey 2011*. Ministry of Health and Population, Kathmandu, New ERA, and ICF International, Calverton, Maryland.

Ministry of National Planning and Economic Development and Ministry of Health (2011) *Myanmar Multiple Indicator Cluster Survey 2009–2010*. Nay Pyi Taw, Myanmar.

Mukherjee, A., Sengupta, M., Hossain, M., Ahamed, S., Das, B. *et al.* (2006) Arsenic contamination in groundwater: a global perspective with emphasis on the Asian scenario. *Journal of Health, Population and Nutrition* 24(2), 142–163.

NIPS/Pakistan and ICF International (2013) *Pakistan Demographic Health Survey 2012–2013*. National Institute of Population Studies (Pakistan) and ICF International, Islamabad.

NITI Aayog (2018) *Report of Working Group I Inventory and Revival of Springs in the Himalayas for Water Security*. National Institution for Transforming India. Available at: http://niti.gov.in/writereaddata/files/document_publication/doc1.pdf (accessed 3 September 2023).

Niu, Y., Xie, G., Xiao, Y., Liu, J., Zou, H. *et al.* (2022) The story of grain self-sufficiency: China's food security and food for thought. *Food and Energy Security* 11(1), e344. DOI: 10.1002/fes3.344

Nyima, Y. and Hopping, K.A. (2019) Tibetan lake expansion from a pastoral perspective: local observations and coping strategies for a changing environment. *Society & Natural Resources* 32(9), 965–982. DOI: 10.1080/08941920.2019.1590667

Pandey, K. (2018) Down to Earth report. Available at: https://www.downtoearth.org.in/news/india/crisis-in-the-himalayas-nearly-50-perennial-springs-in-the-region-have-dried-up-61482 (accessed 3 September 2023).

Peng, H.-Y., Ren, Y., Li, Q. and Wei, J.-H. (2022) Spatial and temporal land use/cover change characteristics of Qinghai-Tibet Plateau. *Journal of Yangtze River Scientific Research Institute* 39(8), 41–49. DOI: 10.11988/ckyyb.20210470

Poudel, S. and Shaw, R. (2015) Demographic changes, economic changes and livelihood changes in the HKH region. In: Nibanupudi, H.K. and Shaw, R. (eds) *Mountain Hazards and Disaster Risk Reduction*. Springer, pp. 105–123.

Rasul, G., Hussain, A., Sutter, A., Dangol, N. and Sharma, E. (2016) *Towards an Integrated Approach to Nutrition Security in the Hindu Kush Himalayan Region*. ICIMOD Working Paper. International Centre for Integrated Mountain Development, Kathmandu.

Rodell, M., Velicogna, I. and Famiglietti, J.S. (2009) Satellite based estimates of groundwater depletion in India. *Nature* 460(7258), 999–1002. DOI: 10.1038/nature08238

Rusk, J., Maharjan, A., Tiwari, P., Karen Chen, T.H., Shneiderman, S. *et al.* (2022) Multi-hazard susceptibility and exposure assessment of the Hindu Kush Himalaya. *Science of the Total Environment* 804(4), 150039. DOI: 10.1016/j.scitotenv.2021.150039

Sharma, D. and Singh, R. (2014) Building urban climate resilience: learning from the ACCCRN experience in India. *International Journal of Urban Sustainable Development* 6(2), 133–153. DOI: 10.1080/19463138.2014.937720

Sharma, E. (2017) China–India collaborative opportunities on science and technology for sustainable development of the Hindu Kush Himalaya. *Science Bulletin* 62(10), 673–674. DOI: 10.1016/j.scib.2017.03.007

Sharma, E., Chettri, N. and Gyamtsho, P. (2006) Advances in community based natural resources management in the Hidnu Kush-Himalayan region. In: Gymtsho, P., Shing, B.K. and Rasual, G. (eds) *Capitalization and Sharing Experiences on the Interaction between Forest Policies and Land Use Pattern in Asia*. International Centre for Integrated Mountain Development (ICIMOD), Kathmandu, pp. 9–23.

Singh, V. and Pandey, A. (2020) Urban water resilience in Hindu Kush Himalaya: issues, challenges and way forward. *Water Policy* 22 (S1), 33–45. DOI: 10.2166/wp.2019.329

Snyder, A. and Iowa, M. (2014) Shortage in the mountains of plenty: water supply in mountain and hill cities throughout the Hindu-Kush Himalayan region. 2014 Borlaug-Ruan International Internship, ICIMOD and the World Food Prize. Available at: https://www.worldfoodprize.org/documents/filelibrary/images/youth_programs/2014_interns/2014_br_research_papers/SnyderAbigail_LONGReport_56ED38F157B76.pdf (accessed 3 September 2023).

UN (2017) *The Sustainable Development Goals Report 2017*. UN DESA, New York. Available at: https://unstats.un.org/sdgs/report/2017/ (accessed 31 January 2024).

UNDP (2020) *A Review of Multidimensional Poverty Index and Uses in Asia-Pacific*. United Nations Development Programme, UNDP Bangkok Regional Hub.

UNICEF (1998) *The State of the World's Children 1998*. United Nations International Children's Emergency Fund, New York.

USAID, WHO, UNICEF, JNFPA and Indian Institute of Health Management Research Jaipur (India) (2010) *Ministry of Public Health, Afghanistan, Afghanistan Mortality Survey 2010*. Government of the Islamic Republic of Afghanistan, Kabul.

Wangchuk, S. and Siebert, S.F. (2013) Agricultural change in Bumthang, Bhutan: market opportunities, government policies, and climate change. *Society & Natural Resources* 26(12), 1375–1389. DOI: 10.1080/08941920.2013.789575

Wester, P., Mishra, A., Mukherjee, A. and Shrestha, A.B. (2019) The Hindu Kush Himalaya assessment: mountains, climate change, sustainability and people. Springer Nature Switzerland AG, Cham, Switzerland, pp. 127–165.

Wester, P., Rathore, B.M.S., Vasily, L.A., Sharma, E. and Molden, D. (2020) The Hindu Kush Himalaya call to action: sustaining mountain environments and improving livelihoods. *Mountain Research and Development* 40(1), 1–4. DOI: 10.1659/MRD-JOURNAL-D-20-00040.1

Xiong, Q., Xiao, Y., Liang, P., Li, L., Zhang, L. *et al.* (2021) Trends in climate change and human interventions indicate grassland productivity on the Qinghai–Tibetan Plateau from 1980 to 2015. *Ecological Indicators* 129(31), 108010. DOI: 10.1016/j.ecolind.2021.108010

Yao, T., Thompson, L., Yang, W., Yu, W., Gao, Y. *et al.* (2012) Different glacier status with atmospheric circulations in Tibetan Plateau and surroundings. *Nature Climate Change* 2, 663–667. DOI: 10.1038/nclimate1580

2 Biodiversity Assessment for Sustainable Conservation in the Hindu Kush Himalayan Region

Zhen Peng[1,*], Allan Degen[2], Devendra Gauchan[3] and Zhanhuan Shang[1]
[1]*College of Ecology, Lanzhou University, Lanzhou, China;*
[2]*Institutes for Desert Research, Ben-Gurion University of the Negev, Israel;*
[3]*Alliance of Bioversity International and CIAT, Lalitpur, Nepal*

Abstract
Protected areas in the Hindu Kush Himalayan (HKH) region cover 40% of the land; however, sustainable biodiversity remains threatened due to climate warming, anthropogenic activities and land-use changes. This chapter reviews the following aspects of the HKH region: (i) status of vascular plants, mammals, birds and reptiles; (ii) risks and threats to biodiversity; (iii) sustainable conservation strategies implemented by eight member states, and the impacts on biodiversity; and (iv) achievements and suggestions for biodiversity conservation.

Keywords: biodiversity hotspots; species extinction; climate change; land-use change; conservation strategies; protected areas

2.1 Introduction

Biodiversity is an important factor in the stability of ecosystems (Cardinale *et al.*, 2012). The Hindu Kush Himalayan (HKH) region is world-renowned for its rich biodiversity of plant (Fig. 2.1) and animal species due to the climatic and ecological gradients (Chettri *et al.*, 2008), and has a number of important eco-regions that have been listed in numerous global conservation priorities (Olson and Dinerstein, 2002). The HKH region is also the source of ten major river systems which provide water for 1.9 billion people. Its ecosystem services support the livelihoods and provide food security and energy for the inhabitants in the region and downstream along the major rivers (IUCN, 2021). The rich biodiversity, six UNESCO World Natural Heritage Sites, more than 1000 languages, diverse cultures and a range of unique traditional knowledge systems are valuable assets (ICIMOD, 2020a).

Resources in the HKH region are vulnerable, especially in the mountains, where biodiversity and ecosystem degradation are widespread. In addition, traditional medicines of the local residents, such as Tibetan medicines based on shamanism, are decreasing and indigenous mountain identities, practices and knowledge are being neglected (ICIMOD, 2020a).

The HKH region contains four of the world's 36 ecological hotspots, all protected by the International Union for Conservation of Nature (IUCN), and include the mountains of Central Asia, the Himalaya, the mountains

*Corresponding author: pengzh20@lzu.edu.cn

DOI: 10.1079/9781800622579.0002

Fig. 2.1. Flowering plants on the Qinghai Tibet Plateau. (Photo: Zhen Peng)

of south-west China, and Indo-Myanmar. The biodiversity in these hotspots is threatened, and there are indications that at least 70% of the native vegetation has been lost. The IUCN describes the four biodiversity hotspots as follows (IUCN: https://www.iucn.org/): (i) the mountains of Central Asia were known to early Persians as the 'roof of the world'. This hotspot holds a large number of endemic plant species, but water stress and civil conflict have placed much of the unique biodiversity under serious threat. In 2019–2024, IUCN funded the mountains of Central Asia $80,000; (ii) the Himalayan biodiversity hotspot is home to the world's highest mountains, including Everest. The abrupt rise of these mountains, from less than 500 m to more than 8000 m, has resulted in a diversity of ecosystems. The region is home to important populations of bird and mammal species, including vultures, tigers, rhinoceroses and wild water buffaloes. Despite the hotspot's remoteness and inaccessibility, the Himalayas are still subjected to human-induced biodiversity loss. In 2005–2010, IUCN funded the Himalayan region $50,000, mainly to protect the eastern Himalayan region; (iii) the mountains of south-west China biodiversity hotspot stretches from south-east Tibet through western Sichuan and extends into central and northern Yunnan. Although only an estimated 8% of the original forest cover remains, this hotspot is home to approximately 50% of the country's birds and mammals and close to 40% of the vascular plants. In 2002–2013, IUCN funded the mountains of south-west China $70,000. The most serious threat to the region is illegal trade in wildlife and infrastructure development; and (iv) in terms of species diversity and endemism, the Indo-Myanmar biodiversity hotspot – which encompasses more than 2 million km^2 of tropical Asia – is one of the most biologically important regions in the world. It is also one of the most threatened, as only 5% of the hotspot's natural habitat remains in relatively pristine condition. There are three funds to protect the Indo-Myanmar hotspot: $90,000 for 2008–2013, $150,000 for 2013–2020 and $100,000 for 2020–2025. The biodiversity hotspots in India and Myanmar are increasingly under threat, and the Critical Ecosystem Partnership Fund (CEPF) for these hotspots has moved from pilot to long-term intervention conservation, integrated into government policies and operational practices.

The comprehensive HKH assessment report (ICIMOD, 2023) stated that a quarter of the endemic species in some parts of the Himalayas could be extinct by 2100; more than 30% of the population in the HKH region suffers from food insecurity and about 50% faces malnutrition. At present, there are 17 sites in the HKH region named on the World Heritage list, and more areas are expected to be added. The global importance of the HKH region under the World Heritage Convention was flagged by the IUCN in its 2021 report (IUCN, 2021), which listed seven broad areas of the HKH that are expected to become new World Natural Heritage Sites, namely Meghalaya subtropical forest ecozone, Shey Phoksundo National Park, Three Rivers Headwaters, Band-e-Amir, and Irrawaddy river corridor. In addition, existing World Heritage Sites could be expanded; for example, the area north of Mount Everest, in China, could be included in the Sagarmatha National Park in southern Nepal.

2.2 Biodiversity in the Hindu Kush Himalayan region

Mountain ecosystems are often rich in rare plant species, which are found throughout the entire HKH region, but many species are isolated at high altitudes (Chettri, 2014; Enquist *et al.*, 2019). The four biodiversity hotspots in the HKH region contain 48,443 plant and animal species (Fig. 2.2) and 1267 endemic species (Fig. 2.3). This region is home to snow leopards (*Uncia uncia*), tigers (*Panthera tigris*), leopards (*Panthera pardus*), clouded leopards (*Neofelis nebulosa*), kiangs (*Equus kiang*), yaks (*Poephagus grunniens*), Tibetan antelopes (*Pantholops hodgsonii*) and Tibetan gazelles (*Procapra picticaudata*) (Chettri, 2014), and is considered a valuable gene bank.

2.2.1 Plant species

There are approximately 10,700 plant species in the HKH region (Jansky *et al.*, 2002; Shinwari and Gilani, 2003), of which more than 2500 species have been reported, including endemic taxa such as *Androsace*, *Viola*, *Rhododendron*, *Saxifraga*, *Gentiana*, *Primula*, *Lentopodium* and *Saussurea* (Shaheen and Shinwari, 2012). In the Lake Karambar area, in north Pakistan, 27 families (108 plant species) were recorded, with an average species diversity, evenness and species richness of 2.18, 0.68 and 29, respectively. The flora of Lake Kalambar is composed mainly of Asteraceae (19%), Leguminosae (13%), Caprifoliaceae (11%), Rosaceae (7%), Primulaceae (7%) and Poaceae (5%) (Shaheen and Shinwari, 2012).

The Over-Aru Wildlife Sanctuary is one of the largest protected areas in the HKH region, with 76 families, 254 genera and 397 plant species. Asteraceae is the dominant family with 57 species, followed by Rosaceae with 28 and Lamiaceae with 27. The dominant plants are *Salix* and *Veronica* with seven species each, followed by

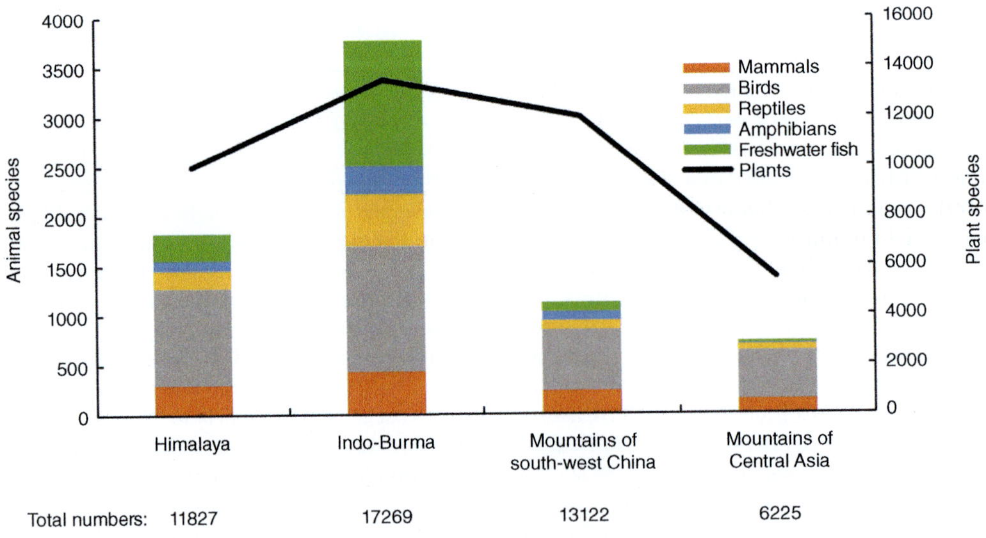

Fig. 2.2. Total plant and animal species in the four biodiversity hotspots in the Hindu Kush Himalayan region. (Data from Xu *et al.*, 2019)

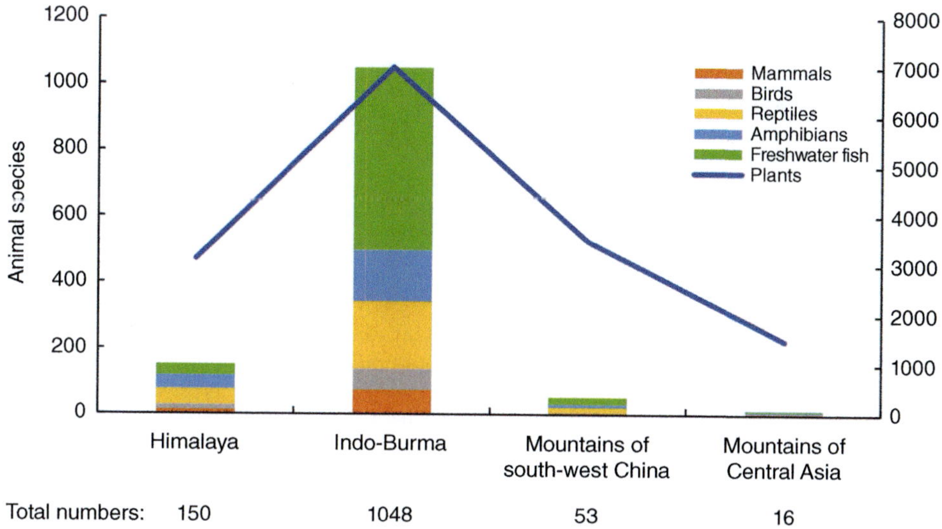

Fig. 2.3. Endemic plant and animal species in the four biodiversity hotspots in the Hindu Kush Himalayan region. (Data from Xu *et al.*, 2019)

Artemisia with six species, while there are 307 species of herbs and 41 species of trees (Islam *et al.*, 2023). In the mountains of south-west China biodiversity hotspot, Hengduan Mountains is a gathering centre of temperate plants, and its vegetation types are divided into southern and northern regions. The southern vegetation is subtropical – warm temperate broad-leaved forest with a mix of evergreen and deciduous species, namely *Castanopsis*, *Cyclobalanopsis*, *Lithocarpus*, *Quercus* (Fagaceae), *Michelia* (Magnoliaceae) and *Schima* (Theaceae) (Yu *et al.*, 1989). The northern vegetation is less dense than in the south, and is composed mainly of conifers; that is, the alpine vegetation in the Qinghai-Tibet cold zone or plateau (Zhang *et al.*, 2009). The Hengduan Mountains are also home to the largest number of seed plant species, with 8439 species belonging to 184 families and 1273 genera, of which 5817 (69%) are herbaceous and 1896 (22%) are shrubs (Yu *et al.*, 2020).

The Himalayas, as an important biogeographical eco-region, have 5468 plant species belonging to 177 families and 1146 genera, with 3681 herbaceous species (67%), 1251 shrub species and 536 tree species (Yu *et al.*, 2020). In the forest vegetation area of the western Himalayas, there is a total of 155 plant species belonging to 49 families and 120 genera. Asteraceae accounts for 12%, followed by Rosaceae with 11%. The distribution pattern of species in the region is uneven, with 50% belonging to only seven families. Of the species, 15% are invasive (Haq *et al.*, 2023). The eastern Himalaya is a geographical convergence of Indo-Malayan, Palearctic and Sino-Japanese regions and is also a crisis eco-region (ICIMOD, 2010), with at least 7500 flowering plant species, 700 orchid species, 58 bamboo species, 64 citrus species, 28 conifer species, 500 moss species, 700 fern species and 728 lichen species. The region is home not only to a large number of rare species but also to some of the most species-rich temperate forests in the world.

The three coniferous forests in Kashmir Himalaya contain 272 vascular plant species in 64 families and 196 genera, including 70 invasive plant species belonging to 27 families and 62 genera. Magnoliophyta, which includes Asteraceae, Apiaceae and Rosaceae, and the genera *Artemisia*, *Commiphora*, *Corydalis* and *Pocillophora*, has 190 taxa, 139 genera and 50 families; Pinophyta has seven taxa, six genera and three families, while Pteridophyta has three taxa, three genera and two families. More than 16% of the taxa are listed in the IUCN Red List, of which 24 taxa are endemic to the Himalayan region. An updated inventory in the Darjeeling Himalayas, India, recorded 239 species of epiphytic vascular plants in 93 genera of 38 families, including holoepiphytes, hemiepiphytes, facultative and accidental epiphytes (Fig. 2.4). Orchidaceae is the largest family with 70 species followed by Polypodiaceae with 42 species (Rai and Moktan, 2022).

2.2.2 Crop wild relatives

Crop wild relatives (CWRs) are wild plant taxa closely related to crops and can contribute genes and traits for crop improvement. Crop wild relatives are a traditional form of ensuring food security and an important measure addressing climate change and population growth as they can contain genes for coping with changes in climate. These genetic resources can be used to improve pest and disease resistance, drought and salt tolerance, and increase yields of crops. The four biodiversity hotspots overlap with the CWR hotspots, with the Indian Myanmar hotspots having the largest overlap area (Vincent *et al.*, 2022). Table 2.1 presents the distribution, area and number of CWRs among the four diversity hotspots in the HKH region.

2.2.3 Vertebrates

Mammals

In Darjeeling-Sikkim Himalaya, India, 173 mammal species from 11 orders and 33 families were recorded, with 25 species listed in the IUCN threatened category, 58 in the Convention on International Trade in Endangered Species of Wild Fauna and Flora (CITES) Appendices, and 112 species in the Wildlife (Protection) Act (Naulak and Pradhan, 2020). The northern part of Pakistan is a high-altitude wetland ecosystem with an area of high biodiversity due to its location on the boundary between the Palearctic and Oriental biogeographical regions (Ficetola *et al.*, 2010). There are 54 species of large and small mammals in the area, consisting of 1 shrew, 10 bats, 18 carnivores, 6 artiodactyls, 3 lagomorphs and 16 rodents, with carnivores and rodents the most diverse groups (Virk *et al.*, 2003). Species such as the markhor (*Capra falconeri*), Asian black bear (*Ursus thibetanus*), brown bear (*Ursus arctos*), alpine musk deer (*Moschus chrysogaster*) and snow leopard (*Uncia uncia*) are listed as endangered on the IUCN Red List (www.iucnredlist.org), while the hairy flying squirrel (*Eupetaurus cinereus*), a small mammal thought to be extinct for 70 years, inhabits the area (Khan and Baig, 2017). Of the 175 vertebrate species in the eastern Himalaya, 45 are threatened (ICIMOD, 2010), of which 14 are endangered, 29 are vulnerable, and two are extremely rare. In addition, 28 endemic mammal species, of which 18 are strictly endemic, inhabit 14 eco-regions. Some endemic species have specific niches, such as the Gee's golden langur (*Trachypithecus geei*), which is limited to semi-evergreen and tropical forests on the northern bank of the Brahmaputra river between the Sankosh and Manas rivers. The pygmy hog (*Sus salvanius*) and hispid hare (*Caprolagus hispidus*) are restricted to

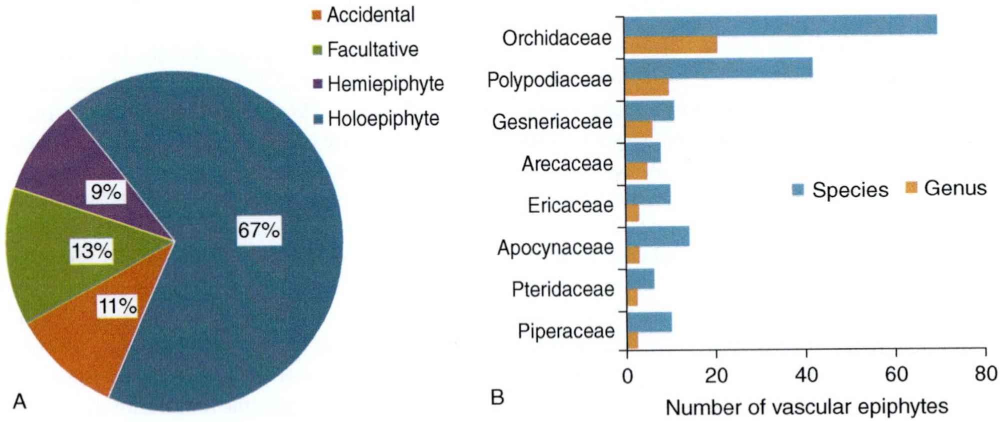

Fig. 2.4. Proportion and number of epiphytic vascular plant species in the Darjeeling Himalayas, India. (From Rai and Moktan, 2022)

Table 2.1. Crop wild relatives (CWRs) among the four diversity hotspots. (From: Vincent *et al.*, 2022)

Hotspot ID	Biodiversity hotspot	Area of biodiversity hotspct/area of overlap with CWR hotspot (kms^2)	Proportion overlap with CWR hotspot	Number of priority CWRs *in situ* conservation sites within biodiversity hotspot	Number of CWRs within biodiversity hotspot	Number of related crops within biodiversity hotspot	Number of CWRs in CWR hotspot and biodiversity hotspot overlap
H22	Mountains of Central Asia	172,672/60,435	35.0	2	96	53	94
H23	Himalaya	185,427/23,197	12.5	1	148	79	139
H24	Mountains of south-west China	20,966/5910	28.2	5	95	54	88
H25	Indo-Burma	118,653/51,234	43.2	7	146	55	139

alluvial grasslands, while the Namdapha flying squirrel (*Biswamoyopterus biswasi*) is restricted to the temperate broadleaved forests.

Birds

There are a total of 151 species of breeding and resident birds on the southern slopes of the central Himalayas in China (within Tibet). Bird diversity is highest in areas with medium altitudes (2400–3600). There are 13–65 bird species in each 600 m (6 sections) between 1800 m and 5400 m (Ding *et al.*, 2021). In the mountainous areas of south-west China, the complex terrain and diverse microclimate have caused a decreasing trend of nucleotide diversity in passerine birds from south-west to north-east, while haplotype diversity is more likely to decrease from west to east (Chang *et al.*, 2022). The northern part of Ailao Mountain in south-west China is a part of the Indo-Myanmar biodiversity hotspot, and is also one of the Important Bird and Biodiversity Areas (IBAs) listed by BirdLife International. The latest estimate of the number of bird species was 462 (Wu *et al.*, 2015). He *et al.* (2018) observed and recorded a total of 139 species of birds, including 125 passerine species, and 14 non-passerine species belonging to Galliformes, Columbiformes, Cuculiformes, Accipitriformes, Piciformes and Psittaciformes. There are 230 bird species in northern Pakistan, with the most species (110) in the Naltar, Gilgit, Hunza wetlands and valleys, followed by 109 in the Central Karakoram National Park (CKNP) (Khan and Baig, 2020). There are 25–47 bird species in a single wetland; for example, 47 species at Lulusar Lake in KPK province, 45 species at Heosar Lake on the Deosai Plateau, 40 species at Rama Lake, 39 species at Shandur Lake and 25 species at Karumber Lake.

Dendup *et al.* (2021) recorded 272 bird species belonging to 143 genera and 59 families in the Jigmedoji National Park in Bhutan. Among them, the chestnut-breasted partridge (*Arborophila mandellii*) and wood snipe (*Gallinago nemoricola*) were vulnerable species, and there were six endangered species. There are differences in the richness index of bird species in different forest ecosystems, with the greatest index in subtropical forests, followed by warm temperate forests, cold temperate forests, cool temperate forests and rhododendron scrub. In the Kailash Sacred Landscape-India, Kumaon Himalaya, 205 species belonging to 52 families were recorded between 2013 and 2016. Among them, there were 6 threatened, 5 endangered and 2 new species that were discovered in the Pithoragarh district. A total of 63 migratory and 81 Himalayan endemic bird species were observed (Arya and Gopi, 2021).

Reptiles

Twenty-three reptile species were recorded in the Hindu Kush Karakoram Himalayan region of northern Pakistan (Khan and Baiq, 2020). The six reptile families Agamidae, Colubidae, Elaphidae, Gekkonidae, Scincidae and Viperidae are the dominant ones in the high-altitude wetlands in northern Pakistan (Khan and Baiq, 2017). In the Himalayas, reptiles are most abundant at an altitude above 3000 m (Ficetola *et al.*, 2010).

2.3 Risks and Threats

The HKH region is being threatened by ecosystem collapse and species loss or even extinction

Table 2.2. Bird populations in eight member countries. (Data from BirdLife International, 2023; https://www.birdlife.org/ [accessed 25 January 2024])

Country	Total	Land birds	Migratory	Breeding endemic	Seabirds	Waterbirds
Afghanistan	393	291	294	1	16	102
Bangladesh	603	456	277	1	17	144
Bhutan	618	548	220	0	6	70
China	1288	1053	599	67	67	213
India	1210	994	467	76	50	200
Myanmar	1034	873	352	10	21	158
Nepal	820	685	341	1	13	135
Pakistan	611	440	355	0	35	165

(Chettri, 2014). Land-use change, pollution and climate change have led to habitat degradation and landscape fragmentation, with rare species particularly affected and natural resources decreasing (Chettri and Sharma, 2016; Enquist *et al.*, 2019). Poverty is widespread in mountainous areas, and when risks to human livelihoods and well-being increase, there is excessive exploitation of natural resources which leads to a vicious cycle of ecosystem degradation and reduction in ecosystem services (Chettri, 2014). Forest fragmentation caused by over-exploitation has been identified as one of the main driving factors in the Himalayas (Ravindranath *et al.*, 2011; Uddin *et al.*, 2015; Xu *et al.*, 2019). Even protected areas (PAs) face livelihood pressures from anthropogenic activities (Gurung *et al.*, 2019).

2.3.1 Climate change

The rate of climate warming in the HKH region is faster than the global rate, and as the altitude increases, the rate of warming increases gradually (Shrestha and Devkota, 2010; Chettri, 2014). This climate change poses a threat to the region's biodiversity, and ecosystem functions and services. The most dramatic impact of climate warming in the HKH region is the melting of glaciers, snow and permafrost, and an increase in melt water, leading to more frequent floodings, landslides and livestock diseases (Kattel, 2022). Increasing trends are being reported in droughts, mountain torrents and rainstorms in Afghanistan; glacial lake eruptions and floods in Nepal; rapidly melting glaciers in Bhutan; floods and hurricanes in Bangladesh; rising temperature and declining groundwater level in India; and floods in Pakistan (Devkota, 2013). Climate change has a strong impact on animal species that have a small population and restricted habitat, and that are limited to mountainous or plain areas (Dutta *et al.*, 2022). Due to climate change, the endemic species of the yellow throated mink (*Martes flavigula*) in the HKH region may be heading towards extinction. Some effects of climate warming are presented in Table 2.3.

2.3.2 Land-use change

Land-use changes in the HKH area are extensive. Between 1990 and 2005, forest cover in the Indus and Ganges river basins decreased by 90% and 85%, respectively, and by 85% and 73% in the Youngster and Yarlung Zangbo river basins, respectively. Land-use change can alter the ecosystem and, ultimately, affect the livelihoods of the 240 million people and habitat of animals in the region. Large animals that lose their habitat are often forced to flee the area. This region harbours two species of geographically separated brown bears – the Tibetan (*Ursus arctos pruinosus*) and Himalayan (*Ursus arctos isabellinus*). It is expected that by 2070 the distribution areas for these bears will shrink by 30–35% and shift to higher altitudes (Dai *et al.*, 2021).

Table 2.3. Changes in the Hindu Kush Himalayan region caused by climate change. (Source: You *et al.*, 2017)

Specific manifestations of climate change	Problems or difficulties caused
Rapid warming in the HKH region	Glacial retreat, snow melt and permafrost degradation
Hydrological	Reduction or pollution of water resources
Trends in climate extremes in the HKH region	Reduced extreme cold days and increased hot weather
The higher the altitude, the faster the temperature increases	Ecological environment in high-altitude areas worsens
Erratic change in precipitation	Warming and wetting of the Qinghai Tibet Plateau
Decrease in wind speed	Not conducive to the reproductive expansion of seed species
Decrease in solar radiation and sunshine duration	Plant phenological changes

2.3.3 Deforestation

Forests are important for animals, as more than twice the number of reptile, bird and mammal species inhabit forests compared to other habitats. However, because of the extensive deforestation, the proportion of animal and amphibian species that are threatened in forest habitats is greater than in non-forest habitats (Cox *et al.*, 2022). Afghanistan loses 3.1% of its forest cover annually, followed by Pakistan, Nepal and Myanmar. The forest cover in Nepal decreased from 34% to 25% between 1990 and 2005 (Khan and Baig, 2020). In the mountainous areas of northern Pakistan, deforestation and land-use change have led to land fragmentation and have encouraged illegal hunting and fishing for livelihood purposes. The biodiversity of the HKH region is seriously threatened, with a large decrease in many previously abundant large mammal species (BirdLife International, 2016; Siraj-UD-Din *et al.*, 2016).

2.3.4 Illegal hunting

The mountainous area is a special and fragile ecosystem, and poverty is widespread. The local people rely heavily on natural resources to sustain their livelihoods. Biodiversity is declining rapidly due, at least in part, to illegal hunting, wildlife trade and habitat destruction (Khan and Baig, 2017), which have led to 32 bird species being threatened in Pakistan (BirdLife International, 2016), and the snow partridge (*Lerwa lerwa*) and Himalayan monal (*Cophorus impejanus*) facing extinction in many wetland areas (Virk *et al.*, 2003). Large mammals such as the snow leopard (*Uncia uncia*) are listed in the Convention on International Trade in Endangered Species of Wild Fauna and Flora Habitat fragmentation because of illegal hunting for body parts and retaliatory killings by livestock owners. Unregulated fishing and rising water temperatures are depleting fish populations in the high-altitude wetlands in northern Pakistan. The number of fish in Nepal's Kaligandaki-Narayani river has decreased significantly, especially over the past 30 years (Gillette *et al.*, 2022).

2.3.5 Urban and agricultural land expansion

The Impervious Surface Areas (ISAs) in the HKH region increased from 1615 km^2 in 1993 to 4315 km^2 in 2018. From the perspective of regional differences, the average annual growth rates of ISAs in Afghanistan, Pakistan and some areas of China from 1993 to 2018 were 0.05, 0.05 and 0.09, respectively, indicating that urban expansion in some areas in China is faster than in other regions. These data indicate that the habitat of plants and animals in the HKH region has been invaded and destroyed, and is decreasing (Chao *et al.*, 2023). However, with climate warming and increased rainfall, the HKH region will have a larger area suitable for crop production.

2.3.6 Hazards of a major epidemic such as COVID-19

The social, economic and environmental impacts of the COVID-19 pandemic have been very severe (ICIMOD, 2020b). The conflict between livelihoods and the protection of natural resources has surfaced and even intensified, especially in protected areas in the HKH region. Major epidemics not only have negative impacts on the management capabilities and budgets, but also damage and reduce the livelihoods of communities. People in mountainous areas who already suffer from hunger and malnutrition are more likely to fall into long-term poverty with economic downturns in agriculture and tourism. The risks of increased poverty, reduced access to education and increased maternal mortality increase. In addition, due to epidemics and the negative effects on livelihoods, illegal hunting and wildlife trade flourish. Maintaining natural ecological integrity through protected areas is crucial for preventing biodiversity loss and reducing the spread of zoonotic diseases.

2.4 Sustainable Protection Policies of Countries and International Organizations

2.4.1 National measures

Without promoting restoration/conservation programmes, there is a serious risk of extinction of over half (57%) the threatened species. On average, there are 54 threatened species in each country that require urgent action. However, countries with rich biodiversity often invest fewer resources in biodiversity research and protection than countries with poor biodiversity (Reed *et al.*, 2020).

2.4.2 International Centre for Integrated Mountain Development (ICIMOD) initiative

The ICIMOD adopted a flagship species approach to conserve biodiversity and habitats in the HKH region. An ecosystem is considered healthy if its flagship species, which normally occupies the top of the food web pyramid, flourish. Conservation approaches shift from species-centred to landscape level (Sharma *et al.*, 2010), with local communities empowered in conservation and management decision-making processes (Phuntsho *et al.*, 2012). In addition, ICIMOD has undertaken the following efforts to protect HKH's biodiversity: (i) considering the importance of biodiversity, key ecological zones and critical habitats for flagship species, four 'HKH trans-Himalayan cross-sections' were proposed, ranging from west to east, dry to wet, and south to north, to address the information gap in the region; (ii) identified six transboundary landscapes (TBLs; Table 2.4) and strengthened co-operation among countries to ensure the integrity of the landscape. TBLs facilitated multi-regional co-operation and have the potential to boost regional co-operation processes, help countries work together to address milestones, targets and impacts related to biodiversity conservation and development; and (iii) a new generation of conservation approaches with economic dimensions: evaluation of biodiversity and its derived products or services.

The World Wildlife Fund (WWF) and ICIMOD selected seven ecosystems in the eastern Himalayas, namely: (i) alpine/sub-alpine ecosystems between the tree line at 4000 m to the snowline at 5500 m; (ii) cool moist forests; (iii) cloud forests at temperature elevations where moisture tends to mix with other particles and remains in the air; (iv) areas with intensive agriculture; (v) freshwater wetlands; (vi) riparian habitats nurtured by silt deposited by overflow rivers; and (vii) ephemeral stream habitats (Chettri *et al.*, 2010).

In 2023, ICIMOD was guided by four strategic goals: (i) serving as a knowledge and innovation centre; (ii) promoting regional co-operation and collaboration; (iii) leading global knowledge and providing a voice for sustainable mountain development; and (iv) ensuring that internal organizations are suitable for the 2030 goals (Fig. 2.5). On the basis of these strategic goals, the four impact pathways were proposed to work together in different combinations. These consist of a series of actions identified in the World Bank's Fifth Medium Term Action Plan (MTAP V), ultimately resulting in three high-level outcomes: (i) policy responses and practical solutions, which will be widely adopted and supported by large-scale investments; (ii) regional co-operation and collaboration that will lead to key policy decisions,

Table 2.4. Details of six transboundary landscapes in the Hindu Kush Himalayan region. (Source: Kotru *et al.*, 2020)

Landscape (west to east)	Participating countries	Key landscape feature	Total area (km^2)	Total population
Hindu Kush Karakoram Pamir Landscape (HKPL)	Afghanistan, China, Pakistan, Tajikistan	High alpine and endangered species	67,506	1,000,000
Kailash Sacred Landscape (KSL)	China, India, Nepal	Sacred landscape with high cultural significance	31,000	1,300,000
Kangchenjunga Landscape (KL)	Bhutan, India, Nepal	Corridors and habitat connectivity	25,085	7,248,293
Landscape Initiative for the far-eastern Himalaya (HI-LIFE)	China, India, Myanmar	Biodiversity hotspots and high endemism	71,452	213,600

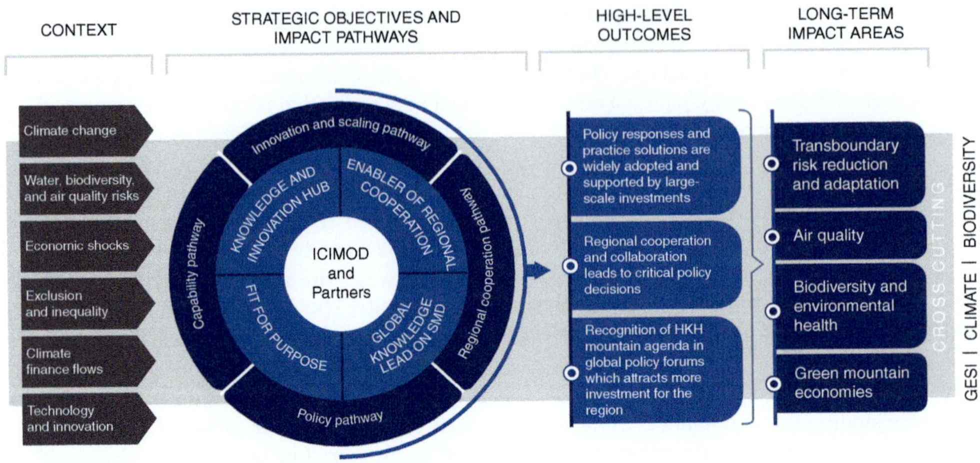

Fig. 2.5. New direction framework for the International Centre for Integrated Mountain Development (ICIMOD). GESI: Gender equality and social inclusion. (From Izabella and Pema, 2023)

including regional institutional mechanisms and protocols; and (iii) recognition of the Hong Kong Mountain Agenda at global policy forums to attract more investment to the region (ICIMOD, 2023).

2.4.3 International Union for Conservation of Nature (IUCN)

The IUCN is committed to protecting global biodiversity by providing funds and technical support to biodiversity hotspots. In addition, IUCN is committed to incorporating more areas of the HKH region into the World Heritage Convention in order to protect this globally unique system (IUCN, 2021). As of 2007, the HKH region had 488 protected areas, comprising 39% of the land (Chettri, 2014). With the strengthening of biodiversity conservation awareness, the number of protected areas continues to increase. The protected areas in the HKH region, which belong to the Chinese administrative region alone, is as high as 36%, followed by India (1.46%) and Nepal (0.58%). The four biodiversity hotspots in the HKH region account for 32% of the area, and the protected areas established in the region reached 62% (Chettri *et al.*, 2008). At present, the HKH region has a total of 575 protected areas, covering 40.2% of the land. In addition, the HKH region has 335 important bird and biodiversity areas (IBAs), 348 key biodiversity areas (KBAs), 12 global top 200 ecological areas and four global biodiversity hotspots. However, the ecological representation of the current protected areas is limited, as 67% of ecological areas, 39% of hotspots, 69% of KBA and 76% of IBA are outside the protected areas. About 47% of the protected areas are small, less than 250 km^2, and lack connection with other protected areas, with the majority distributed in the lower reaches of the Hong Kong river (ICIMOD, 2022). Corridors should be developed among protected areas, and they should be incorporated into broader landscapes through regional co-operation. It is also necessary to evaluate and strengthen the effectiveness and governance of protected area management, and consider other effective protection measures, especially in high-altitude areas.

2.5 Conclusions

2.5.1 Achievements

According to ICIMOD's report, the following achievements have been accomplished through the joint efforts of ICIMOD and the eight member states (Chaudhary *et al.*, 2022):

1. Progress has been made towards achieving the 20 Aichi targets: target 17 has been mostly achieved, 11 targets have been partially achieved, while 8 targets have not been achieved.

Table 2.5. Public attention to biodiversity and its protection in member countries of the Hindu Kush Himalayan region. (Data source: Caetano *et al.*, 2023. LCI, lower confidence interval; UCI, upper confidence interval)

Country	Google use (%)	Biodiversity LCI	Biodiversity median	Biodiversity UCI	Conservation LCI	Conservation median	Conservation UCI
Afghanistan	94.7	0.0000	0.0000	0.0010	−0.0034	0.0000	0.0000
Bangladesh	96.4	0.0000	0.0000	0.0006	−0.0008	0.0000	0.0000
Bhutan	93.9	0.0000	0.0015	0.0021	−0.0070	−0.0048	−0.0032
India	96.9	0.0010	0.0016	0.0020	−0.0008	0.0000	0.0000
Myanmar	92.3	0.0019	0.0026	0.0031	0.0000	0.0000	0.0000
Nepal	95.1	0.0006	0.0013	0.0017	0.0000	0.0000	0.0000
Pakistan	97.3	0.0000	0.0000	0.0010	0.0009	0.0020	0.0025

2. Targets with the best progress are: target 7 (sustainable agriculture, forestry); target 11 (protected areas); target 15 (ecosystem resilience); and target 16 (Nagoya protocol).

3. Of all the targets, about 11% have been completed, 55% are on track; 29% reported 'progress has been made, but at an insufficient rate'; 3% had 'no significant change'; less than 0.5% had a 'shift away from them'; and 2% provided no information.

4. Some of the major gaps and challenges in the region include variations in the scope and level of ambition in terms of aligning national targets with Aichi targets; lack of attention to research results and data on species, ecosystems and drivers of change in the region; and over-dependence on international funding for Convention on Biological Diversity implementation.

2.5.2 Attention to biodiversity

With the proposal of Aichi target 1, human attention and interest in biodiversity increased, but the protection of biodiversity has not been given enough attention. Table 2.5 depicts the interest displayed by countries in the HKH region (excluding China) in biodiversity and its conservation through Google searches. The Myanmar people demonstrate a high interest in biodiversity, while the Pakistani people focus on biodiversity conservation. Compared to other areas, people in the HKH region have a high interest in biodiversity, but a low interest in conservation (Caetano *et al.*, 2023).

2.5.3 Protection suggestions

More efforts are urgently needed to prevent the loss of biodiversity and the extinction of plant and animal species. In addition to understanding the responses of existing protected areas to climate change (Watson *et al.*, 2014), it is necessary to expand habitats and corridors and strengthen linkages among existing protected areas and habitats (Worboys *et al.* 2010).

Comprehensive changes in policies and practices are necessary for biodiversity conservation. Some suggestions are:

1. Long-term monitoring of environmental and social changes in the HKH region, with a focus on impacts of climate, land use and livelihood changes on biodiversity;

2. Assessing the effectiveness of existing protected habitat and ecological areas, and identifying potential threats to ecosystems and livelihoods;

3. Incorporating livelihoods and enterprises in ecological conservation (Panwar, 2023);

4. Enhancing awareness of biodiversity conservation and understanding the vulnerability of mountain ecosystems; and

5. Formulating policies to address threats such as climate and land-use changes, and encouraging co-operation among HKH countries to promote biodiversity values.

References

Arya, S.K. and Gopi, G.V. (2021) An annotated bird checklist of community-managed lands in Kailash Sacred Landscape-India, Kumaon Himalaya. *Check List* 17(2), 365–383. DOI: 10.15560/17.2.365

BirdLife International (2016) Country profile. Available at: http://datazone.birdlife.org/country (accessed 13 September 2023).

Caetano, G.H.O., Vardi, R., Jarić, I., Correia, R.A., Roll, U. *et al.* (2023) Evaluating global interest in biodiversity and conservation. *Conservation Biology* 37(5), e14100. DOI: 10.1111/cobi.14100 (CC by 4.0)

Cardinale, B.J., Duffy, E.J., Gonzalez, A., Hooper, D.U., Perrings, C. *et al.* (2012) Biodiversity loss and its impact on humanity. *Nature* 486, 59–67. DOI: 10.1038/nature11148

Chang, Y., Song, G., Zhang, D., Jia, C., Fan, P. *et al.* (2022) Distribution pattern and driving factors of genetic diversity of passerine birds in the mountains of Southwest China. *Avian Research* 13, 100043. DOI: 10.1016/j.avrs.2022.100043

Chao, Z., Shang, Z., Fei, C., Zhuang, Z. and Zhou, M. (2023) Spatiotemporal analysis of urban expansion in the mountainous Hindu Kush Himalayas region. *Land* 12(3), 576. DOI: 10.3390/land12030576

Chaudhary, S., Uddin, K., Chettri, N., Thapa, R. and Sharma, E. (2022) Protected areas in the Hindu Kush Himalaya: a regional assessment of the status, distribution, and gaps. *Conservation Science and Practice* 4(10), e12793. DOI: 10.1111/csp2.12793

Chettri, N. (2014) Reconciling mountain biodiversity conservation in a changing climate: a Hindu Kush-Himalayan perspective. *Conservation Science* 2, 17–27. DOI: 10.3126/cs.v2i1.13766

Chettri, N. and Sharma, E. (2016) Reconciling the mountain biodiversity conservation and human wellbeing: drivers of biodiversity loss and new approaches in the Hindu-Kush Himalayas. *Proceeding of Indian National Science Academy* 82, 53–73. DOI: 10.16943/ptinsa/2016/v82i1/48378

Chettri, N., Shakya, B., Thapa, R. and Sharma, E. (2008) Status of a protected area system in the Hindu Kush-Himalayas: an analysis of PA coverage. *International Journal of Biodiversity Science and Management* 4(3), 164–178. DOI: 10.3843/Biodiv.4.3:4

Chettri, N., Sharma, E., Shakya, B., Thapa, R., Bajracharya, B. *et al.* (2010) *Biodiversity in the Eastern Himalayas: Status, Trends and Vulnerability to Climate Change*. International Centre for Integrated Mountain Development, Kathmandu.

Cox, N., Young, B.E., Bowles, P., Fernandez, M., Marin, J. *et al.* (2022) A global reptile assessment highlights shared conservation needs of tetrapods. *Nature* 605, 285–290. DOI: 10.1038/s41586-022-04664-7

Dai, Y., Peng, G., Wen, C., Zahoor, B., Ma, X. *et al.* (2021) Climate and land use changes shift the distribution and dispersal of two umbrella species in the Hindu Kush Himalayan region. *Science of the Total Environment* 777, 146207. DOI: 10.1016/j.scitotenv.2021.146207

Dendup, P., Wangdi, L., Jamtsho, Y., Kuenzang, P., Gyeltshen, D. *et al.* (2021) Bird diversity and conservation threats in Jigme Dorji National Park, Bhutan. *Global Ecology and Conservation* 30, e01771. DOI: 10.1016/j.gecco.2021.e01771

Devkota, D.C. (2013) Regional cooperation: key to climate change adaptation in HKH region. *Nepal Journal of Environmental Science* 1(1), 27–35. DOI: 10.3126/njes.v1i1.36545

Ding, Z., Hu, H., Cadotte, M.W., Liang, J., Hu, Y. *et al.* (2021) Elevational patterns of bird functional and phylogenetic structure in the central Himalaya. *Ecography* 44, 1403–1417. DOI: 10.1111/ecog.05660

Dutta, R., Mukherjee, T., Sharief, A., Singh, H., Kumar, V. *et al.* (2022) Climate change may plunder the facultative top predator Yellow-throated Martin from the Hindu-Kush Himalayan region. *Ecological Informatics* 69(1), 101622. DOI: 10.1016/j.ecoinf.2022.101622

Enquist, B.J., Feng, X., Boyle, B., Maitner, B., Newman, E.A. *et al.* (2019) The commonness of rarity: global and future distribution of rarity across land plants. *Science Advances* 5(11), eaaz0414. DOI: 10.32942/osf.io/qfpa9

Ficetola, G.F., Crottini, A., Casiraghi, M. and Schioppa, E.P. (2010) New data on amphibians and reptiles of the northern areas of Pakistan: distribution, genetic variability and conservation issues. *North-Western Journal of Zoology* 6(1), 1–12.

Gillette, D.P., Edds, D.R., Jha, B.R. and Mishra, B. (2022) Thirty years of environmental change reduces local, but not regional, diversity of riverine fish assemblages in a Himalayan biodiversity hotspot. *Biological Conservation* 265(8), 109427. DOI: 10.1016/j.biocon.2021.109427

Gurung, J., Chettri, N., Sharma, E., Ning, W., Chaudhary, R.P. *et al.* (2019) Evolution of a transboundary landscape approach in the Hindu Kush Himalaya: key learnings from the Kangchenjunga landscape. *Global Ecology Conservation* 17, e00599. DOI: 10.1016/j.gecco.2019.e00599

Haq, S.M., Khoja, A.A., Lone, F.A., Waheed, M., Bussmann, R.W. *et al.* (2023) Floristic composition, life history traits and phytogeographic distribution of forest vegetation in the western Himalaya. *Frontiers in Forest and Global Change* 6, 1169085. DOI: 10.3389/ffgc.2023.1169085

He, X., Luo, K., Brown, C., Lin, L. (2018) A taxonomic, functional. and phylogenetic perspective on the community assembly of passerine birds along an elevational gradient in southwest China. *Ecology and Evolution* 8(5), 2712–2720. DOI: 10.1002/ece3.3910

ICIMOD (2007) *Ecological Regions of Hindu Kush Himalayan (HKH) Region*. Regional Database System, ICIMOD, Kathmandu, Nepal.

ICIMOD (2010) *Biodiversity in the Eastern Himalayas: Status, Trends and Vulnerability to Climate Change.* International Centre for Integrated Mountain Development, Kathmandu.

ICIMOD (2011) *Important Habitat Area of Bird Species of Hindu Kush Himalayan (HKH) Region.* Regional Database System, ICIMOD, Kathmandu, Nepal.

ICIMOD (2020a) *The Hindu Kush Himalaya Call for Action Maintaining the 'Pulse of the Planet'*. International Centre for Integrated Mountain Development, Kathmandu.

ICIMOD (2020b) *COVID-19 Impact and Policy Responses in the Hindu Kush Himalaya.* International Centre for Integrated Mountain Development, Kathmandu.

ICIMOD (2022) *Implementation of the Convention on Biological Diversity in the Hindu Kush Himalayan Countries: A Retrospective Analysis of Aichi Targets.* International Centre for Integrated Mountain Development, Kathmandu.

ICIMOD (2023) *Water, Ice, Society, and Ecosystems in the Hindu Kush Himalaya: An Outlook.* International Centre for Integrated Mountain Development, Kathmandu.

Islam, T., Khuroo, A.A. and Nawchoo, I.A. (2023) An annotated checklist of flora of Overa-Aru wildlife sanctuary, Kashmir Himalaya. *Phytotaxa* 599, 1–31. DOI: 10.11646/PHYTOTAXA.599.1.2

IUCN (2021) *Leveraging the World Heritage Convention for Conservation in the Hindu Kush Himalaya.* The International Union for Conservation of Nature, Switzerland.

Izabella, K. and Pema, G. (2023) Moving mountains: a new strategy and action plan for ICIMOD to embrace change and accelerate impact to 2030. *Mountain Research and Development* 43(1), P1–P3. DOI: 10.1659/mrd.2023.00002. (CC BY 4.0)

Jansky, L., Ives, J.D., Furuyashiki, K. and Watanabe, T. (2002) Global mountain research for sustainable development. *Global Environmental Change* 12(3), 231–239.

Kattel, G.R. (2022) Climate warming in the Himalayas threatens biodiversity, ecosystem functioning and ecosystem services in the 21st century: Is there a better solution? *Biodiversity and Conservation* 31(1), 2017–2044. DOI: 10.1007/s10531-022-02417-6

Khan, H. and Baig, S.U. (2017) High altitude wetlands of the HKH Region of Northern Pakistan – status of current knowledge, challenges and research opportunities. *Wetlands* 37(2), 371–380. DOI: 10.1007/s13157-016-0868-y

Khan, H. and Baig, S.U. (2020) Biodiversity conservation in the Hindu Kush-Karakoram-Himalayan mountain region of northern Pakistan: overview of big mammal protection. *Journal of Mountain Science* 17(6), 1360–1373. DOI: 10.1007/s11629-018-5113-0

Kotru, R.K., Shakya, B., Joshi, S., Gurung, J., Ali, G. *et al.* (2020) Biodiversity conservation and management in the Hindu Kush Himalayan region: Are transboundary landscapes a promising solution? *Mountain Research and Development* 40(2), A15–A23. DOI: 10.1659/MRD-JOURNAL-D-19-00053.1 (CC BY 4.0)

Naulak, T. and Pradhan, S. (2020) A checklist of mammals with historical records from Darjeeling-Sikkim Himalaya landscape, India. *Journal of Threatened Taxa* 12(11), 16434–16459. DOI: 10.11609/jott.6062.12.11.16434-16459

Olson, D.M. and Dinerstein, E. (2002) The Global 200: priority ecoregions for global conservation. *Annals of Missouri Botanical Garden* 89(2), 199–224. DOI: 10.2307/3298564

Panwar, R. (2023) Business and biodiversity: achieving the 2050 vision for biodiversity conservation through transformative business practices. *Biodiversity and Conservation* 32(11), 1–7. DOI: 10.1007/s10531-023-02575-1

Phuntsho, K., Chettri, N. and Oli, K.P. (2012) *Mainstreaming Community-based Conservation in a Transboundary Mountain Landscape – Lessons from Kangchenjunga.* International Centre for Integrated Mountain Development, Kathmandu.

Rai, P. and Moktan, S. (2022) An updated checklist of vascular epiphytes in the Darjeeling Himalaya, India. *Check List* 18(6), 1279–1319. DOI: 10.15560/18.6.1279 (CC BY 4.0)

Ravindranath, N.H., Chaturvedi, R.K., Joshi, N.V., Sukumar, R. and Sathaye, J.A. (2011) Implications of climate change on mitigation potential estimates for forest sector in India. *Mitigation and Adaptation Strategies for Global Change* 16(2), 211–227. DOI: 10.1007/s11027-010-9256-8

Reed, J., Oldekop, J., Barlow, J., Carmenta, R., Geldmann, J. *et al.* (2020) The extent and distribution of joint conservation-development funding in the tropics. *One Earth* 3(6), 753–762. DOI: 10.1016/j.oneear.2020.11.008

Shaheen, H. and Shinwari, Z.K. (2012) Phytodiversity and endemic richness of Karambar lake vegetation from Chitral, Hindukush-Himalayas. *Pakistan Journal of Botany* 44(1), 15–20.

Sharma, E., Chettri, N. and Oli, K.P. (2010) Mountain biodiversity conservation and management: a paradigm shift in policies and practices in the Hindu Kush-Himalayas. *Ecological Research* 25(5), 909–923. DOI: 10.1007/s11284-010-0747-6

Shinwari, Z.K. and Gilani, S.A. (2003) Sustainable harvest of medicinal plants at Bulashbar Nullah, Astore (Northern Pakistan). *Journal of Ethnophormacology* 84(2–3), 289–298. DOI: 10.1016/S0378-8741(02)00333-1

Shrestha, A.B. and Devkota, L.P. (2010) *Climate Change in the Eastern Himalayas: Observed Trends and Model Projections: Climate Change Impact and Vulnerability in the Eastern Himalayas.* Technical Report 1. International Centre for Integrated Mountain Development, Kathmandu.

Siraj-UD-Din, M., Minhas, R.A., Khan, M., Khan, M., Ali, U. *et al.* (2016) Conservation status of Ladakh urial (*Ovis vignei vignei* Blyth, 1841) in Gilgit Baltistan, Pakistan. *Pakistan Journal of Zoology* 48(5), 1353–1365.

Uddin, K., Chaudhary, S., Chettri, N., Kotru, R., Murthy, M. *et al.* (2015) The changing land cover and fragmenting forest on the Roof of the World: a case study in Nepal's Kailash sacred landscape. *Landscape Urban Planning* 141, 1–10. DOI: 10.1016/j.landurbplan.2015.04.003

Vincent, H., Hole, D. and Maxted, N. (2022) Congruence between global crop wild relative hotspots and biodiversity hotspots. *Biological Conservation* 265(7401), 109432. DOI: 10.1016/j.biocon.2021.109432

Virk, A.T., Sheikh, K.M. and Marwat, A.H. (2003) *NASSD Background Paper: Biodiversity*. IUCN Pakistan, Northern Areas Progamme, Gilgit.

Watson, J.E.M., Dudley, N., Segan, D.B. and Hockings, M. (2014) The performance and potential of protected areas. *Nature* 515, 67–73.

Worboys, G.L., Francis, W. and Lockwood, M. (2010) *Connectivity Conservation Management: A Global Guide*. Earthscan, London.

Wu, F., Liu, L., Gao, J., Yan, D., Hu, W. *et al.* (2015) Birds of the Ailao Mountains, Yunnan province, China. *Forktail* 31, 47–54.

Xu, J., Badola, R., Chettri, N., Chaudhary, R.P., Zomer, R.J. *et al.* (2019) Sustaining biodiversity and ecosystem services in the Hindu Kush Himalaya. In: Wester, P., Mishra, A., Mukherji, A. and Shrestha, A.B. (eds) *The Hindu Kush Himalaya Assessment*. Springer, Berlin/Heidelberg, Germany, pp. 127–165. (CC BY 4.0)

You, Q., Ren, G., Zhang, Y., Ren, Y., Sun, X. *et al.* (2017) An overview of studies of observed climate change in the Hindu Kush Himalayan (HKH) region. *Advances in Climate Change Research* 8(3), 141–147. DOI: 10.1016/j.accre.2017.04.001 (CC BY 4.0)

Yu, H., Miao, S., Xie, G., Guo, X., Chen, Z. *et al.* (2020) Contrasting floristic diversity of the Hengduan mountains, the Himalayas and the Qinghai-Tibet Plateau sensu stricto in China. *Frontiers in Ecology and Evolution* 8, 136. DOI: 10.3389/fevo.2020.00136

Yu, Y.D., Liu, L.H. and Zhang, J.H. (1989) Vegetation regionalization of the Hengduan Mountain region. *Mountain Research* 7(1), 47–55.

Zhang, D.C., Zhang, Y.H., Boufford, D.E. and Sun, H. (2009) Elevational patterns of species richness and endemism for some important taxa in the Hengduan Mountains, southwestern China. *Biodiversity and Conservation* 18(3), 699–716. DOI: 10.1007/s10531-008-9534-x

3 The Environmental Impact on Livelihood and Food Security in the Hindu Kush Himalayan Region: Current and Future Challenges for Nepal

Madan Koirala[1,*] and Yamuna Ghale[2]

[1]*Central Department of Environmental Science, Tribhuvan University, Nepal;* [2]*Research Lead, Nepal Center for Contemporary Research (NCCR), Civil Homes, Nepal*

Abstract

Most people in mountainous areas are extremely poor. Approximately 25% of the 240 million people in the Hindu Kush Himalayan (HKH) region are below the international poverty line. The percentage of the population in the region with access to basic sanitation services ranges between 37% and 75%, while the population using basic drinking water ranges between 57% and 97%. Consequently, the environmental status, livelihood requirements and basic guarantees of food security for the people are urgent issues in this mountain ecosystem where agriculture is the backbone of the economy. This chapter describes: (i) the environmental issues in Nepal and its dependence on local natural resources for livelihood improvements; and (ii) the areas of high food insecurity concerns and their inter-connectedness with production resources such as biodiversity in Nepal.

Key words: biodiversity, food security, livelihood, climate change

3.1 Contextualization of the Context

Multiple environmental problems exist in the HKH region in relation to food security and livelihood. The main (direct) environment-related drivers of change to mountain sustainability are land use and land-cover change, levels of resource exploitation, forms of pollution, presence of invasive alien species, mountain hazards, and climate change (Wang *et al*., 2019). These are followed by socio-economic drivers such as changing demographic, socio-economic situations, and governance, plus institutional provisions. Economic drivers add further dimensions through economic growth and differentiation, rapid infrastructure development, and urban and peri-urban expansions. Some of the challenges for sustainability are related to weak governance, over-exploitation of natural resources, environmental degradation, unregulated or rapid urbanization, and loss of indigenous and traditional culture.

The importance of agricultural biodiversity encompasses socio-cultural, economic and environmental elements. All domesticated crops and animals result from management of biodiversity, which is constantly responding to new challenges to maintain and increase productivity under constantly varying conditions and population pressures. Agricultural biodiversity is essential to satisfy basic human needs for food and

*Corresponding author: madankoirala@gmail.com

DOI: 10.1079/9781800622579.0003

livelihood security (Khush, 2017). As a general principle, dealing with these inter-relations needs an overarching approach. Through legally binding instruments, followed by the societal realization and practice, such issues can be dealt with. Food security can only be attained through global policy mechanisms and the country's constitutional provisions.

The constitution of Nepal ensures the right to clean environment (article 30), food security, right to food (article 36) and food sovereignty (Nepal Law Commission, 2015) of its citizens. To fulfil these commitments, inter-connections among environmental concerns, and food security and livelihoods are quite evident. The Climate Change Policy of Nepal (Ministry of Forests and Environment, 2019), in its article 8, specifically aims to improve the condition of food security, nutrition and livelihoods by adopting a climate-friendly agricultural system. It also aims to ensure the climate-responsive ecosystem services in the biodiversity, livelihood and watershed conservation sectors. The IPCC report clearly spells out that agriculture-oriented livelihoods are negatively impacted due to climate risks such as landslide frequency (92.6) and intensity (97.3) over a 20-year period between 1996 and 2016 (Birkmann, 2022). Likewise, the Global Hunger Index 2022 (von Grebmer *et al.*, 2022) identifies Nepal falling into the category of mild risk of hunger with a score of 19.1. The Index has identified climate risks as one of the major reasons. Furthermore, Nepal, in tenth position, definitely has concerns over understanding and response measures to ensure food security and livelihoods are minimally impacted by environmental hazards.

Furthermore, climate risks are not neutral and they impact more negatively vulnerable topographies, weak economies and disadvantaged communities who are on the verge of displacement, insecurity, discrimination and exclusion. Nepal is prone to all these risks, impacting food security and threatening livelihoods.

Given the international commitments, national priorities and political commitments as reflected in the political manifestos of different major political parties during the elections for three tiers of government during 2021/22, there are rays of hope towards environmental conservation, food security and improving livelihoods.

3.2 Current Status of Environmental Dimensions

3.2.1 Environment-related issues

An assessment report published on environmental dimensions of HKH, assessing the biodiversity status towards maintaining sustainability, gives the following key messages (Xu *et al.*, 2019):

(a) The mountain ecosystems of the Hindu Kush Himalaya are diverse with one of the highest diversities of flora and fauna providing varied ecosystem services to one fourth of humanity. With four out of 36 global biodiversity hotspots the HKH is a cradle for 35,000+ species of plants and 200+ species of animals. At least 353 new species – 242 plants, 16 amphibians, 16 reptiles, 14 fish, two birds, and two mammals, and at least 61 invertebrates – have been discovered in the Eastern Himalaya between 1998 and 2008, equating to an average of 35 new species finds every year.

(b) The HKH has numerous seeds of good practices in conservation and restoration of degraded habitat along with community development which need upscaling and outscaling. These participatory and community-based approaches have had large ecological, economic, and social positive impacts. Substantial degraded forest areas are regenerating, as decentralized practices reverse deforestation trends. Local communities have gained institutional space to decide for themselves on issues related to forests, income, inclusion and social justice.

(c) Global and regional drivers of change on biodiversity and ecosystem loss are prevalent and increasing in the HKH. These drivers include land use and land cover change, pollution, climate change, invasive species, solid waste, habitat degradation, and over-exploitation of resources, among others, impacting biodiversity, ecosystem services, and human well-being.

The National Climate Change Policy 2019 recognizes that Nepal is fragile in terms of its ecology and its climatic and microclimatic conditions. The same has identified that poverty, social disparities and high dependency on natural resources are making it more sensitive to negative impacts of climate change. Moreover, to achieve the goal of socio-economic prosperity through climate-resilient society, inter-sectoral co-ordination such as between agriculture and food security is crucial (Ministry of Forests and Environment, 2019).

The Nepal National Adaptational Plan (2021–2050), submitted to the United Nations Framework Convention on Climate Change (UNFCCC), has identified eight thematic areas where agriculture and food security are the prime areas of importance in responding to the impact of climate change. Agriculture has contributed 27.65% of the GDP and about 66% of the population worked in the agriculture sector in 2019/20 (Bruci *et al.*, 2021).

Climate change has the potential to prevent the achievement of national goals by negatively impacting agricultural production and nutrition security. Rising temperature, changes in precipitation patterns and an increase in the occurrence of extreme weather events have negatively impacted productivity in the agricultural sector. The sector is vulnerable to climate impacts because of high reliance on small-scale, rain-fed agriculture and dry-land farming (Paudel *et al.*, 2016). The direct economic cost of climate vulnerability in the agriculture sector in 2020 was 1.5–2% of the country's GDP (MoSTE, 2014). Similarly, a study published by the government of Nepal mentions that Nepal suffers high economic cost due to climate variability and extreme events. The economy of Nepal and the livelihoods of its people are dependent on the climate, because a large portion of GDP is associated with climate-sensitive activities (IDS-Nepal, PAC and GCAP, 2014). Current climate variability and extreme events lead to major impacts and costs in Nepal. These are predominantly floods, but also include rainfall variability on agriculture (rain-fed agriculture, soil erosion, droughts etc.). The analysis has used crop models in determining potential impacts of climate change on the agricultural sector in Nepal. The results show a strong distributional pattern over time and across the country, with potentially high impacts in the plain (Tarai) in later years but with potential benefits in the hills and mountains.

In this connection, the overall sectoral development, especially of agriculture and food security, aims to promote national seed-sector development as envisaged in the National Seed Vision 2013–2025, which clearly states environmental concerns. National Agriculture Policy also aims to protect, promote and properly utilize natural and environmental resources and biological diversity. The Agriculture Development Strategy (ADS) for 2015–2035 is considered a main document to enhance agriculture sector development through enhanced productivity, good governance, profitable commercialization and competitiveness (Ministry of Agriculture and Livestock Development, 2015). Nepal, as a signatory to international frameworks and commitments, also has a mandate to fulfil those. The recent UNFCCC declarations at COP 26 in Glasgow (UNFCCC, 2021a) and COP 27 in Sharm el-Sheikh (UNFCCC, 2022b) were expected to focus on mountains and the impact on livelihoods, though they could not table them on the mainstream agenda. Nepal therefore has fewer options to mainstream priority concerns, rather focusing on possible gains from offers related to services- and payment-related provisions.

3.2.2 Natural resources-related concerns

Exploitation of natural resources is one of the major environmental problems and drivers of change in the mountains. Over-exploitation includes the destruction of forest and shrubland for fuelwood and for commercial timber collection, over-harvesting of non-timber forest products (NTFPs) such as medicinal plants, over-grazing, over-fishing, and unsustainable and/or illegal mining (Shrestha and Dhillion, 2003; Shrestha and Bawa, 2015).

Generally, demands for natural resources increase in tandem with local and regional socio-economic development. Demand pressures are further exacerbated by population growth. Together, these two factors are the main (ultimate) driving forces leading to over-exploitation in HKH (Poudel and Shaw, 2015). In the medium to long term, resource over-exploitation always results in the degradation of ecosystems, often followed by loss of genetic diversity and the extinction of species (Kaur *et al.*, 2012). A decline in the provision of ecosystem services, including the loss of biodiversity, ultimately has disastrous effects on human well-being.

Agriculture is the direct or indirect source of livelihood for over 70% of the population of the HKH, and a substantial contributor to national incomes (Tiwari, 2000). Changing precipitation patterns, reduced run-off in the major river basins, and increasing temperatures will simultaneously put additional pressure on available water resources and increase agricultural water demand for rain-fed and irrigated crops, thus increasing competition for water for agriculture, industry and human consumption (Hanjra and Qureshi, 2010). Agriculture in this region is mostly rain-fed (about 60%; World Bank, 2016) and therefore vulnerable to changes in timing and frequency of precipitation. Besides water availability, crop yield depends on a number of biophysical processes and variables (such as thermal stress, humidity, solar radiation, nitrogen stress, ozone and the fertilization effect of CO_2) and their complex, non-linear interactions (Challinor *et al.*, 2009). Extremes in floods and droughts through much of the upcoming century may destroy the food production base of the region (Bruinsma, 2003). Fischer *et al.* (2002) expected a temperature increase of 1.5–2.5°C to lead to a decline in the agricultural productivity of crops such as rice, maize and wheat. Climate changes are predicted to reduce the livelihood assets of rural people, alter the path and rate of national economic growth and undermine regional food security due to changes in natural systems and impacts on infrastructure.

Nepal occupies only 0.1% of the global area, but harbours over 3% and 1% of the world's known flora and fauna, respectively. Of the global biodiversity, Nepal holds 9.3% birds, 4.5% mammals, 4% butterflies, 2.7% flowering plants, 1.6% reptiles, 1% amphibians and 1% freshwater fishes (www.ntnc.org.np).

The abundance of biodiversity is very much acknowledged by the national policies and programmes. The Agrobiodiversity Policy 2006 has identified the role of agrobiodiversity in relation to environment, food systems, associated indigenous knowledge and skills, as well as equitable benefit sharing out of it. Agrobiodiversity, indeed, has a crucial role in protection and sustenance of the natural environment, promoting resilient farming systems and a diversified food system that nurtures soil, as well as being a rich source of nutritious foods both in domesticated and wild forms. Moreover, agrobiodiversity is a key agent to reveal the cultural aspects of Nepalese life where different cropping seasons, plants and animals are worshipped, and where different dishes prepared according to the cultural norms and values are consumed (Ghale, 2021a). In the context of climate change and possible negative impacts in the food system, agrobiodiversity is the major catalyst to develop resilient production, balance weather conditions, and have major input in research (genetic materials) and thinking for alternative food systems. It has thus been identified by global actors concerned about the environment, climate change and food security who opt for localized good systems (Ghale, 2023).

Biodiversity loss

Nepal's biodiversity is threatened by multiple factors. Loss and degradation of natural habitats, such as forests, grasslands and wetlands, due to expansion of settlements, agriculture and infrastructure; over-exploitation; invasion by alien species; and pollution of water bodies remain the predominant threats to natural ecosystems. Rapid expansion of hybrid varieties and improper use of insecticides and pesticides are the major threats to agrobiodiversity. Climate change can have profound impacts in future, particularly in the mountains. Most threats continue to increase. Moreover, interaction of multiple threats is speculated to cause increased pressures, thereby leading to further decline, degradation and loss of habitats (Government of Nepal, 2014).

The main threats to biodiversity at different biodiversity levels are habitat loss and deforestation; high livestock grazing intensity, mainly in protected areas and rangelands; over-harvesting of selected NTFPs, invasion by invasive alien plant species (IAPs); forest fires; improper use of pesticides; urbanization; and loss of traditional practices, mainly in agrobiodiversity; and climate change in mountain biodiversity.

It is believed that about nine species of birds, 11 species of mammals and seven species of medicinal herbs have already vanished from the country in just a quarter of a century. With their habitat taken down in the name of deforestation and urbanization, species like musk deer, Asiatic elephants, gaur, snow leopards etc. are on the verge of extinction. As per the data of the National Conservation status of Nepal's fauna, out of 878 bird species recorded in Nepal,

167 are facing the threat of extinction. About 27 mammals, nine birds and three reptiles are categorized as endangered in Nepal. According to Nepal's biodiversity data 2007, the country is losing most of the magnificent species rapidly. Due to illegal trade, many of the medicinal species like *Swertia chiraita, A. aconitum, Nardostachys jatamasi, Raulfia serpentine, Cordyceps sinensis* etc. are threatened with extinction (Nishan, 2019).

3.2.3 Livelihoods-related issues

Sustainable Rural Livelihoods: Practical Concepts for the 21st Century (Chambers and Conway, 1992), published by the Institute of Development Studies, introduces the concept of sustainable livelihoods (SL). This foundational publication was a working paper that Chambers co-authored with Gordan Conway entitled 'Livelihood in its simplest sense is a means of gaining a living'. In a fuller 'working definition':

> A livelihood comprises the capabilities, assets (stores, resources, claims and access) and activities required for a means of living: a living is sustainable which can cope with and recover from stress and shocks, maintain or enhance its capabilities and assets, and provide sustainable livelihood opportunities for the next generation; and which contributes net benefits to other livelihoods at the local and global levels and in the short and long term.

Chambers and Conway mention social and environmental sustainability issues related to livelihood. However, over the years human-induced global environmental changes are leaving more scars on the livelihood of humans compared to social ones. An avenue has been opened recently after the breakthrough decision made by the UNFCCC COP27 in 2022 to respond the loss and damage caused by climate change-induced disasters to vulnerable countries (UNFCCC, 2022a). Nepal not only holds an abundance of biodiversity, but it is also rich in social, cultural, ethnic and climatic diversities. The social statistics of Nepal show that the inhabitants are diverse in nature by their social and cultural identities. Of the total population of approximately 30 million, there is 51.04% female and 48.96% male, with a gender ratio of 95:91. Geography-wise, Terai, hills and mountains hold 53.61%, 40.31% and 6.08%, respectively. Likewise, Nepal has 125 castes and ethnicities with the highest percentage Chhetri (16.6%) and the lowest Rai (2.3%). Ghale (2021b) conceptualizes that there is a strong nexus between food security, climate change and livelihood vulnerability, especially in the mountains, due to their geographic fragility as mentioned in Fig. 3.1. The nexus shows that there is a correlation between climate change, especially in the fragile ecosystem of the mountains, and its negative impact on food security and livelihoods.

Regmi and Sapkota (2021) have reported various impacts of climate change on forests, and loss of and damage to species. There is a growing incidence of animal death, injury and wild animals being swept away in and around Tarai protected areas, especially in Chitwan National Park during heavy rainfall and flash floods. During rainy seasons, rhinoceros calves and other animals are highly susceptible to flash flooding. Forests in Nepal are permanent livelihood sources and safety nets for over 65% of the population. The impact of climate change on the livelihoods of forest-dependent populations and communities has become severe for women, indigenous people, Dalit and poor households. They are the ones who mostly rely on forest resources for their survival, income and employment opportunities. These social groups are already discriminated against for accessing quality forest resources. The diminished availability of forest products will further aggravate access to quality forest resources (Khanal *et al.*, 2019). The usual working hours for women of western Nepal are reported to be over 18 hours a day, mostly for collecting fuelwood, grass and fodder (Sugden *et al.*, 2014).

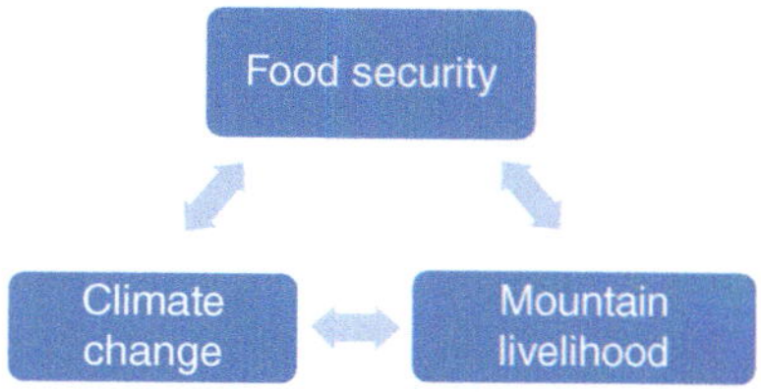

Fig. 3.1. Inter-relation between climate change, mountain livelihood and food security.

3.2.4 Inclusion-related issues

The sufferings of people in different geographies and associated climatic vulnerabilities are different. For example, erratic rainfall and snowlessness, avalanches in the mountains, landslides, forest fires in the hills, and floods in the Terai are some of the major incidents that are directly related to environment, natural-resource management, farming practices and demographic settlements. They therefore impact on people differently at different times and in different seasons according to their coping strengths. They create an opportunity for the people to make best use of their indigenous knowledge, skills and experiences to develop resilience and cope with the negative impacts. The beliefs and practices of the particular communities are embedded in their cultures and rituals, castes and ethnicities. Women are the repository of such knowledge and skills as they are the most permanent dwellers in that specific context. However, the decision-making platforms are not female and women are not getting enough space to voice their needs, priorities and demands to be reflected in the policies, programmes and plans (Ghale, 2021b). Their suffering is more evident and more pronounced between resource-poor and rich people. Inclusion of people from different geographic, social, cultural and gender dimensions is crucial to make targeted interventions to combat negative impacts of climate change, protect the environment and improve food security and livelihoods.

3.3 Food Security and Livelihoods

The *The Hindu Kush Himalaya Assessment Mountains, Climate Change, Sustainability and People* report (Rasul *et al.*, 2019) lists key findings, especially on food and nutrition security:

1. Food and nutrition insecurity remains a serious challenge in the Hindu Kush Himalaya (HKH) region; more than 30% of the population suffers from food insecurity and around 50% face some form of malnutrition, with women and children suffering the most. The insecurity is more severe in remote mountain areas. Challenges to food security in the mountain areas differ from those in the plains due to inaccessibility, fragility, seasonality, limited economic opportunities, poor market access, and harsh biophysical conditions.
2. The causes of food and nutrition insecurity in the HKH are multi-faceted and complex, and influenced by a range of factors including high poverty, natural resource degradation, climate change, low level of market development, uncertain food support, and inadequate policy and institutional support.
3. Traditional food systems in mountains are currently under threat from rapid socio-economic and environmental changes including changing dietary habits, changes towards mono-cropping and commodity crops, loss of water sources, soil degradation, and decline in market value. Mountain agriculture is becoming relatively less competitive and the youth are increasingly abandoning agricultural livelihoods, leading to decreased food production and adversely affecting local food systems.

Regmi and Sapkota (2021) produce a long list of observed impacts on agriculture and food security: key extreme events and hazards impacting the agriculture sector include droughts, floods, hailstorms, cold waves, heatwaves, thunderstorms, and insect and disease pests. Climate-related extreme events and hazards impact large tracts of agricultural land in different crop-growing seasons. For instance, between 2002 and 2009, the largest area affected was rice, followed by maize, and the biggest impact was experienced in 2002, followed by 2006 and 2004. Among all hydro-meteorological hazards in Nepal, drought, a slow-onset widespread disaster, ranks top in terms of impact on crops. Drought and slow onset of monsoon act to delay planting, which in turn leads to poor crop germination/flowering, and fruit setting, and eventually declining productivity. Between 1971 and 2007, nearly 850,000 ha of crops were lost due to weather- and climate-related events. Droughts accounted for 38.9% of lost crops and floods for 23.2%. In 2006, drought contributed to the loss of 11% rice yield and 7% wheat yield in Nepal. The continuous rainfall between 11 and 14 August 2017 flooded 35 districts (18 districts severely affected) and inundated about 80% of the land along the Tarai region. Fifty-eight per cent of the total loss and damage caused by the flood was on agriculture, livestock and irrigation infrastructure. Hailstorms, cold waves, and heatwaves adversely affect crop yield, mainly in the mid-hills and the Tarai region. Hailstorms damage cereal crops, vegetables and fruits at all stages, thus causing a yield decline. Cold waves affect crops, particularly belonging to the solanaceous family – potatoes, tomatoes and chillies. Insects and disease pests are also major hazards affecting agriculture. Recent analysis revealed that the changes in maximum and minimum temperatures and precipitation have a direct contribution to the yield of major crops. The existing literature on projection and scenario analysis shows that climate change will have a substantive impact on crop yield and production. An IFAD report shows a 6.77% decline in rice yield and a 5.66% increase in wheat yield by 2050 and a 12.90% decline in rice yield and 9.77% decline in wheat yield by 2080, in comparison to 2011, due to changing temperature and precipitation.

Climate change has also impacted agricultural biodiversity. Depending on its nature, extent and intensity, sometimes whole crops, varieties or genes are wiped out from a locality. It is reported that farmers cannot harvest anything at all when droughts affect growth and reproduction, floods inundate entire land for days, insects and disease damage crops, and weeds outperform crops. However, positive impacts have been observed in higher-altitude regions on new crops, vegetables and fruit trees such as maize, tomato, chilli, mango and lychee.

Livestock, poultry and fisheries are impacted by rising temperatures, negatively affecting weight gain, reproduction, breeding patterns, feed intake and conversion efficiency, heat stress, morbidity, vector- (e.g. ticks and flies) borne diseases and parasitic diseases. Temperature rises have led to deterioration in the quality of water (Regmi and Sapkota, 2021).

3.3.1 Conceptual understanding of food security, right to food, food sovereignty and food governance

The constitution of Nepal has articulated provision of food security, right to food and food sovereignty in a very progressive manner. Article 36 – Rights Relating to Food – mentions: '(i) Every citizen shall have the right relating to food; (ii) Every citizen shall have the right to be safe from the state of being in danger of life from the scarcity of food; and (iii) Every citizen shall have the right to food sovereignty in accordance with law.' Likewise, Article 42.4, under Right to Social Justice, states: 'Every farmer shall, in accordance with law, have the right to have access to land for agro-activities, select and protect local seeds and agro-species which have been used and pursued traditionally'. Article 51.e, under Policies of the State, relating to Agriculture and Land Reforms, specifies: (i) 'To introduce scientific land reforms having regard to the interests of the farmers, while ending the dual ownership existing in the land; (ii) To enhance product and productivity by carrying out land pooling, while discouraging absentee land ownership; (iii) To carry out land management and commercialization, industrialization, diversification and modernization of agriculture by pursuing land-use policies to enhance agriculture product and productivity, while protecting and promoting the rights and interests of the farmers; (iv) To make proper use of land, while regulating and managing it on the basis of, inter alia, productivity, nature of land and ecological balance; (v) To make provisions for appropriate prices of agricultural inputs and agricultural products and for market access' (Nepal Law Commission, 2015).

In alignment with the constitution of Nepal, the government of Nepal has promulgated the Right to Food and Food Sovereignty Act 2018. However, studies show that the legal documents in most cases do not correspond to the principles of conceptual understanding and theoretical orientations as shown in Fig. 3.2.

3.3.2 Overall trend of food security and trend of livelihood status

According to the *World Data Atlas on Nepal Agriculture* (2021), the food production index of Nepal in 2021 was 122.3. Between 1972 and 2021, the index grew substantially from 24.5 to 122.3, rising at an increasing annual rate and reaching a maximum rate of 22.10 in 1983, then decreasing to 1.51 in 2021 (Knoema, 2023). Six major cereals – paddy, maize, wheat, millet, buckwheat and barley – contribute to food security in Nepal from a production perspective. However, there is relatively low productivity of

Major theories	Major focuses
Household Food Security	Individual's intake requirements and intra-household issues
Entitlement	People's entitlement to secure food
Capability	State and people's ability to respond to the fulfillment
Governance	Mechanisms, processes and institutions
Citizen Participation	Legitimacy, political credibility of the process and increased public ownership

Concepts	Pillars
Food Security	Availability, Accessibility, Affordability, Stability and Utilization
Right to Food	Respect, Fulfil, Protect
Food Sovereignty	Political, Economic, Social, Ecological Dimensions
Food Security Governance	Accountability, Participation, Transparency, Effectiveness and Efficiency, Rule of Law

Fig. 3.2. Theories and concepts related to food security. (From Ghale, 2021a)

the crops due to the dominance of traditional farming; a significant quantity is still supplemented through external imports. Over recent years, an increasing trend in vegetable, fruit, fish and milk production is an indication of increasing food security. The *World Hunger Report 2022* scores Nepal moderate, with a score of 19.1, and ranks it 81 among the 121 countries. The hunger status is measured with four elements to calculate the hunger index: (i) under-nourishment; (ii) child wasting; (iii) child stunting; and (iv) child mortality.

Though the overall food and nutrition situation is improving, women and disadvantaged people are still suffering. The report shows that stunting rates range from 22.6% to 22.9% in Gandaki and Bagmati provinces, respectively, in the central region of the country, to more than double that figure (47.8%) in Karnali province in the west of the country. Likewise, children consuming foods rich in iron and iron supplements are low across provinces. Provinces 1 and 2 have the lowest percentage in the case of women and children, respectively. Women and children in Provinces 1 and 2 were also found to be the most anaemic, which could be due to poor iron intake. Moreover, food security is a challenge among the Terai/Madhesh and Muslim communities in Madhesh province. This confirms that food insecurity is a manifestation of a combination of technical, social, cultural and religious norms, beliefs and practices among different communities in certain demographies. There is an absence of mechanisms to assess and monitor the status of right to food and food sovereignty along with conceptual lack of clarity among the key actors, especially of the three tiers of government on food security governance. The policies and regulatory provisions were changing over time in terms of their content, scope and comprehension among the three tiers. Since the summary of the *Global Hunger Report*, the UN Global Food System Summit also highlighted that conflicts, climate extremes and COVID-19 had an effect on food security. It therefore urges local action as a mechanism to strengthen inclusive, resilient and sustainable food systems and realizes the need to transform food systems at all levels (KC, 2019).

3.3.3 Crisis and impacts on food security

The *National Climate Change Policy* (Government of Nepal, 2021) identified that there were three major shocks that impacted production, productivity and food security: (i) lockdowns against COVID-19 created difficulties in the supply of agricultural inputs like fertilizers, seeds, pesticides and agricultural implements and affected the price chain by increasing the price of agricultural produce such as food grains, fresh vegetables and fruits; (ii) the supply of production materials including feeds, medicines and vaccines required for livestock and fisheries was also affected. There has been delay in providing output-based incentives on foundation seeds and the construction of granaries at local levels has not been implemented in time; and (iii) loan investment from co-operatives has declined by about 2.0%. Identification and data collection of poor households conducted at the local level has also been affected. It clearly means that these kind of shocks, either of natural disasters or crisis, impact negatively on food security and livelihoods directly.

3.3.4 Global commitments and obligations

There are global instruments that have a mandate to deal with the issues related to biodiversity, climate, trade, agriculture and food security. Among them, the most relevant ones to Nepal are the International Treaty on Plant Genetic Resources for Food and Agriculture (ITPGRFA), the Convention on Biological Diversity (CBD), the World Trade Organization (WTO), the International Labour Organization (ILO) and the International Panel on Climate Change (IPCC).

3.5 Conclusion and the Way Forward

- Environment, livelihood and food security are inter-related components of a common system. Three of the components need to be looked at from a complementary perspective and cannot be dealt with independently or in isolation.
- The overall scenarios of global and national food insecurity reveal that the situation is likely to worsen in the current global crises caused by conflict, climate change and the economic fallout due to COVID-19. It shows that the vulnerability of populations facing food insecurity may rise.
- Inequalities of wealth and income, gender-based discrimination, discrimination against population groups and other forms of inequalities are increasing. Therefore, Leave No One Behind (LNOB) not only entails reaching the poorest of the poor, but requires combating discrimination and rising inequalities and their root causes.
- A holistic approach to food security, right to food and food sovereignty are not yet fully materialized because of lack of proper understanding of the essence of the three intertwined concepts. Recent laws and policies related to food security, right to food and food sovereignty are supportive to promote rights-based public policies but still lack operationalization in fulfilling needs, demands and priorities of the public.
- The need to devise inclusive governance and accountability should be at the centre of efforts to transform food systems ensuring citizens' participation, action and oversight, and considering the local context.
- Necessity to scale up resources for humanitarian needs while transforming food systems to make them resilient to shocks.
- The need to actualize the 3Cs (Co-ordination, Co-operation, Collaboration) of the constitution among the different actors and programmes is of utmost importance (Fig. 3.3).

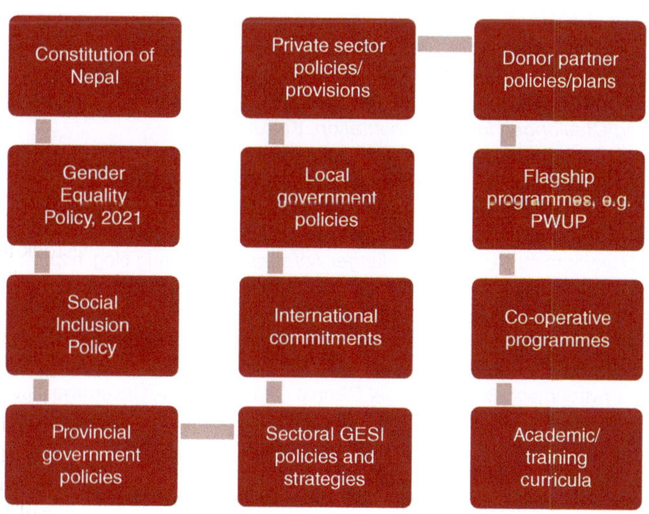

Fig. 3.3. Different actors for addressing issues of environment, livelihood and food security. GESI: gender equality and social inclusion. (From Ghale, 2023)

References

Birkmann, J., Liwenga, E., Pandey, R., Boyd, E., Djalante, R. *et al.* (2022) Poverty, livelihoods and sustainable development. In: Pörtner, H.-O., Roberts, D.C., Tignor, M., Poloczanska, E.S., Mintenbeck, K. *et al.* (eds) *Climate Change 2022: Impacts, Adaptation and Vulnerability. Contribution of Working Group II to the Sixth Assessment Report of the Intergovernmental Panel on Climate Change*. Cambridge University Press, Cambridge, UK and New York, pp. 1171–1274. DOI: 10.1017/9781009325844.010

Bruci, E., Islami, B., Kamberi, M., Bego, F., Deda, M. *et al.* (2021) *Third National Communication of the Republic of Albania under the United Nations Framework Convention on Climate Change*. Ministry of Environment Reports, Kathmandu.

Bruinsma, J. (2003) *World Agriculture: Towards 2015/2030*. Food and Agriculture Organization of the United Nations (FAO), Rome.

Challinor, A.J., Ewert, F., Arnold, S., Simelton, E. and Fraser, E. (2009) Crops and climate change: progress, trends, and challenges in simulating impacts and informing adaptation. *Journal of Experimental Botany* 60(10), 2775–2789. DOI: 10.1093/jxb/erp062

Chambers, R. and Conway, G. (1992) *Sustainable Rural Livelihoods: Practical Concepts for the 21st Century*. IDS Discussion Paper no. 296, Institute of Development Studies, Sussex, UK.

Fischer, G., Shah, M. and van Velthuizen, H. (2002) Climate change and agricultural vulnerability. IIASA, Vienna. Available at: http://adapts.nl/ perch/resources/climateagri.pdf (accessed 21 July 2023).

Ghale, Y. (2021a) Food security governance and right to food: understanding state-people relationship in Nepal. PhD thesis, Agriculture and Forestry University, Rampur, Chitwan, Nepal.

Ghale, Y. (2021b) Food security, climate change, and mountain livelihoods: understanding nexus. Paper presented in a virtual national symposium on *Mountain Voices for Sustainable Human Future: Constraints and Opportunities*, organized by KIRDARC. Focus International and the Mountain Foundation Nepal, 15 September 2021.

Ghale, Y. (2023) Review of the GHI report in the Nepalese context. Presentation in the Global Hunger Index Report launching programme organized by Welt Hunger Hilfe and NGO Federation of Nepal, 27 January 2023, Kathmandu.

Government of Nepal (2014) *Nepal Biodiversity Strategy and Action Plan 2014–2020*. Ministry of Forests and Soil Conservation, Kathmandu.

Government of Nepal (2021) *National Climate Change Policy*. Ministry of Forests and Environment, Kathmandu.

Hanjra, M.A. and Qureshi, M.E. (2010) Global water crisis and future food security in an era of climate change. *Food Policy* 35(5), 365–377. DOI: 10.1016/j.foodpol.2010.05.006

IDS-Nepal, PAC and GCAP (2014) *Economic Impact Assessment of Climate Change in Key Sectors in Nepal*.

Kaur, R., Joshi, S.P. and Srivastava, M.M. (2012) Natural resource degradation in three sub-watersheds of river Tons, Uttarakhand, India. *Tropical Ecology* 53(3), 333–343.

KC, H. (2019) *Federal Government's Policies, Priorities and Strategies for Food and Nutrition Security: Opportunities and Challenges for Implementation*. Kathmandu.

Khanal, P., Wagle, B.H., Upadhaya, S., Ghimire, P. and Acharya, S. (2019) Perceived climate change impacts and adaptation strategy of indigenous community (Chepang) in rural midhills of Nepal. *Forestry: Journal of Institute of Forestry Nepal* 16, 48–61. DOI: 10.3126/forestry.v16i0.28353

Khush, G.S. (2017) *The Importance of Biodiversity to Food and Agricultural Systems across the Globe. Is the Sustainable Development Goal to End Hunger Achievable?* Borlaug blog. https://www.worldfoodprize.org/index.cfm/88533/18098/the_importance_of_biodiversity_to_food_and_agricultural_systems_across_the_globe (accessed 4 April 2024).

Ministry of Agriculture and Livestock Development (2015) *Agriculture Development Strategy (2015–2035), Parts 1 and 2*. Government of Nepal, Kathmandu.

Ministry of Forests and Environment (2019) *National Climate Change Policy*. Government of Nepal, Kathmandu.

MoSTE (2014) *Economic Impact Assessment of Climate Change in Key Sectors in Nepal*. IDS Nepal, Global Climate Action Partnership and Practical Action, Kathmandu.

Nepal Law Commission (2015) Constitution of Nepal.

Nishan, K.C. (2019) *Biodiversity Loss in Nepal*. TUNZA Eco Generation. https://tunza.eco-generation.org/ambassadorReportView.jsp?viewID=48145&searchName= (accessed 4 April 2024).

Paudel, B., Zhang, Y.-L., Li, S., Liu, L., Wu, X. *et al.* (2016) Review of studies on land use and land cover change in Nepal. *Journal of Mountain Science* 13(4), 643–660. DOI: 10.1007/s11629-015-3604-9

Poudel, S. and Shaw, R. (2015) Demographic changes, economic changes and livelihood changes in the HKH. In: Nibanupudhi, H.K. and Shaw, R. (eds) *Mountain Hazards and Disaster Risk Reduction*. Springer, Japan, pp. 105–123.

Rasul, G., Saboor, A., Tiwari, P.C., Hussain, A., Ghosh, N. *et al.* (2019) Food and nutrition security in the Hindu Kush Himalaya: unique challenges and niche opportunities. In: Wester, P., Mishra, A., Mukherji, A. and Shrestha, A. (eds) *The Hindu Kush Himalaya Assessment Mountains, Climate Change, Sustainability and People*. ICIMOD, HIMAP, Springer (Open).

Regmi, B. and Sapkota, R. (2021) *Vulnerability and Risk Assessment and Identifying Adaptation Options: Summary for Policy Makers*. Government of Nepal, Ministry of Forests and Environment, Kathmandu.

Shrestha, P.M. and Dhillion, S.S. (2003) Medicinal plant diversity and use in the highlands of Dolakha district, Nepal. *Journal of Ethnopharmacology* 86, 81–96.

Shrestha, U.B. and Bawa, K.S. (2015) Harvesters' perceptions of population status and conservation of Chinese caterpillar fungus in the Dolpa region of Nepal. *Regional Environmental Change* 15(8), 1731–1741. DOI: 10.1007/s10113-014-0732-7

Sugden, F., de Silva, S., Clement, F., Maskey-Amatya, N., Ramesh, V. *et al.* (2014) A framework to understand gender and structural vulnerability to climate change in the Ganges River Basin: lessons from Bangladesh, India and Nepal. IWMI Working Paper 159. DOI: 10.5337/2014.230

Tiwari, P.C. (2000) Land-use changes in Himalaya and their impact on the plains ecosystem: need for sustainable land use. *Land Use Policy* 17(2), 101–111. DOI: 10.1016/S0264-8377(00)00002-8

UNFCCC (2022a) COP27 reaches breakthrough agreement on new 'Loss and Damage' fund for vulnerable countries. Available at: https://www.powershiftafrica.org/storage/publications/Cop27%20six-point%20plan_1667424915.pdf (accessed 21 July 2023).

UNFCCC (2022b) *Six-Point Action Agenda for Realizing Sharm El Sheikh as An African Cop27 for Africa*. Available at: https://www.powershiftafrica.org/storage/publications/COP%2027%20Assessment_1689600881.pdf (accessed 21 July 2023).

UNFCCC (2022c) Report of the Conference of the Parties serving as the meeting of the parties to the Paris Agreement on its third session, Glasgow, 31 October–13 November 2021. Available at: https://www.powershiftafrica.org/storage/publications (accessed 21 July 2023).

von Grebmer, K., Bernstein, J., Resnick, D., Wiemers, M., Reiner, L., Bachmeier, M. *et al.* (2022) 2022 Global Hunger Index: Food Systems Transformation and Local Governance. Welthungerhilfe, Bonn; and Concern Worldwide, Dublin.

Wang, Y., Wu, N., Kunze, C., Long, R. and Perlik, M. (2019) Drivers of change to mountain sustainability in the Hindu Kush Himalaya. In: Wester, P., Mishra, A., Mukherji, A. and Shrestha, A.B. (eds) *The Hindu Kush Himalaya Assessment – Mountains, Climate Change, Sustainability and People*. Springer Nature, Cham, Switzerland.

World Bank (2016) *World Bank data*. Available at: http://data.worldbank.org/ indicator/ (accessed 21 July 2023).

World Data Atlas (2021) Nepal agriculture. Nepal Food Production Index. https://knoema.com/atlas/Nepal/Food-production-index (accessed 4 April 2024).

Xu, J., Badola, R., Chettri, N., Chaudhary, R.P., Zomer, R.J. *et al.* (2019) Sustaining biodiversity and ecosystem services in the Hindu Kush Himalaya. In: Wester, P., Mishra, A., Mukherji, A. and Shrestha, A.B. (eds) *The Hindu Kush Himalaya Assessment: Mountains, Climate Change, Sustainability and People*. Springer, Berlin/Heidelberg, Germany, pp. 127–165.

4 Challenges of Ecological Conservation and Restoration under the COVID-19 Global Pandemic in the Hindu Kush Himalayan Region

Muhammad Khalid Rafiq[1,*] and Awais Iqbal[2]

[1]*Rangeland Research Institute, National Agricultural Research Center, Islamabad, Pakistan;* [2]*College of Ecology, Lanzhou University, Lanzhou, China*

Abstract

The Hindu Kush Himalayan (HKH) region extends across eight countries, from Afghanistan in the west to Myanmar in the east, crossing Pakistan, India, China, Nepal, Bhutan and Bangladesh. It is the source of water for ten major Asian river systems and hosts diverse ecosystems, biodiversity and cultures. The HKH region is the poorest and largest mountain system in the world, characterized by harsh climate, poor living conditions, poor sanitation and high probability of quickly-spreading diseases. COVID-19 exacerbated the risks affecting the HKH region. This chapter focuses on challenges in the HKH region in terms of diseases, namely COVID-19, and discusses: (i) the influence, scope and impact of COVID-19; (ii) the medical and health services to combat COVID-19 and other diseases; (iii) the environmental consequences after COVID-19; and (iv) potential pathways for recovery from COVID-19 in the HKH region.

Keywords: pandemic, medical and health services, ecological restoration pathways

4.1 Introduction

4.1.1 The Hindu Kush Himalayan region: an overview

The Hindu Kush Himalayan (HKH) region extends from 15.95° N to 39.31° N latitudes, and 60.85° E to 105.04° E longitudes, and encompasses mountains of eight south- and east-Asian countries: Afghanistan, Pakistan, China, India, Nepal, Bhutan, Bangladesh and Myanmar, from west to east (Table 4.1). The region is characterized by rugged high-altitude landscapes, and has the largest area covered by glaciers and permafrost outside the polar regions. It is, therefore, often referred to as the 'third pole'. The HKH region, including the Tibetan Plateau, plays an important role in global atmospheric circulation that influences two major weather systems: monsoons and westerlies. It is a geomorphologically unique region with the Himalayas acting as a geological buffer protecting the lush green southern slopes from the dry and arid winds of the north and, thus, supports a unique mosaic of biota and eco-regions.

*Corresponding author: khalid_parc@hotmail.com

DOI: 10.1079/9781800622579.0004

Table 4.1. The Hindu Kush Himalayan (HKH) region: size of the country in area and percentage in HKH region. (From Sharma and Partap, 1994)

Country	HKH area (km^2)	Percentage of country in HKH
Afghanistan	390,475	**60**
Bangladesh	13,189	9
Bhutan	38,394	100
China	1,647,725	17
India	482,920	14
Myanmar	317,640	47
Nepal	147,181	100
Pakistan	404,195	51
Total	**3,441,719**	

4.1.2 Main features of the Hindu Kush Himalayan mountains

The HKH region covers approximately 4.3 million km^2 and is rich in biological and cultural diversity along with a complex terrain (Xu *et al.*, 2019). The mountains of the HKH have been blessed with soil diversity across the altitudes and, thus, possess a large number of microclimates under different ecological gradients. HKH maintains vegetation from tropical to alpine in a short distance across changing elevation levels, which is dependent on climatic variation, altitude, latitudinal gradients and soil type (Sharma *et al.*, 2010). This biophysical diversity set up the foundation for a rich biodiversity and species evolution (Xu *et al.*, 2019). The region also includes all, or parts of, major global biodiversity hotspots, namely Himalayan mountains, mountains of south-west China, Indo-Burma, and Central Asian mountains (Chettri *et al.*, 2010; Mittermeier *et al.*, 2011), which contain many threatened and endemic species. Moreover, the region has more than 60 eco-regions of which 12 are part of the Global 200 eco-regions (Olson and Dinerstein, 2002), and more than 39% of the HKH is a protected area network (Chettri *et al.*, 2008; Xu *et al.*, 2019). More than 50% of the animal species are under pressure and are decreasing in population (Xu *et al.*, 2019).

The HKH region is a source of livelihood for 240 million people and provides drinking water to 1.9 billion people downstream (Sharma *et al.*, 2019). The region provides food, timber for building material, fuelwood, fibre, medicinal plants, and water for drinking and irrigation. There is evidence that the mountain ecosystems also play a role in climate regulation, carbon storage and the maintenance of aesthetic, cultural and spiritual values (Sharma *et al.*, 2015; Chaudhary *et al.*, 2017; Xu *et al.*, 2019). The diverse vegetation cover in forests and agro forestry systems stabilize watersheds, reduce flooding and landslides, and regulate continuous flows of water over the year. However, in recent decades, both the quality and quantity of ecosystem services in the region have been degrading (Xu *et al.*, 2019).

4.1.3 River basins of importance in the HKH region

The total geographical area of the river basins in the HKH is 2.79 million km^2, including downstream areas with basins covering 8.99 million km^2. From Afghanistan in the west to Myanmar in the east, the region extends 3500 km over all or parts of eight countries. The region has about 60,000 km^2 of glaciers (Bajracharya and Shrestha, 2011). The HKH mountains are the source of ten major river systems – the Amu Darya, Indus, Ganges, Brahmaputra (Yarlungtsanpo), Irrawaddy, Salween (Nu), Mekong (Lancang), Yangtse (Jinsha), Yellow River (Huanghe) and Tarim (Dayan), which provide water, ecosystem services and the basis for livelihoods for about 210 million people. The people in the river basins depend on the water for irrigation, drinking, sanitation and industrial use. The rivers also recharge and nurture some of the richest ecosystems and farmlands in the world. The HKH mountains have been sites of adaptation, mitigation and resilience of the local people, who have maintained a rich cultural identity, food security and biogenetic diversity.

4.1.4 Ecosystem services provided by the HKH mountains

The HKH region is extremely rich in ecosystem goods and services including water, biodiversity, medicinal and aromatic plants, wild food crops, and ecotourism resources. The natural resources provide the basis for a substantial part of the region's total gross domestic product besides supporting the subsistence of more than a billion people. The global importance of ecosystem resources in the HKH region has been recognized (Sanlaville and Prieur, 2005). The rapid economic growth in China and India can be attributed partially to the availability of a strong natural resource base, especially water and industrial raw materials. Conserving and regulating water, preserving soil, supporting forest management and biodiversity conservation, providing fodder and forage resources for pastoral development, maintaining good environmental conditions for agriculture development, producing timber and non-timber forest products, and promoting adventure and nature-based tourism are some of the most prominent ecosystem services in the region. Researchers (Lu *et al.*, 2004) estimated the annual ecosystem service value of the Himalayas at US$150–170 billion. Climate change, especially global warming, is already having pronounced negative effects on these services; notably, the shrinking of alpine lakes; reduction of river flow; glacial retreat; decrease of water availability; degradation of permafrost, forests and grasslands; desertification; water and soil losses; and degradation of rangeland resources. Mountains play a vital role in natural water purification and retention in groundwater, ice, snow, lakes, forested watersheds and streams. They also regulate and modulate atmospheric and climate balances beyond geographical boundaries. For example, the cryospheric environment of 'water-ice-air-ecosystem-human' systems in the HKH region is believed to influence and respond to the global and regional environmental change processes and mechanisms including the Asian monsoon systems. Mountain soils are important reservoirs for water, carbon and nutrients that enhance soil fertility. Mountain forests (28% of the world's forests) are also highly important as protective shields against natural hazards; they help ensure slope stability and prevent or reduce erosion, landslides and avalanches. Furthermore, tropical mountain forests have high genetic diversity and serve as important wildlife habitat. Together with highland wetlands, mountain forests play a key role in biospheric carbon storage (IPCC, 2007). The HKH region has vast forests, range and shrublands, and therefore plays a major role in carbon storage.

Fresh water

Water and water cycles are the most important ecosystem goods and services in the HKH mountains. This is why they are also referred to as the 'water tower of Asia' (Viviroli *et al.*, 2007). The ten major rivers originating in the HKH region supply fresh water to about 1.5 billion people. Recent estimates suggest that snow cover in the region averages 760,000 km^2 (Gurung *et al.*, 2011). The region also has a large number of freshwater lakes and aquifers.

Biological diversity

The HKH mountains are a treasure-trove of biological and agricultural diversity, including food, medicinal and aromatic plants, mushrooms, fibres such as cashmere, and mountain crops such as amaranths, buckwheat and different types of millet that are in demand downstream and in global markets. The mountains are also a source of timber, firewood, fruits, vegetables, forage and non-timber forest products. The vast and diverse gene pool, especially wild relatives of important crops in the HKH region, is vital for development.

Rangelands

One of the ecosystems – range and pasture lands (Stoddart *et al.*, 1975) – accounts for over 60% of the surface of the HKH. Over 100 million people are dependent, either directly or indirectly, on rangeland resources for animal husbandry and collecting rangeland products for their livelihoods. Furthermore, HKH rangelands are habitats for globally important fauna and flora and contain many biodiversity hotspots and a great number of protected areas, especially in the Tibetan Plateau. Climate-induced negative changes are presenting threats to rangeland ecosystem services across the HKH. In China, more than 90% of rangelands are degraded with about a quarter under severe degradation or threat of desertification (Yi and Sharma, 2009). This has resulted in sharp declines in fodder productivity and livestock-carrying capacity and has also endangered the social-economic well-being of pastoral communities.

Cultural services

Mountain areas support a large number of cultures. Many of these have rich traditional agricultural knowledge that promotes sustainable production systems. However, these lifestyles are severely threatened by climate and other changes. For many of these cultures, mountains play an important spiritual role, as they are living forces, sources of power and symbols of the sacred. Additionally, the mountains and mountain cultures attract tourists, who in turn can contribute towards improving local livelihoods. Every year, hundreds of thousands of tourists visit the unique HKH landscapes for rest, recreation and relief from summer heat and urban pollution.

4.2 COVID-19 Pandemic Crisis across the Hindu Kush Himalayan Region

As the world comes to grips with the public health implications of the global COVID-19 pandemic, it is also clear that responses to the pandemic have led to severe global socio-economic disruption and economic losses in local and national economies, affecting production, industries, services and jobs. Its severe impact on the poor and vulnerable is widening inequality gaps, and many who had made gains will slide back into poverty.

The people and mountain environments of the HKH region have not been isolated from the impacts of COVID-19. The social, economic and environmental impacts have been harsh and have exacerbated old risks and vulnerabilities as well as highlighting new ones. The pandemic struck at a time when HKH countries were enjoying economic growth. With the pandemic, HKH countries have felt the sharp pain of economic decline and skyrocketing unemployment. Decreased mobility of workers has impacted the remittance economy, and the lockdown has challenged social and family norms.

Mountain-based economies such as agriculture and tourism have collapsed, and mountain people are highly at risk of sliding into chronic poverty. The looming threat of severe hunger and malnutrition has intensified, as mountain farming systems were already declining before the pandemic. Gender, social and economic inequalities have widened, accompanied by more poverty, constricted access to education, and higher risks of rising maternal child mortality and morbidity rates. With their rich natural resources, mountain environments are particularly sensitive to the likelihood of increased resource extraction, illegal wildlife trade, and zoonotic diseases. Under various scenarios, it is predicted that over half of the additional poor will be in South Asia (Sumner *et al.*, 2020), a number that will increase if China and Myanmar are included. Although each country has individual specificities, the eight countries in the HKH have been profoundly impacted.

The HKH region is a critically important global asset given its water, food, energy, biodiversity, forest and carbon resources. The region is also home to diverse cultures, languages, religions and traditional knowledge (Sharma *et al.*, 2019). The current crisis has caused disruptions to sustainable development and resilience in the HKH, with clear impacts not only in the hills and mountains but also in downstream areas. The HKH region is experiencing intolerable pollution and biodiversity loss, while land-use changes have contributed to habitat loss, expanding the human–wildlife interface and facilitating the spread of diseases such as COVID-19.

4.2.1 Food security, nutrition and health dimensions of COVID-19

The COVID-19 crisis is not only a public health shock but also a global economic crisis. The global economy has absorbed a major disruption and might be heading towards a recession.

4.2.2 Food and nutritional security perspectives

In the HKH, the decline in agrobiodiversity is high, and traditional nutritious food crops such as buckwheat, barley, millet and beans are gradually disappearing (Adhikari *et al.*, 2019). The region is already vulnerable to food insecurity and malnutrition due to difficult terrain, climate change, high poverty rates, seasonality, abandonment of cultivable land and inadequate access to safe drinking water and health services (Rasul *et al.*, 2019). In the region, one third of the population is food insecure, and half is facing malnutrition. This is coupled with a very high prevalence of stunting, wasting and underweight in children (< 5 years of age) and micronutrient deficiencies in women. Some mountain areas, such as eastern Afghanistan, Meghalaya state in India, Chin and Rakhine states in Myanmar, the high mountains of Nepal, and Balochistan province in Pakistan, are facing extreme malnutrition (Rasul *et al.*, 2019). Against this backdrop, the additional shocks from the COVID-19 pandemic have further deepened existing challenges of food insecurity and malnutrition in the region with disruptions to agricultural production, food and input supply chains, trade, income-generating activities and access to food (FAO *et al.*, 2020). People in the mountains, who already face intense poverty and lack local employment opportunities, are extremely vulnerable. Since temporary food insecurity can have a profound impact on current and future well-being, it is important to assess immediate impacts and likely risks to food security and nutrition in the HKH region to identify the necessary short-, medium- and long-term actions that are needed.

Food availability

Most hilly and mountainous areas in the HKH depend heavily on plain areas to meet their food demand. For example, in Pakistan, around 30–60% of the food demand in mountain areas is met by food from plain areas. The lockdown situation has severely affected food availability due mainly to disruptions in food supply chains. At the national level, the HKH governments have adequate food reserves, but the supply of food to remote areas is a big challenge. Even within mountain areas, transportation of local agricultural products has been disrupted and food cannot reach markets or collection hubs. Return of migrant workers has also added to the pressure on food reserves in mountain areas. Even while some returnees begun leasing and planting on fallow land, continued challenges in accessing markets affect income from agriculture. Harvesting of winter food crops has also been affected in mountainous and hilly regions. For instance, in the double-cropping zones of Pakistan's Gilgit-Baltistan, harvesting of wheat and barley has been delayed, and this has implications for availability of locally produced food.

Food accessibility

In the mountains, both physical and economic access to food has been disrupted by the lockdowns. Panic buying and hoarding depleted locally available food items, leaving poor and lower-middle-class people with inadequate physical access to required food (Naz, 2020). An estimated 30–50% of HKH households dependent on tourism, small businesses and daily wage work (Hussain *et al.*, 2016) had income halted with lockdown; migrant workers lost their jobs temporarily or permanently, and the return of these workers affected the income of 15–20% remittance-dependent households, particularly in mountain areas of Pakistan and Nepal. This decline in household income has led to a serious decline in food purchasing power of the people in mountains.

Food utilization and health services

Dietary diversity is very important for nutrition and health, particularly of children and women. In the HKH region, dietary diversity is already low due to a gradual decline in the cultivation of traditional food crops such as barley, millet, buckwheat and sorghum (Adhikari *et al.*, 2019). The lockdowns are causing a further decline in dietary diversity because supply chains have been disrupted. In general, health systems in the region are fragile and in mountain areas, there is inadequate access to health services, inadequate public health infrastructure, and inadequate transport infrastructure for mountain people to reach the health facilities. In Afghanistan, for instance, 55% of children are deprived of adequate immunization services, and almost 40% of women are without proper antenatal care. In Balochistan province of Pakistan, 70% of children do not have all basic vaccinations. Although there are great differences across and within countries, around 20–50% of HKH households do not have adequate water and soap in their homes. In Pakistan and Afghanistan, all polio-eradication activities, including national vaccination campaigns and house-to-house surveillance have been suspended for six months. Hospitals and doctors are discouraging people with other health issues from accessing hospitals and meeting with doctors. New public health crises can emerge across the region if governments and health service providers cannot maintain other preventative health services such as immunization and curative treatment of acute and chronic diseases.

4.2.3 Environmental perspectives and COVID-19 in the Hindu Kush Himalayan region

Ecosystem services loss, encroachment of human settlements leading to human–wildlife interactions, and trade in live animals contribute to greater risk of human exposure to novel diseases and zoonotic transmission of diseases (Pongsiri *et al.*, 2009). As the pandemic situation continues to unfold, the complex interactions among social, economic and environmental factors require deeper analyses to identify effective interventions and to anticipate potential consequences, especially for the biodiversity-rich HKH ecosystems.

The pandemic has reinforced the critical importance of access to clean water and sanitation (WHO, 2020) and clean air for public health. However, the percentage of the HKH population with hand-washing facilities is relatively low; for instance, in Bangladesh it is a mere 35% and in India and Pakistan it is 60% (Matheswaran, 2020). Although air quality overall has improved with reduced vehicular and industrial pollutants, poor households may be experiencing increased and more concentrated exposure to indoor air pollutants due to the use of low-quality fuels for cooking and heating. Fragile HKH ecosystems require that environmental sustainability be a central tenet in all recovery pathways, and the longer-term response to the COVID-19 pandemic could be the catalyst for societal transformation towards sustainability and a means to achieve the 1.5°C climate target as set out in the Paris Agreement.

Ecosystems, landscapes and human–wildlife livestock interactions

Fragile HKH ecosystems have been under threat from environmental degradation and biodiversity loss and are likely to face more pressure due to increasing population as migrants return to villages (Maxwell *et al.*, 2016) The immediate need for food security and rapid economic recovery may propel over-harvesting of non-timber forest products (NTFP) and non-sustainable agricultural practices. The pandemic has increased human–wildlife interaction, causing a rise in incidents of poaching (Nag, 2020), with easier access to wildlife-rich ecosystems, and wildlife in urban settlements (Narayani, 2020). As vehicular and human traffic has decreased, the wildlife–livestock interface has also increased in transboundary and urban areas. The increased interaction between wildlife and people has exposed risks of zoonotic disease transfer (Pongsiri *et al.*, 2009) and risks of attack.

4.2.4 Potential pathways for recovery against COVID-19

Over the past decades, the environment and the air and water supporting agriculture and livelihoods in the HKH region have been compromised by intolerable levels of pollution. Biodiversity reduction and land-use changes have contributed to habitat loss, expanding the human–wildlife interface and facilitating the spread of zoonotic diseases such as COVID-19. In addition, the HKH is a climate change hotspot, subject to dramatic losses of glaciers, and increases in weather extremes leading to increased frequency and intensity of natural disasters. A business-as-usual post-pandemic recovery that relies on fossil fuel and inequitable growth will accelerate the rate of climate change and exacerbate the vulnerability of mountain communities. An inclusive and climate-resilient recovery along a trajectory towards prosperity will provide sustainable opportunities for growth, employment generation and well-being and must include: scaled-up social protection systems; investments aligning financial services with an inclusive and climate-responsive recovery; investments in nature-based solutions; strengthened entrepreneurial ecosystems; investments in resilient infrastructure to encourage low-carbon growth; and a transformation of labour markets to create sustainable employment opportunities.

4.3 A Systematic Framework for Promoting the Achievement of Sustainable Development Goals in the Post-pandemic Era

To address the fragile foundations, uneven progress and degraded co-operation context towards sustainable development, ecological restoration and to help cope with the multiple challenges facing sustainable development progress before and amid COVID-19, it is recommended that post-pandemic restoration adopt a systematic framework of 'Classification–Co-ordination–Collaboration' (Fu *et al.*, 2020).

4.3.1 Classification

Classification helps rationalize and identify the current development status, clarify the expected goals for sustainability, and select the appropriate development priorities and paths aligned with the national conditions. It is necessary for governments to adopt multiple solutions that deal with COVID-19's adverse effects while maintaining positive momentum (Ranjbari *et al.*, 2021). Achieving herd immunity, in which most of the population is vaccinated, remains an urgent requirement. Improving healthcare systems, ensuring resource security, and restoring livelihoods and production are also critical. Some Sustainable Development Goals (SDGs) are more urgent amid COVID-19 (e.g. medical and healthcare systems, food and water security) and, under the ongoing pandemic and economic recession, cash-strapped governments need to identify priorities among diverse SDG targets according to challenges and national development needs. Based on the principles of cost-benefit and social-ecological analysis, governments and experts need to determine win-win solutions (e.g. nature-based solutions) that minimize trade-offs, promote synergies and address multi-dimensional development issues, so as to meet their most urgent needs and to systematically promote the SDGs.

Classifying standards for SDGs

COVID-19 has amplified deficiencies in the foundations of development, and inadequate resilience and recovery potential. It is difficult for less-developed countries to achieve SDGs that are measured against a unified global standard, which therefore frustrates their efforts towards making progress. To implement targeted optimization measures for the SDGs, it would be helpful for the UN to classify different countries/regions, considering their differences in SDG baseline levels, resources, infrastructure, technology and socio-economic status. More realistic and practicable SDG actions require common but differentiated targets for countries/regions at different levels of development. Countries within similar categories can establish national plans to take concerted action in terms of policy, finance, trade, infrastructure and values.

4.3.2 Co-ordination and co-operation

Co-ordination includes domestic and international support and inter- and intra-departmental co-ordination for both short-term recovery and long-term development.

Domestic and international co-ordination

The global experience of COVID-19 suggests that stronger domestic and international co-ordination is needed for achieving sustainable development and ecological restoration. On one hand, countries need to establish their own management systems to ensure the security of basic resources, such as improving infrastructure and logistics and strengthening domestic production. On the other hand, international organizations need to strengthen their co-ordination capacity to improve international dialogue, build platforms for bilateral and multi-lateral investments and trade, and co-ordinate market rules for basic resources, especially given the additional stresses brought about by the pandemic and associated impacts. Co-ordination between different administrative departments is also very important. To set the goals back on the path to recovery, both short- and long-term issues must be identified and addressed with targeted measures. In the short term, the pandemic has negatively impacted goals related to reducing poverty and securing food, water and energy, which must be addressed urgently. However, in the longer term, improving infrastructure, promoting scientific and technological innovation, and optimizing the relationship between people and nature are imperatives for the realization of the SDGs. Accordingly, in formulating socio-economic development policies, both short-term recovery and long-term development must be addressed by researchers and policy-makers.

Co-operation

Co-operation includes strengthening economic co-operation, promoting technological innovation, and jointly building a global culture of sustainable development.

References

Adhikari, L., Tuladhar, S., Hussain, A. and Aryal, K. (2019) Are traditional food crops really 'future smart foods?' A sustainability perspective. *Sustainability* 11(19), 5236. DOI: 10.3390/su11195236

Bajracharya, S.R. and Shrestha, B.R. (2011) *The Status of Glaciers in the Hindu Kush-Himalayan Region*. International Centre for Integrated Mountain Development (ICIMOD), Kathmandu.

Chaudhary, R.P., Bhattarai, S.H., Basnet, G., Bhatta, K.P., Uprety, Y. *et al.* (2017) Traditional practice and knowledge of indigenous and local communities in Kailash Sacred Landscape, Nepal. ICIMOD Working Paper 2017/1. Kathmandu.

Chettri, N., Shakya, B., Thapa, R. and Sharma, E. (2008) Status of protected area system in the Hindu Kush Himalaya: an analysis of PA coverage. *International Journal of Biodiversity Science and Management* 4(3), 164–178. DOI: https://doi.org/10.3843/Biodiv.4.3:4

Chettri, N., Sharma, E., Shakya, B., Thapa, R., Bajracharya, B. *et al.* (2010) Biodiversity in the Eastern Himalayas: status, trends and vulnerability to climate change. *Climate Change Impact and Vulnerability in the Eastern Himalayas*. Technical Report 2. ICIMOD, Kathmandu.

FAO, IFAD, WFP and the World Bank (2020) Joint statement on COVID-19 impacts on food security and nutrition. Available at: https://www.worldbank.org/en/news/statement/2020/04/21/joint-statement-on-covid-19-impacts-on-food-security-and-nutrition (accessed 29 September 2023).

Fu, B., Zhang, J., Wang, S. and Zhao, W. (2020) Classification–coordination–collaboration: a systems approach for advancing Sustainable Development Goals. *National Science Review* 7, 838–840. DOI: 10.1093/nsr/nwaa048

Gurung, D.R., Kulkarni, A.V., Giriraj, A., Aung, K.S., Shrestha, B. and Srinivasan, J. (2011) Changes in seasonal snow cover in Hindu Kush-Himalayan region. *The Cryosphere Discussions* 5(2), 755–777.

Hussain, A., Rasul, G., Mahapatra, B. and Tuladhar, S. (2016) Household food security in the face of climate change in the Hindu-Kush Himalayan region. *Food Security* 8, 921–937. DOI: 10.1007/s12571-016-0607-5

IPCC (2007) *The Scientific Basis: Contribution of Working Group I to the Fourth Assessment Report of the Intergovernmental Panel on Climate Change*. Cambridge University Press, Cambridge, UK.

Lu, H., Wang, X., An, Z., Miao, X., Zhu, R. *et al.* (2004) Geomorphologic evidence of phased uplift of the northeastern Qinghai-Tibet Plateau since 14 million years ago. *Science in China Series D: Earth Sciences* 47(9), 822–833.

Matheswaran, K. (2020) Access to water and COVID-19: seven measures countries in Asia can take now. Stockholm Environment Institute. Available at: https://www.sei.org/perspectives/access-to-clean-water-is-vital-in-the-fight-against-covid-19-here-are-seven-measures-that-countries-in-asia-can-take-now/ (accessed 29 September 2023).

Maxwell, S.L., Fuller, R.A., Brooks, T.M. and Watson, J.E. (2016) Biodiversity: the ravages of guns, nets and bulldozers. *Nature* 536(7615), 143–145. DOI: 10.1038/536143a

Mittermeier, R.A., Turner, W.R., Larsen, F.W., Brooks, T.M. and Gascon, C. (2011) Global biodiversity conservation: the critical role of hotspots. In: Zachos, F.E. and Habel, J.H. (eds) *Biodiversity Hotspots: Distribution and Protection of Conservation Priority Areas*. Springer, Berlin Heidelberg, pp. 3–22.

Nag, O.S. (2020) Wildlife poaching surges across the world as the COVID-19 pandemic rages on. World Atlas. Available at: https://www.worldatlas.com/news/wildlife-poaching-surges-across-the-world-as-the-covid-19-pandemic-rages-on.html (accessed 28 September 2023).

Narayani, P.A. (2020) Feeding animals in the time of COVID-19. *The Hindu*. Available at: https://www.cargill.com/feedingintelligence/covid19 (accessed 28 September 2023).

Naz, L. (2020) COVID-19 and panic buying. News International. Available at: https://www.thenews.com.pk/tns/detail/635420-covid-19-and-panic-buying (accessed 28 September 2023).

Olson, D.M. and Dinerstein, E. (2002) The Global 200: priority ecoregions for global conservation. *Annals of the Missouri Botanical Garden* 89, 199–224. DOI: 10.2307/3298564

Pongsiri, M.J., Roman, J., Ezenwa, V.O., Goldberg, T.L. and Koren, H.S. (2009) Biodiversity loss affects global disease ecology. *Bioscience* 59, 945–954. DOI: 10.1525/bio.2009.59.11.6

Ranjbari, M., Saidani, M., Esfandabadi, Z.S., Peng, W., Lam, S.S. *et al.* (2021) Two decades of research on water management in the circular economy: insights from bibliometric text mining and content analyses. *Journal of Cleaner Production* 314, 128009.

Rasul, G., Saboor, A., Tiwari, P.C., Hussain, A., Ghosh, N. *et al.* (2019) Food and nutrition security in the Hindu Kush Himalaya: unique challenges and niche opportunities. In: Wester, P., Mishra, A., Mukherji, A.

and Shrestha, A. (eds) *The Hindu Kush Himalaya Assessment: Mountains, Climate Change, Sustainability and People*. Springer Nature, Cham, Switzerland, pp. 301–338.

Sanlaville, P. and Prieur, A. (2005) Asia, Middle East, coastal ecology and geomorphology. In: Schwartz, M.L. (ed.) *Encyclopedia of Coastal Science*. Springer, Dordrecht, The Netherlands, pp. 71–83.

Sharma, B., Rasul, G. and Chettri, N. (2015) The economic value of wetland ecosystem services: evidence from the Koshi Tappu Wildlife Reserve, Nepal. *Ecosystem Services* 12, 84 –93.

Sharma, E., Chettri, N. and Oli, K.P. (2010) Mountain biodiversity conservation and management: a paradigm shift in policies and practices in the Hindu Kush-Himalayas. *Ecological Research* 25(5), 909–923. DOI: 10.1007/s11284-010-0747-6

Sharma, E., Molden, D., Rahman, A., Khatiwada, Y.R., Zhang, L. *et al.* (2019) Introduction to the Hindu Kush Himalaya assessment. In: Wester, P., Mishra, A., Mukherji, A. and Shrestha, A. (eds) *The Hindu Kush Himalaya Assessment: Mountains, Climate Change, Sustainability and People*. Springer Nature, Cham, Switzerland, pp. 1–16.

Sharma, P. and Partap, T. (1994) Population, poverty, and development issues in the Hindu Kush-Himalayas: development of poor mountain areas. ICIMOD, Kathmandu, pp. 61–78.

Stoddart, D.R. (1975) 'That Victorian Science': Huxley's physiography and its impact on Geography. *Transactions of the Institute of British Geographers* 17–40.

Sumner, A., Hoy, C. and Ortiz-Juarez, E. (2020) Estimates of the impact of COVID-19 on global poverty. UNUWIDER, April, 800–809. Available at: https://blogs.worldbank.org/opendata/updated-estimates-impact-covid-19-global-poverty-turning-corner-pandemic-2021 (accessed 28 September 2023).

Viviroli, D., Durr, H.H., Messerli, B., Meybeck, M. and Weingartner, R. (2007) Mountains of the world, water towers for humanity: typology, mapping and global significance. *Water Resources Research* 43(7).

WHO (2020) Water, sanitation, hygiene, and waste management for the COVID-19 virus: interim guidance. Available at: https://www.who.int/publications/i/item/WHO-2019-nCoV-IPC-WASH-2020.4 (accessed 28 September 2023).

Xu, J., Badola, R., Chettri, N., Chaudhary, R.P., Zomer, R. *et al.* (2019) Sustaining biodiversity and ecosystem services in the Hindu Kush Himalaya. In: Wester, P., Mishra, A., Mukherji, A. and Shrestha, A. (eds) *The Hindu Kush Himalaya Assessment: Mountains, Climate Change, Sustainability and People*. Springer Nature, Cham, Switzerland, pp. 127–165.

Yi, S. and Sharma, E. (2009) Climate change and the Hindu Kush-Himalayan Rangelands. ICIMOD, Kathmandu.

5 Ecological Conservation and Restoration in Afghanistan

Awais Iqbal[1] and Zhanhuan Shang[1,*]
[1]*College of Ecology, Lanzhou University, Lanzhou, China*

Abstract
Afghanistan spans three of the world's eight biogeographical territories: the Afrotropic, the Palaearctic, and the Indomalayan. Because of its unique topography, temperature and geology, Afghanistan is considered one of the most ecologically productive and biologically distinct countries in south Asia. Both the people and environment of Afghanistan have suffered greatly from decades of war and violence, and natural disasters like drought and earthquakes. Most of its people are subjected to oppression and continue to live in conditions at or below the poverty line. The extraction of natural resources without proper controls hastens the degradation of the environment and exacerbates the effects of natural disasters. Furthermore, the country's political instability places the wildlife and territories in unprecedented peril. Research with domestic and international organizations and a new paradigm to restore the balance between economic goals and ecological needs are needed to preserve Afghanistan's ecosystem.

Keywords: assessments, biodiversity, resources, sustainability

5.1 Introduction

Afghanistan, a country in Central Asia, spans ~652,677 km^2 and features a wide range of ecosystems, including forests, woodlands and deserts, in addition to extraordinary mountainous landscapes. Most of the Afghan population views the biodiversity of flora and fauna as a source of sustenance. In addition, its natural resources are abundant and diverse, ranging from coal, oil and natural gas to precious stones and rare earth elements. The rich and most important ecological regions of Afghanistan are presented in Fig. 5.1. However, much of the information on Afghanistan's biota is outdated and no longer dependable. Since the beginning of conflict in 1978, little information has been contributed to the field of biodiversity conservation. Recent research suggests that Afghanistan's biodiversity has deteriorated substantially over the past three decades.

Afghanistan has 34 provinces, seven of which are in the Palaearctic realm, while only a small fraction of the lower Kabul river valley is of Indo-Malayan origin (Fig. 5.2). A recent classification divides Afghanistan into 15 smaller eco-regions, of which four are deemed critical/endangered, eight are deemed vulnerable, and only two are considered comparatively stable and intact. The species diversity of all eco-regions

*Corresponding author: shangzhh@lzu.edu.cn

DOI: 10.1079/9781800622579.0005

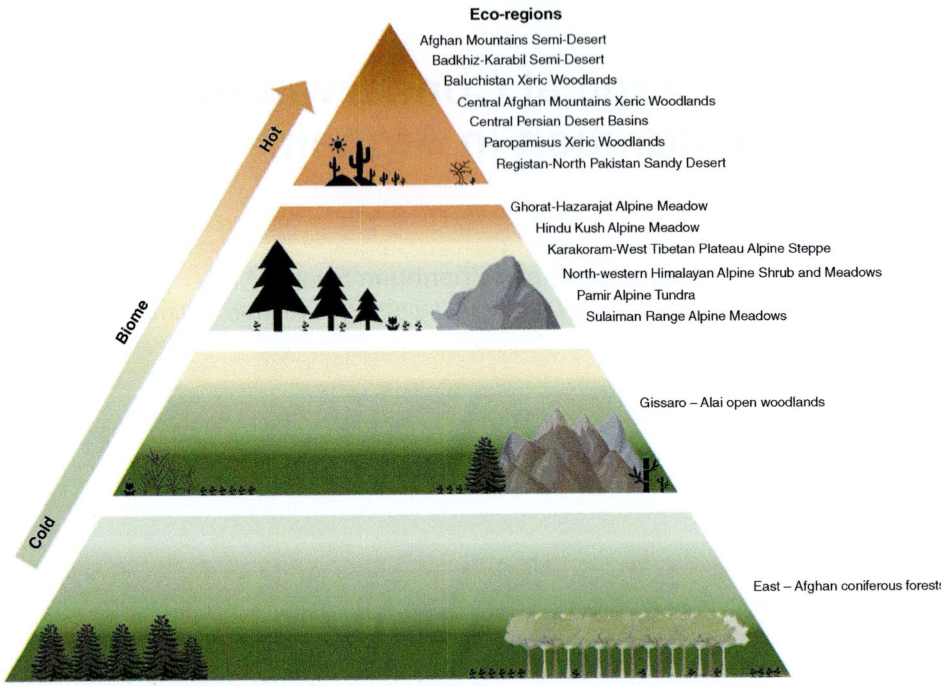

Fig. 5.1. Ecological regions and biomes of Afghanistan. (Adapted from Rolando Y. Wee, www.WorldAtlas.com, 2019)

has been drastically diminished through overgrazing, fuel collection and exploitation of large herbivorous animals. Afghanistan is home to 118–147 species of mammals, 472–510 species of birds, 92–112 species of reptiles, 6–8 species of amphibians, 101–139 species of fish, 245 species of invertebrates and 3500–4000 species of vascular plants. The variation in numbers results from taxonomic uncertainty and the questionable veracity of some records. According to the International Union for Conservation of Nature (IUCN) Red List, 39 species and 8 subspecies are threatened with extinction globally. No formal assessment of endangered species at the national level has been conducted, but numerous species, particularly large mammals, are unquestionably threatened with extinction in Afghanistan. The number of endemic vertebrate species in Afghanistan is only seven; however, estimates for endemic plant species range from 20 to 30. Afghanistan is an arid country; therefore, the few remaining wetland areas are important to biodiversity (IUCN, 2023).

Afghanistan was one of the most important centres for the origin and development of crop plants for humans. Consequently, Afghan farmers cultivate numerous local landraces of wheat and other commodities. Afghanistan is home to nine local breeds of sheep, in addition to eight breeds of cattle and seven breeds of goats. Approximately 5% of Afghanistan's land mass was once covered by deciduous and evergreen tree forests, now restricted to the monsoon-affected eastern region. Recent surveys indicate that just 5% of the original forests still remains. In the past, approximately 38% of Afghanistan was composed of open woodlands dominated by pistachio and juniper. Satellite image analysis of selected areas in 2002 revealed that tree density was too low to detect, indicating that this once-extensive ecosystem in Afghanistan is on the verge of extinction. The vast areas of sub-alpine vegetation were likely once steppe grassland, but are now almost exclusively cushion shrubland. Afghanistan has 15 protected areas, according to the World Database of Protected Areas (UNEP-WCMC, 2006), of which seven were granted formal recognition by the government of Afghanistan in the 1970s and are recognized by IUCN (one national park of Category II and

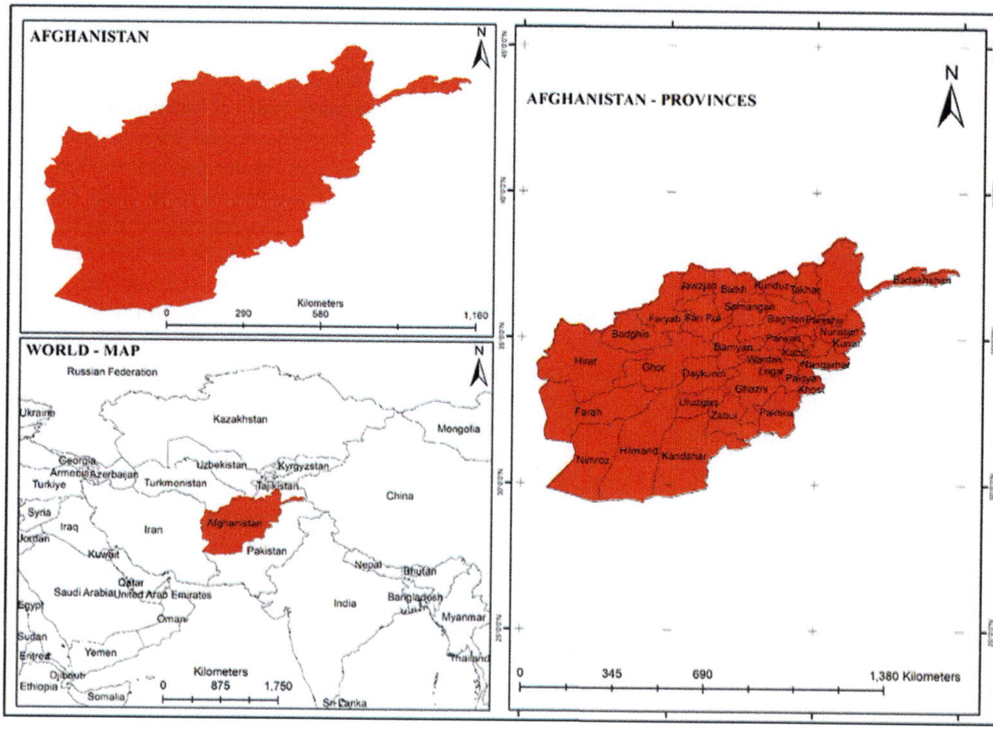

Fig. 5.2. Map of Afghanistan and its provinces.

six wildlife or waterfowl reserves of Category IV) (UNEP-WCMC, 2006). However, the current government does not recognize any; none has official boundaries, and none is managed as a protected area.

5.2 Ecological Degradation: Causes and Problems

Decades of conflict and violence, coupled with desertification and earthquakes, have devastated Afghanistan and its once-pristine and biologically diverse natural environment. The involvement of the United Nations Environment Programme (UNEP) began with a comprehensive post-conflict environmental assessment conducted in collaboration with the Afghan government and Afghan environmental experts. The 2003 report highlighted widespread and severe land and resource degradation, such as lowered water tables, desiccation of wetlands, pervasive loss of vegetative cover, erosion and extinction of wildlife (UNEP, 2021). For instance, the Shah Foladi protected area is located in the Koh-e-Baba mountain range of Bamyan province in the Central Highlands region of Afghanistan. It is an essential area for biodiversity, with approximately 5000 native taxa, of which approximately 1200 are endemic. The rural highland communities utilize shrubs and vegetation from the rangelands as fuelwood. These plants are gathered in large quantities and are frequently uprooted, thereby disrupting the structure of the plant community and the biodiversity, as well as eroding the soil, which can increase the likelihood of natural disasters such as floods and avalanches. Endangered woody taxa include *Acantholimon*, *Astragalus* and *Artemisia*. Bamyan is home to 68 national and 17 provincial endemics out of 110 species (UNEP, 2021). The unsustainable collection of fuelwood raises the risk of extinction, jeopardizes ecological services and reduces future fuel stocks for populations.

According to the Fifth National Report to the Convention on Biological Diversity (CBD) of the United Nations (UN), the primary causes of ecological degradation and ecological threats to the country are the conversion of land for agricultural and housing purposes, illegal hunting, deforestation, over-grazing, shrub collection, dryland farming, water diversion and climate change (UNEP, 2014). In the past three decades, the gravity of these threats has increased. For example, deforestation persists today, and it is estimated that firewood harvesting for the Kabul market destroys 10,000 ha of oak forest and 15,000 ha of juniper forest annually in the Paktia and Khost provinces. In addition to other ecological disturbances, more than 75% of the total land area in Afghanistan's northern, western and southern regions is already affected by desertification, placing the country in the very high risk category for desertification, according to the United Nations Framework Convention on Climate Change (UNFCC) and the United States Department of Agriculture (USDA) global studies (Convention on Biological Diversity, 2023). Because of security and access issues, the illegal export of timber to neighbouring countries is substantial but incalculable. According to national experts, catastrophic landslides, rockslides, mudslides, floods and flash floods caused by spring rainfall and snowmelt have become more frequent in the past 50 years. Water schemes and regular droughts have unquantified effects on wetland and river biodiversity. Numerous large mammals and birds in Afghanistan face great threats from hunting, trapping and habitat devastation. In the past, waterfowl hunting was widespread, and large mammals were hunted for sport illegally. In a context where many residents depend on natural resources and biodiversity for their livelihoods and economic survival, the main underlying threat is the timing in implementing conservation programmes. Human population growth is the greatest threat to biodiversity conservation and quality of life.

5.3 Ecological Risk Profile and Assessment in Afghanistan

Risk knowledge is essential for effective disaster and climate risk management. Integrating risk information into development planning, public policy and investments, and ensuring the resilience of new and existing reconstruction against natural hazards and climate change is crucial for protecting lives and livelihoods. Afghanistan faces ecological risks, but the country's most frequent and intensive natural hazards are earthquakes, floods, flash floods, landslides, avalanches and droughts. Afghanistan ranks second among low-income nations, trailing only Haiti in the number of natural disaster-related deaths between 1980 and 2015. There are 1150 deaths per million inhabitants in Afghanistan, with 50% resulting from geophysical and weather-related causes (Penuel *et al.*, 2013). The population living in hazardous areas and the damage caused by these hazards are summarized in Table 5.1. With funding from the government of Japan and the Global Facility for Disaster Reduction and Recovery (GFDRR), and close collaboration with the Afghanistan National Disaster Management Agency, the World Bank generated an extensive multi-hazard risk assessment at the national level, including detailed analyses for certain areas.

5.3.1 Floods

Flooding is the most common natural disaster in Afghanistan. The country is prone to river inundation due to its steep headwater slopes. Most river flooding is caused by heavy precipitation and rapid snowmelt; most river sources are in the mountains and are nourished by snow and glaciers. Lack of vegetation and deforestation also contribute to inundation in mountainous

Table 5.1. Economic damage and impact of natural hazards of Afghanistan. (Adapted from Ranghieri *et al.*, 2017)

Hazard	Economic damage (US$)	Impact
Earthquake	80 million	560 fatalities on average per year
Flood	54 million	Over 800,000 people exposed
Drought	280 million	US$280 agriculture damage per year
Landslide	n.a.	130,000 private buildings exposed
Avalanche	n.a.	10,000 kilometres of road exposed

regions. Urban flooding is a concern in major cities and is frequently caused by intense local precipitation and inadequate drainage. Hairatan (north) and the Helmand basin (west) are prone to flooding. Population-wise, Kabul province is the most affected, with an average of over 2000 deaths annually. The commercial (40%) and residential (33%) sectors account for the preponderance of expected flood damage (Khalid, 2018). Approximately 100,000 individuals are impacted annually by inundation, and the damage could exceed US$500 million (Convention on Biological Diversity, 2023). The number of people affected by inundation annually could more than double by 2050 due to climate change and socio-economic development. Inadequately constructed flood protection infrastructure, the absence of early-warning systems, and the expansion of human settlements in flood-prone areas contribute substantially to flood risk. In 14 northern provinces, 90,000 people were impacted by flooding in May 2014, and 20,000 were displaced, while the floods caused more than $100 million in damage (Khalid, 2018; Convention on Biological Diversity, 2023).

Assessments

- Flood protection in cities and rural regions could be improved by strengthening dykes and modifying homes.
- A single new embankment in Kabul could prevent $600,000 in flood damage each year.
- Building a local earth embankment along the river in the Kunduz agricultural area, the Phuli Khumri and Fayzabad, could decrease the risk of floods.

5.3.2 Earthquake

Afghanistan is a part of the world where tectonic plates are moving constantly. Every year, Afghanistan is hit by earthquakes ranging in intensity from moderate to strong, often with devastating results. There is a high risk of earthquakes in the north-eastern part of Afghanistan, especially along the borders with Uzbekistan, Tajikistan and Pakistan. Over 10,000 people have died as a result of earthquakes in the last 30 years. Based on a probabilistic analysis of a stochastic hazard catalogue, it is predicted that over $500 million in damage and 8500 deaths will result from a massive earthquake in Kabul province (Samim and Zhiquan, 2020; Convention on Biological Diversity, 2023). Due to the concentration of assets and population in the province, Kabul has the greatest average estimated damage of all regions in Afghanistan, equal to $17 million every year. Large landslides are another common consequence of earthquakes that occur in mountainous areas. Since 1900, earthquakes have killed nearly 2.7 million people, which is more than any other natural disaster. Over the last 30 years, more than 10,000 people have died and more than 250,000 have been affected by earthquakes. In October 2015, an earthquake in the Hindu Kush mountains killed 11 people, harmed more than 58,000 and left more than 35,000 people homeless. At least 1200 people died, and more than 100,000 were affected when an earthquake hit the Hindu Kush mountains in March 2002 (Khalid, 2018).

Assessments

- Building new structures to make them resistant to earthquakes and reinforcing existing buildings can greatly reduce the number of buildings that collapse, and the number of deaths caused by earthquakes.
- Constructing all Afghan schools to be earthquake-resistant would reduce the potential financial losses by 60% and mortality by 90%.
- Building new roads that withstand earthquakes is more feasible than repairing or rebuilding old roads, and keeping money reserved for repairing destruction.

5.3.3 Drought

Afghanistan has repeated droughts that differ in length and intensity. Drought threatens livelihoods, income and agriculture, which employs over 85% of the population. Additionally, severe and protracted droughts have grave consequences for food security. There are fewer droughts in the north-east than in the Helmand basin. Future climate change-related precipitation losses across most of Afghanistan's sub-basins are anticipated to increase drought risk. Drought affects millions of people and damages farmland, production, households and hydropower. Droughts cost the agricultural industry $280 million annually, on average, and, due to climate change, may cost as much as $3 billion annually (Khalid, 2018).

Assessments

- More effective water management could alleviate water scarcity during droughts.
- Irrigation systems and canals could provide water for the agricultural sector.
- Increasing agricultural output and decreasing dependency on water supply could be possible through diversifying livelihoods and enhancing the efficiency of agricultural techniques, such as planting crops and using water-saving irrigation technologies.

5.3.4 Landslide

The risk of landslides spreads from the north-east to the west across Afghanistan's hilly terrain, which makes up more than 60% of the country (Ranghieri *et al.*, 2017). Landslides are fairly prevalent in Afghanistan due to the hilly terrain and unstable soil, and they can be caused by both excessive rainfall and earthquakes. The highest concentration of people vulnerable to landslides reside in Badakhshan, Daykundi and Ghor, and the most at-risk assets are in Badakhshan, Takhar and Ghor (Ranghieri *et al.*, 2017). Landslide risk to settlements and infrastructure can be exacerbated by deforestation, land-use change and increase in heavy rainfall. The availability of nets and drainage pipes to prevent damage from landslides is limited in Afghanistan. More than 3 million people, 400 schools and 300 hospitals, totalling over $6 billion in value, are at risk from landslides in Afghanistan. Moreover, population increase alone might lead to a doubling of landslide casualties by 2050 (Ranghieri *et al.*, 2017; Convention on Biological Diversity, 2023).

Assessments

- Buildings, roads and utility lines should not be constructed in high-risk landslide areas.
- Early detection of landslides is possible with local monitoring of slopes with many slow-moving landslide hazards.
- Reinforcing landslide-prone slopes could lessen the likelihood of landslides.

Small landslides can be prevented from damaging assets by implementing passive protection measures, including boulder barriers, metal containment nets, and boulder-gathering channels at the base of the hill.

5.3.5 Avalanche

Avalanches of snow in the Hindu Kush mountains of Afghanistan cause damage to infrastructure, people and settlements. A common concern associated with avalanche events is the difficulty in reaching and assisting remote populations. The three regions most affected by avalanches in terms of population and gross domestic product (GDP) are Badakhshan, Kabul and Daykundi. Rising temperatures due to climate warming may lessen the risk of avalanches in low-lying areas. However, due to heavy snowfall in the middle of winter, high-altitude regions like the Salang Pass may experience avalanches more frequently. There are 2 million individuals at risk, while over 10,000 km of road and $4 billion in assets are vulnerable to avalanches (Ranghieri *et al.*, 2017; Convention on Biological Diversity, 2023). In Badakhshan alone, $990 million in assets are at risk of avalanches. Moreover, over 300 people were killed in a series of avalanches and flash floods in 2015, the majority of whom were in Panjshir Valley. It also impacted more than 153,000 lives between 2000 and 2015 (Ranghieri *et al.*, 2017; Convention on Biological Diversity, 2023). Avalanche early-warning systems are being tested in the Salang Pass but not elsewhere. There are usually no avalanche mitigation strategies in place to safeguard settlements and infrastructure.

Assessments

- Avalanches can be prevented by retention structures such as snow-supporting structures, snowdrift fences, artificial avalanche releases, avalanche dams, avalanche deflection dams, and avalanche galleries or snow sheds.

- In avalanche-prone areas, buildings and structures can be strengthened to withstand the impact of avalanches.

5.4 Relationship between People's Livelihoods and the Environment

Ecosystems sustain life and serve as the cornerstone of human socio-economic development by providing a range of services that are crucial for social welfare and livelihoods (Costanza *et al.*, 1998; Skewes *et al.*, 2016). In Afghanistan, ecosystem services are of great importance and urgently need the attention of policy-makers to safeguard them in the face of rapid socio-economic development. Nearly 80% of Afghans rely on the local ecosystem for their livelihoods due to the country's slow industrial growth (Saidajan, 2012). Additionally, the agro-ecosystem services sector alone employs 40–60% of the labour force and accounts for a quarter of the GDP. Grasslands and forests contribute to local livelihoods by safeguarding agricultural soils and water supplies and providing products and services to energy, building and other industries (Milbrandt, 2009). For instance, grassland covers roughly 47% of the geographical area and provides approximately 75% of the feed for livestock (Milbrandt, 2009). The forests, which comprise over 2% of the country's total area, produce non-timber goods and timber that are valued highly worldwide (Gouhari *et al.*, 2021). However, decision-makers have historically undervalued ecosystem services despite their significant economic benefits, largely because of a lack of information and research (Gouhari *et al.*, 2021). Rapid population expansion, especially after 2000, and worsening climate conditions, exacerbate the situation by increasing the strain on the ecosystem. Currently, 38% of Afghanistan's total land area is endangered, 61% is considered vulnerable (UNEP, 2014), and 70% of Afghanistan's forested lands have been classified as vulnerable (UNEP, 2014).

Afghanistan frequently experiences severe weather or climatic events that result in large economic losses and fatalities (Kreft and Eckstein, 2013; EM-DAT, 2019). Because their farms are situated in marginal areas (such as slopes, deserts and floodplains), smallholder farmers are very vulnerable to climate change (Donatti *et al.*, 2019). The main factors contributing to the vulnerability of smallholder farmers in the area are poor livelihood strategies, low-income agriculture, and a lack of institutional, financial and technical support. The greatest threats are pests and disease outbreaks, along with irregular rainfall and high temperatures brought on by climate change. Given their low initial costs and minimal ongoing maintenance requirements, a few low-cost and local approaches, such as enhancing farmer extension services, implementing small-scale local infrastructure projects, bolstering informal safety nets, and safeguarding natural ecosystems, could be viable, cost-effective and sustainable choices (Omerkhil *et al.*, 2020).

The principal climate change-related elements impacting water quality are the hydrological cycle and fluctuating air temperature (Simonovic, 2017). Rapid urbanization, industrialization, high population density, poor education and inappropriate treatment all contribute to the decline in water quality. In addition, concerns about water scarcity, outdated infrastructure, rising water demand, and uneven water supply are widespread across the nation. The release of effluents and other pollutants into the Kabul river causes sublethal organic pollution. Along with human and livestock waste, the effluents from numerous leather-processing plants and other industries are destroying the ecology of the river. The prevalence of fecal coliforms and other microbial diseases poses a serious hazard to the health of those who live near the river. Additionally, the role of climate in influencing the dynamics of human diseases has come under close scrutiny in recent years (McMichael *et al.*, 2006), as evidenced by a number of studies that show the potential for both global and regional effects of climate change on population health (Gray *et al.*, 1996; Patz *et al.*, 2005). For instance, diarrhoea is the second most common disease (after respiratory diseases) in a nation where many endemic infections are present, many of which have high fatality rates (Dalgleish *et al.*, 2017). In 2013, diarrhoea was estimated to be the cause of 20% of reported deaths among children five years old and 11% of all reported deaths in health facilities in Afghanistan (Anwar and Burnham, 2016). These authors reported that aridity and mean daily temperature were correlated positively with cases of diarrhoea. For every 1°C increase in mean daily temperature and 0.01-unit change in the aridity index, the risk of diarrhoea increased by 0.7% and 4.79%, respectively (Anwar and Burnham, 2016). The findings of this study demonstrated the critical impact of climate on diarrhoea in Afghanistan, enabling decision-makers to consider climate change implications in public health evaluations.

Climate change may have a negative feedback loop on health and livelihoods (Connolly-Boutin and Smit, 2016). As the number of 'hot' days increases, it may be difficult to work throughout the day in the spring and summer. Increased heat stress will also affect chronic illnesses, such as heart disease and respiratory disorders (particularly in urban places like Kabul, which is already quite polluted). The range of vector-borne diseases such as malaria will also extend as temperatures rise, which will increase the strain on the population (Stanke *et al.*, 2013). Malnutrition may rise (without concerted humanitarian action) if food prices rise and domestic food production decreases due to climate change, which will directly influence health indicators. Since lower incomes impact access to sufficient food and healthcare services, this will negatively affect people's productivity, creating a vicious cycle. Drought is likely to become a regular occurrence by 2030, having a serious impact on semi-arid/arid regions that rely heavily on rain-fed agriculture (Wei *et al.*, 2021). The efforts to develop peace in Afghanistan face obstacles due to the growing frequency of droughts and the depletion of natural resources. Poppy production tends to increase during droughts, tension and violence over increasingly scarce natural resources are on the rise, and people may become involved in violent conflict in their search for alternative sources of income. Prolonged violence in the country has severely limited efforts to develop the healthcare system, provide vaccinations and respond to disasters, and, thus, affect the livelihood of the country. Migration from rural to urban regions is predicted to rise as people search for alternative livelihoods as natural resources deteriorate and ecosystems collapse.

5.5 Ecological Conservation, Restoration Strategies and Policies

To rebuild the degraded ecology, foreign and domestic government agencies and international organizations are pursuing ecological restoration efforts in Afghanistan. Afghanistan has seen a protracted civil war and political unrest for the past 40 years, and the direct and indirect effects of conflict on biodiversity are obvious and long-lasting (Hanson *et al.*, 2009; Buzzard, 2020). Despite this, the Environment Law, passed by the post-conflict Afghan parliament in 2007, demonstrated the then government's commitment to environmental conservation. The status, demographic health and distribution of wildlife species (primarily mammals and birds) in Afghanistan have been studied by the Wildlife Conservation Society's Afghanistan programme and others. These wildlife species include the globally threatened snow leopard (*Panthera uncia*), Persian leopard (*Panthera pardus saxicolor*), Marco Polo sheep, musk deer (*Moschus cupreus*), Bactrian deer (*Cervus elaphus bactrianus*), Asiatic black bear (*Ursus thibetanus*), markhor (*Capra falconeri*), ibex (*Capra sibirica*) and associated species in the Wakhan corridor, the Hazarjat Plateau and Nuristan regions (Maheshwari, 2022). The findings show that, despite 40 years of conflict and troubled governance, and contrary to the narrative provided in many images shown in the global media on the extensive damage to natural habitats caused by inherent warfare and the illegal wildlife trade, Afghanistan is a stronghold of Asia's biodiversity, including species that are vulnerable to extinction. The Ecology and Conservation Organization of Afghanistan (ECOA) was established in 2010 as an Afghan-owned and -managed local non-profit organization with the overarching objectives of achieving environmental conservation and rehabilitation, and poverty alleviation through encouraging and supporting sustainable community-based development (Ecology and Conservation Organisation of Afghanistan, 2023). ECOA aims to prevent the degradation of Afghanistan's natural environment and ensure a sustainable future for its people by conserving biological diversity, ensuring the sustainable use of renewable natural resources, reducing pollution and wasteful consumption, and providing sustainable livelihoods. ECOA has overseen numerous initiatives in Afghanistan, some of which are listed in Table 5.2. In a similar vein, the National Environmental Protection Agency (NEPA) of Afghanistan was founded in 2005, the year in which President Hamid Karzai wrote and signed the country's

Table 5.2. The initiatives of the Ecology and Conservation Organization of Afghanistan. (From Ecology and Conservation Organisation of Afghanistan, 2023)

Projects	Province	Duration	Budget ($)	Donors	Summary
Koh-e-Baba mountain area site planning and conservation	Bamyan	Jan 2011 – Dec 2012	70,000	World Food Programme	Improved planning and management of the selected area. Over 100,000 trees were planted in the 7 villages.
Clean cookstoves and protection of environment in the Koh-e-Baba Mountains	Bamyan	Jan 2012 – Dec 2012	37,469	Embassy of Finland	Protection of land degradation from fuel wood cutting and to potentially save lives. Green energy products (clean cookstoves, clean ovens, solar water heaters and briquette/efficient fuel makers) were distributed among the women to improve health and take better care of their environment.
Mountain landscape community conservation and education centre	Bamyan	March 2012 – Dec 2012	10,000	Ruffords Small Grants Programme	This project increased environmental awareness in local communities on the importance of conservation and nature protection in the poverty-stricken Koh-e-Baba area and throughout Afghanistan, by linking to the national Protected Areas plan.
Establishing of eco-conservation centre in Koh-e-Baba mountains	Bamyan	Jan 2012 – Aug 2012	12,000	Agha Khan Foundation	Ecologically sound conservation centres-cum-guesthouses were established in five of the communities in the Koh-e-Baba protected area. These act as places, outside of the mosque where educational workshops and community meetings can happen.
Training in biodiversity data collection	Bamyan	Jan 2012 – Aug 2012	40,000	Millennium Seed Bank	Teaching and capacity building of DAIL, NEPA, university professors and students in rigorous biodiversity data collection. The data collected from the fieldwork contributed to the Koh-e-Baba protected area management plan and fed into the National Biodiversity Action Plan.
In situ biodiversity conservation strategic plan	Bamyan	Jan 2012 – Sept 2012	25,000	United Nations Environment	The project was implemented together with the National Environment Protection Agency (NEPA) and Ministry of Agriculture Irrigation and Livestock (MAIL) to gather information on biodiversity in the Koh-e-Baba Mountains, a very important area for biodiversity in the Central Highlands.

Continued

Table 5.2. Continued.

Projects	Province	Duration	Budget ($)	Donors	Summary
Environmental education	Nationwide	Apr 2013 – Jul 2013	5000	British Ecological Society	Development and design of environment education materials promoting biodiversity conservation in Afghanistan
Women, environment and natural resources	Bamyan	May 2013 – May 2014	79,577	Fund for Local Cooperation Embassy of Finland in Kabul and Linda Norgrove Foundation	An upgrade and stage–2 project of 'Clean cookstove project' focusing on the improvement of stove performance and manufacturing process of low-cost sustainable irrigation systems. The number of villages involved in this project was 18.
Securing sustainability through community-based rangeland conservation, environmental education and livelihood production	Bamyan	June 2013 – Dec 2014	50,000	UNDP/GEF-SGP (United Nations Development Programme/Global Environment Facility - Small Grants Programme)	The project aimed to help local communities spread awareness about traditional rangeland management, build institutional capacity through establishing rangeland user associations and support rangeland management through training and workshops.
Eco-DRR (Ecosystem based Disaster Risk Reduction), using water conservation planning and community environment management for Koh-e Baba	Bamyan	May 2013 – May 2015	75,000	UNEP SSFA (United Nations Environmental Programme Small Scale Funding Agreement)	The aim of this project was to increase the resilience of local communities against natural disasters whilst also providing multiple benefits including poverty reduction, biodiversity conservation, sustainable livelihood development and carbon management.
Beekeeping, education, empowerment and sustainability	Bamyan	June 2013 – Dec 2014	163,320	New Zealand Aid and DFID (Department of International Development, UK), facilitated by Prime Consulting International	The project brought the sustainable and profitable practice of beekeeping to rural villages across the Bamyan and Yawkalang district. It was particularly aimed at women, who can practise beekeeping around their houses and provide additional income for their family.
LANSA	Bamyan	2016 – 2017	39,647	Leveraging Agriculture for Nutrition in South Asia (LANSA)	The project focused on investigating how agriculture and agri-food systems can be better designed to advance nutrition, focusing on policies, interventions and strategies that can improve nutritional status in South Asia.

Greening Baba schools	Baba Mountains	2017 – 2018	30,000	UN Environment	The Green Schools project was built upon gender-sensitive and community-based practices in order to create a safe, eco-conscious and participatory environment for children to grow up and learn about ecosystem conservation, development and climate.
Darwin Initiative Afghanistan	Bamyan	2016 – 2019	185,290	Royal Botanic Garden Edinburgh (RBGE)	The project supported the National Environmental Protection Agency of Afghanistan (NEPA) and the Conservation Organization for Afghan Mountains to engage with the rural, mountainous communities of the Central Highlands region in Bamyan, to prevent environmental degradation by reducing fuelwood collection through more efficient and sustainable fuel interventions that reduce indoor air pollution in the form of smoke.

first environmental law. The National Assembly then amended it, and, in early 2007, the final version was ultimately promulgated. Legislations define the functions and jurisdictions of NEPA, which is responsible for formulating and enforcing environmental regulations in Afghanistan (NEPA, 2023).

The first National Biodiversity Strategies and Action Plan (NBSAP; 2014-2017) of Afghanistan includes 11 preliminary targets on: protected areas, conserving selected taxonomic groups; conserving genetic diversity of valuable species and maintaining associated indigenous and local knowledge; sustainably managing sources from which biodiversity-based products are derived; decreasing the rate of loss and degradation of natural habitats; alien species; climate change and pollution; ecosystem capacity and sustainable livelihoods; traditional knowledge and the rights of International Law Commissions (ILCs); access and benefits sharing; and financial and technological resources (Convention on Biological Diversity, 2023). These targets have been linked to fulfilling the Aichi Biodiversity Targets and contain timelines and implementation responsibilities. The NBSAP is integrated with several other strategies, including the Environment Sector Strategy and the National Priority Programme 1 – National Water and Natural Resources Development Programme. In addition, the UNFCCC – National Adaptation Plan of Action (2009) – references biodiversity and assists ecosystems and rural communities extensively to adapt to climate change. The Environment Law (2007) assigns the NEPA with overall responsibility for NBSAP implementation in collaboration with the Ministry of Agriculture, Irrigation, and Livestock (MAIL) and other government organizations, and with assistance from international organizations and non-governmental organizations (NGOs). Moreover, implementation of the NBSAP at the provincial level will be facilitated by environmental plans, of which biodiversity is a subcomponent, contained in the National Development Plan (2015–2019). A comprehensive list of indicators will be developed for the updated NBSAP (2015–2019) to monitor activities (Convention on Biological Diversity, 2023).

Since 2002, UNEP has been developing environmental foundations for sustainable growth in Afghanistan. The engagement of UNEP began with a comprehensive post-conflict environmental assessment in collaboration with the Afghan government and environmental experts. The 2003 report highlighted widespread and severe land and resource degradation, such as lowered water tables, desiccation of wetlands, pervasive loss of vegetative cover, erosion, and heavy loss of wildlife (UNEP-WCMC, 2006). This prompted the Afghan government to request technical assistance to establish a national environmental agency. Since then, the focus of UNEP assistance has shifted to restoring governance structures and addressing urgent urban environmental and natural resource management issues. The Afghanistan programme of UNEP focuses on building environmental resilience and sustainability throughout the country by strengthening environmental governance and building institutions; providing technical assistance in fulfilling the administrative obligations of each ratified convention; implementing robust knowledge management and environmental outreach activities; developing community-based natural resources management; and preserving the country's biodiversity (UN Environment Programme, 2023). A concerted effort is essential for ongoing accomplishments, utilizing the technical and policy directions of agencies such as the NEPA and implementing agencies such as the MAIL. However, due to the return of the Taliban, the Afghanistan office of UNEP ceased at the end of 2021.

The World Bank (WB) in Afghanistan, in close co-ordination with the international community, development partners and donors to the Afghanistan Reconstruction Trust Funds (ARTF), supports the Afghan people and provides vital health, education and livelihood assistance (World Bank, 2011). A summary of the initiatives carried out by the WB is provided in Table 5.3. The International Centre for Integrated Mountain Development (ICIMOD) is an inter-governmental organization whose mission is to protect the Hindu Kush Himalayan (HKH) region across eight nations. Afghanistan is a founding member of ICIMOD, which works under MAIL to achieve its primary goals of alleviating poverty and conserving the environment in the HKH. The ICIMOD has implemented a number of regional programmes in the HKH region of Afghanistan, including adaptation and resilience development, transboundary landscapes, river

Table 5.3. The initiatives of the World Bank in collaboration with the national and international organizations in Afghanistan. (From World Bank, 2023).

Project	Objective	Status	Implementing agency	Duration	Financing type	Project cost ($)
Emergency food security	To restore production of food security crops for the targeted smallholder farmers	Active	FAO	June 2022 – June 2024	Grants	195,000,000
Afghanistan community resilience and livelihoods project	To provide short-term livelihood opportunities and deliver urgent essential services in rural and urban areas	Active	UNOPS	April 2022 – April 2024	Grants	265,000,000
Afghanistan water, sanitation, hygiene and institutional support project	To improve access to and quality of water supply in selected cities and to strengthen the capacity of AUWSSC to deliver sustainable services in order to contribute to national efforts to manage COVID-19 and other disasters	Active	AUWSSC	Dec. 2020 – Jan. 2026	Grants	200,000,000
Emergency agriculture and food supply project	In response to COVID-19, the proposed project development objective is to support critical food-supply chains and create short-term economic opportunities	Closed	MAIL	Aug. 2020 – Aug. 2022	Grants	100,000,000
Afghanistan gas project (AGASP)	To facilitate a sustainable supply of gas through targeted investments in gas infrastructure and enhanced gas-sector governance	Active	MMP	Nov. 2019 – Oct. 2024	IDA	n.a.
Afghanistan: national horticulture and livestock productivity project	To promote adoption of improved production practices by target farmers, with gradual roll-out of farmer-centric agricultural services systems and investment support	Closed	MAIL	April 2013 – Dec. 2020	Grants	114,600,000
Afghanistan: sustainable development of natural resources project II	To assist the Ministry of Mines (MoM) and the National Environmental Protection Agency (NEPA) in further improving their capacities to effectively regulate Afghanistan's mineral resource development in a transparent and efficient manner, and foster private-sector development	Closed	MMP	May 2011 – March 2017	IDA	52,000,000
Technical assistance for water-sector capacity building	To build GoA's (Government of Afghanistan) capacity to progressively undertake strategic basin planning and to improve project preparation for water resources development	Closed	n.a.	March 2009 – March 2013	Grants	12,000,000

Continued

Table 5.3. Continued.

Project	Objective	Status	Implementing agency	Duration	Financing type	Project cost ($)
Food crisis response project	To enhance wheat and other cereal production by supporting small-scale irrigation at community level through increase of irrigated land area and capacity building of communities to implement and maintain irrigation sector sub-projects that address community needs	Closed	n.a.	Aug. 2008 – Sep. 2010	n.a.	8,000,000
Emergency irrigation rehabilitation (additional financing)	n.a.	Closed	n.a.	June 2008 – n.a.	IDA	28,000,000
Emergency irrigation rehabilitation (supplemental)	n.a.	Closed	n.a.	April 2007 – n.a.	IDA	25,000,000
Afghanistan: horticulture and livestock productivity project	To assist producer households in adopting improved practices so as to increase horticulture and livestock productivity and production in focus areas	Closed	n.a.	May 2006 – Dec. 2012	Grants	31,000,000
AF: urban water sector project	To assist the recipient to develop the capacity of the AUWSSC for operational management and investment planning and implementation	Closed	n.a.	May 2006 – June 2014	IDA	18,000,000
ARTF - rural water supply and sanitation project	(i) To improve the health of rural communities in Afghanistan by increasing awareness through integration of health and hygiene education with the provision of safe and sustainable water supply and sanitation services; (ii) To strengthen and build the capacity of government (central and provincial) for sector development and that of NGOs, the private sector and the communities to scale up provision of sustainable water supply and sanitation facilities	Closed	n.a.	Dec. 2005 – Dec. 2009	Grants	7,650,000

ARTF - Afghanistan short-term urban water supply and sanitation project	(i) To provide sustainable, improved water supply and sanitation services to the urban segment of the Afghan population; (ii) To build the technical and institutional foundation for the medium-term UWSS programme; (iii) To ensure continuity of investment by benefiting from the implementation arrangements already in place and fully operational	Closed	n.a.	Dec. 2010 – n.a.	Grants	41,000,000
AF: emergency irrigation rehabilitation project	The project development objective is to assist the borrower in restoring irrigated agricultural production in rural areas through improved, reliable water supply to rehabilitated irrigation schemes	Closed	n.a.	Dec. 2003 – Dec. 2011	IDA	40,000,000
Kabul water supply and sewerage project	n.a.	Closed	n.a.	July 1978 – Dec. 1983	n.a.	n.a.
Fruit and vegetable export project	n.a.	Closed	n.a.	March 1978 – June 1985	n.a.	n.a.
Khanabad irrigation project (02)	n.a.	Closed	n.a.	March 1978 – June 1983	n.a.	n.a.
Livestock project (02)	n.a.	Closed	n.a.	June 1976 – Dec. 1982	n.a.	n.a.
KHANABAD IRRIG SUP FIN	n.a.	Closed	n.a.	June 1976 – n.a.	n.a.	n.a.
Kabul water supply and sanitation project	n.a.	Closed	n.a.	June 1975 – Dec. 1982	n.a.	n.a.
Livestock project	n.a.	Closed	n.a.	March 1973 – Sept. 1981	n.a.	n.a.
Khanabad irrigation project	n.a.	Closed	n.a.	May 1971 – Dec. 1981	n.a.	n.a.
Kabul water supply improvement project	To improve quality and quantity of water supplied to selected areas in Kabul and improve the operational performance of AUWSSC	Dropped	AUWSSC	n.a.	n.a.	n.a.
Agro-water and climate resilience project	To improve land and water productivity, and strengthen climate resilience in agricultural systems in selected watersheds and other irrigated areas of Afghanistan	Dropped	MAIL	n.a.	n.a.	n.a.

AUWSSC, Afghanistan Urban Water Supply and Sewerage Corporation; MAIL, Ministry of Agriculture, Irrigation and Livestock; UWSS, Urban Water Supply System; NGOs, Non-governmental organizations; MMP, Ministry of Mines and Petroleum; UNOPS, United Nation Office for Project Services; FAO, Food and Agriculture Organization; n.a., not available.

basins and cryosphere, and mountain environment regional information systems (MENRIS) (World Bank, 2023). Despite these efforts, the return of the Taliban has resulted in the emergence of Afghanistan as an epicenter for environmental degradation. Whether or not the Taliban will prioritize the conservation of the environment, as the former administration was doing in conjunction with both domestic and foreign organizations, is yet unknown.

5.6 State-of-the-art Perspectives and Recommendations

Due to its extremely diverse geography, climate and geology, Afghanistan has habitats spanning three of the world's eight biogeographical zones (Afrotropic, Palaearctic and Indomalayan). This habitat diversity, combined with its central location among biological zones, makes it one of south Asia's most ecologically productive and biologically diverse nations. Conversely, Afghanistan is reportedly one of the largest sources of natural minerals in the world, including one of the world's largest untapped high-grade copper sources (Bhattacharjee *et al.*, 2010). General criteria for assuring sustainable development in Afghanistan would be the provision of education at all levels, the establishment of a long-term database and the implementation of a comprehensive strategy for exploiting natural resources. There is also a need for proper planning and strategies to ensure that the biocultural diversity and social fabric of the region are not compromised by development. After the Taliban regained power in 2021, the country's priorities shifted to stabilizing the government, preventing civil war, protecting the borders, reviving the economy, and averting a famine because of reliance on foreign aid. Conservation and UN groups, both local and foreign conservationists, operating in Afghanistan, face a difficult decision under the Taliban administration. If they leave, the programmes they have established in the last 20 years will be discontinued. If they attempt to continue the programmes, they are aware that conservationists from Afghanistan and other countries may face risks. As a global community, we must not give up on Afghanistan but, instead, call on the Taliban government to ensure the safety of Afghan conservationists, reinstate the rule of law, including laws regarding environmental protection and protected areas, and comply with community-based sustainable natural resource management. In addition, we must advocate for a secure environment in which to conduct ecological studies in collaboration with national and international scientific organizations to generate credible data to protect Afghanistan's ecological integrity.

References

Anwar, M.Y. and Burnham, G. (2016) Tendances de l'incidence des maladies infectieuses chez les enfants en Afghanistan à l'heure de l'élargissement des services de santé publique. *Eastern Mediterranean Health Journal* 22(11), 778–785. DOI: 10.26719/2016.22.11.778

Bhattacharjee, Y., Medlin, J. and Mirzad, S. (2010) Amid war, appraising the mineral wealth of Afghanistan. *Science* 328(5986), 1620. DOI: 10.1126/science.328.5986.1620

Buzzard, P.J. (2020) The Snow Leopard Project and other adventures in warzone conservation. *Oryx* 54(3). DOI:10.1017/S0030605320000113

Connolly-Boutin, L. and Smit, B. (2016) Climate change, food security, and livelihoods in sub-Saharan Africa. *Regional Environmental Change* 16, 385–399. DOI: org/10.1007/s10113-015-0761-x

Convention on Biological Diversity (2023) Country profiles: Afghanistan. Available at: https://www.cbd.int/countries/profile/?country=af (accessed 17 June 2023).

Costanza, R., d'Arge, R., de Groot, R., Farber, S., Grasso, M. *et al.* (1998) The value of the world's ecosystem services and natural capital. *Ecological Economics* 25(1), 3–15. DOI: 10.1016/s0921-8009(98)00020-2

Dalgleish, T., Williams, J.M.G., Golden, A.-M.J:, Perkins, N., Barrett, L.F. *et al.* (2017) Reduced specificity of autobiographical memory and depression. *Journal of Experimental Psychology: General* 136(1), 23–42. DOI: 10.1037/0096-3445.136.1.23

Donatti, C.I., Harvey, C.A., Martinez-Rodriguez, M.R., Vignola, R. and Rodriguez, C.M. (2019) Vulnerability of smallholder farmers to climate change in Central America and Mexico: current knowledge and research gaps. *Climate and Development* 11(3), 264–286. DOI: 10.1080/17565529.2018.1442796

Ecology and Conservation Organisation of Afghanistan (2023) For sustainable livelihoods and green future of Afghanistan. Available at: http://www.myafghanmountains.org (accessed 17 June 2023).

EM-DAT (2019) The international disasters database. Available at: https://www.emdat.be (accessed 17 June 2023).

Gouhari, S., Forrest, A. and Roberts, M. (2021) Cost-effectiveness analysis of forest ecosystem services in mountain areas in Afghanistan. *Land Use Policy* 108, 105670. DOI: 10.1016/j.landusepol.2021.105670

Gray, J.S., Depledge, M. and Knap, A. (1996) Global climate controversy. *The Journal of the American Medical Association* 276(5), 372–373. DOI: 10.1001/jama.1996.03540050032016

Hanson, T., Brooks, T.M., Da Fonseca, G.A.B., Hoffmann, M., Lamoreux, J.F. *et al.* (2009) Warfare in biodiversity hotspots. *Conservation Biology* 23(3), 578–587. DOI: 10.1111/j.1523-1739.2009.01166.x

IUCN (2023) Resources. Available at: https://www.iucn.org/resources#resource_types-resource_type_publication (accessed 17 June 2023).

Khalid, A.M. (2018) Disaster risk management in Afghanistan: exploring the possible implications of climate change. Central European University, Vienna. Available at: https://cn.bing.com/search?q=Khalid%2C+A.+M.+(2018).+Disaster+risk+management+in+Afghanistan%3A+exploring+the+possible+implications+of+climate+change.+Central+European+University%3A+Vienna%2C+Austria.&PC=U316&FORM=CHROMN (accessed 17 June 2023).

Kreft, S. and Eckstein, D. (2013) Global Climate Risk Index 2014. Available at: http://germanwatch.org/en/download/8551.pdf (accessed 17 June 2023).

Maheshwari, A. (2022) Biodiversity conservation in Afghanistan under the returned Taliban. *Nature Ecology and Evolution* 6(6), 342–343. DOI: 10.1038/s41559-021-01655-1

McMichael, A.J., Woodruff, R.E. and Hales, S. (2006) Climate change and human health: present and future risks. *The Lancet* 367(9513), 859–869. DOI: 10.1016/S0140-6736(06)68079-3

Milbrandt, A. (2009) Assessment of biomass resources in Liberia. Available at: https://doi.org/10.2172/1004791 (accessed 17 June 2023).

NEPA (2023) Available at: https://nepa.gov.af/indexen (accessed 17 June 2023).

Omerkhil, N., Chand, T., Valente, D., Alatalo, J.M. and Pandey, R. (2020) Climate change vulnerability and adaptation strategies for smallholder farmers in Yangi Qala District, Takhar, Afghanistan. *Ecological Indicators* 110, 105863. DOI: 10.1016/j.ecolind.2019.105863

Patz, J.A., Campbell-Lendrum, D., Holloway, T. and Foley, J.A. (2005) Impact of regional climate change on human health. *Nature* 438(7066), 310–317. DOI: 10.1038/nature04188

Penuel, K.B., Statler, M. and Hagen, R. (2013) Encyclopedia of crisis management. In: Emrich, C.T. (eds) *NatCatSERVICE (Munich Re)*. Sage Publications, Inc.

Ranghieri, F., Fallesen, D.M.G., Jongman, B., Balog-Way, S.A.B., Mashahid, S.S. *et al.* (2017) Disaster risk profile, Afghanistan. Available at: https:// www.gfdrr.org/en (accessed 17 June 2023).

Saidajan, A. (2012) Effects of war on biodiversity and sustainable agricultural development in Afghanistan. *Journal of Developments in Sustainable Agriculture* 7(1), 9–13. DOI: 10.11178/jdsa.7.9

Samim, S.A. and Zhiquan, H. (2020) Assessment of food security situation in Afghanistan. *SVU-International Journal of Agricultural Sciences* 2(2), 356–377. DOI: 10.21608/svuijas.2020.45765.1044

Simonovic, S.P. (2017) Bringing future climatic change into water resources management practice today. *Water Resources Management* 31(10), 2933–2950. DOI: 10.1007/s11269-017-1704-8

Skewes, T.D., Hunter, C.M., Butler, J.R.A., Lyne, V.D., Suadnya, W. *et al.* (2016) The Asset Drivers, Well-being Interaction Matrix (ADWIM): a participatory tool for estimating future impacts on ecosystem services and livelihoods. *Climate Risk Management* 12(C), 69–82. DOI: 10.1016/j.crm.2015.08.001

Stanke, C., Kerac, M., Prudhomme, C., Medlock, J. and Murray, V. (2013) Health effects of drought: a systematic review of the evidence. *PLOS Currents* 5. DOI: 10.1371/currents.dis.7a2cee9e980f91ad7697b570bcc4b004

UNEP (2014) *Fifth National Report on the Implementation of the Convention on Biological Diversity*. Available at: https://www.unep.org/resources/report/fifth-national-report-implementation-convention-biological-diversity (accessed 17 June 2023).

UNEP (2021) Afghanistan. Available at: https://www.unep.org/explore-topics/disasters-conflicts/where-we-work/afghanistan (accessed 17 June 2023).

UNEP (2023) Publications & data. Available at: https://www.unep.org/publications-data (accessed 17 June 2023).

UNEP-WCMC (2006) Available at: https://www.unep-wcmc.org/en (accessed 17 June 2023).

Wei, W., Zhang, H., Zhou, J., Zhou, L., Xie, B. *et al.* (2021) Drought monitoring in arid and semi-arid region based on multi-satellite datasets in northwest, China. *Environmental Science and Pollution Research* 28(37), 51556–51574. DOI: 10.1007/s11356-021-14122-y

World Bank (2011) Afghanistan Reconstruction Trust Fund. Available at: https://www.wb-artf.org (accessed 17 June 2023).

World Bank (2023) The World Bank in Afghanistan. Available at: https://www.worldbank.org/en/country/afghanistan/ (accessed 17 June 2023) (CC By 4.0).

6 Ecological Conservation and Restoration in Bangladesh

Srijana Khanal[1,*], Madan Koirala[2] and Zhanhuan Shang[1]
[1]*College of Ecology, Lanzhou University, Lanzhou, China;* [2]*Central Department of Environmental Science, Tribhuvan University, Kathmandu, Nepal*

Abstract
High population has put pressure on natural resources in Bangladesh, leading to widespread ecological degradation and a decline in production yield. The degradation and climate change have reduced biodiversity, damaged ecosystem services, caused habitat loss and pollution, and had negative effects on the livelihoods of the people. This chapter describes systematically the causes, problems and risk assessments of ecological degradation in Bangladesh and examines the link between people's livelihoods and ecological services. Additionally, it includes the ecological conservation and restoration strategies or policies of the government, international organizations and local communities and provides suggestions for future improvements.

Keywords: ecosystem services, livelihoods, restoration strategies

6.1 Introduction

With a population of more than 60 million people, Bangladesh is one of the most densely populated countries in the world (Kim *et al.*, 2017). Bangladesh is a country with fragile ecosystems that are degrading at a rapid rate. The causes of ecological degradation are diverse and complex, and can be broadly categorized as natural, anthropogenic and policy-related. The natural causes of ecological degradation are mainly floods, cyclones and droughts (Sultana *et al.*, 2022), while climate change exacerbates the frequency and severity of these disasters.

Most of the people are poor, and depend on local ecosystem services for their livelihoods (Smith *et al.*, 2021). The large population has exerted huge pressure on natural resources, leading to habitat loss and pollution, and, ultimately, negative impacts on ecosystem services, livelihoods and biodiversity. Consequently, ecological conservation and restoration are gaining much attention (Fig. 6.1) as approximately 75% of the population depend on natural resources for their livelihoods, and these natural resources are in a serious state of degradation.

Natural resources are not being managed effectively, and maintaining the biological diversity is always a concern (Rahman *et al.*, 2010). Ecological conservation and restoration are recent international movements that address global policies for climate, biodiversity and sustainable development (Telesetsky *et al.*, 2016). The government of Bangladesh has committed to the conservation and restoration of natural resources guided by international policies, laws,

*Corresponding author: Srijana.khanalgautam@gmail.com

DOI: 10.1079/9781800622579.0006

Fig. 6.1. Map of Bangladesh showing administrative divisions. (From GISGeography.com; https://gisgeography.com/bangladesh-map/ (accessed 28 August 2023)

strategies, treaties and conventions. The Bangladesh National Conservation Strategy approved the main governing document as guidelines and placed it in its constitution (Rashid, 2013). The goal is to foster restoration through the conservation, development and enhancement of natural resources (Lali *et al.*, 2023). National parks, ecologically critical areas (ECAs), wildlife sanctuaries, game reserves, botanical gardens and preservation plots have been established, and gene banks and arboretums have been created (Sharma, 2011). National Adaptation Plans and Nationally Appropriate Mitigation Actions design projects to mitigate and adapt to climate change, reduce disaster risks and greenhouse gas (GHG) emissions, and support sustainable activities (Narayan *et al.*, 2016). The fourth assessment report of the Intergovernmental Panel on Climate Change (IPCC) stressed that attention must be given to climate change, the negative impacts of the rise in sea level, rapid increase in temperature, irregular weather conditions, erratic rainfall patterns, floods, and hazardous tropical cyclones (Islam, 2014).

Apart from government-led initiatives, non-governmental organizations (NGOs) and community-based organizations (CBOs) cooperate to preserve and restore the ecosystems in Bangladesh (Khan *et al.*, 2022). Such issues as climate change, biodiversity conservation and sustainable development are addressed. Firstly, this chapter discusses the causes, impacts and risk assessments of ecological degradation in Bangladesh. Secondly, it provides conservation and restoration strategies or policies initiated by the government, international organizations and local communities. Finally, suggestions for sustainable ecological restoration are discussed.

6.2 Causes, Problems and Risk Assessment of Ecological Degradation

Climate change has caused a rise in sea level, an increase in the frequency and intensity of cyclones and erratic rainfall patterns, displacement of people and increased food insecurity. The inadequate waste management in Bangladesh results in the dumping of waste in bodies of water, which leads to pollution, health problems and environmental degradation (Basak *et al.*, 2013; Ahmed, 2015; Delaporte and Maurel, 2018). The growing demand for fish in Bangladesh has led to over-fishing, which has depleted fish stocks and threatened the livelihood of fishing communities (La Rosa *et al.*, 2021). Anthropogenic activities such as deforestation, over-fishing, over-grazing and pollution are major causes of ecological degradation. The rapid growth of population and urbanization has exerted pressure on natural resources (Rahman *et al.*, 2019). In addition, poor policies contribute to environmental degradation by allowing unsustainable practices to continue or by failing to provide adequate protection of natural resources (Miah *et al.*, 2014).

Ecological degradation and erratic precipitation patterns have decreased the availability of fresh water in the country, as groundwater resources are being depleted and surface water is being polluted (Rahman and Alam, 2022). Food insecurity is widespread among millions of people, mainly those who depend on agriculture for their livelihood. Health risks due to water-borne diseases, air pollution and exposure to harmful chemicals are rampant. The incidence of respiratory illnesses, and diarrhoeal diseases are of particular concern. These problems and the loss of ecosystem services have a detrimental effect on the livelihoods of the local residents, and also exacerbate further environmental degradation (Parvin *et al.*, 2022). The destruction of forest and wetlands reduces biodiversity including endangered species. Many species like the tiger are at risk due to habitat loss and cause human–wildlife conflicts.

Forest land has been disappearing at a rapid rate worldwide due to deforestation and forest destruction (Richards and Friess, 2016). Deforestation has reduced the forest area in Bangladesh by close to 80% over the years and, from 2002 to 2022 the country lost approximately 228 kha of tree cover, a 12% decrease, and 124 megatons of CO_2 emissions (Watch, 2022). Expansions of agriculture, urbanization and infrastructure development are the main reasons for the large areas of forest cover that have been lost. Forests in Chittagong Hill cover 40% of the country's forests; however, they are seriously degraded because of swidden agriculture on the forest area, illegal cutting down of trees, invasion of alien species and conversion of forests into industrial plantations (Ahammad *et al.*, 2023).

Similarly, mangrove forests are being degraded from over-exploitation, deforestation, land recycling, decreased water flow and pollution (Reddy *et al.*, 2016). The Sundarbans mangrove ecosystem, the largest mangrove forest, is the habitat of the royal Bengal tiger (*Panthera tigris tigris*) and many other species of mammals, birds, reptiles and marine life (Fig. 6.2). It is in danger of degradation due to intensive shrimp farming, pollution and over-exploitation of local natural resources, while lack of ecosystem knowledge and sea-level rise are accelerating the degradation (Iqbal and Mozahid, 2022). Collection of nypa (*Nypa fructans*) leaves is a major livelihood source for the people living near the Sundarbans reserve forest, as they are used as

Fig. 6.2. Sundarban royal Bengal tiger (*Panthera tigris tigris*). (From Titumir, 2021)

thatching material and are a source of income. However, it was concluded that nypa leaf collection is not a good sustainable livelihood strategy for the forest-dependent communities (Islam *et al.*, 2020).

Wetlands are essential for the purification of water, flood control and carbon sequestration, and are also the habitat of many plant and animal species (Aziz and Naima, 2021), but they are under threat, mainly due to human activities. The destruction of the aquatic-terrestrial transitional zone, changes in land use, excessive exploitation of natural resources, decreased upstream flow, sea-level rise and over-fishing are the main causes (Alam, 2014).

In Bangladesh, out of 64 districts, 20 are at risk of riverbank erosion, which damages the physical, economic, social and political conditions, and impacts the lives and livelihoods of the riverine households (Lein, 2000; Ahmed, 2015). Destruction of the riverbank occurs gradually and has long-term effects which are often non-recoverable. Each year about 8700 ha of land are lost from riverbank erosion that displaces close to 200,000 people (Hoq *et al.*, 2021; Mamun *et al.*, 2022). Most of the households on the riverbank (e.g. char areas) are poor and marginalized from the mainland people due to the poor communication networks (Sarker *et al.*, 2003).

The destruction of natural habitats and overuse has reduced biodiversity, which results in the reduction of species and genetic diversity, and the disruption of ecosystem functions. Soil erosion due to deforestation, improper agricultural practices and land degradation reduces soil fertility, productivity and stability, leading to reduced crop yields and biodiversity and increased poverty and vulnerability to natural disasters like floods and landslides (Alam, 2009).

Risk assessment is an essential tool to identify the potential impacts of ecological degradation in Bangladesh. It is a technique which measures and assesses the negative impact of disturbances like anthropogenic stresses, pollution and natural disasters (Wu *et al.*, 2021). Risks due to ecological degradation are assessed by analyzing specific consequences of various scenarios that include the loss of key ecosystem services, effects on human health and livelihoods, and the potential for further environmental degradation.

6.3 The Link between People's Livelihoods and Ecology

There is an inter-relationship among ecosystem services, livelihoods and people's well-being in Bangladesh, as most people are dependent on natural resources for their income and food security (Walelign and Jiao, 2017). However, these resources are under threat due to climate change, disasters, deforestation and over-exploitation (Ayeb-Karlsson *et al.*, 2016). Thus, it is essential to understand how people cope with these challenges and how they can enhance their resilience and sustainability. The country is moving towards an urbanized future, but ecosystem services continue to be critical for the well-being of the populations in rural areas.

The loss of biodiversity in Bangladesh affects ecosystem services such as pollination, nutrient cycling and water purification, and leads to negative effects on agriculture, fisheries and tourism, and, ultimately, on the country's economy. There is a loss of employment opportunities and restricted economic growth. The country should prioritize sustainable-development strategies including the protection of natural resources, and the implementation of environmentally friendly policies and practices. The development of alternative livelihoods for communities that rely on natural resources is essential. It is crucial to make the public aware of the importance of environmental protection and to promote community participation in conservation efforts. Bangladesh suffered from 19 droughts between 1960 and 1991, which affected 47% of the country (Islam, 2009).

Agriculture is the backbone of the country's economy, as it provides livelihood for over 60% of the population (Shelley *et al.*, 2016). The fertile alluvial soil and water resources provide ideal conditions for rice cultivation (Mallick, 2019), which is the staple food of the country, and is produced by millions of farmers. Other crops such as jute, tea and sugarcane are also important sources of income. For many people who live in rural and remote areas, agriculture is the only option for food and income (Leighton *et al.*, 2011). Lessening the high dependency on agriculture and balancing farm and non-farm activities (Ahmed *et al.*, 2018) are aims of the present government. Forests provide livelihoods and security for the farming households suffering from poverty (Pouliot *et al.*, 2012).

Coastal communities in Bangladesh have to contend with hydro-meteorological disasters including cyclones, tidal surges, floods, drought and saline water intrusion, which have a direct impact on agriculture (Leighton *et al.*, 2011). These agriculture-dependent people have taken out informal and formal loans, which have driven them further into poverty (Lázár *et al.*, 2015). Many of these people are forced to change their occupations, diversify their income sources, and self-organize to recover. In communities in the Chittagong Hills Tracts (CHT), the people depend on cash income from fuelwood, bamboo, fruits, honey and medicinal plants for their subsistence. Forest conservation and management should take responsibility for the needs and preferences of the local people and the ecological functions of the forest (Jannat *et al.*, 2018). Here, rain-fed rice is the main crop in the monsoon season, while irrigation is essential in the dry season for growing rice, pulses, oil seeds and vegetables. Thus, their agriculture is dependent upon environmental aspects like timing and intensity of the monsoon, and the availability of freshwater for irrigation.

A climate-resilient and sustainable future for communities in vulnerable countries starts through resilient livelihoods. In Sundarbans, there are many ecosystem services for the local residents depending on forests, including fishermen, crabbers, honey hunters and nypa leaf collectors. Climate change poses the main threat to the livelihoods of these people by destroying the productivity of the forest resources. People affected by these changes are forced to change their occupations (Ela *et al.*, 2021). For instance, some honey collectors engage in fishing that results in over-fishing of resources, and, uninten tionally, affects the ecosystem and livelihood of the fishermen. The government of Bangladesh has an information gap in the context of climate change and often does not advise the people of impending dangers. It is an essential condition to find solutions to the livelihood threats, shocks and disasters.

Local knowledge, social memory and community networks are important to foster resilience (Uddin *et al.*, 2021). However, indigenous knowledge is neglected while western models are implemented in Bangladesh, which are not usually appropriate for the conditions.

6.4 Ecological Conservation and Restoration Strategies or Policies

Bangladesh has diverse flora and fauna, but their habitats are threatened by anthropogenic and natural disturbances. There are different ecological conservation and restoration strategies and policies which are essential for local communities, the government and international organizations to implement (Cliquet and Akhtar-Khavari, 2019). The concepts on ecological conservation differ between local communities and institutions; co-operation between them could safeguard the ecosystem. It is essential for the local people to protect themselves from external pressures regarding their livelihoods and to make decisions in utilizing natural resources. For instance, the Village Development Committees (VDC) made public functioning at the community level, as it is meant to represent the interests of the entire resident population (Carloni and Crowley, 2005). The community needs strong traditional authorities and decision-making abilities and should develop strong networks for information exchange and exposure with other people (Bevir, 2012).

The government of Bangladesh implemented the Sundarbans Management Plan to protect biodiversity, manage resources sustainability and address the effect of climate change on the ecosystem (Mahmood *et al.*, 2021). The maintenance of the ecosystem is vital for the survival of the endangered Bengal tiger and the tiger's prey species (Seidensticker and 'Abdulḥaʼī, 1983). The National Biodiversity Strategy and Action Plan was developed by the government of Bangladesh to promote the conservation and sustainable usage of biodiversity. The plan was urgent as threats to the flora and fauna were increasing due to the destruction of habitats and ecosystems. For instance, in the context of forest biodiversity, planting in forest areas increased the coverage, but natural forests declined over the years. The plan consisted of different measures including the protection of habitats, the regulation of trade of endangered species and the promotion of sustainable agriculture. Bangladesh committed to implement the three main objectives which were the conservation of biodiversity, sustainable use of its components and fair and equitable sharing of genetic resources (Mohamed, 2015; Miah *et al.*, 2023).

The International Principles and Standards for the Practice for Ecological Restoration accounted for the complex ecosystem dynamics, and navigated trade-offs associated with land management priorities and decisions (Gann *et al.*, 2019). The United Nations (UN) declared six practical strategies to fulfil restoration aspirations such as incorporating holistic actions, encouraging traditional ecological knowledge, collaborating with allied movements and organizations, advancing soil microbiome science and technology, providing training and capacity-building opportunities for community people, and emphasizing the relation between ecosystem health and human health (Aronson *et al.*, 2020).

The mangrove afforestation programme was initiated through the collaboration of the United Nations Development Program (UNDP) and the Global Environment Facility (GEF) in Bangladesh. This programme included the planting of new mangrove forests to protect the coastline from erosion, the provision of habitats for marine species and the mitigation of the effects of climate change. The green measures were seen to be low-cost rather than human-made solutions for dealing with climate change. Thus, the mangrove reforestation project helped to strengthen resilience to climate change (Lovelock *et al.*, 2022) by protecting and restoring greenery ecosystems like forests and wetlands and utilizing nature for infrastructures like dams and seawalls.

Local people were involved in programmes like the Community-Based Adaptation to Climate Change, which promoted the use of sustainable agriculture, protection of biodiversity and the establishment of community-managed natural resource systems. Community people were involved in raising motivation through awareness, capacity building, providing alternative income-generating solutions, and promoting the efficient use of wetland resources (Reid, 2016; Sovacool, 2018). This project included conservation measures such as swamp protection and reforestation, recovery of wetlands, establishment of fish protection areas, and species conservation. The Nishargo Support Project (2003–2008) was initiated through Bangladesh Forest Department with USAID's support, and introduced co-management approaches in protected areas (DOE, 2008). The Coastal and Wetland Biodiversity Management Project (CWBMP) was directed by the Department of the Environment from 2003 to 2011, and focused mainly on the management of select places with the involvement of community people. It focused on unsustainable exploitation of biodiversity resources, clearance of forests and watersheds, transformation of wetlands and forest ecosystems, and illegal poaching of animals. Capacity building of the local people on natural resources conservation, organizational management and livelihood improvement were given priority (Sunny *et al.*, 2020). The effect of livestock grazing and human activities were monitored in forest areas, and the protective effect of the international ecological network was assessed. However, the site covered only a few national species of community interest (Zisenis, 2017).

Women in rural Bangladesh are engaged in crop cultivation, harvesting, seed preservation, livestock and poultry rearing, and aquaculture. They were offered training, education-awareness and livelihood activities for safeguarding biodiversity (Yousuf Haroon and Kibria, 2017), and are involved in activities to ensure that they receive equitable benefits (NBSAP 2016–2021). The Community-Based Adaptation (CBA) approach was promoted by different international organizations like UNDP, CARE, IIED, IUCN, EEF and local NGOs for conserving threatened biodiversity. The aim was to empower the most vulnerable groups, mainly women, to participate in identifying and implementing adaptation strategies depending on local knowledge, needs, preferences and capacities (Jahan, 2022).

In addition, the Bangladesh government and international organizations promoted ecotourism for conservation and generating income by developing ecotourism facilities, promoting sustainable tourism, and establishing community-managed tourism programmes. These strategies and policies demonstrated the commitment of the Bangladesh government to protect the ecological diversity and promote sustainable development.

6.5 Suggestions for Sustainable Ecological Construction in the Future

Most under-developed countries like Bangladesh are vulnerable to natural disasters and climate change due to their geographical location and poverty. Consequently, they are initiating programmes to support sustainable development.

Different nature-based solutions provide cost-effective options for addressing these challenges, although policy-makers lack evidence of their effectiveness (Seddon *et al.*, 2020). In Bangladesh, it is essential to protect, restore and manage mangroves, forests and wetlands, and conserve agricultural land (Smith *et al.*, 2021).

Bangladesh has sustainable building materials such as bamboo, mud and thatch. Locally available material reduces environmental pollution due to transportation and supports the local economy. Ecological restoration and protective measures should be considered in projects like the construction of corridors to minimize the impacts on protected areas (Jin *et al.*, 2021). Passive design strategies should consider orientation, shading and natural ventilation to decrease the energy consumption in buildings (Ahsan, 2009). Bangladesh has abundant renewable energy resources like solar, wind and hydropower, which can be used to decrease the dependence on fossil fuels and mitigate climate change.

There are heavy rainfalls during the monsoon season that cause floods and damage to infrastructure. Harvesting rainwater can be a sustainable solution for managing excessive amounts of water and reducing the strain on water supply. The National Water Management Plan should balance drinking water and the environment, especially during drought (Rahaman *et al.*, 2019). For drinking-water purposes, the Managed Aquifer Recharge (MAR) should analyze the physical and chemical properties and the bacteria content to assess the quality.

Green spaces are vital for ecosystem services, but increased urbanization in Bangladesh has reduced green spaces. Conservation and the development of green spaces are effective for reducing friction between environmental quality and urbanization (Siddique *et al.*, 2022). By prioritizing green spaces in construction projects, developers create healthier and more sustainable environments for residents. Thus, sustainable design and construction of buildings and infrastructures which minimize the negative aspects on the environment and improve positive social and economic benefits should be encouraged.

The adoption of green building standards and certification systems like leadership in Energy and Environmental Design, and Green Rating for Integrated Habitat Assessment should be encouraged as they promote: (i) energy effectiveness, water conservation, waste disposal, and indoor environmental quality for sustainable ecological conservation; and (ii) lower-carbon and climate-resilient construction using alternative cement and concrete materials, managing thermal insulation and ventilation, and installing solar panels and biogas plants which incorporate green roofs and walls.

Natural regeneration is important to preserve and manage biodiversity, and can be modified to address societal needs. The modified ecosystem can be designed, at least partly, through natural regeneration, which decreases the impact of floods and landslides while benefiting crops by improving water-holding capability of the soil. Integration of ecological restoration and conservation principles should be implemented to plan and design urban and rural landscapes like restoring wetlands and mangroves. It is essential to create urban forests and parks, protect biodiversity corridors and apply ecosystem-based adaptation approaches.

6.6 Conclusions

The biodiversity and unique ecosystems in Bangladesh are a major global concern. Ecological conservation and restoration measures are applied to conserve natural resources and improve the livelihood of the people. Anthropogenic activities such as deforestation, over-fishing, over-grazing and pollution are the major causes of the reduction in ecosystem services, while inappropriate or poorly planned policy intervention exacerbates the severity of the problems. The transformation of forests into cultivated land due to high population leads to over-exploitation of natural resources and environmental degradation. Natural disasters like floods, cyclones and droughts damage ecosystems, while climate change causes the rise of sea levels and increased frequency and intensity of cyclones. The reduction in ecosystem services has negative impacts on pollination, nutrient cycling and water purification, which is detrimental to agriculture, fisheries, tourism and public health. Risk assessment could be an effective tool to identify the potential impacts of ecological degradation. The ecology is closely linked with the livelihood of local people, as most people are dependent on the natural resources for their income, food security and well-being. The alluvial soil is very fertile and ideal for rice cultivation, the main crop in the country.

Different ecological conservation and restoration strategies are being implemented by local communities, the government and international organizations to protect and maintain the country's rich biodiversity and ecosystem services. The government of Bangladesh has made a detailed management plan to conserve forests, protect biodiversity, manage resources and pinpoint the effects of climate change on the ecosystem. The National Biodiversity Strategy and Action Plan was developed to promote the sustainable usage of natural resources and prevent the loss and degradation of biodiversity. The International Principles and Standards for Practice designed strategies to restore the ecosystem, while the United Nations and Global Environment Facility initiated programmes to mitigate the effect of climate change.

There are plans for future sustainable ecological development in Bangladesh, some of which are nature-based. These activities include: (i) ecological restoration, conservation measures and the use of local materials; (ii) passive design strategies using orientation, shading and natural ventilation to decrease energy consumption in buildings. Renewable energy can decrease the dependency on fossil fuels and help mitigate the impact of climate change; (iii) rainwater harvesting during monsoon season to store water that can be used during droughts; and (iv) developing or maintaining green spaces through restoration and conservation to reduce pollution in a rapidly urbanizing country. In summary, these activities could benefit the ecological conservation and restoration in Bangladesh.

References

Ahammad, R., Hossain, M.K., Sobhan, I., Hasan, R., Biswas, S.R. *et al.* (2023) Social-ecological and institutional factors affecting forest and landscape restoration in the Chittagong Hill Tracts of Bangladesh. *Land Use Policy* 125, 106478. DOI: 10.7202/051336ar

Ahmed, I. (2015) *People of Many Rivers: Tales from the Riverbanks*, The University Press Limited, Dhaka, Bangladesh.

Ahmed, M., Bhandari, H., Gordoncillo, P., Quicoy, C. and Carnaje, G. (2018) Factors affecting extent of rural livelihood diversification in selected areas of Bangladesh. *SAARC Journal of Agriculture* 16, 7–21. DOI: 10.3329/sja.v16i1.37419

Ahsan, T. (2009) Passive design features for energy-efficient residential buildings in tropical climates: the context of Dhaka, Bangladesh. Master's thesis in Environmental Research. KTH-Royal Institute of Technoloy, Sweden at the Division on Environmental Strategies Research (FMS).

Alam, G. (2009) *Environmental Pollution of Bangladesh: Its Effect and Control.* ICME2009, Dhaka, Bangladesh.

Alam, M.Z. (2014) Status of biodiversity at wetland ecosystem of Mohangonj Upazila in Netrakona district. *Advances in Ecology* 2014. DOI: 10.1155/2014/642450

Aronson, J., Goodwin, N., Orlando, L., Eisenberg, C. and Cross, A.T. (2020) A world of possibilities: six restoration strategies to support the United Nation's Decade on Ecosystem Restoration. *Restoration Ecology* 28, 730–736. DOI: 10.1111/rec.13170

Ayeb-Karlsson, S., Van Der Geest, K., Ahmed, I., Huq, S. and Warner, K. (2016) A people-centred perspective on climate change, environmental stress, and livelihood resilience in Bangladesh. *Sustainability Science* 11, 679–694. DOI: 10.1007/s11625-016-0379-z

Aziz, A. and Naima, U. (2021) Rethinking digital financial inclusion: evidence from Bangladesh. *Technology in Society* 64, 101509. DOI: 10.31235/osf.io/7sr5c

Bevir, M. (2012) *Governance: A Very Short Introduction*. Oxford University Press, Oxford, UK.

Carloni, A.S. and Crowley, E.L. (2005) *Rapid Guide for Missions: Analysing Local Institutions and Livelihoods: Guidelines*. Food and Agriculture Organization of the United Nations, Rome.

Cliquet, A. and Akhtar-Khavari, A. (2019) *Ecological Restoration and International Law*. Oxford University Press, Oxford, UK.

Delaporte, I. and Maurel, M. (2018) Adaptation to climate change in Bangladesh. *Climate Policy* 18, 49–62. DOI: 10.1080/14693062.2016.1222261

Ela, M.Z., Rabya, T., Khan, L., Rahman, M.H., Shovo, T.E.A. *et al.* (2021) Climate change and livelihood vulnerabilities: the forest resource-dependent communities of the Sundarbans of Bangladesh.

In: Alam, G.M.M., Erdiaw-Kwasie, M.O., Nagy, G.J. and Filho, W.L. (eds) *Climate Vulnerability and Resilience in the Global South: Human Adaptations for Sustainable Futures*. Springer Nature, Cham, Switzerland, pp. 341–352.

Gann, G.D., McDonald, T., Walder, B., Aronson, J., Nelson, C.R. *et al.* (2019) International principles and standards for the practice of ecological restoration. *Restoration Ecology* 27, S1–S46. DOI: 10.1111/rec.13771

Hoq, M.S., Raha, S.K. and Hossain, M.I. (2021) Livelihood vulnerability to flood hazard: understanding from the flood prone Haor ecosystem of Bangladesh. *Environmental Management* 67, 532–552. DOI: 10.1007/s00267-021-01441-6

Iqbal, M.H. and Mozahid, M.N. (2022) Valuing conservation of the Sundarbans mangrove forest ecosystem. *Trees, Forests and People* 9, 100278. DOI: 10.1016/j.tfp.2022.100278

Islam, M. (2009) Sustainable development framework: Bangladesh perspective. *Bangladesh Journal of Public Administration* 1–12. DOI: 10.36609/bjpa.v18i1.194

Islam, M.N., Dana, N.H., Rahman, K.S., Hossain, M.T., Ahmed, M.U. *et al.* (2020) *Nypa fruticans* Wurmb leaf collection as a livelihoods strategy: a case study in the Sundarbans impact zone of Bangladesh. *Environment, Development and Sustainability* 22, 5553–5570. DOI: 10.1007/s10668-019-00438-w

Islam, M.S. (2014) Green banking in Bangladesh: How much implemented by the practitioners? *Abasyn University Journal of Social Sciences* 7. e-ISSN: 2278-487X.

Jahan, I. (2022) Scope of public participation for the protection of biodiversity in Bangladesh: some legal reflections. *Environmental Policy and Law* 52, 55–68. DOI: 10.3233/EPL-210113

Jannat, M., Hossain, M.K., Uddin, M.M., Hossain, M.A. and Kamruzzaman, M. (2018) People's dependency on forest resources and contributions of forests to the livelihoods: a case study in Chittagong Hill Tracts (CHT) of Bangladesh. *International Journal of Sustainable Development & World Ecology* 25, 554–561. DOI: 10.1080/13504509.2018.1434571

Jin, Y., Li, A., Bian, J., Nan, X., Lei, G. *et al.* (2021) Spatiotemporal analysis of ecological vulnerability along Bangladesh-China-India-Myanmar economic corridor through a grid level prototype model. *Ecological Indicators* 120, 106933. DOI: 10.1016/j.ecolind.2020.106933

Khan, N.A., Choudhury, J.K., Rashid, A.M., Siddique, M.R.H. and Sinha, K. (2022) Co-management practices by non-government organizations (NGOs) in selected coastal forest zones of Bangladesh: a focus on sustainability. *Sustainability* 14, 14885. DOI: 10.3390/su142214885

Kim, Y., Smith, J.B., Mack, C., Cook, J., Furlow, J. *et al.* (2017) A perspective on climate-resilient development and national adaptation planning based on USAID's experience. *Climate and Development* 9, 141–151. DOI: 10.1080/17565529.2015.1124037

La Rosa, D., Pauleit, S. and Xiang, W.N. (2021) Unearthing time-honored examples of nature-based solutions. *Socio-Ecological Practice Research* 3, 329–335. DOI: 10.1007/s42532-021-00099-y

Lali, O., Nanuli, K. and Maka, S. (2023) Green economy directives of the European Union: problems and challenges. *European Science Review* 58–64. DOI: 10.29013/ESR-23-3.4-58-64

Lázár, A.N., Clarke, D., Adams, H., Akanda, A.R., Szabo, S. *et al.* (2015) Agricultural livelihoods in coastal Bangladesh under climate and environmental change: a model framework. *Environmental Science: Processes & Impacts* 17, 1018–1031. DOI: 10.1039/C4EM00600C

Leighton, M., Shen, X. and Warner, K. (2011) *Climate Change and Migration: Rethinking Policies for Adaptation and Disaster Risk Reduction*. Publication series of United Nations University. UNU-EHS, Bonn, Germany.

Lein, H. (2000) Hazards and forced migration in Bangladesh. *Norsk Geografisk Tidsskrift-Norwegian Journal of Geography* 54, 122–127. DOI: 10.1080/002919500423735

Lovelock, C.E., Barbier, E. and Duarte, C.M. (2022) Tackling the mangrove restoration challenge. *PLOS Biology* 20, e3001836. DOI: 10.1371/journal.pbio.3001836

Mahmood, H., Ahmed, M., Islam, T., Uddin, M.Z., Ahmed, Z.U. *et al.* (2021) Paradigm shift in the management of the Sundarbans mangrove forest of Bangladesh: issues and challenges. *Trees, Forests and People* 5, 100094. DOI: 10.1016/j.tfp.2021.100094

Mallick, B. (2019) The nexus between socio-ecological system, livelihood resilience, and migration decisions: empirical evidence from Bangladesh. *Sustainability* 11, 3332. DOI: 10.3390/su11123332

Mamun, A.A., Islam, A.R.M.T., Alam, E., Chandra Pal, S. and Alam, G.M.M. (2022) Assessing riverbank erosion and livelihood resilience using traditional approaches in northern Bangladesh. *Sustainability* 14, 2348. DOI: 10.3390/su14042348

Miah, M.D., Akther, S., Shin, M.Y. and Koike, M. (2014) Scaling up REDD+ strategies in Bangladesh: a forest dependence study in the Chittagong Hill Tracts. *Forest Science and Technology* 10, 148–156. DOI: 10.1080/21580103.2014.889045

Miah, M.R., Hasan, M.M., Parisha, J.T., Alam, M.S.E., Sayok, A.K. *et al.* (2023) Enhancing national park information knowledge to improve biodiversity conservation in Bangladesh: a study on policy perspectives. *International Journal of Plant Research* 13, 1–23. DOI: 10.5923/j.plant.20231301.01

Narayan, S., Beck, M.W., Reguero, B.G., Losada, I.J., Van Wesenbeeck, B. *et al.* (2016) The effectiveness, costs and coastal protection benefits of natural and nature-based defences. *PLOS ONE* 11, e0154735. DOI: 10.1371/journal.pone.0154735

Parvin, F., Haque, M.M. and Tareq, S.M. (2022) Recent status of water quality in Bangladesh: a systematic review, meta-analysis and health risk assessment. *Environmental Challenges* 6, 100416. DOI: 10.1016/j.envc.2021.100416

Pouliot, M., Treue, T., Obiri, B.D. and Ouedraogo, B. (2012) Deforestation and the limited contribution of forests to rural livelihoods in West Africa: evidence from Burkina Faso and Ghana. *Ambio* 41, 738–750. DOI: 10.1007/s13280-012-0292-3

Rahaman, M.F., Jahan, C.S. and Mazumder, Q.H. (2019) Rainwater harvesting: practiced potential for integrated water resource management in drought-prone Barind tract, Bangladesh. *Groundwater for Sustainable Development* 9, 100267. DOI: 10.1016/j.gsd.2019.100267

Rahman, M.M. and Alam, K. (2022) CO_2 emissions in Asia-Pacific region: Do energy use, economic growth, financial development, and international trade have detrimental effects? *Sustainability* 14, 5420. DOI: 10.3390/su14095420

Rahman, M.M., Rahman, M.M. and Islam, K.S. (2010) The causes of deterioration of Sundarban mangrove forest ecosystem of Bangladesh: conservation and sustainable management issues. *Aquaculture, Aquarium, Conservation & Legislation* 3, 77–90.

Rahman, S., Islam, M.S., Khan, M.N.H. and Touhiduzzaman, M. (2019) Climate change adaptation and disaster risk reduction (DRR) through coastal afforestation in South-Central Coast of Bangladesh. *Management of Environmental Quality: An International Journal* 30, 498–517. DOI: 10.1108/MEQ-01-2018-0021

Rashid, H.E. (2013) Bangladesh: National Conservation Strategy. *Strategies for Sustainability: Asia.* Routledge, London.

Reddy, C.S., Pasha, S.V., Jha, C., Diwakar, P. and Dadhwal, V. (2016) Development of national database on long-term deforestation (1930–2014) in Bangladesh. *Global and Planetary Change* 139, 173–182. DOI: 10.1016/j.gloplacha.2016.02.003

Reid, H. (2016) Ecosystem and community-based adaptation: learning from community-based natural resource management. *Climate and Development* 8, 4–9. DOI: 10.1080/17565529.2015.1034233

Richards, D.R. and Friess, D.A. (2016) Rates and drivers of mangrove deforestation in Southeast Asia, 2000–2012. *Proceedings of the National Academy of Sciences* 113, 344–349. DOI: 10.1073/pnas.1510272113

Sarker, M.H., Huque, I., Alam, M. and Koudstaal, R. (2003) Rivers, chars and char dwellers of Bangladesh. *International Journal of River Basin Management* 1, 61–80. DOI: 10.1080/15715124.2003.9635193

Seddon, N., Chausson, A., Berry, P., Girardin, C.A., Smith, A. *et al.* (2020) Understanding the value and limits of nature-based solutions to climate change and other global challenges. *Philosophical Transactions of the Royal Society B* 375, 20190120. DOI: 10.1098/rstb.2019.0120

Seidensticker, J. and Abdulḥai̇̄l, M.A. (1983) *The Sundarbans Wildlife Management Plan: Conservation in the Bangladesh Coastal Zone: A World Wildlife Fund Report*. International Union for Conservation of Nature and Natural Resources, Gland, Switzerland.

Sharma, R.A. (2011) Co-management of protected areas in South Asia with special reference to Bangladesh. *Asia-Pacific Journal of Rural Development* 21, 1–28. DOI: 10.1177/1018529120110101

Shelley, I.J., Takahashi-Nosaka, M., Kano-Nakata, M., Haque, M.S. and Inukai, Y. (2016) Rice cultivation in Bangladesh: present scenario, problems, and prospects. *Journal of International Cooperation for Agricultural Development* 14, 20–29. DOI: 10.50907/jicad.14.0_20

Siddique, M.A.M., Uddin, A., Rahman, S.M.A., Rahman, M., Islam, M.S. *et al.* (2022) Microplastics in an anadromous national fish, Hilsa shad Tenualosa ilisha from the Bay of Bengal, Bangladesh. *Marine Pollution Bulletin* 174, 113236. DOI: 10.1016/j.marpolbul.2021.113236

Smith, A., Tasnim, T., Irfanullah, H.M., Turner, B., Chausson, A. *et al.* (2021) Nature-based solutions in Bangladesh: evidence of effectiveness for addressing climate change and other sustainable development goals. *Frontiers in Environmental Science* 511. DOI: 10.3389/fenvs.2021.737659

Sovacool, B.K. (2018) Bamboo beating bandits: conflict, inequality, and vulnerability in the political ecology of climate change adaptation in Bangladesh. *World Development* 102, 183–194. DOI: 10.1016/j.worlddev.2017.10.014

Sultana, N., Rahman, M.M. and Khanam, R. (2022) Environmental kuznets curve and causal links between environmental degradation and selected socioeconomic indicators in Bangladesh. *Environment, Development and Sustainability* 24, 5426–5450. DOI: 10.1007/s10668-021-01665-w

Sunny, A., Alam, R., Sadia, A., Miah, Y., Hossain, S. *et al.* (2020) Factors affecting the biodiversity and human well-being of an ecologically sensitive wetland of North Eastern Bangladesh. *Journal of Coastal Zone Management* 23, 471. DOI: 10.35841/2473-3350.23.1-471

Telesetsky, A., Cliquet, A. and Akhtar-Khavari, A. (2016) *Ecological Restoration in International Environmental Law*. Taylor & Francis, London.

Titumir, R.A.M. (2021) *Numbers and Narratives in Bangladesh's Economic Development*. Springer, Singapore.

Uddin, M.S., Haque, C.E., Khan, M.N., Doberstein, B. and Cox, R.S. (2021) Disasters threaten livelihoods, and people cope, adapt and make transformational changes: community resilience and livelihoods reconstruction in coastal communities of Bangladesh. *International Journal of Disaster Risk Reduction* 63, 102444. DOI: 10.1016/j.ijdrr.2021.102444

Walelign, S.Z. and Jiao, X. (2017) Dynamics of rural livelihoods and environmental reliance: empirical evidence from Nepal. *Forest Policy and Economics* 83, 199–209. DOI: 10.1016/j.forpol.2017.04.008

Watch, G.F. (2022) Forest monitoring designed for action. Available at: https://www.globalforestwatch.org/ (accessed 28 September 2023).

Wu, Z., Lin, C., Shao, H., Feng, X., Chen, X. *et al.* (2021) Ecological risk assessment and difference analysis of pit ponds under different ecological service functions: a case study of Jianghuai ecological economic zone. *Ecological Indicators* 129, 107860. DOI: 10.1016/j.ecolind.2021.107860

Yousuf Haroon, A. and Kibria, G. (2017) Wetlands: biodiversity and livelihood values and significance with special context to Bangladesh. *Wetland Science: Perspectives from South Asia*, 317–346. DOI: 10.1007/978-81-322-3715-0_17

Zisenis, M. (2017) Is the Natura 2000 network of the European Union the key land use policy tool for preserving Europe's biodiversity heritage? *Land Use Policy* 69, 408–416. DOI: 10.1016/j.landusepol.2017.09.045

7 Ecological Conservation and Restoration in Bhutan

Furbe Lama[1,*], Allan Degen[2] and Zhanhuan Shang[1]
[1]College of Ecology, Lanzhou University, Lanzhou, China; [2]Institutes for Desert Research, Ben-Gurion University of the Negev, Israel

Abstract
Bhutan is recognized as a hotspot of biodiversity and an eco-friendly country. Forests cover 70% of the total land area and are important for the livelihood of the people and economy of the country. However, climate change, steep and rugged terrain, urbanization, infrastructure development and natural disasters are degrading the landscape and environment. This chapter reviews the ecology, conservation policies and livelihoods of the country. Implementation of national policies, laws, international treaties, agreements and declarations on the conservation and management of ecosystems could be the key for sustainable conservation and restoration of those systems. Focus should be placed on the preservation of ecosystem services, socio-ecological projects, diversification of livelihoods and sustainable food security. Ecotourism could be important for employment opportunities and the economy, for diversification of livelihoods and for improving socio-economic status.

Keywords: biodiversity, degradation, ecosystem services, environmental management, Himalaya

7.1 Introduction

7.1.1 Background

Bhutan is a small Himalayan country covering 38,394 km^2 and spanning 300 km east to west and 170 km north to south, with an elevation range between 150 m and 7500 m above sea level (masl). It is characterized by complex topography consisting mostly of steep and high mountains and diverse landscapes (Bhutan Glacial Lake Inventory, 2021). The ecological zones of Bhutan can be divided into three: (i) southern belt – a subtropical area, ranging in altitude between 150 masl and 2000 masl; (ii) central belt – steep mountainous region with river valleys, ranging in altitude between 2000 masl and 4000 masl; and (iii) northern belt – high Himalayan mountains over 4000 masl.

About 69% of the population live in rural areas and depend on subsistence agriculture (Yeshey *et al.*, 2023); however, poverty is widespread (Cosma *et al.*, 2023). Most of the forest land is controlled by the central government and only 12 ha are private forest (Feuerbacher *et al.*, 2021). Forest covers 70% of the land of which 46% is dominated by broad-leaved forest, 14% by mixed conifer, 3% by chir pine, 3% by blue pine forest, and 4% by alpine scrub (Rai and Phuntsho, 2016). Potentially, 11.3% of the forest is suitable for production, 5.8% for commercial use and 5.8% for rural use (Tshewang *et al.*, 2021). Due to the rugged mountain terrain, only 2.75% of

*Corresponding author: lama2021@lzu.edu.cn

DOI: 10.1079/9781800622579.0007

the area is cultivated (Tshewang *et al.*, 2021). Bhutan is home to many wildlife species and, with the support of the international community, has strong programmes on conservation. The mountain ecosystem is very fragile and vulnerable due to climate change and the lack of adaptation and management strategies (IPCC, 2013). Ecosystem degradation reduces ecosystem services, habitats and biodiversity, and fragments landscapes.

Ecosystems are threatened by poverty, vast resource extraction, landscape fragmentation and widespread development (Cosma *et al.*, 2023). Poaching of animals is widespread as the country lies next to China, India and Nepal (Felbab-Brown, 2018). The keystone and flagship species like the West Bengal tiger (*Panthera tigris tigris*), snow leopard (*Uncia uncia*), white-bellied heron (*Ardia insignis*) and black-necked crane (*Grus nigricollis*) are at severe risk (Tshewang *et al.*, 2018). Poorly constructed infrastructure causes landslides, erosion and degradation, especially in the Dzongkhags district (Wangda *et al.*, 2018). Climate change not only affects the pristine environment but also threatens the livelihood of people and sustainable development (Tshewang *et al.*, 2021). Hydropower exported to India is considered the backbone of the economy but resources of the hydropower might be impacting climate change (MOEA, 2008).

7.1.2 Ecology and biodiversity of Bhutan

Bhutan is part of eastern Himalaya, situated at the Indo-Malayan and Palearctic biogeographic realms (Penjor *et al.*, 2021). It has three global biodiversity hotspots, 60 eco-regions, 330 important bird areas, 53 plant areas and 29 Ramsar sites (Table 7.1; Royal Government of Bhutan, 2014). Bhutan initiated the first conservation programmes in the 1960s with five-year development plans, and developed a Public Environmental Expenditure Review (PEER) to devote a substantial portion of public expenditure to environment-related objectives (Royal Government of Bhutan, 2014). The protected areas system of Bhutan consists of five national parks, four wildlife sanctuaries, one strict nature reserve and eight biological corridors (Dorji *et al.* 2020). Climatic variation supports a wide range of forest types and vegetation, and the country is known as a hub of endangered species (Wangyal *et al.*, 2022).

According to the vegetation, there are three types of floral zones in Bhutan: (i) alpine: over 4000 masl; (ii) temperate: 2000–4000 masl; and (iii) subtropical: 150–2000 masl (Feuerbacher *et al.*, 2021). About 5600 species of seed plants, including 105 endemic species, and 350 fungi and 287 lichen species were identified in Bhutan. Close to 200 mammal species, including 27 globally threatened species (Table 7.2); 700 bird species, including 18 globally threatened species (Table 7.3; Royal Government of Bhutan, 2014); 61 amphibian species; 124 reptile species; and 900 butterfly species were recorded (Wangyal *et al.*, 2022). Bhutan also has a diverse aquatic ecosystem with 91 freshwater native fish species. Agriculture is traditional and subsistence at 150–4600 masl. Agriculture biodiversity is diverse with 100 agricultural species, 384 varieties of rice landraces, 36 varieties of wheat, 32 varieties of barley, 36 varieties of millet, 105 varieties of maize, 11 varieties of bitter buckwheat and 22 varieties of amaranth. In addition, a total of 230 species of wild crops belonging to 120 genera and 51 families occur in Bhutan (Royal Government of Bhutan, 2014).

The aquatic ecosystem of Bhutan consists of rivers, lakes, marshlands and hot springs. Bhutan has adequate water resources because of the many glaciers, glacier lakes, watersheds, and high precipitation (Bhutan Glacial Lake Inventory, 2021). There are four major river basins, namely: Drangme Chu (the largest), Amo Chu, Puna Tsang Chu and Wang Chu, which are estimated to drain one third of the area of the country. There are many small- and medium-sized lakes across the country, and 2674 glacial lakes of which 562 are associated with glaciers and 24 are at high risk of outburst (Ukita *et al.*, 2011). There are 63 marshlands, which are water-logged areas rich in biota, and a suitable habitat for birds, amphibians and reptiles. The Phobjikha Valley marshland provides a suitable habitat for the globally threatened black-necked cranes (*Grus nigricollis*). Ten hot springs are active in Bhutan, including Gasa, Duenmang, Dhur and Chubu (Royal Government of Bhutan, 2014), which are popular for therapeutic treatment of bone and skin diseases. Six major agro-ecological zones in Bhutan are presented in Table 7.4.

Bhutan has a rich heritage due to, in part, the strong biodiversity conservation. The distinct customary practices and traditions associated

Table 7.1. Protected areas in Bhutan. (From Royal Government of Bhutan, 2014; used with permission from National Biodiversity Center, Bhutan)

SN	Name of protected area	Notification year	Gazettment year	Establishment year	Zhongkhags (districts)	Area (km^2)
1	Wangchuck Centennial Park	2008	2008	2008	Gasa, Wangduephodrang, Bumthang, Trongsa, Lhuentse	4914
2	Jigme Dorji National Park	1993	1995	1995	Punakha, Gasa, Thimphu, Paro	4316
3	Jigme Singye Wanchuck National Park	1993	1995	1995	Trongsa, Wangdue, Sarpang, Tsirang, Zhemgang	1730
4	Bumdeling Wildlife Sanctuary	1993	1995	1998	Trashiyangtse, Lhuntse, Mongar	1521
5	Royal Manas National Park	1993	1995	1996	Sarpang, Zhemgang	1057
6	Thrumshingla National Park	1993	1999	2000	Bumthang, Lhuntse, Mongar, Zhemgang	905
7	Sakten Wildlife Sanctuary	1993		2003	Trashigang, Samdrup Jongkhar	741
8	Toorsa Strict Nature Reserve	1993			Haa	610
9	Khaling Wildlife Sanctuary	1993			Samdrup Jongkhar	335
10	Phibsoo Wildlife Sanctuary	1993			Sarpang, Dagana	269

Table 7.2. Globally threatened mammal species in Bhutan. (From Royal Government of Bhutan, 2014; used with permission from National Biodiversity Center, Bhutan)

SN	Species	Common name	Global threat category
1	*Sus salvanius*	Pygmy Hog	Critically endangered
2	*Trachypithecus geei*	Golden Langur	Endangered
3	*Trachypithecus pileatus*	Capped Langur	Endangered
4	*Cuon alpinus*	Dhole/Wild Dog	Endangered
5	*Ailurus fulgens*	Red Panda	Endangered
6	*Panthera tigris tigris*	Bengal Tiger	Endangered
7	*Uncia uncia*	Snow Leopard	Endangered
8	*Elephas maximus*	Asian Elephant	Endangered
9	*Rhinoceros unicornis*	One-horned Rhinoceros	Endangered
10	*Bubalus bubalis*	Asiatic Water Buffalo	Endangered
11	*Caprolagus hispidus*	Hispid Hare	Endangered
12	*Plantanista gangetica*	Ganges River Dolphin	Endangered
13	*Macaca assamensis*	Assamese Macaque	Vulnerable
14	*Melursus ursinus*	Sloth Bear	Vulnerable
15	*Ursus thibetanus laniger*	Himalayan Black Bear	Vulnerable
16	*Moschus chrysogaster*	Himalayan Black Deer	Vulnerable
17	*Lutrogale perspicillata*	Smooth-coated Otter	Vulnerable
18	*Prionailurus viverrinus*	Fishing Cat	Vulnerable
19	*Pardofelis marmorata*	Marbled Cat	Vulnerable
20	*Neofelis nebulosa*	Clouded Leopard	Vulnerable
21	*Catopuma temmincki*	Asiatic Golden Cat	Vulnerable
22	*Cervus duvauceli*	Swamp Deer	Vulnerable
23	*Bos gaurus*	Gaur	Vulnerable
24	*Capricornis sumatrensis*	Serow	Vulnerable
25	*Budorcas taxicolor*	Takin	Vulnerable
26	*Myotis sicarius*	Mouse-eared Bat	Vulnerable
27	*Rattus sikkimensis*	Sikkim Rat	Vulnerable

Table 7.3. Globally threatened bird species in Bhutan. (From Royal Government of Bhutan, 2014; used with permission from National Biodiversity Center, Bhutan)

SN	Species	Common name	Global threat category
1	*Gyps bengalensis*	White-rumped Vulture	Critically endangered
2	*Ardea insignis*	White-bellied Heron	Critically endangered
3	*Sarcogyps calvus*	Red-headed Vulture	Critically endangered
4	*Aythya baeri*	Baer's Pochard	Critically endangered
5	*Arborophila mandellii*	Chestnut-breasted Partridge	Vulnerable
6	*Tragopan blythii*	Blyth's Tragopan	Vulnerable
7	*Aceros nipalensis*	Rufous-necked Hornbill	Vulnerable
8	*Apus acuticauda*	Dark-rumped Swift	Vulnerable
9	*Grus nigricollis*	Black-necked Crane	Vulnerable
10	*Gallinago nemoricola*	Wood Snipe	Vulnerable
11	*Haliaeetus leucoryphus*	Palla's Fish-eagle	Vulnerable
12	*Sitta formosa*	Beautiful Nuthatch	Vulnerable
13	*Prinia cinereocapilla*	Grey-crowned Prinia	Vulnerable
14	*Aquila clanga*	Greater Spotted Eagle	Vulnerable
15	*Aquila heliaca*	Eastern Imperial Eagle	Vulnerable
16	*Mulleripicus pulverulentus*	Great Slaty Woodpecker	Vulnerable
17	*Chlamydotis undulata*	Houbara Bustard	Vulnerable
18	*Saxicola insignis*	White-throated Bushchat	Vulnerable

Table 7.4. Agro-ecological zones of Bhutan. (From Royal Government of Bhutan, 2014; used with permission from National Biodiversity Center, Bhutan)

SN	Agro-ecological zone	Altitude (masl)	Rainfall (mm)	Farming system and major crops
1	Alpine	3600–4600	< 650	Semi-nomadic people, yak herding, dairy products, barley, buckwheat, mustard and vegetables
2	Cool temperature	2600–3600	650–850	Yak, cattle, sheep, horses, dairy products, barley, wheat, potatoes, buckwheat, mustard, temperate fruits and vegetables
3	Warm temperature	1800–2600	650–850	Rice on irrigated land, mixed cropping with wheat, mustard, barley, potatoes; temperate fruits, vegetables
4	Dry subtropical	1200–1800	850–1200	Maize, rice, millet, pulses, fruit trees, vegetables, wild lemons, grass, livestock
5	Humid subtropical	600–1200	1200–2500	Irrigated rice, mustard, wheat, pulses, vegetables, tropical fruit trees
6	Wet subtropical	150–600	2500–5500	Humid zones-irrigated rice rotated with mustard, wheat, pulses and vegetables, tropical trees

with biodiversity are testimony to the harmonic and vibrant relationship that exists between nature and culture (Samal and Dash, 2023). The country is rich in a diverse array of vascular plants. There are 411 pteridophyte species in 27 families, and more than 200 species of medicinal plants. The non-vascular species include sphagnum mosses, liverworts and hornworts (Royal Government of Bhutan, 2017), with 282 moss species in 156 genera. Approximately 350 species of fungi, of which 53 are edible mushroom species, and 1000 lichen species with 27 lichenicolous fungi, have been identified (Royal Government of Bhutan, 2017). Close to 100 species of insect-fungi, which are complex associations of fungi and insects have been observed.

There are no forests above 4000 masl but there are alpine meadows, where medicinal plants and herb species, and mammals such as the snow leopard (*Uncia uncia*), blue sheep (*Pseudo nayaur*), musk deer (*Moschus chrysogaster*) and lynx (*Felis lynx*) are found. Fir forests occur above 3000 m, with the plant species *Abies densa, Juniperus, Taxus* and *Larix* and the animal species black bear (*Ursus thibetanus*), langur (*Trachypithecus pileatus*), red panda (*Ailurus fungens*), clouded leopard (*Neofelis nebulosa*) and golden cat (*Catopuma temminckii*). Mixed conifer forests occur between 2500 m and 3500 m with floral species spruce, juniper, fir, rhododendron, *Betula* spp. and *Taxus* spp. Broadleaf mixed conifer forests occur between 2000 m and 2500 m with floral species *Quercus, Persea* and *Castanopsis*. Broadleaf forests occur between 1000 m and 2000 m with floral species *Schima, Quercus* and *Castanopsis* and faunal species sambar deer (*Cervus duvaucelii*), water buffalo (*Bubalus bubalis*) and golden langur (*Trachypithecus pileatus*). Tropical lowland forests occur below 700 m with the plant species *Shorea robusta, Bombax ceiba* and *Daubanga grandiflora*.

7.2 Ecological Degradation in Bhutan: Cause, Problems and Risk Assessment

Although Bhutan has a long history of environmental conservation, there are many stresses on the ecosystems. The topographical complexity and extreme altitudinal variation affect the weather patterns, which create unique microclimates and a wide assembly of flora and fauna (Penjor *et al.*, 2021). The principal drivers of environmental degradation are climate change, population growth, anthropogenic activities, habitat alterations and land-use conversions (Kristensen, 2004), while the ecosystem is also affected by forest fires, over-grazing, invasion of alien plant species, pests and diseases (Penjor *et al.*, 2021). Land-use changes are due to modern developments, including hydropower, transmission lines, and roads and mining (Wangyal *et al.*, 2022). Growth rate of the urban population in Bhutan is 5.7%, which was the highest among the south-Asian countries during the period between 2000 and 2010, and is increasing pressure on the environment with an acute demand for buildings and roads, while generating waste (Tshewang *et al.*, 2021).

7.2.1 Forest degradation and poaching

Forests are a major source of fuelwood for cooking and heating in Bhutan, with more than 94% of rural households fully dependent on it (WMD, 2017). Forests are being degraded due to illegal extraction of timber and non-wood forest products, poaching and converting forest land into cultivated and urban land. Forest products, including timber, are used for industry, road construction and buildings (MOAF, 2010). Between 2010 and 2017, 25,261 m^3 of wood were extracted, including 11,444 m^3 for household use and 13,817 m^3 for industrial purposes (MOAF, 2010). In 2018, there were 924 cases of illegal timber extraction and 40 cases of poaching (DFPS, 2018). Land conversion leads to habitat loss and fragmentation, reduced biodiversity and human–wildlife conflicts (Feuerbacher *et al.*, 2021). A major loss of forest areas is attributed to rapid expansion of road networks (Penjor *et al.*, 2021). Poaching of top predators has affected prey species which triggered stresses on the ecology (Tshewang *et al.*, 2021). About 270 poaching cases were encountered between 2013 and 2014 for body parts of Himalayan black bear, musk deer, common leopard, elephants, tigers, pheasants, Asiatic golden cats, clouded leopards and pangolins (DFPS, 2018). A total of 113,378 ha of forest were converted to agricultural land, roads, buildings, and hydropower and transmission lines from 2013 to 2017 (Table 7.5; Royal Government of Bhutan, 2014). Six large hydropower projects required 2276 ha of land and about 272 ha of forest (WMD, 2017).

Forest ecosystem services are threatened by the exploitation of forest resources, livestock grazing, soil erosion and climate change (Feuerbacher *et al.*, 2021). Deforestation and forest degradation leads to the loss of soil, while snowmelt leads to flooding in the lower part (Wangchuk *et al.*, 2021). Ecological degradation has negative effects on the economy and productivity (Wangyal *et al.*, 2022).

7.2.2 Land conversion and land-use change

Agriculture is important for socio-economic development, but unsustainable agriculture practices have negative effects on cropland (Penjor *et al.*, 2021). Farming on steep slopes, without slope stabilization, causes the loss of fertile soil, while chemical fertilizers and pesticides reduce soil fertility and cause ecological degradation

Table 7.5. State forest land allotted in hectares for different purposes from 2008 to 2013. (From Royal Government of Bhutan, 2014; used with permission from National Biodiversity Center, Bhutan)

SN	Activity	2008/09	2009/10	2010/11	2011/12	2012/13	Total	Percentage
1	Government institutions	333	318	178	466	212	1507	9.64
2	Government reserved forest	136	101	137	192	228	794	5.08
3	Kidu land	19.4	14.2	21.1	26.3	128.8	210	1.34
4	Land exchange	32.4	18.2	15.0	5.67	5.67	76.6	0.49
5	Lease	934	376	1951	895	478	4714	30.2
6	Lhakhang	10.9	8.51	6.48	3.24	3.65	32.8	0.21
7	Roads	307	651	906	516	2310	4690	30.0
8	Satshab	130	104	162	117	103	616	3.94
9	Transmission lines	754	614	789	307	521	2984	19.1
	Total	**2656**	**2204**	**4166**	**2529**	**3990**	**15624**	**100.00**

(FRMD, 2018). Agricultural land decreased from 7.85% to 2.75% from 1995 to 2016, as land was converted to urban towns (Rai and Phuntsho, 2016). In total, 36,298 ha of forest were converted to agricultural land from 2000 to 2015. Wetland agricultural land increased by 8303 ha, dryland agriculture by 25,691 ha and orchards by 2303 ha (DFPS, 2018). But during this period, rural-urban migration converted 24,631 ha of abandoned agricultural land to forest land (FMRD, 2018), which indicates that there may be growth of second- and third-generation forests in the future. Protected areas have been managed effectively, but non-protected areas were exposed to habitat alterations and human-induced changes (Williams *et al.*, 2017). Farm roads increased from 1658 km to 5363 km from 2008 to 2017, mainly at the expense of wilderness and forests (Wangda *et al.*, 2018). Mining extracted 45.9 million metric tons of minerals from 2008 to 2014, which were exported mainly to Bangladesh and provided 4.22% to the GDP of the overall economy (Tshewang *et al.*, 2021). Extraction of coal, stone, iron, talc and quartzite damaged the ecological zones that led to the degradation and deterioration of habitats, biomes and soil, and pollution of water and air (MOAF, 2010; Tshewang *et al.*, 2021). The herders in the northern region of the country, where free range and migratory grazing are practised, are dependent mainly or wholly on livestock for their livelihoods. Over-grazing of livestock can cause degradation of forests and rangelands (Wangda *et al.*, 2018).

7.2.3 Forest fires

Forest fires are a major cause of ecosystem degradation, disruptions of forest ecosystem dynamics, shaping vegetation distribution and loss of biodiversity (Fig. 7.1; Royal Government of Bhutan, 2014). Besides destroying trees, forest fires reduce soil moisture content, which causes an increase in soil erosion, and reduces the carbon storage capacity, carbon cycling, and soil stability (Vilà Vilardell *et al.*, 2020). Forest fires also contribute to global warming due to carbon emissions and also affect soil porosity, structure and hydraulic conductivity. Of vital interest, forest fires reduce livelihood sources of the local population (Wang *et al.*, 2021).

There were 234 fires from 2013 to 2017, which affected 37,921 ha of forest land (WMD, 2017). The 803 fires from 1993 to 2004 destroyed 275,642 ha of forest (Wang *et al.*, 2021), while 255 fires from 2012 to 2017 destroyed 43,712 ha (Wang *et al.*, 2021). The middle belt of the country, between 100 m and 4962 m, is dominated by chirpine and blue pine forests, which are the most fire-prone forests. Blue pine forests cover 3.7% of the forest area and are

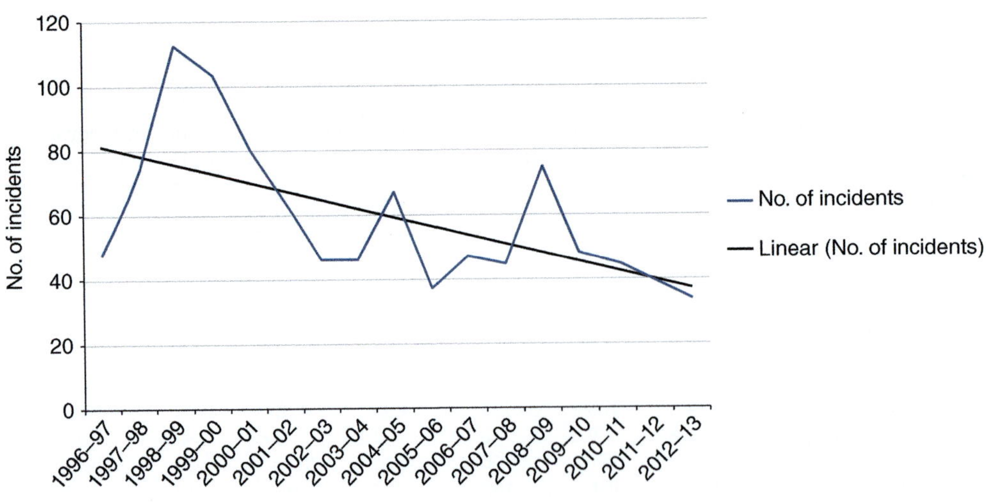

Fig. 7.1. Forest fire trends from 1996 to 2013 in Bhutan. (From Royal Government of Bhutan, 2014; used with permission from National Biodiversity Center, Bhutan)

economically very important (Villa-Vilardell *et al.*, 2020). Mongar Dzongkhag (district) is the most fire-prone area, followed by Wangduephordrang, Trashingang, Bumthang, Thimpu and Paro Dzongkhags (Tshering *et al.*, 2020). Main causes of forest fires are slash-and-burn agriculture, burning of agriculture residue and seasonal events (Wang *et al.*, 2021). Coniferous forests are affected by forest fires due to electrical short circuits along the power transmission lines (Tshering *et al.*, 2020).

7.2.4 Habitat destruction and unsustainable use of resources

Over-harvesting of forest products is one of the major causes of biodiversity loss (Feuerbacher *et al.*, 2021). Therefore, limits for the use of forest resources should be enforced strictly. Forests are used mainly for timber, which is legally harvested by Natural Resources Development Corporation Limited (NRDCL). Over-grazing in rangelands leads to soil erosion, change in vegetation, and loss of habitats and plant species. Humans are often in conflict with wildlife and some species like Asian elephant, tiger and mongoose select habitat sites far from human settlements. Illegal trade of forest products and wildlife, and illegal collection of medicinal herbs and wood threaten the biodiversity (Fig. 7.2; Penjor *et al.*, 2021).

7.2.5 Urbanization

Urbanization, industrialization and socio-economic development increase pollution (Choden *et al.*, 2020). Solid wastes are a major source of land pollution; domestic sewage, oil residues and effluents from industries and automobiles are a major source of water pollution (Uddin *et al.*, 2007); while emission of carbon dioxide from industries, vehicles and burning are a major source of air pollution (Kamei *et al.*, 2021). Land is being converted into urban settlements, industrial areas and roads and, consequently, changing the energy flow of natural ecosystems (Lin and Wang, 2023; Tian *et al.*, 2023). Indoor air pollution is becoming a major problem, and causes respiratory and diarrhoeal diseases (Tshewang *et al.*, 2021).

The composition of urban wastes includes medical, electronic, plastic, municipal and organic wastes (MOAF, 2010). The annual quantity of waste of the four major cities, namely Thimpu, Paro, Phuntsholing and Gelephu, is 14,491 tons (MOAF, 2010). Quantity of waste is increasing due to the socio-economic development in the country. But the wastes are not properly managed due to lack of segregation systems at source, lack of scientific technology and poorly managed landfills (MOAF, 2010). Also, there are no controlling measures for impact of waste on the environment – water, air, land and human health (Tshewang *et al.*, 2021).

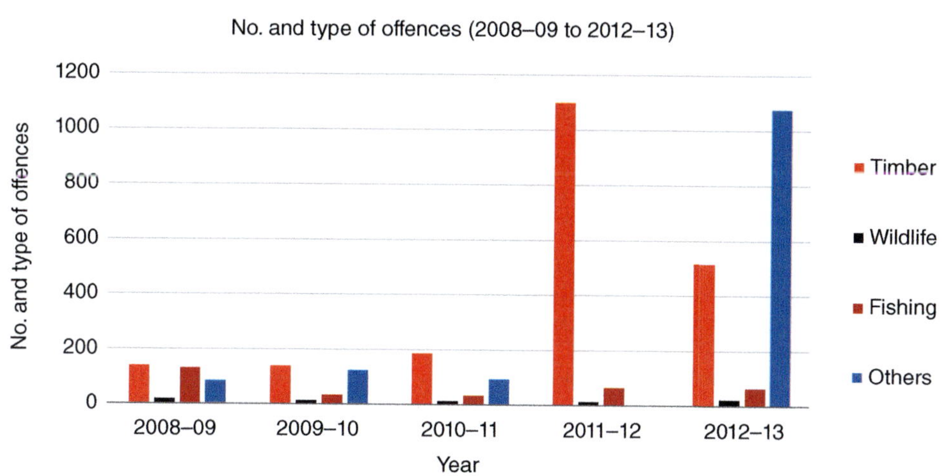

Fig. 7.2. Illegal activities in Bhutan forests. (From Royal Government of Bhutan, 2014; used with permission from National Biodiversity Center, Bhutan)

7.2.6 Invasive and alien species

Invasive species cause major problems for biodiversity conservation in Bhutan. The invasions are caused by increased land degradation due to over-grazing, deforestation and climate change (Tshewang *et al.*, 2021). According to the World Conservation Union, invasive species are the second greatest threat to biodiversity after habitat destruction (Tshewang *et al.*, 2021), but there still is no systematic inventory of invasive alien species in Bhutan, nor on the damage they cause the ecosystem and biodiversity (Table 7.6; Royal Government of Bhutan, 2014). According to a

Table 7.6. List of Invasive alien plant species in Bhutan. (From Royal Government of Bhutan, 2014; used with permission from National Biodiversity Center, Bhutan)

SN	Botanical name	Common name	Family	Native range/ country
1	*Acanthospermum hispidum*		Asteraceae	Brazil
2	*Acmella uliginosa*		Asteraceae	South America
3	*Ageratina adenophora*		Asteraceae	Mexico
4	*Alternanthera pungens*		Amaranthaceae	Central and South America
5	*Amaranthus spinosus*		Amaranthaceae	Tropical Americas
6	*Argemone mexicana*	Mexican poppy	Asteraceae	Mexico
7	*Axonopus compressus*		Poaceae	Southern US, Mexico and Brazil
8	*Bidens pilosa*	Beggar's tick	Asteraceae	Tropical Americas
9	*Cassia tora*		Leguminosae	Not known
10	*Chenopodium ambrosioides*		Chenopodiaceae	Central America
11	*Chromolaena odoratum*	Baby tea, bitter bush	Asteraceae	Tropical South America
12	*Convolvulus arvensis*		Convolvulaceae	Europe
13	*Conyza bonariensis*		Asteraceae	Not known
14	*Conyza canadensis*		Asteraceae	North and Central America
15	*Conyza floribunda*		Asteraceae	South America
16	*Cosmos bipinnata*	Cosmos	Asteraceae	Tropical Americas
17	*Crassocephalum crepidioides*		Asteraceae	Not known
18	*Cryptomeria japonica*	Japanese cedar	Taxodiaceae	Japan
19	*Datura stramonium*	Thorn apple	Solanaceae	Americas
20	*Drymaria cordata*		Caryophyllaceae	Tropical Americas
21	*Eichhornia crassipes*	Water hyacinth	Pontedriaceae	South Amerca
22	*Euphorbia heterophylla*		Euphorbiaceae	Mexico
23	*Galinsoga parviflora*		Asteraceae	Tropical Americas
24	*Ipomonea purpurea*	Morning glory	Convolvulaceae	Tropical America
25	*Lantana camara*	Lantana, shrub verbena	Verbenaceae	Tropical America
26	*Lepidium virginicum*	Pepper weed	Brassicaceae	United States
27	*Mikania micrantha*	American rope	Asteraceae	United States
28	*Mimosa pudica*	Sensitive plant	Leguminosae	South America
29	*Nicandra physalodes*	Apple of Peru	Solanaceae	Peru
30	*Opuntia vulgaris*	Prickly pear	Cactaceae	South America
31	*Oxalis latifolia*		Leguminosae	Mexico
32	*Parthenium hysterophorus*	Parthenium weed	Asteraceae	Mexico
33	*Paspalum conjugatum*		Poaceae	Not known
34	*Paspalum distichum*		Poaceae	Tropical America
35	*Pennisetum clandestinum*	Kinkuyo grass	Poaceae	East Africa
36	*Robinia pseudacacia*	Black locust	Fagaceae	South-eastern USA
37	*Scoparia dulcis*	Sweet broom weed	Scrophulariaceae	Neo-Tropics
38	*Sida acuta*		Malvaceae	Central America

Continued

Table 7.6. Continued.

SN	Botanical name	Common name	Family	Native range/ country
39	*Stellaria media*	Chickweed	Caryophyllaceae	Europe
40	*Tagetes minuta*	Mexican marigold	Asteraceae	South America
41	*Tagetes patula*	French marigold	Asteraceae	Americas
42	*Thuja orientalis*	Chinese Arbor-vitae	Cupressaceae	North-eastern China
43	*Tithonia diversifolia*		Asteraceae	Mexico and Central America

pilot study of the National Biodiversity Center (NBC), there are more than 40 invasive alien species, including the eight major species which are most damaging (Royal Government of Bhutan, 2017), but the Global Invasive Species Database recorded 46 invasive species, including 11 alien species (Tshewang *et al.*, 2021). The most invasive species in Bhutan have been recorded in agriculture, horticulture and floriculture, and were introduced through travel, trade and tourism (Tshewang *et al.*, 2021). Clearly distinct invasive alien species in Bhutan include *Ageratina adenophora, Trifolium repens, Eichhornia crassipes, Chromolaena odoratum* and *Potamogeton distinctus.* Alien species can threaten species diversity, alter soil properties, degrade wildlife forage, change land use, alter fire regimes and threaten endangered species (Tshewang *et al.*, 2021). *Potamogeton distinctus*, an invasive alien species due to climate change, can reduce the rice yield by 35% (Tshewang *et al.*, 2021).

7.2.7 Climate change

Climate warming is triggering cascading effects on local livelihoods, biodiversity, water availability and monsoons (Villa-Vilardell *et al.*, 2020). Between 1996 and 2018, the maximum temperature increased by 1.3°C while the minimum temperature decreased, which made the days warmer and the nights colder. In addition, the increased temperature in the Himalayas was greater than in the lowlands (NCHM, 2019). Increased warming has been forecasted for the period 2021–2100 under two scenarios: representative concentration pathway (RCP) 4.5, which projected an increase of 0.8–2.8°C and RCP 8.45, which projected an increase of 3.2–5.6°C (NCHM, 2019). The changes could impact the climate-sensitive sectors such as agriculture, forestry, tourism and hydropower (Bhutan National Adaptation Program of Action, 2006). Warming caused, at least in part, glacial lake outburst floods (GLOF) events in 1957, 1960, 1968 and 1994, which affected the downstream environment and settlements (NCHM, 2019). A GLOF in 1994, which originated from Luggye Tsho lake, killed 21 people and damaged 91 houses and 1781 acres of land (NCHM, 2019). Torrential monsoons in 2016 triggered landslides and flash floods in the southern parts of the country, which displaced more than 100 households, damaged bridges and blocked roads (NCHM, 2019).

Rising temperature and declining precipitation are altering the habitats of plants and animals (Penjor *et al.*, 2021) and reducing biodiversity. Climate warming increases risk of forest fires, outbreaks of pests and diseases, and habitat fragmentation. Other impacts include melting snow, drying wetlands, habitat loss of agricultural species, and invasion of alien species (Choden *et al.*, 2021). Some species shift their range due to climate change, while some wild native species face extinction (Penjor *et al.*, 2021). Although the growth rate is declining, the population of Bhutan is projected to increase. Poor rural populations are dependent on natural resources for their livelihood and, therefore, unsustainable harvesting of forest products could exacerbate forest degradation (Wangyal *et al.*, 2022).

7.3 Policies and Strategies for Ecological Conservation

Bhutan has promoted environmental and ecological conservation with the promulgation of policies, acts and other legal frameworks (Branch, 2019; Penjor *et al.*, 2021). Bhutan has initiated conservation policies since 1993, and in 1995 passed the Protected Areas (PA) and Forest and Nature Conservation Acts. Protected areas in Bhutan cover 19,750 km^2 or 51% of the country's area. In 2008, the constitution of Bhutan approved that a minimum of 60% of the country should be maintained as forest land. Article 5.1 of the constitution states that 'Every Bhutanese is a trustee of the Kingdom's natural resources and environment' (Royal Government of Bhutan, 2014).

To prevent the spread of pests, internally and internationally, Bhutan enacted the Plant Quarantine Act (1993), which controls the widespread movement of plants in the country (Gawel and Ahsan, 2014). The Forest and Nature Conservation Act of Bhutan (1995) promoted the management of forests, wildlife conservation, community forests and protected areas (Feuerbacher *et al.*, 2021). The Pesticide Act (2000) was passed for the promotion and encouragement of organic agriculture and integrated pest management (IPM), while the Seeds Act (2000) aimed to prevent new diseases in agriculture and promote the trade of seeds (Gawel and Ahsan, 2014). The Environmental Assessment Act (2000) aimed to ensure environmental sustainability (Wangdi *et al.*, 2013), while the Bhutan Water Policy (2003) was established to integrate management of all water resources through watershed conservation and treatment, and to conserve soil and forests (Feuerbacher *et al.*, 2021). The Biodiversity Act of Bhutan (2003) was promulgated for the sustainable utilization of biological resources, preservation of traditional knowledge and conservation of biodiversity to ensure the protection and equitable benefits to the local community (Wangdi *et al.*, 2013). The Land Act (2007) was approved to lease state land to conserve grasslands and forests (Gawel and Ahsan, 2014). The National Environment Protection Act (2007) was established to promote equitable sharing of costs and benefits of conservation among stakeholders, to conserve the forest and natural resources and to promote clean energy technologies (Wangdi *et al.*, 2013). The Local Government Act (2009) empowered the local government to regulate environmental conservation policies, rules and laws; protect forest products; promote clean air and community forests; prevent noise pollution; and conserve medicinal herbs; while the Waste Prevention and Management Act of Bhutan (2009) was passed for waste management with adjustments for the fragile ecological systems (Gawel and Ahsan, 2014). The Economic Development Policy (2010) created a broad range of economic growth opportunities by recognizing the country as 'Brand Bhutan' along with environmental conservation products (Gawel and Ahsan, 2014), and by promoting biodiversity, genetic resources and indigenous knowledge. This policy is more effective for the promotion of green technology, micro-hydro, solar, wind, biomass and energy conservation programmes. The Biosecurity Policy of the Kingdom of Bhutan (2010) was established to protect the biodiversity from invasive plant species, toxic chemicals, genetically modified organisms, and pests and diseases (Gawel and Ahsan, 2014).

Bhutan National Strategy for Community Forestry (2010) empowered the communities with sustainable management of community forests to receive socio-economic and environmental benefits by generating opportunities with timber and increasing the rural economy (BNSC, 2010). The Water Act of Bhutan (2011) aimed to ensure that water resources are conserved and protected, while being managed in a socially equitable, economically efficient, and environmentally sustainable manner (Wangdi *et al.*, 2013). The National Forest Policy (2011) was passed to manage forest biodiversity sustainably (Wangdi *et al.*, 2013) by focusing on a science-based participatory approach in forest governance, environment friendly technologies and waste minimization (Wangdi *et al.*, 2013). The Food and Nutrition Security Policy of Bhutan (2014) was formulated to promote food security and resilience and to contribute to the national happiness of Bhutan by providing a healthy environment and safe and nutritious food. With the establishment of different national-level policies and laws, Bhutan has strong international co-operation for ecological management and biodiversity management (Gawel and Ahsan, 2014). These policies enhance scientific forest management, provide guidelines and facilitate the community group formation opportunities to achieve bottom-up strategy and transparent accountability (MOAF, 2011).

Bhutan has strengthened its passion for biodiversity conservation and has made commitments to the many regional and international environmental agreements and treaties. To eliminate poaching and illegal wildlife crime, Bhutan ratified the Convention on International Trade in Endangered Species of Wild Fauna and Flora (CITES). In addition, Bhutan is a member of the tiger conservation alliance (along with China, Cambodia, India, Indonesia, Nepal, Bangladesh, Malaysia, Laos, Vietnam, Myanmar, Russia and Thailand), which was established with an aim to prevent poaching of tiger, rhino and elephant. Bhutan follows a zero-tolerance poaching policy and vows to strengthen and track illegal activities (Tshewang *et al.*, 2021). The international agreements and treaties in which Bhutan participates are presented in Table 7.7 (Royal Government of Bhutan, 2014).

Government institutions were established to co-ordinate, plan, regulate and implement programmes, policies and laws for the promotion of sustainable utilization of natural resources and sustainability of the environment. The Ministry of Agriculture and Forest (MoAF) protects natural resources and agriculture and is responsible for the planning and implementation of legal frameworks, policies and other laws related to forests, agriculture and livestock, and biodiversity conservation. The MoAF also promotes the social and economic well-being of local peoples through sustainable natural resources and food availability (Royal Government of Bhutan, 2017), and public awareness and education of wildlife resources. The National Biodiversity Centre (NBC) is mandated to co-ordinate the implementation of biodiversity conservation programmes (Royal Government of Bhutan, 2014), and to provide guidelines for sustainable use of natural resources by ensuring a balance between their use and conservation. The Department of Agriculture (DoA) is mandated to enhance food security and improve management of field and horticulture crops and medicinal plants (Royal Government of Bhutan, 2017), and is responsible for new technologies, pest management and assessment of farm and market economies. The Department of Livestock (DoL) is responsible for livestock health and production, and implementation of technologies and services. The National Environment Commission (NEC) is a high-level, multi-sectorial group chaired by the Prime Minister and is an independent authority of the highest decision-making body for environmental concerns (Gawel and Ahsan, 2014). It is responsible for the implementation of policies, regulations and directives related to environmental conservation and management. The Department of Local Governance (DLG) is a branch of the government that is responsible for the development of local government affairs, guidelines, frameworks and delivery of public services (Gawel and Ahsan, 2014). In the non-governmental sector,

Table 7.7. Bhutan's membership of environment-related treaties and agreements. (From Royal Government of Bhutan, 2014; used with permission from National Biodiversity Center, Bhutan)

SN	Treaties/agreement	Ratification
1	Nagoya Protocol on Access to Genetic Resources and the Fair and Equitable Sharing of Benefits Arising from their Utilization to the Convention on Biological Diversity	September 2013
2	RAMSAR Convention on Wetlands	January 2012
3	South Asian Wildlife Enforcement Network (SAWEN)	January 2010
4	Vienna Convention for the Protection of the Ozone Layer	April 2004
5	Montreal Protocol on Substances that Deplete Ozone Layer	April 2004
6	International Treaty on Plant Genetic Resources for Food and Agriculture (ITPGRFA)	September 2003
7	UN Convention to Combat Desertification (UNCCD)	August 2003
8	Cartagena Protocol on Biosafety to the UN Convention on Biological Diversity	September 2002
9	Kyoto Protocol to the UN Framework Convention on Climate Change	August 2002
10	Convention on International Trade in Endangered Species of Wild Fauna and Flora (CITES)	August 2003
11	UNESCO World Heritage Convention	October 2001
12	UN Framework Convention on Biological Diversity	August 1995
13	UN Framework Convention on Climate Change (UNFCCC)	August 1996
14	International Plant Protection Convention	June 1994
15	UN Convention on the Law of Sea	December 1982

the Royal Society for Protection of Nature (RSPN) promotes environmental education, conservation of endangered species, sustainable livelihoods, and research (Gawel and Ahsan, 2014), and is registered as a Public Benefit Organization under the Civil Society Organization authority of Bhutan. It is involved in environmental research on solid-waste management and climate change (Wangdi *et al.*, 2013).

7.4 Suggestions for Sustainable Ecosystems in the Future

The Global Sustainable Development Report (GSDR, 2019) stated that urgent action must be taken to mitigate the impacts of climate change, inequalities, biodiversity loss, and increasing waste produced (Branch, 2019). Broad discussions are needed on inter-disciplinary, multi-disciplinary and trans-disciplinary issues to solve the complex conservation problems (Biedenweg *et al.*, 2023). The Convention on Biodiversity set a new global framework to address conservation challenges of balancing demand and supply. Landscape management, protected areas, community-based natural resource management, species conservation, and dissemination of science-based knowledge and skills should be priorities for the sustainability of ecosystems.

Natural disasters, pests and diseases, and forest fires, should be controlled by protected area management strategies and forest management regimes like community forests, heritage forests, non-timber forest products and wetland management (DFPS, 2017). Sustainable livelihoods, watershed management, and forest and water-quality improvement should be promoted in the non-protected areas (Tshering *et al.*, 2011). Anthropogenic pressures on the environment need to be mitigated with legal policies, plans and improved technologies at the community level (NECS, 2012). Utilization of local and traditional indigenous knowledge for non-timber forest production should be enhanced with the adoption of sustainable harvesting and innovative management practices (Tshewang *et al.*, 2021). Steep slopes and fragile landscape should be protected and improved through well-planned planting of crops and trees (Tshewang *et al.*, 2021). There should be a focus on conservation of keystone and flagship species for the preservation of ecosystem health (Tshewang *et al.*, 2018).

Bhutan is a net-zero country in carbon emission. It is estimated that the forest areas are able to store 521 million tons of biomass carbon and 188 million tons of soil organic carbon (SOC) (FRMD, 2018); therefore, Bhutan should promote global carbon trading. Bhutan should adhere to emission reduction projects for climate change mitigation by adoption of Clean Development Mechanism (CDM) of United Nations Framework Convention on Climate Change (UNFCCC) (MOEA, 2008). Environmental impact assessments at every stage of mining should be implemented to control environmental pollution (Tshewang *et al.*, 2021). Mining sites should be restored and rehabilitated by the reintroduction of vegetation and animal species (Tshewang *et al.*, 2021). The impact of pesticides and chemical fertilizer needs proper management to ensure environmental sustainability for future generations. Bhutan has committed to limiting pesticides and pest control (Wangda *et al.*, 2018). Proper waste management should be developed to ensure a non-polluted environment for the local inhabitants and tourists (Samal and Dash, 2023). Waste treatment policies should be promoted with bottom-up management strategies.

The status of conservation and restoration can be measured by the quality of forests, crop yields, landscapes, quality of species, economic benefits, and human well-being (Vittini *et al.*, 2023). Although there are many forest policies, holistic management strategies and comprehensive planning at local level should be employed (WMD, 2017) to minimize forest degradation (Tshewang *et al.*, 2021). Widespread growth of invasive alien species (IAS) needs to be controlled by proper and cost-effective biological controls (Tshewang *et al.*, 2021). Controlled burning and advanced forest fire detection technologies, like airborne optical and thermal remote sensing, should be promoted to minimize the damage caused by forest fires (Tshewang *et al.*, 2021). Use of modern firefighting equipment and techniques, training community members, fire-prone mapping, and long-term forest management plans should be implemented (Villa-Vilardell *et al.*, 2020). Research should assess the impacts of over-grazing and strategies should be developed for grassland regeneration and restoration (Wangda *et al.*,

2018). Pasture development, rotational grazing and stall-feeding programmes should be considered in alpine meadows for rangeland management (Tshewang *et al.*, 2021).

Bhutan has hydropower potentiality of 27,000 MW, and, till 2019, had installed hydroelectric plants with a capacity of 2326 MW. This produces the cheapest electricity supply, and Bhutan will be adding more hydroelectric plants, with some of the electricity to be exported to India. Clean energy use such as electric vehicles and mass transportation systems should be promoted to replace fossil fuels. Mapping the major sources of pollution, creation of green jobs, public awareness, and local-level emission policies are effective strategies for air pollution control (Tshewang *et al.*, 2021). Bhutan has been promoting environmental awareness and education since the late 1980s (Royal Government of Bhutan, 2014). Awareness programmes regarding biodiversity conservation, climate change, water-saving strategies, agriculture and organic production can be crucial for ecological restoration and conservation (Wangdi *et al.*, 2013). The synergistic effect of climate change and land use should be evaluated for the conservation of landscapes and facilitation of species movement (Penjor *et al.*, 2021). Public awareness campaigns, school programmes, celebration of environment day, biodiversity day, national forest day, and media broadcasting should be organized for the promotion of biodiversity conservation. Workshops, meetings and declarations should be made with all relevant stakeholders to spread environmental education and awareness (Choden *et al.*, 2020). Understanding biodiversity conservation and government accountability is important for sustainable ecosystems (Royal Government of Bhutan, 2017).

A substantial proportion of Bhutan has been declared protected areas, which are integral parts of the community (Royal Government of Bhutan, 2017). Protected areas can maintain the balance between preservation, conservation and sustainable utilization of the natural resources (Choden *et al.*, 2020), while community-based conservation and landscape level management ensures that the communities are beneficiaries of conservation. The management of protected areas can protect wildlife habitats and monitor species dynamics (Royal Government of Bhutan, 2017). Species-based approach to conservation is a widely adopted practice in conserving ecologically, culturally and economically important species and their habitats (Royal Government of Bhutan, 2017). Innovative methods should be employed for the welfare of animal species and their co-existence with people. Allocation of sufficient funds, awareness of species conservation and enforcement of laws for wildlife crimes should be implemented to protect endangered species.

Restoration and rehabilitation of degraded ecosystems should be prioritized with afforestation, river, wetland and catchment area restoration; sustainable forest, watershed, and grassland management; and the development of climate resiliency. Forests should be protected from the conversion of forest to agricultural and urban lands (Penjor *et al.*, 2021). Better understanding of genetic diversity could develop more resilient species. Genetic-level diversity assessment is needed for the conservation of varieties of species. In Bhutan, genetic diversity assessment is practised in some livestock and crop varieties (Royal Government of Bhutan, 2014), while the introduction of exotic species is also attempted. Proper research and preservation of genetic diversity of crops and livestock should be documented for the conservation of biodiversity and balancing ecological functions. Subsidies for forestry extensions, agricultural enterprises, environmental education, and promotion of organic production should be encouraged. Traditional indigenous knowledge, in combination with laws and regulations, should be used for biodiversity conservation (Choden *et al.*, 2021); for harvesting and use of medicinal herbs; for biological and natural resources; and for managing forest fires, invasive species, water, and human–wildlife conflict. Gender equality and social inclusion (GESI) is important in recognizing the critical roles of women, locals and indigenous and marginalized peoples in environmental conservation (Verma and Ura, 2022). Engaging and empowering women and disadvantaged groups should be encouraged to strengthen their status in the community (Verma and Ura, 2022). The government should support the empowerment of women through gender-specific interventions and address their inclusion during the development, implementation and evaluation of policies and programmes.

Due to its unique landscape, rich biodiversity, cultural heritage and religious practices, Bhutan has a very high potential to develop ecotourism. Promotion of nature-based tourism, including trekking, bird watching, ecological adventures and hiking strengthens the local economy by diversifying livelihoods and providing income for the local community. In addition, community led ecotourism could be beneficial for ecological conservation and for the economy through the development of small-scale enterprises (Samal and Dash, 2023).

The participation of the community is critical for the success of ecological management and conservation governance of natural resources (Vittini *et al.*, 2023). For ecological management, cultural values of the society, income generation, and environmental conservation should be considered (Feuerbacher *et al.*, 2021). Urban forests should be promoted as they can play a vital role in conservation of native species (Yang *et al.*, 2022). Funding is needed to safeguard ecosystems which provide diversified sources of income for conservation (Wangyal *et al.*, 2022), and for reducing emissions from forest degradation and deforestation (REDD) (Uddin *et al.*, 2007). Assessment of ecosystem service links between rural life and forest, food availability, cultural values, and security of livelihood are beneficial for landscape and resource management planning (Choden *et al.*, 2020). Conservation planning should be well documented to achieve sustainable management of natural resources and be a part of social behaviour and social values of the local community (Vittini *et al.*, 2023).

7.5 Conclusions

The pristine environment of Bhutan is being threatened by global warming, greenhouse gases, population growth, habitat loss, urbanization, infrastructure development, land fragmentation and anthropogenic activities. Ecological degradation is leading to food insecurity, natural disasters, social conflicts and ecosystem imbalance. The local population is dependent on natural resources for their livelihoods, and, consequently, the ecosystem services must be protected. The commitment of the Bhutanese towards environmental conservation and restoration is mandated by policies, laws, rules and legislation, and, in addition, Bhutan is co-operating in global agreements, treaties and declarations. Landscape management, protected-area management, community-based natural-resource management, ecotourism, species conservation, scientific technology, financial management, and monitoring and evaluation systems are being implemented for ecological conservation and restoration.

References

Bhutan Glacial Lake Inventory (2021) Cryosphere Services Division, National Center for Hydrology and Meteorology. Royal Government of Bhutan.

Bhutan National Adaptation Program of Action (2006) National Environment Commission of Bhutan, RGOB. Available at: http://unfccc.int/resource/docs/napa/btn01.pdf (accessed 20 September 2023).

Biedenweg, S., Anderson, L., Hatfield, S.C., Hollender, R., Kintner, L. *et al.* (2023) Seeking consilience: traditional ecological knowledge and western social science contributions to orca conservation knowledge. *Journal for Nature Conservation* 72, 1–10. DOI: 10.1016/j.jnc.2023.126364

Branch, M.P. (2019) The future is now. *American Scientist* 99(2), 163–164. DOI: 10.1511/2011.89.163

Choden, K., Keenan, R.J. and Nitschke, C.R. (2020) An approach for assessing adaptive capacity to climate change in resource dependent communities in the Nikachu watershed, Bhutan. *Ecological Indicator* 114, 1–13. DOI: 10.1016/j.ecolind.2020.106293

Choden, K., Nitschke, C.R., Stewart, S.B. and Keenan, R.J. (2021) The potential impacts of climate change on the distribution of key tree species and *Cordyceps* in Bhutan: implications for ecological functions and rural livelihoods. *Ecological Modelling* 455, 1–11. DOI: 10.1016/j.ecolmodel.2021.109650

Cosma, S., Rimo, G. and Cosma, S. (2023) Conservation finance: What are we not doing? A review and research agenda. *Journal of Environmental Management* 336, 1–14. DOI: 10.1016/j.jenvman.2023.117649

DFPS (2017) *Forest and Nature Conservation Rules and Regulations*. Department of Forest and Park Services, Bhutan.

DFPS (2018) *Forestry Facts and Figures*. Department of Forest and Park Services, Ministry of Agriculture and Forest, RGOB.

Dorji, T., Sheldon, F. and Linke, S. (2020) Fulfilling nature needs half through terrestrial-focused protected areas and their adequacy for freshwater ecosystems and biodiversity protection: a case from Bhutan. *Journal for Nature Conservation* 58, 1–6. DOI: 10.1016/j.jnc.2020.125894

Felbab-Brown, V. (2018) *Wildlife and Drug Trafficking, Terrorism, and Human Security*. PRISM, Institute for National Strategy Security, National Defense University.

Feuerbacher, A., Rai, A., Lofgren, H., Sander, K. and Grethe, H. (2021) Policies to reconcile forest conservation and rural development: a pathway to bridge the forest transition in Bhutan. *Land Use Policy* 109, 1–14. DOI: 10.1016/j.landusepol.2021.105647

FRMD (2018) Proposed forest references emission level and forest reference level: submission for technical assessment to UNFCCC. Forest Resources Management Division (FRMD), Ministry of Agriculture and Forests, RGOB. Available at: https://doi.org/10.1111/cobi.12827 (accessed 20 September 2023).

Gawel, A. and Ahsan, I. (2014) Review and compendium of environmental policies and laws in Bhutan – input to the Asian Judges Network on Environment. Asian Development Bank, Mandaluyong City, Phillippines. Available at: https://www.adb.org/sites/default/files/publication/150136/review-compendium-environmental-policies-and-laws-bhutan.pdf (accessed 20 September 2023).

GSDR (2019) The Future is now – science for achieving sustainable development. *American Scientist* 99(2), 163–164. DOI: 10.1511/2011.89.163

IPCC (2013) *The Physical Science Basis*. IPCC fifth assessment report on climate change. Cambridge University Press, Cambridge, UK and New York.

Kamei, M., Wangmo, T., Leibowicz, B.D. and Nishioka, S. (2021) Urbanization, carbon neutrality, and gross national happiness: sustainable development pathways for Bhutan. *Cities* 111, 1–10. DOI: 10.1016/j.cities.2020.102972

Kristensen, P. (2004) The DPSIR framework: workshop on a comprehensive detailed assessment of the vulnerability of water resources to environmental change in Africa using river basin approach. Workshop, 27–29 September, pp. 1–10. UNEP Headquarters, Nairobi. Available at: http://enviro.lclark.edu:8002/rid=1145949501662_742777852_522/DPSIR Overview.pdf (accessed 20 September 2023).

Lin, X. and Wang, Z. (2023) Landscape ecological risk assessment and its driving factors of multi-mountainous city. *Ecological Indicators* 146(3), 109823, 1–15. DOI: 10.1016/j.ecolind.2022.109823

MOAF (2010) Bhutan National Strategy for Community Forestry (BNSC). Royal Government of Bhutan.

MOAF (2011) *The National Forest Policy of Bhutan*. Department of Forestry and Park Services, Ministry of Agriculture and Forests, RGOB.

MOEA (2008) *Hydropower Development Master Plan of Bhutan*. Ministry of Economic Affairs, RGOB, Thimpu. Available at: https://www.moea.gov.bt/wp-content/uploads/2017/07/Hydropower-Policy.pdf (accessed 20 September 2023).

NCHM (2019) *Analysis of Historical Climate and Climate Projection for Bhutan*. National Centre for Hydrology and Meteorology (NCHM), RGOB, Thimphu.

NECS (2012) *Bhutan: A National Strategy and Action Plan for Low Carbon Development*. National Environment Commission Secretariat (NECS), RGOB.

Penjor, U., Wangdi, S., Tandin, T. and Macdonald, D.W. (2021) Vulnerability of mammals communities to the combined impacts of anthropic land-use and climate change in the Himalayan conservation landscape of Bhutan. *Ecological Indicators* 1–10, 121. DOI: 10.1016/j.ecolind.2020.107085

Rai, A. and Phuntsho, P. (2016) *Land Use and Land Cover Assessment of Bhutan*. Department of Forests and Park Services, RGOB, Thimpu.

Royal Government of Bhutan (2014) National biodiversity strategies and action plan of Bhutan. National Biodiversity Center, Ministry of Agriculture and Forests, Royal Government of Bhutan. Available at: https://www.cbd.int/nbsap/ (accessed 20 September 2023).

Royal Government of Bhutan (2017) *Biodiversity Statistics of Bhutan*. National Biodiversity Centre, Ministry of Agriculture and Forests, Royal Government of Bhutan.

Samal, R. and Dash, M. (2023) Ecotourism, biodiversity conservation and livelihoods: understanding the convergence and divergence. *International Journal of Geoheritage and Parks* 11, 1–20. DOI: 10.1016/j.ijgeop.2022.11.001

Tian, M., Gao, J., Song, G. and Feng, C. (2023) Prioritizing ecological conservation strategies to enhance national ecological connectivity in China. *Ecological Indicators* 150, 1–11. DOI: 10.1016/j.ecolind.2023.110273

Tshering, K., Gilmour, D., Tshering, C., Meijboom, M., Tenzin, J. *et al.* (2011) *A Roadmap for Watershed Management in Bhutan*. Watershed Management Division, Department of Forest and Park Services, Ministry of Agriculture and Forests, RGOB.

Tshering, K., Thinley, P., Tehrany, M.S., Thinley, U. and Shabani, F. (2020) A comparison of the qualitative analytic hierarchy process and the quantitative frequency ratio techniques in predicting forest fire-prone areas in Bhutan using GIS. *Forecasting* 2, 36–58. DOI: 10.3390/forecast2020003

Tshewang, U., Morison, J.G. and Tobias, M.C. (2018) *Bionomics in the Dragon Kingdom: Ecology, Economics and Ethics in Bhutan*. Springer, New York. Available at: http://www.springer.com/series/15408 (accessed 20 September 2023).

Tshewang, U., Morison, J.G. and Tobias, M.C. (2021) *Bhutan: Conservation and Environmental Protection in the Himalayas*. Springer, New York. Available at: https://doi.org/10.1007/978-3-030-57824-4 (accessed 20 September 2023).

Uddin, S.N., Taplin, R. and Yu, X. (2007) Energy, environment and development in Bhutan. *Renewable and Sustainable Energy Reviews* 11, 2083–2103. DOI: 10.1016/j.rser.2006.03.008

Ukita, J., Narama, C., Tadono, T., Yamanokuchi, T., Tomiyama, N., Kawamoto, S. and Nishimura, K. (2011) Glacial lake inventory of Bhutan using ALOS data: methods and premilinary results. *Annals of Glaciology* 52(58), 65-71. https://doi.org/10.3189/172756411797252293

Verma, R. and Ura, K. (2022) Gender differences in gross national happiness: analysis of the first nationwide wellbeing survey in Bhutan. *World Development* 150, 1–17. DOI: 10.1016/j.worlddev.2021.105714

Vilà Vilardell, L., Keeton, W.S., Thom, D., Gyeltshen, C., Tshering, K. *et al.* (2020) Climate change effects on wildfire hazards in the wildland–urban interface: blue pine forests of Bhutan. *Forest Ecology and Management* 461. DOI: 10.1016/j.foreco.2020.117927

Vittini, M.S., Valenzuela, V.G. and Ramirez, K. (2023) Social perceptions and conservations in protected areas: taking stock of the literature. *Land Use Policy* 131, 1–11. DOI: 10.1016/j.landusepol.2023.106696

Wang, S.W., Lim, C. and Lee, W. (2021) A review of forest fire and policy response for resilient adaptation under changing climate in the Eastern Himalaya region. *Forest Science and Technology* 17(4), 180–188. DOI: 10.1080/21580103.2021.1979108

Wangchuk, J., Choden, K., Sears, R.R., Baral, H., Yoezer, D. *et al.* (2021) Community perceptions of ecosystem services from commercially managed forests in Bhutan. *Ecosystem Services* 50, 1–12. DOI: 10.1016/j.ecoser.2021.101335

Wangda, P., Dukpa, C., Wangmo, K., Penjor, S. and Tenzin, K. (2018) Bhutan RNR statistics 2017: renewable natural resources statistics. Division Directorate Service, RGOB, Thimpu. Available at: https://doi.org/10.13140/RG.2.2.14859.34084 (accessed 20 September 2023).

Wangdi, T., Lhendup, P. and Wangdi, N. (2013) An analysis of forestry policy, acts and rules of Bhutan to mainstream climate change adaptation. Regional Climate Change Adaptation Knowledge Platform for Asia, Partner Report series No. 13. Stockholm Environment Institute, Bangkok. Available at: www.asiapacificadapt.net (accessed 20 September 2023).

Wangyal, J.T., Bower, D., Vernes, K. and Thinley, P. (2022) Employing citizen science to understand amphibian and reptile diversity and distribution in the Himalayan Kingdom of Bhutan. *Global Ecology and Conservation* 37, 1–14. DOI: 10.1016/j.gecco.2022.e02157

Williams, S., William, K.S., Lewis, B.P. and Hill, R. (2017) Population dynamics and threats to an apex predator outside protected areas: implications for carnivore management. *Royal Society Open Science* 4(4). DOI: 10.1098/rsos.161090

WMD (2017) Drivers of deforestation and forest degradation in Bhutan: a synthesis report for REDD+ policymakers. Watershed Management Division (WMD), Ministry of Agriculture and Forest, Bhutan. Available at: http://www.era-mx.org/biblio/Drivers of deforestation and forest degradation.pdf (accessed 20 September 2023).

Yang, J., Wang, Z., Zheng, Y. and Pan, Y. (2022) Shifts in plant ecological strategies in remnant forest patches along urbanization gradients. *Forest Ecology* 524, 1–8. DOI: 10.1016/j.foreco.2022.120540

Yeshey, Keenan, R.J., Ford, R.M. and Nitschke, C.R. (2023) How does conservation land tenure affect economic impacts of wildlife: an analysis of subsistence farmers and herders in Bhutan. *Trees, Forests, and People* 11, 1–13. DOI: 10.1016/j.tfp.2023.100

8 Ecological Conservation and Restoration on the Tibetan Plateau in China

Mei Huang[1,*], Allan Degen[2] and Zhanhuan Shang[1]
[1]*College of Ecology, Lanzhou University, Lanzhou, China;* [2]*Institutes for Desert Research, Ben-Gurion University of the Negev, Israel*

Abstract
Climate change and human activities are the main causes of ecological degradation on the Qinghai-Tibetan Plateau (QTP) in China. The ecosystems of the QTP are very fragile, in particular the grasslands and forests, and are threatened by degradation, soil erosion and desertification. In the past 30 years, China has implemented major ecological projects, including grassland and forest protection and restoration, water- and soil-loss mitigation, and desertification control. By initiating ecological projects and improving compensation policies, the livelihoods and living environment of the QTP are improving, thereby promoting the development of local ethnic industries and ecotourism, and increasing the *per capita* income of residents. The sustainable development of the QTP requires the co-ordination and unity of ecological, economic and social functions of the ecosystem.

Keywords: ecological protection and restoration, Qinghai-Tibetan Plateau, sustainable development, livelihood

8.1 Introduction

The Qinghai-Tibetan Plateau (QTP) is approximately 4400 m above sea level (masl), and covers 3.08 million km^2. It extends from the foot of the Kunlun and Qilian mountains in the north to the southern foot of the Himalayas in the south, and from the Hindu Kush mountains and Pamir Plateau in the west to the eastern foot of the Hengduan mountains in the east (Zhang *et al.*, 2021; Dong *et al.*, 2022). The QTP covers an area of 2.58 million km^2 in China, which is close to 84% of the total area of the QTP, while 61% of the QTP is located in Tibet and Qinghai province and, to a lesser extent, in Gansu, Sichuan, Yunnan and Xinjiang (Zhang, 2019). The QTP is characterized by complex topography, generally sloping from north-west to south-east, low air temperature with an average annual temperature of 1.37°C, low oxygen content, high solar radiation of 5400–8000 MJ m^{-2} yr^{-1} and strong winds. The soil types are diverse and are prone to erosion.

The grasslands of the QTP cover approximately 1.28 million km^2, accounting for one third (33.2%) of the total grasslands area in the country, and has 30 vegetation types out of 42 types in China and the world. Alpine meadows and alpine typical steppe have the largest areas, accounting for 45.4% and 29.1% of the total grasslands area, respectively, while alpine meadow steppe, alpine desert steppe, alpine desert, montane meadow and temperate steppe account for 4.4%, 6.8%, 4.6%, 5.5% and 1.3%, respectively (Dong, 2023). The grasslands in the Tibet Autonomous

*Corresponding author: huangm2018@lzu.edu.cn

 Sustainable Ecological Restoration and Conservation in the Hindu Kush Himalayan Region (Z. Shang *et al.*)
DOI: 10.1079/9781800622579.0008

Region account for 63.9% of the total grasslands area on the Qinghai-Tibetan Plateau, followed by Qinghai province with 22.1%, Sichuan province with 10.9%, Gansu province with 2.0% and Yunnan province with 1.1%. The flora and fauna of the QTP are similar to the northern mid-latitude regions (Dong, 2023).

The QTP has high biodiversity, with 210 mammal and 532 bird species, and 5000 flowering and 12,000 plant species, including endemic species (Miehe *et al.*, 2019). The QTP includes two key eco-regions and is the key hotspot of alpine biodiversity in the world (Dong, 2023). The ecosystems of the QTP are fragile (Zhan *et al.*, 2021) and the quality and functions affect ecological security, biodiversity, water supply, climate stability and carbon (C) balance. Understanding the factors and restoration measures of grassland degradation is critical for planning future sustainability in the QTP (Dong *et al.*, 2020a).

The current policies on the Qinghai-Tibetan Plateau advocate livestock reduction, sedentarization, land privatization and grazing exclusion, thereby reducing the mobility and flexibility of herders, which can have serious consequences for the development and degradation of grasslands (Qiu, 2016). Approximately 2 million km^2 of grasslands have been severely degraded in the QTP over the past 20 years, resulting in a 30% decrease in productivity (Liu *et al.*, 2018). The most severe degradation has occurred in Tibet, followed by Gansu and Qinghai provinces (Hao *et al.*, 2008).

There is a complex entanglement of fine minerals, plant roots and nutrients in the mattic layer, which holds the majority of soil ecological functions and ecosystem services in both topsoil and subsoil. This layer creates a barrier for ecological security (Zhang *et al.*, 2023a), and its destruction and functional degradation results in detrimental consequences. For example, the loss of the protective function of the mattic layer enhances soil erosion, greenhouse gas emissions and permafrost thawing (Zhang *et al.*, 2023a). The common degradation causes are: (i) polygonal crack patterns and the drifting downwards of polygonal sods enhanced by needle ice; (ii) sods resting like stepping-stones on a deflation-pavement, otherwise covered with alpine steppe plants; and (iii) patch-wise dieback of the ground in front of the burrows of small mammals (e.g. *Ochotona curzoniae* Hodgson, pika) (Miehe *et al.*, 2019). Low temperature or dryness in winter can lead to changes in soil volume. When the root mat reaches a certain thickness, with a large portion of dead roots, tensions resulting from the volume changes lead to the formation of polygonal cracks. Polygons are progressively separated while drifting downhill with gravity on the steep incline (Miehe *et al.*, 2019).

8.2 Causes of Grassland Degradation in the Qinghai-Tibetan Plateau

Vegetation on the QTP tends to exhibit an early greening period, long growing period, high coverage and productivity, and enhanced C sink function. The overall vegetation on the QTP is improving, while other areas have deteriorated (Zhang *et al.*, 2015). The areas of forests, grasslands and wetlands are closely related to human activities such as timber harvesting, afforestation and livestock grazing. The coupling of internal driving factors of ecosystem vulnerability with external driving factors has led to the degradation of grasslands on the QTP (Dong *et al.*, 2020a). The internal driving factors of grassland degradation can be excessive compensatory growth (Zhang *et al.*, 2020), and the external driving factors can be divided into the following categories: (i) natural factors (climate change, topographic factors, freeze-thawing); (ii) over-grazing by livestock; (iii) rodent and insect pests; (iv) deforestation; and (v) harvesting of medicinal plants and reclamation of wasteland (Chen *et al.*, 2013; Chen *et al.*, 2020; Dong *et al.*, 2020a; Wang *et al.*, 2022). Overall, human activities such as overgrazing have a greater impact on grasslands degradation in the eastern regions with better hydrothermal conditions, while climate change has a greater impact in the western regions (Wang *et al.*, 2022). Degraded grasslands decrease vegetation coverage, above-ground biomass, species diversity and the proportion of edible forage; are prone to the invasion of poisonous weeds, insects and rodents; and alter the composition of plant and microbial communities (Wang and Yang, 2004; Wang *et al.*, 2022).

8.2.1 Climate change

About one third of protected areas have increased the average daily temperature by 1.5°C in the last 60 years (Shrestha *et al.*, 2021). Climate changes have been evident across the QTP since the beginning of the 1980s, where the temperature has increased by 0.2°C per decade, with the northern part experiencing the greatest warming (Li *et al.*, 2010), with a rate as high as 0.08°C yr^{-1} (Chen *et al.*, 2013). The increase in temperature has resulted in substantial permafrost thawing and glacier retreat (Li *et al.*, 2010; Chen *et al.*, 2013). The annual amount and frequency of precipitation have increased in the north-western QTP but have decreased slightly in the south-eastern part (Wang *et al.*, 2022). Overall, the QTP is displaying a trend of warming and increased precipitation (Li *et al.*, 2010; Wang *et al.*, 2022). The C sequestration of the alpine grasslands ranged from 26.4 to 79.2 Tg C $year^{-1}$ between 1982 and 2018, with an increase of 1.14 Tg C $year^{-1}$. The C sequestration capacity of alpine grasslands on the plateau has undergone persistent enhancement under a warmer and moister climate (Wang *et al.*, 2023).

Continued climate warming can lead to grasslands degradation (Bardgett *et al.*, 2021). The impact of warming on plant productivity is closely related to grasslands type, warming range, ambient temperature and soil moisture. For example, warming generally increases net primary productivity by ~14% in alpine meadows but decreases productivity by ~6% in alpine steppe (Wang *et al.*, 2022). Warming can decrease plant richness by ~8%, equivalent to the disappearance of 1.08 plant species per m^2, but does not affect the Shannon-Wiener diversity (Wang *et al.*, 2022). The total N deposition rate in the QTP was estimated to be less than 10 kg N ha^{-1} yr^{-1}, with a maximum deposition of 36 kg N ha^{-1} yr^{-1} from 1990 to 2003 (Lü and Tian, 2007). The effects of N deposition on biodiversity are severe and are unlikely to be quickly reversed (Payne *et al.*, 2017). N enrichment reduced plant richness by ~22%, but increased Shannon-Wiener diversity (Wang *et al.*, 2022).

8.2.2 Anthropogenic activities

Human activities can alter the response of alpine grasslands to climate change on the QTP (Wei *et al.*, 2020). However, human activity intensity decreased by ~16% from 2000 to 2017 across the alpine grasslands, which might be driven by recent ecological conservation policies, especially reduction in livestock numbers (Li *et al.*, 2021). In general, there is high human activity across the QTP, but low and medium activity in the north and south sections (Li *et al.*, 2021; Fig. 8.1).

The intensity of human activity was correlated positively with grazing intensity (Li *et al.*, 2021). Under the traditional grazing management system, the grasslands have been in a long-term negative state of 'over-grazing–grasslands degradation', which restricts the development of alpine grasslands and animal husbandry and the improvement of local people's livelihoods. Grasslands degrade when the stocking rate exceeds the level that the land can support (Maestre *et al.*, 2022). From 2000 to 2015, the actual carrying capacity of the QTP increased from 145 million to 158 million sheep units, and the theoretical carrying capacity increased from

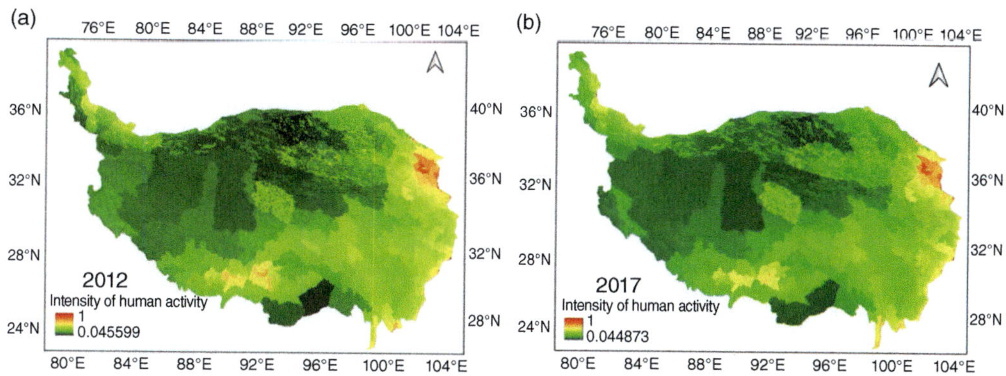

Fig. 8.1. Intensity of human activity on the Qinghai-Tibetan Plateau in 2012 and 2017. (Data from National Tibetan Plateau Data Center; Zhang, 2019).

86 million to 94 million sheep units. The actual carrying capacity was more than 1.6 times that of the theoretical carrying capacity, and the proportion of counties experiencing over-grazing reached 81%. Over-grazing on the QTP was most serious in 1990, and then gradually declined. From a spatial perspective, over-grazing is greatest in Xinjiang and least in Yunnan (Mo, 2020).

The effect of grazing on grasslands

The magnitude and directions of grazing impacts on vegetation and soil depend on the environment, and vary with soil conditions, vegetation characteristics, livestock type, herbivore diversity, grazing strategies (e.g. continuous versus rotational grazing), and grazing intensity and duration (Bai and Cotrufo, 2022; Wang *et al.*, 2022). The feeding preference of livestock alters the dominance of plant species, reducing competition among species, and thus alters the composition of the plant community. The proportion of bare land and erosion index were correlated positively with grazing pressure (Harris *et al.*, 2016).

Continuous livestock grazing reduces plant cover, diversity and productivity, and thus root inputs and plant- and microbial-mediated SOC formation, while stimulating losses through microbial turnover and erosion caused by increased compaction and reduced cover (Bai and Cotrufo, 2022). In the QTP, light grazing did not affect plant richness, Shannon-Wiener diversity, soil-water content, soil-bulk density, nutrients, and microbial biomass C and N. Moderate grazing increased plant richness and Shannon-Wiener diversity but decreased soil organic C and total N (Zhang *et al.*, 2023b). Heavy grazing led to soil erosion and fecal N leaching in alpine grasslands ecosystems (Dong *et al.*, 2022), which, ultimately, reduced C storage regardless of grassland ecosystem types (Deng *et al.*, 2023). Severe mechanical interference (e.g. trampling) coupled with excessive nutrient input (dung, urine) led to strong changes and dieback of above-ground vegetation (Miehe *et al.*, 2019). In addition, plant compensatory growth led to a decrease in root biomass, a loss of nutrients in the ecosystem, destruction of soil physical structure, and soil erosion under over-grazing conditions (Zhang *et al.*, 2020). The regrowth and regeneration of grasslands in response to grazing intensity (i.e. grazing optimization) depended on grassland type and grazing duration (Deng *et al.*, 2023). Over-grazing is associated closely with grassland degradation, while moderate grazing is a sustainable practice for animal production as it maintains ecological functions for alpine grasslands on the QTP (Zhang *et al.*, 2023b). However, there is uncertainty in estimating degradation from grazing due to inconsistent definitions, unclear baselines, and varying standards and indicator systems, as well as the merging of different spatial scales (Miehe *et al.*, 2019; Fig. 8.2).

The effect of infrastructure construction on grassland

With the support and promotion of government policies, many herders have turned from nomadic to settled or semi-settled grazing. The continuous high-intensity grazing near settlements has led to severe grasslands degradation. In addition, the increase in population and infrastructure construction (e.g. railways, power poles, highways and bridges) on the QTP has restricted the grasslands area and exacerbated grasslands degradation (Fayiah *et al.*, 2020; Wang *et al.*, 2022). The development of tourism, which has increased the number of people, has also damaged the grasslands.

Others factors

Illegal hunting of large animals (e.g. *Pantholops hodgsonii*, *Moschus moschiferus*, *Vulpes ferrilata* and *Panthera uncia*) and birds has led to the disruption of the food chain in the plateau ecosystem, while the increase in the number of burrowing animals has exacerbated grasslands degradation (Li *et al.*, 2013). The excessive harvesting by farmers and herders of wild plant species, such as *Fritillaria* spp., has led to a sharp decline in some important economic plant resources, with some even reaching the brink of extinction (Zhang *et al.*, 2015).

8.3 Risk Assessment of Qinghai-Tibetan Plateau

Global climate change poses challenges to the stability of ecosystems such as grasslands, forests and wetlands on the QTP. The main reasons for degradation are discussed below.

Fig. 8.2. Turf degradation in *Kobresia pygmaea* pastures: (a) degeneration of alpine meadow hillside to form black soil mountain; (b) polygonal separation of the felty root mat; (c) downslope drift of the sods; (d) pika increase habitat diversity mainly through their digging – the excavated soil covers the ground and root mats; (e) damage of grass felt layer by pika; (f) erosion of felty root mat follows, forming patches of *Kobresia*; and (g) mattic layer disappears, and poisonous weeds spread. (Photos: Mei Huang)

8.3.1 Managing degraded grasslands is a difficult task

The natural restoration of alpine meadows and alpine grasslands is poor, requiring active human restoration. However, restored grasslands are prone to secondary degradation and invasion of poisonous plants, and return of rodents and pests. The 'black soil beach' and 'black soil slope' (severely degraded grasslands) are still widely distributed in Tibet and Qinghai, with the black soil beach covering 11 million ha.

8.3.2 Poor quality of forest ecosystems

The artificial forests on the QTP have simple stand structures composed of single-tree species, high forest density, and intense competition among trees. The biodiversity of artificial forests is poor, there is widespread serious degradation, large areas of low-quality forests are present in some areas, and tree diseases such as pine needle disease, spruce needle cast and *Arceuthobium sichuanense* are causing considerable damage.

8.3.3 Increased risk of wetland and glacier degradation

Temperature is the decisive factor of glacier shrinkage in Qinghai-Tibetan Plateau (Yao *et al.*, 2019). In the case of global warming of 1.5°C in the future, the temperature of the Qinghai-Tibetan Plateau and the adjacent areas will increase by 2.1°C, resulting in the glaciers on the Qinghai-Tibetan Plateau and the adjacent areas will decrease by 49%, 51% and 64% under the RCP4.5, RCP6.0 and RCP8.5 scenarios, respectively (Kraaijenbrink *et al.*, 2017). The melting and retreat of glaciers and snow on mountains has increased, and the thawing of frozen layers has accelerated, leading to enhanced peri-glacial processes such as slope instability, debris flow, and thermal karst, which have weakened the water storage function in the Three Rivers source region.

8.3.4 Damage due to soil erosion and desertification

Soil erosion and desertification are expanding because of wind, water and freezing-thawing. The QTP has 46 million ha of moderate soil erosion, 47 million ha of moderate sandy desertification land, and 0.427 million ha of moderate rocky desertification land.

8.3.5 Increase of invasive alien species

Due to the impact of biological invasion, irrational release and the introduction of foreign species, indigenous biodiversity on the QTP is at risk. Major invasive species such as *Spodoptera frugiperda*, *Pomacea canaliculata* and *Solenopsis invicta* have caused great harm, and the distribution of poisonous alien species (*Trachemys scripta* and *Lithobates catesbeiana*) have been detected.

8.4 Ecological Conservation and Restoration Strategies and Policies

The restoration and management of degraded grasslands, the co-ordination of the relationship between ecological, economic and social functions of the alpine grasslands ecosystem, and the sustainable utilization of alpine grasslands are related to ecological security, regional stability and economic development. The management of alpine ecosystems should be based on ecological theory to achieve sustainable utilization of grasslands resources and to protect and restore grasslands ecosystems (Sun *et al.*, 2019). The white paper 'The Construction of Ecological Civilization on the Qinghai Tibet Plateau' (State Council Information Office of the People's Republic of China, 2018) mentioned that, from 1982 to 2009, the grasslands coverage and net primary productivity on the QTP displayed an overall increasing trend. The area with increased grasslands coverage accounted for 47% of the total grasslands area, and increased net primary productivity accounted for more than 32%. In the past decade, the grasslands ecosystem has been stable and improving.

8.4.1 Implement a series of ecological engineering programmes

A series of programmes, including 'Rodent Control', 'Grain for Green', 'Black Soil Patch Degraded Grassland Control', 'Grazing Exclusion by Fencing', 'Ecological Migration' and grassland fire prevention programmes, have been initiated since 2000 to restore degraded grasslands in the QTP (Zhen *et al.*, 2018; Wang *et al.*, 2022). From 1990 to 2020, the 'Return Grazing Land to Grassland Protection and Construction of Engineering' programme was implemented on an area of 250,000 km^2; the 'Grasslands Pests and Mice Governance' project covered 201,000 km^2; the 'Forest Protection and Construction of Artificial Afforestation' project covered 18,513 km^2; the 'Natural Forest Protection' project, concentrated in the southern and south-east regions and part of the Qilian mountain counties, covered 11,274 km^2; the 'Soil and Water Loss' project, mainly in the mountain valley area, Tibet, one river and two streams area and Three Rivers south-east of some counties, covered 737.8 km^2; and the 'Desertification Land Area' project, focusing on one river and two streams in Tibet, middle river valley, and river south-west of Three Rivers, covered 637.3 km^2 (Zhao *et al.*, 2023).

In addition, Qinghai province put forward five objectives in 2021: to stabilize the ecological protective screen in the Three Rivers source region; to build an ecological protective screen in the southern foot of Qilian mountains; to construct an ecological community in Hehuang Valley; to construct an ecological zone in Tsaidam basin; and to construct an ecosphere in Gonghe basin. Several protective measures are examined below.

Grazing exclusion by fencing and rotational (rest) grazing

Since the 1990s, grazing exclusion by fencing, which prohibits mobility and changes the land-use form, has been employed widely to restore degraded grasslands (Xu and Huntsing, 2022). The effect of grazing exclusion depends mainly on the degree of degradation, years of fencing, and vegetation types (Sun *et al.*, 2020). After 39 years of grazing exclusion, below-ground biomass increased but soil nutrient content decreased (Cao *et al.*, 2021). Recently, however, there has been increasing criticism of long-term fencing, as ecological consequences emerge. For example, long-term grazing exclusion decreased species richness of vegetation and soil seed banks (Saatkamp *et al.*, 2018), caused a large amount of litter to gather, affected the material and energy circulation of the ecosystem, and reduced grasslands resources (Sun *et al.*, 2020). Consequently, the period of fencing should be determined before implementation. It is suggested that the fencing policy should be tailored to local conditions, by taking into account the degree of degradation, natural resilience, livestock carrying capacity, and herders' grazing traditions and knowledge (Xu and Huntsing, 2022).

Seasonal rotational grazing includes two options: spring and autumn rotational grazing. Grazing could stop during the annual greening and flowering periods of the grasslands to ensure growth and reproduction of the pasture (Bai *et al.*, 2016). To avoid grasslands degradation, artificial grasslands in warm and cold seasons should be grazed at 40–60% and 70–90% of the biomass, respectively (Wang *et al.*, 2013). Grazing regimes (i.e. intensity and duration applied) must consider specific grasslands characteristics to ensure stable productivity and optimal impacts on ecosystem C stocks (Deng *et al.*, 2023). Rotational grazing systems promoted the functional genes of soil microorganisms regulating C and N cycles, which were conducive to nutrient cycling in alpine grasslands (Dong *et al.*, 2020b).

Planting artificial grasslands

Severe or extremely degraded grasslands have poor self-renewal ability due to low soil nutrient and seed activity, and, therefore, natural restoration is unreliable. Establishing artificial grasslands is the best way to restore degraded grasslands, as it provides both ecological and economic benefits for herders and achieves a win-win situation (Shang *et al.*, 2014). The grasslands area in China is approximately 265 million ha, and the artificial grasslands are approximately 0.581 million ha (Central Government of the People's Republic of China, 2021). Current research on the technology of artificial grasslands in China focuses mainly on seed selection, supplementary sowing, water and fertilizer management, controlled grazing, and mowing silage. Artificial grasslands should be established in large-scale degraded grasslands or grasslands with low vegetation coverage to avoid damaging the local natural grasslands.

Cold- and drought-resistant plant species should be selected for artificial grasslands due to the large temperature difference between day and night on the QTP and the strong radiation. Eight native plant species are important forage resources in alpine areas, including *Poa crymophila*, *Elymus nutans*, *Avena chinensis*, *Puccinellia tenuiflora* cv. Tongde, *Bromus inermis*, *Elymus dahuricus* cv. Aba, *Medicago sativa* and *Medicago ruthenica* (Zhou *et al.*, 2022a, b). In Tibet and Qinghai, *Bromus inermis*, *Elymus nutans*, *Elymus dahuricus* cv. Aba and *Poa crymophila* are planted widely, and the area planted with grasses is much larger than that of legumes. The mixed sowing of legumes (*Medicago sativa* and *Vicia sativa*) and grasses increased productivity and soil nutrient content (Adnane *et al.*, 2015). The planting of *Festuca sinensis* in Xinjiang produced a high yield, which also reduced the forage shortage.

In addition, semi-artificial grasslands can be established by re-seeding on moderately degraded grasslands. Re-seeding can be employed in combination with shallow tillage and loose harrowing without damaging native vegetation. Re-seeding technology includes mainly the selection of species, and determination of plots, period, method, dose, depth and post-sowing management. According to the degree of grasslands degradation, appropriate combinations of shallow tillage, irrigation, fertilization and other restoration measures should be used. The seeding rate should be determined based on the thousand seed weight, germination rate and degree of degradation.

Control of rodents in grasslands

Rodent species in grasslands include those from the orders Rodentia and Mammalia; for example, *Myospalax aspalax*, *Myospalax baileyi*, *Ochotona curzoniae* and *Marmota himalayana* (Hong *et al.*, 2014). There are 2183–4423 rodent burrows per ha and about 10% of plant biomass is consumed by pika (Fayiah *et al.*, 2020). In 2020, rodents inhabited 3.59×10^7 ha of grasslands in China, or 8.97% of the total grasslands area, with a serious damage area of 2.21×10^7 ha, or 62% of the grasslands rodent area (Yang *et al.*, 2022). The area damaged by *Ochotona curzoniae* and *Myospalax baileyi* on the QTP accounted for more than 50% of the total rodent area in China (Hong *et al.*, 2014; Wang *et al.*, 2016).

Rodents increased the heterogeneity of plant communities, changed the rate of plant species turnover, and reduced the availability of forage and water storage by digging burrows, creating bare spots, redistributing soil and selective feeding (Li *et al.*, 2013; Miehe *et al.*, 2019; Chen *et al.*, 2022). Overall, rodents and herbivores compete for plant resources, reducing the feed availability for grazing livestock (Wang *et al.*, 2016).

The control policy of 'prevention first and comprehensive control' and the working principle of 'territorial management, group responsibility' have been adopted, forming a working policy of 'overall planning, unified prevention rule, state support, local people participation' and a technology roadmap of 'biological control first, supplemented by chemical and physical measures' (Hong *et al.*, 2014). Since the 1960s, rodenticides have been used widely to eradicate the plateau pika, resulting in ecological disasters and environmental contamination. Recently, ecological strategies that encourage the rodents' natural enemies, such as foxes and eagles, to control the pika population have proved effective. The ranking of various control measures is as follows: biological agents (e.g. botulinum toxin) > recruited eagles > chemical agents > artificial captures > wild fox.

Establish nature reserves

More than 2500 reserves have been established in China, and the number of reserves in the east and south far exceeds the number in the west. Reserves in western China suffer from climate change but are less affected by anthropogenic activities than those in the east (Shrestha *et al.*, 2021). The establishment of protected areas has an 'initial containment and partial reversal' effect in suppressing grasslands degradation. As of 2018, 155 nature reserves, including 41 national and 64 provincial, have been established on the QTP. These reserves cover 0.82 million ha or 31.6% of the total area of the QTP and 57.6% of the total area of China's land nature reserves, and include unique and fragile ecosystems and rare species (The State Council Information Office of the People's Republic of China, 2018). The reserves vary in size, with Qiangtang the largest at 29.8 million ha, Lalu wetland the smallest at 1220 ha, and nine reserves exceeding 0.1 million ha (Qin *et al.*, 2022).

The reserves increased net primary productivity and the normalized difference vegetation index (NDVI), protecting the animals and plants as well as their environment on the QTP. For example, the number of black-necked cranes (*Grus nigricollis*) is increasing in the national nature reserves and accounts for about 80% of the global population. From 2000 to 2016, the number of Tibetan antelopes increased from

~60,000 to more than 200,000, while the number of wild yaks on the Qiangtang Plateau increased from ~6000 to more than 10,000. Some rare and endangered species, such as *Rhinopithecus strykeri*, *Cervus wallichii* and *Perisoreus internigrans*, were also observed in reserves.

Improve laws and regulations

Policies and regulations related to the establishment of ecological civilization on the QTP are being improved. Tibet, Qinghai, Sichuan, Gansu and Yunnan have proposed local regulations and implementation methods related to ecological civilization construction based on the actual situation of the plateau. The Tibet Autonomous Region initiated the 'Implementation Opinions on Building an Important National Ecological Safety Barrier and Accelerating the Construction of Ecological Civilization', the 'Opinions on Building a Beautiful Tibet', and the 'Tibet Autonomous Region Environmental Protection Assessment Measures'; Qinghai proposed the 'Overall Plan for the Construction of the Ecological Civilization System in Qinghai Province', the 'Regulations on the Promotion of Ecological Civilization Construction in Qinghai Province', and the 'Action Plan for Establishing a National Ecological Civilization Pilot Zone in Qinghai Province'; Sichuan proposed the 'Regulations on the Management of Nature Reserves in Sichuan Province'; Gansu proposed the 'Regulations on the Management of Gansu Qilian Mountain National Nature Reserve'; Yunnan initiated the 'Plan for the Protection and Construction of Ecological Security Barriers in the "Two Rivers" Basin of Diqing Prefecture' and the 'Action Plan for Biodiversity Protection in Northwest Yunnan'.

In October 2021, the Central Committee of the Communist Party of China issued the 'Qinghai-Tibetan Plateau Ecological Environment Protection and Sustainable Development Plan', proposing the establishment of an ecological environment protection system for the QTP. On 26 April 2023, the 13th National People's Congress passed the 'Law of China on Ecological Protection of the Qinghai Tibet Plateau (Draft)'. The draft aims to establish the basic principles of ecological protection in the QTP, improve the management system, build an ecological security pattern, strengthen ecological protection and restoration, strengthen ecological risk prevention and control, and improve guarantees and supervision.

8.5 The Link between People's Livelihood and Ecology

Livelihoods depend mainly on environmental resources and services, while environmental sustainability affects the range and quality of livelihood opportunities (Li *et al.*, 2023). The QTP grasslands create existence values (intrinsic benefits for the global society), option values (retaining future opportunities), direct values (local benefits in social, economic and environmental dimensions), and indirect values (associated with tourism and agriculture). The pastoralism rooted in the QTP grasslands sustains approximately 150 million people, 18% of the total global grasslands population (Dong, 2023). The implementation of a series of ecological engineering projects has improved the ecological and living environment. The average annual water conservation capacity (WCC) of the QTP reaches 34.5 million m^3. After the completion of the first phase of the ecological engineering project in the Three Rivers source region, the annual average WCC of the forest ecosystem increased by 15.6%. The QTP has lesser human activity, fewer and lower concentrations of pollutants, and the air quality is less affected by human activities than other areas. With the continuous promotion of green energy, the construction of ecologically friendly towns, and the comprehensive improvement of the rural environment, the air quality on the QTP has improved and, at present, the region is one of the 'cleanest' on Earth. Since 2009, the country has invested 6.294 billion yuan in supporting domestic sewage treatment facilities and sewage pipeline projects in towns on the QTP, and in improving the local urban sewage treatment capacity and the living environment. The economic development of the QTP adheres to ecological friendliness, resource conservation and intensification, and strives for green development.

8.5.1 Creating an ecological grassland and animal husbandry technology system

There are two main problems in grasslands management in China: (i) unreasonable allocation of grasslands ecological and production functions, severe degradation of natural grasslands, and low productivity; and (ii) artificial grasslands are on a small scale, with low yield, poor quality and short utilization period, with low coupling between grasslands and livestock industry, and high dependence of meat and milk production on cultivated land and imported forage (Bai *et al.*, 2016). The shortage of forage and poor nutritional quality during the cold season result in substantial weight loss of livestock on the QTP. The imbalance of forage availability and needs for livestock constrain the production of livestock and income of herders (Shang *et al.*, 2014). There are three major benefits in establishing high-quality artificial grasslands: (i) improves the quality of forage and forage reserves, reduces livestock weight loss during harsh winters, and increases the income of herders; (ii) drives the development of enterprises and reduces feed costs for livestock; and (iii) promotes ecological restoration of degraded grasslands. A suitable regional ecological grasslands industry model of 'enterprise + base + herders' has been gradually formed in the farming-pastoral zone of the QTP, i.e. a forage industry development model integrating 'land transfer in farming-pastoral zone – establishment of high-quality and high-yield artificial grasslands – processing of high-quality forage – guarantee supply of forage – balanced livestock nutrition – disaster resistance and forage storage' (Xu *et al.*, 2020).

8.5.2 Developing distinctive, ethnic and cultural industries

Distinctive, ethnic and cultural industries are the key to the sustainable economic and cultural development of the QTP (Shang *et al.*, 2014). For this to occur, it is important to strengthen the standardized and green production of agricultural and livestock products so that they are of high quality and are distinctive; expand and strengthen the organic industries; establish production bases for yaks, Tibetan sheep, plateau summer vegetables, cold-water fish, barley, rapeseed and high-quality forage; create many national areas for agricultural and livestock products; and improve the entire industrial chain of green and organic agricultural and livestock products.

Since 2004, the country has invested more than 3 billion yuan in implementing over 450 special agricultural and animal husbandry industry projects in Tibet, and fostered more than 100 leading enterprises, increasing the income of farmers and herders by 1.182 billion yuan, and benefiting 1.754 million people. At present, ten agricultural and pastoral geographical indication protection products, such as Pali yaks, Gamba sheep, Longzi black barley, Chayu kiwifruit and Bomi rhizome gastrodiae, have been certified by national departments. Qinghai province focuses on creating four '10 billion yuan' industries, including grain and oil cultivation (e.g. *Brassica rapa*, *Hordeum vulgare* and *Zea mays*), livestock and poultry breeding (e.g. yak and Tibetan sheep), fruit and vegetables, *Fructus lycii* and buckthorn. By the end of 2015, Qinghai had 1879 family farms and ranches, 8876 co-operatives and 451 leading enterprises in agriculture and animal husbandry. The proportion of speciality crop products reached 85%, and the percentages of improved varieties of crops, livestock and poultry, and aquatic products reached 96%, 62% and 95%, respectively. Gansu established a base for barley, plateau quality rapeseed and Chinese medicinal plants, and accelerated the development of speciality plantations and economic fruit. Ganzi and Aba Tibetan autonomous prefectures in Sichuan implemented an organic combination of ecological civilization construction and ecological agriculture, creating two 'million mu' (0.133 million ha) industrial bases for distinctive agriculture, animal husbandry, fruit and medicinal industries. Diqing Tibetan autonomous prefecture in Yunnan has 90,700 ha of plateau agricultural planting area, with a production value of 1.9 billion yuan since 2017 (The State Council Information Office of the People's Republic of China, 2018).

Yaks are important to the inhabitants of the QTP, as butter and cheese from milk can be sold for cash or bartered for other necessities. The long hair of yaks is sold or used for making ropes, blankets, bags and tents, while the fine, inner wool is used for making clothing and blankets. Yak hides are sold or used for leather products, and their dung (Fig. 8.3) can be used as fuel and fertilizer (Dong, 2023).

Fig. 8.3. Dried yak dung to be used for cooking and heating. (Photo: Allan Degen)

8.5.3 Developing ecotourism

Ecotourism on the QTP has demonstrated strong vitality. The *Blue Book of Ecological Development* points out that ecotourism relies on: (i) natural resources: for example, geomorphic tourism (high mountains, ice margin, limestone and volcano), forest ecotourism (Xianmi, Bassoon Lake and Sejila National Forest Park), grasslands ecotourism (Caiqutang, YamdrokTso and Zoige grasslands) and wildlife observations (*Equus kiang*, *Rhinopithecus* and *Grus nigricollis*); and (ii) traditions and culture, such as Tibetan culture (Tibetan medicine, murals and religion) and folk culture (Jiangzi Dama Festival and Tibetan Family Music). The development of tourism has driven enterprises of catering, accommodation, transportation and cultural entertainment, and promoted the protection of cultural heritage, the production of traditional handicrafts, and the development of ethnic products. Tourism has become important for the QTP for green growth and for increased income for residents.

The statistical bulletin released by Qinghai revealed that Qinghai received ~51 million tourists from home and abroad in 2019, an increase of 21% compared to the previous year; and tourism revenue reached ~56 billion yuan, an increase of 20.5% compared to the previous year. From 2011 to 2019, the annual number of tourists in the Tibet autonomous region increased from 8.7 million to 40.1 million, with an average annual growth rate of close to 22%. In 2020, the tourism revenue in Gansu was ~77 billion yuan and in Qinghai it was ~33 billion yuan, while in 2021, the tourism revenue in Aba prefecture was ~35 billion yuan and in Diqing prefecture, it was ~9.5 billion yuan (Shi, 2023). Overall, with the improvement of the ecological environment on the QTP, tourism revenue has been increasing steadily.

8.5.4 Improving the ecological compensation policy

The ecological compensation system is an important measure taken to protect the ecological environment. The state established ecological compensation in the QTP, including transfer payments for key ecological functional areas, compensation for forest ecological benefits, subsidies and rewards for grasslands ecological protection, and compensation for wetlands ecological benefits. From 2008 to 2017, the central government allocated ~16 and ~8.3 billion yuan in transfer payment funds for key ecological functional zones in Qinghai and Tibet, respectively, covering 77 key ecological counties and all national-level prohibited development zones in the two provinces. From 2011 to 2015, the state allocated ~10.9 billion yuan in subsidies and reward funds for grasslands ecological protection in Tibet. Since 2015, Tibet has invested 85 million yuan to help herders mitigate losses caused by wildlife accidents. Since 2013, the central government has allocated a total of 16.4 billion yuan to Qinghai and has implemented ecological benefit compensation projects for grasslands, forests and wetlands. To promote the organic combination of ecological protection and poverty alleviation, Qinghai launched the ecological public welfare management system, and allocated 1800 yuan per month in salary subsidies, with an annual subsidy fund of 880 million yuan. The subsidies and reward funds of Diqing were ~1.1 billion yuan from 2009 to 2017, while in Ganzi and Aba they were 284 and 154 million yuan, respectively in 2017 (The State Council Information Office of the People's Republic of China, 2018).

The grasslands ecological subsidy policy was formulated to improve the grasslands vegetation status, but the allocation of funds is spatially heterogeneous, with more funds allotted to Qinghai than Tibet, and more to pastoral counties than to semi-pastoral counties (Ma *et al.*, 2021). The 'six in one' integrated ecological compensation for alpine grasslands was meant to guide herders to manage and use grasslands efficiently (Fu *et al.*, 2012). However, the effectiveness of the subsidy on grasslands protection was reduced or neutralized by other socio-economic and climatic factors, and, consequently, subsidy policies should also consider other factors related to market and climate conditions in the future (Ma *et al.*, 2021).

8.6 Suggestions for Sustainable Development in the Future

The sustainable utilization of grasslands on the QTP requires the co-ordination and unification of ecological, economic and social functions of the grasslands. There should be reasonable utilization of the natural resources of the QTP, adherence to the principle of 'ecological priority', and proper allocation of the grasslands to production and ecological functions. Furthermore, there should be adherence to the principle of 'functional replacement' of replacing some of the production functions of natural grasslands with artificial grasslands to improve the quantity and quality of forage for livestock. The principle of 'adapting to local conditions' should be adopted to determine the appropriate stocking rate, the proper proportion of natural and artificial grasslands, and the suitable types and varieties of livestock according to local conditions (Bai *et al.*, 2016). Based on the degradation and economic situation, grasslands should be divided into ecological protection, ecological protection-production development, and production development priority areas. The degraded grasslands should be restored by classification, subdivision and zoning (Dong *et al.*, 2015). With the balance of grasslands and livestock as the core, the grazing system of 'delayed grazing in spring, nomadic in summer, rotational grazing in autumn, and free grazing in winter' should be practised (Cao *et al.*, 2018).

The allocation of grasslands and livestock resources should be optimized to realize the coupling development of regional functions. Pastoral areas should actively promote seasonal animal husbandry models, farming-pastoral zones should develop ecological grass–animal husbandry industries, and valley agricultural areas should employ development models such as the combination of planting and breeding, and the coupling of agriculture and animal husbandry (Xu *et al.*, 2020). The development of renewable energy should be prioritized to actively promote the clean and low-carbon transformation of energy, and establish an energy system dominated by renewable energy and carbon neutrality.

Ecological risk prevention, warning mechanisms and control systems should be improved. A prediction system for the impact of climate change on the ecosystem, biodiversity and natural disasters on the QTP should be established and action plans for adapting to or mitigating climate change should be formulated. Systems to prevent and remove invasive alien species should be strengthened.

The vertical ecological compensation system should be improved. Regions that provide more and higher-quality ecological products should receive more compensation. The establishment of a horizontal ecological protection compensation system with cost sharing, benefit sharing, and co-operative governance should be examined. Inter-regional ecological compensation in areas with relationships between ecological product suppliers and users should be promoted, so that users of ecological products pay and suppliers benefit. In ecological compensation, the local government should be held accountable and the implementation of all transactions should be strengthened. Only with good grasslands quality can herders receive ecological compensation (Wang *et al.*, 2016).

If temperatures continue to rise in the future, the permafrost melting rate may accelerate and carbon emissions may increase, thereby weakening the carbon sequestration capacity of the alpine grasslands on the QTP (Wang *et al.*, 2023). It is important to focus on high-resolution, multi-sphere earth system models, based on an optimized coupling between observation and modelling, to achieve sustainable development of the QTP (Yao *et al.*, 2023).

References

Adnane, B., Mainassara, Z.-A., Mohamed, F., Farissi, M., Lazali, M. *et al.* (2015) Physiological and molecular aspects of tolerance to environmental constraints in grain and forage legumes. *International Journal of Molecular Sciences* 16, 18976–19008.

Bai, Y. and Cotrufo, M.F. (2022) Grassland soil carbon sequestration: current understanding, challenges, and solutions. *Science* 377(6606), 603–608. DOI: 10.1126/science.abo2380

Bai, Y.F., Pan, Q.M. and Xing, Q. (2016) Fundamental theories and technologies for optimizing the production functions and ecological functions in grassland ecosystems. *Chinese Science Bulletin* 61(2), 201–212 (in Chinese).

Bardgett, R.D., Bullock, J.M., Lavorel, S., Manning, P., Schaffner, U. *et al.* (2021) Combatting global grassland degradation. *Nature Reviews Earth & Environment* 2(10), 720–735. DOI: 10.1038/s43017-021-00207-2

Bargaz, A., Zaman-Allah, M., Farissi, M., Lazali, M., Drevon, J.-J. *et al.* (2015) Physiological and molecular aspects of tolerance to environmental constraints in grain and forage legumes. *International Journal of Molecular Sciences* 16(8), 18976. DOI: 10.3390/ijms160818976

Cao, G., Lin, L., Zhang, F., Li, Y., Du, Y. *et al.* (2018) Long term ecological research and experimental demonstration provide theoretical and technical support for adaptive management of alpine grasslands. *Bulletin of Chinese Academy of Sciences* 33, 1115–1126. DOI: 10.16418/j.issn.1000-3045.2018.10.013 (in Chinese).

Cao, J., Wei, C., Adamowski, J.F., Biswas, A., Li, Y. *et al.* (2021) On China's Qinghai-Tibetan Plateau, duration of grazing exclosure alters R:S ratio, root morphology and attending root biomass. *Soil and Tillage Research* 209(1–2), 104969. DOI: 10.1016/j.still.2021.104969

Central Government of the People's Republic of China (2021) Major data bulletin of the third National Land Survey. Available at: https://www.gov.cn/xinwen/2021-08/26/content_5633490.htm (accessed 24 April 2024).

Chen, H., Zhu, Q., Peng, C., Wu, N., Wang, Y. *et al.* (2013) The impacts of climate change and human activities on biogeochemical cycles on the Qinghai-Tibetan Plateau. *Global Change Biology* 19(10), 2940–2955. DOI: 10.1111/gcb.12277

Chen, H., Ju, P., Zhang, J., Wang, Y., Zhu, Q. *et al.* (2020) Attribution analyses of changes in alpine grasslands on the Qinghai-Tibetan Plateau. *Chinese Science Bulletin* 65, 2406–2418. DOI: 10.1360/TB-2019-0619 (in Chinese).

Chen, Y., Yang, H., Bao, G., Pang, X. and Guo, Z. (2022) Effect of the presence of plateau pikas on the ecosystem services of alpine meadows. *Biogeosciences* 19, 4521–4532. DOI: 10.5194/bg-2021-313

Deng, L., Shangguan, Z., Bell, S.M., Soromotin, A.V., Peng, C. *et al.* (2023) Carbon in Chinese grasslands: meta-analysis and theory of grazing effects. *Carbon Research* 2(1), 19.

Dong, Q., Ma, Y., Xu, C., Shi, J., Wang, Y. *et al.* (2015) Study of classification and gradation, restoration of black-soil beach degraded grassland in the headwaters of Three Rivers. *Acta Agrestia Sinica* 23, 441–447 (in Chinese).

Dong, S. (2023) Revitalizing the grassland on the Qinghai-Tibetan Plateau. *Grassland Research* 1–10. DOI: 10.1002/glr2.12055

Dong, S., Shang, Z., Gao, J. and Boone, R.B. (2020a) Enhancing sustainability of grassland ecosystems through ecological restoration and grazing management in an era of climate change on Qinghai-Tibetan Plateau. *Agriculture Ecosystems and Environment* 287, 106684. DOI: 10.1016/j.agee.2019.106684

Dong, S., Li, Y., Ganjurjav, H., Gao, Q., Gao, X. *et al.* (2020b) Grazing promoted soil microbial functional genes for regulating C and N cycling in alpine meadow of the Qinghai-Tibet Plateau. *Agriculture Ecosystems and Environment* 303, 107111. DOI: 10.1016/j.agee.2020.107111

Dong, S., Shang, Z., Gao, J. and Boone, R. (2022) Enhancing the ecological services of the Qinghai-Tibetan Plateau's grasslands through sustainable restoration and management in era of global change. *Agriculture Ecosystems and Environment* 326, 107756. DOI: 10.1016/j.agee.2021.107756

Fayiah, M., Dong, S., Khomera, S.P., Rehman, S.A.U., Yang, M. *et al.* (2020) Status and challenges of Qinghai-Tibet Plateau's grasslands: an analysis of causes, mitigation, measures, and way forward. *Sustainability* 12, 1099. DOI: 10.3390/su12031099

Fu, W., Zhao, J. and Du, G. (2012) Qinghai-Tibet Plateau alpine grassland ecological compensation mechanism research. *Ecological Economy* 10, 153–157, 168 (in Chinese).

Hao, L., Wang, J.A. and Zhang, H. (2008) Comprehensive assessment and diagnosis of the health of animal husbandry ecosystems in north China pastures. *Acta Ecologica Sinica* 28, 1456–1465 (in Chinese).

Harris, R.B., Samberg, L.H., Yeh, E.T., Smith, A.T., Wang, W. *et al.* (2016) Rangeland responses to pastoralists' grazing management on a Tibetan steppe grassland, Qinghai Province, China. *The Rangeland Journal* 38(1), 1–15. DOI: 10.1071/RJ15040

Hong, J., Yun, X., Lin, J. and Zhang, H. (2014) Analysis and prevention and control of rodent damage in natural grasslands of China. *Chinese Journal of Grassland* 36, 1–4 (in Chinese).

Kraaijenbrink, P.D.A., Bierkens, M.F.P., Lutz, A.F., Immerzeel, W.W. (2017) Impact of a global temperature rise of 1.5 degrees Celsius on Asia's glaciers. *Nature* 549, 257–260.

Li, L., Yang, S., Wang, Z., Zhu, X. and Tang, H. (2010) Evidence of warming and wetting climate over the Qinghai-Tibet Plateau. *Arctic Antarctic and Alpine Research* 42(4), 449–457. DOI: 10.1657/1938-4246-42.4.449

Li, M., Zhang, X., Wu, J., Ding, D., Niu, B. *et al.* (2021) Declining human activity intensity on alpine grasslands of the Tibetan Plateau. *Journal of Environmental Management* 296, 113198. DOI: 10.1016/j.jenvman.2021.113198

Li, T., Singh, R.K., Cui, L., Xu, Z., Liu, H. *et al.* (2023) Navigating the landscape of global sustainable livelihood research: past insights and future trajectory. *Environmental Science and Pollution Research.* DOI: 10.1007/s11356-023-29567-6

Li, X.-L., Gao, J., Brierley, G., Qiao, Y.-M., Zhang, J. *et al.* (2013) Rangeland degradation on the Qinghai-Tibet Plateau: implications for rehabilitations. *Land Degradation and Development* 24(1), 72–80. DOI: 10.1002/ldr.1108

Liu, S., Zamanianb, K., Schleussa, P.-M., Zarebanadkoukic, M. and Kuzyakova, Y. (2018) Degradation of Tibetan grasslands: consequences for carbon and nutrient cycles. *Agriculture Ecosystems and Environment* 252, 93–104. DOI: 10.1016/j.agee.2017.10.011

Lü, C. and Tian, H. (2007) Spatial and temporal patterns of nitrogen deposition in China: synthesis of observational data. *Journal of Geophysical Research* 112(D22), D22S05. DOI: 10.1029/2006JD007990

Ma, H., Lin, H., Xiong, X. and Han, X. (2021) The impact of the grassland ecological reward policy on grassland vegetation in Qinghai-Tibet Plateau. *Acta Agrestia Sinica* 29, 545–554 (in Chinese).

Maestre, F.T., Le Bagousse-Pinguet, Y., Delgado-Baquerizo, M., Eldridge, D.J., Saiz, H. *et al.* (2022) Grazing and ecosystem service delivery in global drylands. *Science* 378(6622), 915–920. DOI: 10.1126/science.abq4062

Miehe, G., Schleuss, P., Seeber, E., Babel, W., Biermann, T. *et al.* (2019) The *Kobresia pygmaea* ecosystem of the Tibetan highlands – origin, functioning and degradation of the world's largest pastoral alpine ecosystem. *Science of the Total Environment* 648, 754–771. DOI: 10.1101/135558

Mo, X. (2020) Sustainable livestock carrying capacity and overgrazing rate of grassland over Qinghai-Tibet plateau since 1980. Available at: https://doi.org/10.11888/Socioeco.tpdc.270347 (accessed 15 September 2023).

Payne, R.J., Dise, N.B., Field, C.D., Dore, A.J., Caporn, S.J.M. *et al.* (2017) Nitrogen deposition and plant biodiversity: past, present, and future. *Frontiers in Ecology and the Environment* 15(8), 431–436. DOI: 10.1002/fee.1528

Qin, L., Zhu, Y., Ren, Y., Li, B., Fu, M. *et al.* (2022) Differences in management capacity of national nature reserves on the Qinghai-Tibet Plateau and the impact on conservation effectiveness. *Biodiversity Science* 30, 22419 (in Chinese).

Qiu, J. (2016) Trouble in Tibet. *Nature* 529, 142–145. DOI: 10.1038/529142a

Saatkamp, A., Henry, F. and Dutoit, T. (2018) Vegetation and soil seed bank in a 23-year grazing exclusion chronosequence in a Mediterranean dry grassland. *Plant Biosystems: An International Journal Dealing with all Aspects of Plant Biology* 152(5), 1020–1030. DOI: 10.1080/11263504.2017.1407375

Shang, Z., Gibb, M.J., Leiber, F., Ismail, M., Ding, L. *et al.* (2014) The sustainable development of grassland-livestock systems on the Tibetan plateau: problems, strategies and prospects. *The Rangeland Journal* 36(3), 267–296. DOI: 10.1071/RJ14008

Shi, B. (2023) Typical of the Qinghai-Tibet plateau area living related data sets (2010–2019). National Tibetan Plateau Data Center. Available at: https://doi.org/10.11888/HumanNat.tpdc.300257 (accessed 15 September 2023).

Shrestha, N., Xu, X., Meng, J. and Wang, Z. (2021) Vulnerabilities of protected lands in the face of climate and human footprint changes. *Nature Communications* 12(1), 1632. DOI: 10.1038/s41467-021-21914-w

The State Council Information Office of the People's Republic of China (2018) Ecological Progress on the Qinghai-Tibet Plateau. White Paper. Available at: http://www.scio.gov.cn/zfbps/ndhf/2018n/202207/t20220704_130607.html (accessed 17 April 2024).

Sun, J., Zhang, Z. and Dong, S. (2019) Adaptive management of alpine grassland ecosystems over Tibetan Plateau. *Pratacultural Science* 36(4), 933–admane938.

Sun, J., Liu, M., Fu, B., Kemp, D., Zhao, W. *et al.* (2020) Reconsidering the efficiency of grazing exclusion using fences on the Tibetan Plateau. *Science Bulletin* 65(16), 1405–1414. DOI: 10.1016/j.scib.2020.04.035

Wang, D. and Yang, L. (2004) *Grassland Ecology and Management Utilization*. Chemical Industry Press, Beijing (in Chinese).

Wang, D., Li, X., Pan, D. and De, K. (2016) The ecological significance and controlling of rodent outbreaks in the Qinghai-Tibetan Grasslands. *Journal of Southwest University for Nationalities (Natural Science Edition)* 42(3), 237–245. DOI: 10.11920/xnmdzk.2016.03.001 (in Chinese).

Wang, J., Zhang, X., Chen, B., Shi, P., Zhang, J. *et al.* (2013) Causes and restoration of degraded alpine grassland in northern Tibet. *Journal of Resources and Ecology* 4(1), 43–49. DOI: 10.5814/j.issn.1674-764x.2013.01.006

Wang, P., Wolf, S.A., Lassoie, J.P., Poe, G.L., Morreale, S.J. *et al.* (2016) Promise and reality of market-based environmental policy in China: empirical analyses of the ecological restoration program on the Qinghai-Tibetan Plateau. *Global Environmental Change* 39, 35–44. DOI: 10.1016/j.gloenvcha.2016.04.004

Wang, Y., Lv, W., Xue, K., Wang, S., Zhang, L. *et al.* (2022) Grassland changes and adaptive management on the Qinghai-Tibetan Plateau. *Nature Reviews Earth and Environment* 3, 668–683. DOI: 10.1038/s43017-022-00330-8

Wang, Y., Xiao, J., Ma, Y., Ding, J., Chen, X. *et al.* (2023) Persistent and enhanced carbon sequestration capacity of alpine grasslands on Earth's Third Pole. *Science Advances* 9(20), eade6875. DOI: 10.1126/sciadv.ade6875

Wei, D., Zhao, H., Zhang, J., Qi, Y. and Wang, X. (2020) Human activities alter response of alpine grasslands on Tibetan Plateau to climate change. *Journal of Environmental Management* 262, 110335. DOI: 10.1016/j.jenvman.2020.110335

Xu, T., Zhao, X., Zhang, X., Wang, X., Geng, Y. *et al.* (2020) Sustainable development of ecological grass-based livestock husbandry in Qinghai-Tibet Plateau alpine area: principle, technology and practice. *Acta Ecologica Sinica* 40, 6324–6337 (in Chinese).

Xu, W. and Huntsing, L. (2022) Minding the boundary: social-ecological contexts for fence ecology and management. *Frontiers in Ecology and the Environment* 20(7), 405–412. DOI: 10.1002/fee.2500

Yang, D., Yue, F., Li, X., Xu, Z., Wang, Z. *et al.* (2022) Current research status and hot spots of grassland rodent pests based on CiteSpace. *Acta Agrestia Sinica* 30, 670–681 (in Chinese).

Yao, T., Yu, W., Wu, G., Xu, B., Zhao, H. *et al.* (2019) Glacier anomalies and relevant disaster risks on the Tibetan Plateau and surroundings. *Chinese Science Bulletin* 6, 2770–2782.

Yao, T., Thompson, L., Chen, D. and Piao, S. (2023) Reflections and future strategies for third pole environment. *Nature Reviews Earth and Environment* 3(10), 608–610. DOI: 10.1038/s43017-022-00342-4

Zhan, J., Teng, Y. and Liu, S. (2021) 1 km grid data set of ecological vulnerability in agricultural and pastoral areas of Qinghai Tibet Plateau. Available at: https://doi.org/10.11888/Ecolo.tpdc.271117 (accessed 15 September 2023).

Zhang, G., Yang, F. and Long, H. (2023a) Save the life-sustaining mattic layer on the Qinghai-Tibetan Plateau. *The Innovation* 4(3),100418. DOI: 10.1016/j.xinn.2023.100418

Zhang, X., Yang, Y., Piao, S., Bao, W., Wang, S. *et al.* (2015) Ecological changes in the Qinghai-Tibet Plateau. *Chinese Science Bulletin* 60, 3048-3056 (in Chinese).

Zhang, Y. (2019) Integration dataset of Tibet Plateau boundary. Available at: https://doi.org/10.11888/Geogra.tpdc.270099 (accessed 15 September 2023).

Zhang, Y., Ganjurjav, H., Dong, S. and Gao, Q. (2020) Excessive plant compensatory growth: a potential endogenous driver of meadow degradation on the Qinghai-Tibetan Plateau. *Ecosystem Health and Sustainability* 6(1), 1816500. DOI: 10.1080/20964129.2020.1816500

Zhang, Y., Li, B., Liu, L. and Zheng, D. (2021) Redetemine the region and boundaries of Tibetan Plateau. *Geographical Research* 40, 1543–1553 (in Chinese).

Zhang, Z., Zhao, Y., Lin, H., Li, Y., Fu, J. *et al.* (2023b) Comprehensive analysis of grazing intensity impacts alpine grasslands across the Qinghai-Tibetan Plateau: a meta-analysis. *Frontiers in Plant Science* 13, 1083709. DOI: 10.3389/fpls.2022.1083709

Zhao, H., Wei, D., Wang, X., Hong, J., Wu, J., *et al.* (2023) Three decadal large-scale ecological restoration projects across the Tibetan Plateau. *Land Degradation and Development* 35, 22–32.

Zhao, H. and Wang, X. (2023) The Qinghai-Tibet plateau atlas of key ecological engineering layout. Available at: https://doi.org/10.11888/HumanNat.tpdc.300293 (accessed 15 September 2023).

Zhen, L., Du, B., Wei, Y., Xiao, Y. and Sheng, W. (2018) Assessing the effects of ecological restoration approaches in the alpine rangelands of the Qinghai-Tibetan Plateau. *Environmental Research Letters* 13(9), 095005. DOI: 10.1088/1748-9326/aada51

Zhou, H., Su, W., Zhou, B., Shi, M. and Zhao, H. (2022a) The zonation dataset of native forage planting suitability in Qinghai (2021). Available at: https://doi.org/10.11888/Terre.tpdc.272696 (accessed 15 September 2023).

Zhou, H., Shi, M., Zhou, B., Su, W. and Sun, W. (2022b) The zonation dataset of native forage planting suitability in Tibet (2021). Available at: https://doi.org/10.11888/Terre.tpdc.272697 (accessed 15 September 2023).

9 Ecological Conservation and Restoration in the Indian Himalaya Region

Huma Ali[1], Muhammad Rafiq[2] and Zhanhuan Shang[1,*]
[1]College of Ecology, Lanzhou University, Lanzhou, China; [2]Department of Microbiology, Balochistan University of Information Technology, Engineering and Management Sciences, Quetta, Pakistan

Abstract

The Indian Himalaya region (IHR) is unique among the mountainous ecosystems in the world. Not only are these mountains important in providing water to a large portion of the Indian subcontinent population, but they also support a diverse range of flora and fauna. Although the region contains a wealth of natural resources, the majority of the people are poor and live at or below the sustainable level. The uncontrolled removal of natural resources is hastening the destruction of the surrounding ecosystems and exacerbating the impacts of natural disasters. A new paradigm must be enforced to restore an equilibrium between economic goals and ecological needs, while also paying attention to socio-cultural norms.

Keywords: ecological degradation, ecosystem restoration, Indian Himalaya region, livelihood

9.1 Introduction

The Indian Himalaya region (IHR) consists of the states of Jammu and Kashmir, Himachal Pradesh, Uttarakhand, Sikkim, Arunachal Pradesh, Nagaland, Manipur, Mizoram, Tripura, Meghalaya, and two partially hill states, Assam and West Bengal (Aayog, 2017) (Fig. 9.1). It lies between the Indus and Brahmaputra rivers and from the southern foothills (*Siwaliks*) to the northern Tibetan Plateau. The IHR is more than 2500 km long, covers an area of roughly 530,000 km^2 (Farming *et al.*, 2014) and contains a wide variety of landscapes (Trans-Himalay) (Kaur *et al.*, 2022a). Geological characteristics separate the Himadri (Greater Himalaya), the Himanchal (Lesser Himalaya) and the Siwaliks (Outer Himalaya) into three main geographical divisions that are continuous throughout the length of the IHR (Valdiya, 2002).

Strategically, the IHR is indispensable for the protection of the entire northern border of the country (Aayog, 2018). The extensive forests in the Himalayan region are important for several reasons, including national security (Matthew, 2012). They encompass approximately 65% of the land, which is close to 35% of the world's forested land and 46% of the world's very outstanding forested land (de Camino, 2000). The Himalayas store approximately 12,000 km^3 of freshwater in 9000 glaciers and high-altitude lakes (Lynas, 2008). Millions of people in the Great Plains depend on the IHR for water and their livelihoods (Cunfer, 2005). The ecosystems of the IHR are described in Fig. 9.2.

*Corresponding author: shangzhh@lzu.edu.cn

DOI: 10.1079/9781800622579.0009

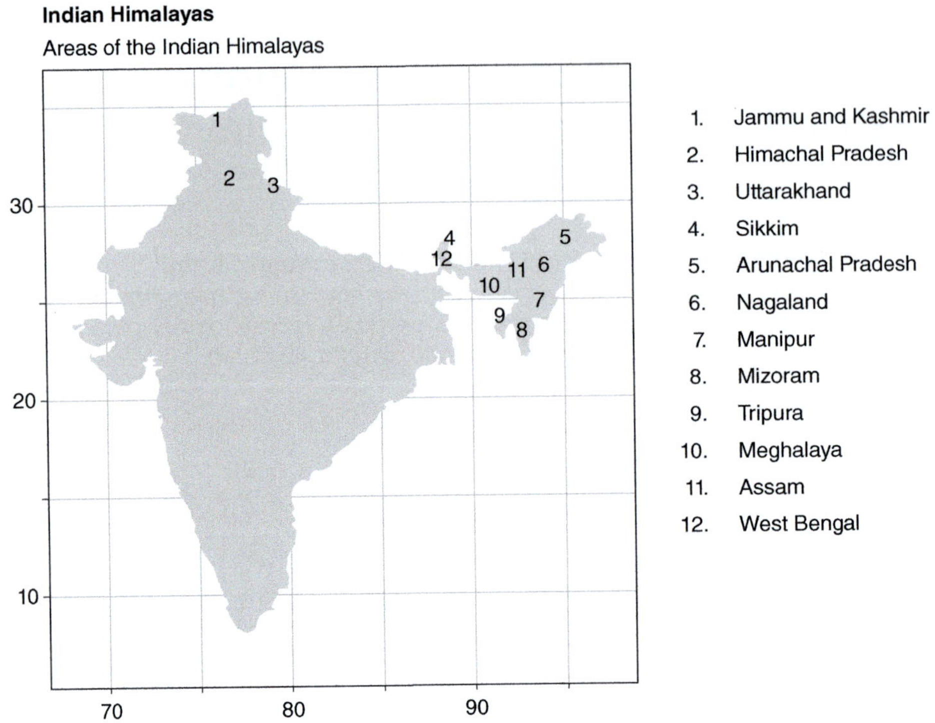

Fig. 9.1. Map of Indian Himalayan region (IHR).

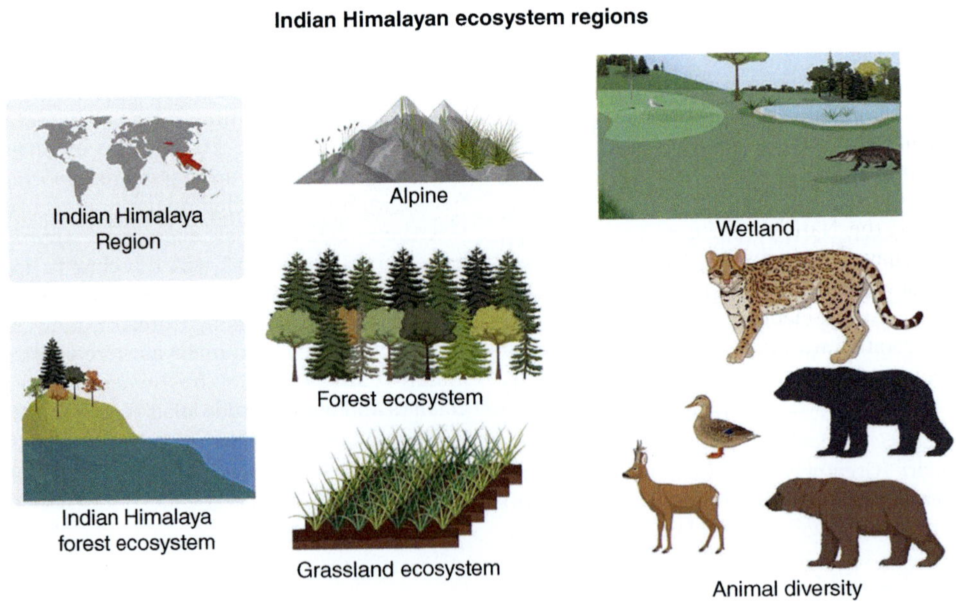

Fig. 9.2. Overview map of Indian Himalayan ecosystems.

The risk of landslides in the IHR has increased over the last several decades, mostly as a result of more frequent and heavy downpours (Kumar *et al.*, 2017). However, little focus has been placed on determining the origin of events that occur before and after the Indian summer monsoon (ISM), which can be devastating (Chen *et al.*, 2014). Severe precipitation causes landslides and rockfalls with the onset or retreat of the ISM in the north-west Himalayan region (Yu *et al.*, 2021). For example, in September 2014, an extraordinary rainfall of 488 mm in 24 hours occurred during the waning phase of the ISM (Bhellum, 2020) causing a landslide and widespread destruction in the Udhampur district of Jammu and Kashmir. Based on geological studies, bedding planes, that is strata of sedimentary rock, in the region are composed of an inconsistent succession of solid, packed sandstone and fragile clay stone, which is the source of the area's planar weakness (Roy and Jakhar, 2002). Collapse along the planar weakness that dips down the valley slope demolished the Sadal village after the heavy rainfall (Dahal, 2016). According to measurements of climatic variables recorded by the Tropical Rainfall Measuring Mission (TRMM), extreme rainfall events do not follow a regular cycle (Yu *et al.*, 2021). During the ISM's withdrawal, the area had heavy precipitation in 2005, 2009, 2011, 2012 and 2014 (i.e. closely spaced and clustered heavy rainfall episodes), and extraordinary heavy rainfalls at the onset of the ISM in 2003 and 2013 (Kumar *et al.*, 2017).

Global warming and climate variability are damaging social and economic systems in mountain, riverine, forest and wetland biomes (Heide, 2012). Consequently, the Indian government launched the National Mission for Sustaining the Himalayas Ecosystem (NMSHE), an initiative being developed by the Indian government's Department of Science and Technology (DST) in collaboration with the Swiss Agency for Development and Cooperation in the Indian Himalayas Climate Adaptation Programme (IHCAP) (Kaur *et al.*, 2022b). The project's official title is 'Capacity Development on Climate Change', and its purpose is to assess the risks in the Indian Himalayas by bringing together representatives from various government agencies via a series of workshops (Fisher *et al.*, 2016). The goal of the project is to persuade the 12 states of the IHR to adopt a unified and integrated approach in combatting disasters (Bergmann, 2016). By combining their efforts, regions can better assess their vulnerability, adaptive capacity and resilience in the face of climate change (Pandey and Jha, 2012). The Indian Institute of Technology Guwahati (IIT Guwahati) and the Indian Institute of Technology Mandi (IIT Mandi) are responsible for carrying out the project, with support from the Indian Institute of Science in Bangalore (IISc) (Laltlanmawii, 2011).

The IHR, a place of tremendous natural beauty and culture, is confronted with deforestation, habitat degradation and the effect of unregulated tourism (Wang *et al.*, 2019). These challenges have far-reaching implications on the fragile ecological balance, local residents' livelihoods, and the tourist industry's long-term viability (Aayog, 2018). Consequently, ecological protection and restoration programmes have taken priority in the IHR (Kumar *et al.*, 2021). Due to reasons such as population expansion, resource demands and unsustainable land-use practices, deforestation has arisen as a serious problem (Pandit and Kumar, 2013). Agriculture, infrastructure expansion and wood extraction have caused the shrinking of forests, resulting in the loss of vital animal habitats (Chatterjee *et al.*, 2006). This loss of forest cover not only increases soil erosion, water shortages and landslide danger but it also directly affects the lives of local communities that rely on forest resources for subsistence and revenue (Ives and Messerli, 1989).

Furthermore, changing environmental conditions in the IHR impact the tourist sector, which has thrived in this area because of the magnificent landscapes, rich cultural legacy and numerous options (Aayog, 2018). Uncontrolled tourism-related activities, such as poorly planned infrastructure building, insufficient waste management and increased foot traffic, have taken a toll on the fragile ecosystems (Wang *et al.*, 2019). These activities cause environmental degradation and pollution, thus jeopardizing the tourist industry's long-term survival (Chakraborty, 2021). To maintain the sustainability and mutual benefits of tourism in the IHR, a careful balance must be reached between the economic benefits and the preservation of natural and cultural resources (Rana and Bisht, 2023). Ecological concerns must be addressed, and conservation and restoration measures must be promoted (Kattel, 2022). By doing so, it is feasible to conserve the unique biodiversity, important ecosystem services and the interests of local populations that depend on these resources (Rawal *et al.*, 2021). It is critical to address the linked concerns of deforestation and habitat degradation, and the consequences of tourism. A comprehensive strategy will help preserve the region's natural and cultural legacy, ensuring a sustainable future that benefits the environment and the well-being of local residents (Choudhury *et al.*, 2022).

9.2 Indian Himalaya Ecological Degradation

The northern section of the IHR, which separates it from the Tibetan Plateau, is plagued by difficulties and inefficient management (Saikawa *et al.*, 2019). Serious transformations are required. Tourism drives India's economy, especially treks in the Himalayas (Bhalla and Chowdhary, 2022), but much waste is left unattended. The increasing urbanization is causing widespread litter (Thakur *et al.*, 2021) due to poor public infrastructure, low air temperature and high altitude (Thakur *et al.*, 2021). Air pollution, often generated by uncontrolled industries, affects the tourist sector (Maikhuri *et al.*, 2017). The idea of using the Himalayas for hydroelectric power has been raised, but this would displace inhabitants, and have an impact on the ecosystem (Bizikova *et al.*, 2013). Dams in geologically unstable areas can erode soil, and garbage from the region can flow into larger bodies of water (Bartarya and Valdiya, 1989). Snow and ice are melting in the Himalayas as a result of climate change, but the water flowing downhill is becoming contaminated (National Research, 2012).

Waste management is desperately needed, but there is a knowledge vacuum in infrastructure planning and trash removal (Dar *et al.*, 2021). Locals have few options for trash disposal. Dumping or burning the garbage harms both the environment and human health (Marsh and Shalvey, 2018). Moreover, owing to the harsh climate, biodegradation solutions are not possible (Maitra and Krishan, 2005). The possible solution to these problems appears to be working as a team (Ives and Messerli, 1989). Waste management and climate change are not limited to the Himalayas in India, but to the entire Himalayas, so all the host countries from these fragile mountains should work together by utilizing government aid and discussing these issues to protect the environment of the Himalayas (Blaikie and Muldavin, 2004). Ecological conservation and restoration initiatives in the IHR are described in Table 9.1.

9.2.1 Causes of ecosystem degradation

The Indian Himalayas are the most vulnerable and fragile mountains in the world, subject to extreme climate change (Chettri *et al.*, 2010). The poor soil, steep slopes and frequent monsoon rains render the mountains prone to erosion (Mishra *et al.*, 2022). The topsoil erodes into the river and is swept away to sea, while deforestation and unsuitable land-use practices hasten the process (Thompson and Warburton, 1985).

Steep slopes and thin soil

Due to the region's steep slopes and extreme land relief, ecological conservation and restoration initiatives in the IHR are confronted by major obstacles. Soil erosion, landslides and water supply issues are all exacerbated by these geographical features (Narain and Singh, 2019). Steep slopes are detrimental for plant and animal life due to soil erosion and low fertility (Tse-Ring *et al.*, 2010). Major causes of landslides include deforestation and poor land management, which, in turn, cause environmental degradation and human hazards (Prasad *et al.*, 2016). These biophysical elements upset the ecological equilibrium and contribute to a loss of plant life, biodiversity and water quality (Nath *et al.*, 2021). Sustainable land-use methods such as terracing, afforestation and erosion control measures are necessary to be implemented (Sur *et al.*, 2021). Promoting both reforestation and sustainable agriculture requires educating the public on the value of good land management and environmentally responsible land use (Kumar, 2023). Because of the rocky landscape, the soil is thin and lacks essential nutrients and water, rendering it prone to erosion, which is a major contributor to ecological deterioration in the IHR (Blaikie, 2016). Aquatic ecosystems and water quality are also impacted negatively by sedimentation and erosion (Meraj *et al.*, 2022). Erosion may be

Table 9.1. Ecological conservation and restoration initiatives in the Indian Himalayan region.

Initiative	Description	References
Project Snow Leopard	A conservation effort in the Himalayas focused on saving endangered snow leopards and their habitats. It emphasizes community-based conservation activities and the promotion of sustainable livelihoods for local populations.	McCarthy and Chapron (2003)
Great Himalayan National Park Conservation Area	This UNESCO World Heritage Site is noted for its diverse biodiversity. Conservation activities include habitat preservation, community involvement and tourist development, all while assuring resource sustainability.	Nandy *et al.* (2015)
Chipko movement	A grass-roots movement that began in Uttarakhand in the 1970s for forest conservation. The movement advocates reforestation, biodiversity protection and long-term livelihoods. It was essential in boosting environmental awareness and strengthening local populations.	Shiva and Bandyopadhyay (1986)
Valley of Flowers National Park	This national park in Uttarakhand is notable for its varied alpine flora. Human activities, invasive species management and biodiversity monitoring are all part of conservation initiatives.	Kala (2005)
Project Hangul	The Kashmir stag, or hangul, is an endangered deer species found in Jammu and Kashmir. Habitat restoration, anti-poaching activities and community participation are examples of conservation strategies.	Aziz *et al.* (2010)
Ecological Restoration in river basins	Several programmes, including reforestation, soil and water conservation, and the promotion of sustainable agriculture are aimed at recovering damaged river basins in the Himalayan region. These initiatives seek to save watersheds, increase water supply and promote ecosystem health.	Prasad *et al.* (2021)

mitigated and water retention can be enhanced by the use of conservation techniques such as terracing, contour plowing and the use of plant barriers (Acharya and Kafle, 2009). However, extensive education and training in sustainable practices are necessary for long-term restoration efforts. Collaboration across many agencies, organizations and communities increases the likelihood of successful restoration.

Landslides, floods and fluvial erosion

Ecological conservation and restoration efforts in the IHR are complicated by frequent geomorphological occurrences such as landslides, flash floods and river erosion (Dikshit *et al.*, 2020). Vegetation loss, altered drainage patterns and habitat fragmentation are only some of the environmental effects of these occurrences (Bhandari, 2006). Intense rainfall events and steep terrain can cause flash floods, which, in turn, can degrade ecosystems (Wang *et al.*, 2019). Moreover, landslides damage human settlements, infrastructure and livelihoods. Besides glacial erosion, fluvial erosion is a major geomorphological process in the Himalayan river systems (Kotlia and Joshi, 2013), causing riverbank erosion, channel extensions and downstream sedimentation. Loss of riparian vegetation, changes in the hydrological cycle and detrimental effects on the aquatic ecosystem are all consequences (Jain and Singh, 2020). Fluvial erosion can be mitigated by riverbank stabilization, afforestation and sustainable river management practices (Debnath *et al.*, 2023), and also by slope stabilization, reforestation, early-warning systems for landslides, retaining walls, proper land use, and floodplain restoration. Community involvement and capacity building that take into account the relationship between geomorphological processes, land use and ecosystem health are also essential (Bhattacharyya *et al.*, 2016).

Active tectonic earthquakes

Because of the movement of the Indian and Eurasian tectonic plates colliding, the Himalayas are a seismically active region with many faults (Bartarya and Valdiya, 1989). Infrastructures and natural ecosystems are damaged by these earthquakes, causing biodiversity loss and ecological system disruptions (Kattel, 2022). Knowledge of active faults, location of vulnerable locations and implementation of suitable land-use practices are all essential for seismic risk management (Kumar *et al.*, 2022b). Buildings can be constructed to be more resilient to earthquakes (Ruchi and Fatima, 2021). Participation from local residents and their indigenous knowledge of the area's conditions are important in the conservation efforts and the creation of early-warning systems (Paudel and Billon, 2020). Ecosystems, and community resilience and restoration, can be improved, and vulnerability to earthquakes can be lessened (Malik and Mohanty, 2007).

9.2.2 Problems

Climate change, the extinction of wildlife, and the breakdown of essential infrastructure pose a danger to local biodiversity, environmental stability and quality of life.

Climate shift

Accelerated glacier retreat is due mainly to a continual rise in air temperature (Shrestha and Aryal, 2011). The Himalayas are home to the greatest number of glaciers outside the polar regions and, therefore, also contain enormous reserves of fresh water (Shrestha and Aryal, 2011). Changes in freshwater flow brought on by ongoing climate change can have severe consequences for human well-being, agricultural productivity, and biodiversity (Tse-Ring *et al.*, 2010). The Himalayas are very susceptible to changes in temperature and precipitation (Sharma and Goyal, 2020). Freshwater sources, aquatic habitats, agriculture and hydropower generation are all impacted by melting glaciers and changing hydrological patterns (Sahu *et al.*, 2020). Climate change can reduce biodiversity and has already been linked to the extinction of several species (Chandra *et al.*, 2022). Animal behaviour, migration routes, breeding windows and habitat are all influenced by climate change (Telwala *et al.*, 2013), which threatens endangered species (Sharma *et al.*, 2016). Reducing greenhouse gas emissions, adopting sustainable practices and constructing climate-resilient infrastructures are all essential in mitigating the detrimental impacts of climate change (Pandey and Jha, 2012). Effective conservation and restoration cannot be implemented without first monitoring the effects of climate change and identifying the areas most at risk (Mishra and Mainali, 2017).

Habitat destruction

Ecological conservation and restoration efforts should focus on reversing the effects of habitat degradation, which is a major danger to biodiversity (Dar and Khuroo, 2020). The delicate ecosystems of the IHR have been severely damaged by anthropogenic activities (Nandy *et al.*, 2015). Habitat loss has been caused by deforestation, urbanization, agricultural expansion, infrastructure construction and population growth (Tiwari, 2008). Endemic plant and animal species are particularly at risk from habitat destruction (Tse-Ring *et al.*, 2010; Kattel, 2022). Sustainable land-use strategies, forest conservation, protected area networks, wildlife corridors, public participation, strict enforcement of environmental laws and better monitoring systems are all necessary to combat habitat degradation (Hilty *et al.*, 2020). Abiding by these measures could protect the Himalayan ecosystems and biodiversity for the long term. The biodiversity hotspots in the IHR are presented in Fig. 9.3.

Endangered species

Poaching is the biggest problem for wildlife in the area, particularly for endangered species like

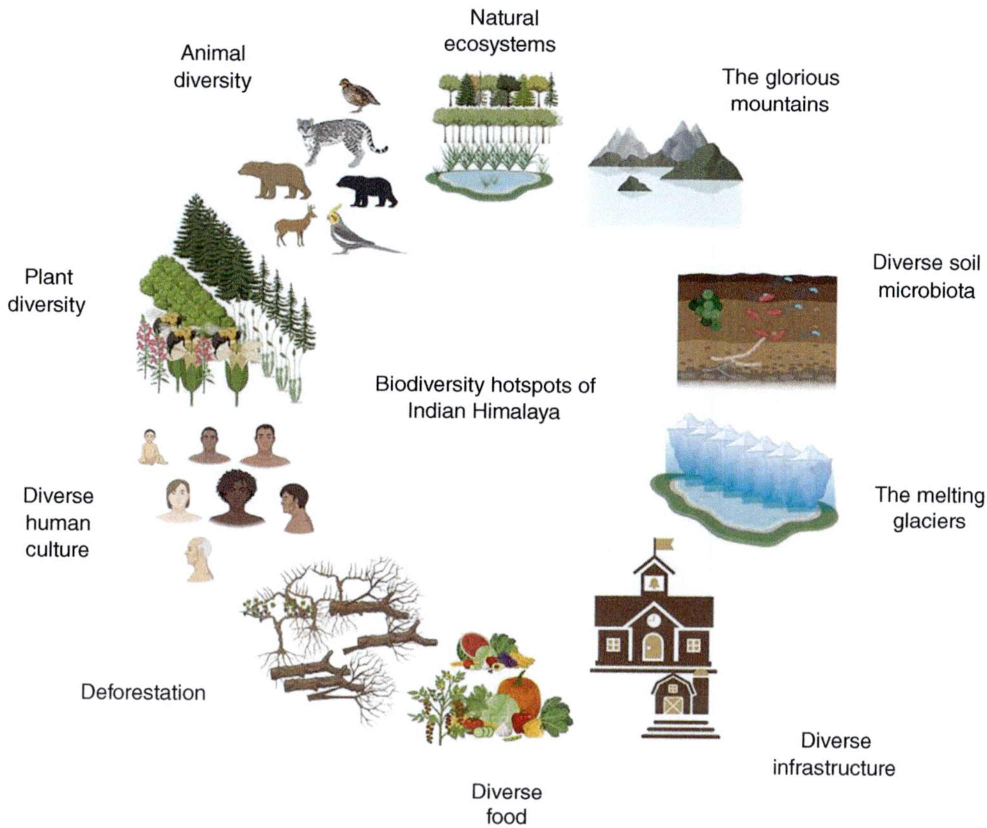

Fig. 9.3. Biodiversity hotspots of the Indian Himalayan region.

rhinoceroses, elephants, tigers, snow leopards, musk deer and black bears (Choudhury, 2007; Lyngdoh *et al.*, 2014). Loss of habitat, fragmentation, hunting and climate change are other factors that threaten endangered species (Sharma and Batish, 2022). Humans and animals are in competition for land and other resources, with humans taking retaliatory measures against elephants and rhinoceroses for raiding crops, and against tigers and snow leopards for killing cattle (Choudhury, 2004). Key methods to preserve endangered species in the Himalayas include the creation of protected areas, habitat preservation, reduction of human–wildlife interaction, introduction of anti-poaching measures, involvement of local people, education on the species' ecological value, and promotion of international collaborations (Zahoor *et al.*, 2021). The endangered species in the IHR are presented in Table 9.2.

Development and infrastructure

With progress comes a greater need for energy. Hydroelectric power is generated in the eastern Himalaya, and it is predicted that energy needs will increase greatly in the future (Rasul, 2014). Submerging arable land and the destruction of biodiversity hotspots could result from the construction of dams without proper environmental impact assessment (Sati, 2015). The construction of reservoirs could lead to the flooding of valleys and damage to nearby villages (Maikhuri *et al.*, 2017). Furthermore, dams can pose a threat to fish populations and aquatic ecosystems (Lakra *et al.*, 2011). Development typically has negative effects on wildlife habitats (Srivathsa *et al.*, 2023). Deforestation and the modification of natural landscapes are occurring at an alarming rate in the Himalayas as a consequence of the fast development of cities, transportation

Table 9.2. Endangered species in the Himalaya region.

Species name	Scientific name	Conservation status	Habitat preference	Threats	Conservation measures	References
Snow leopard	*Panthera uncia*	Endangered	Alpine and sub-alpine	Poaching, habitat loss, human–wildlife conflict	Protected areas, anti-poaching efforts, community-based conservation	Ghoshal *et al*. (2019)
Himalayan musk deer	*Moschus chrysogaster*	Endangered	Forests and alpine meadows	Hunting, habitat degradation	Protected areas, habitat restoration, anti-poaching initiatives	Ilyas (2015)
Red panda	*Ailurus fulgens*	Endangered	Temperate forests	Habitat loss, poaching, climate change	Protected areas, habitat conservation, community-based initiatives	Pradhan *et al*. (2001)
Himalayan monal	*Lophophorus impejanus*	Vulnerable	Temperate forests	Habitat loss, illegal hunting	Conservation education, protected areas, habitat restoration	Rai *et al*. (2020)
Western tragopan	*Tragopan melanocephalus*	Endangered	Temperate forests	Habitat loss, hunting	Protected areas, habitat restoration, anti-poaching efforts	Ramesh *et al*. (2009)

systems, hydropower projects and tourism (Lorenz and Dittmer, 2021). Species that depend on a particular ecosystem for survival are particularly vulnerable to the effects of habitat loss (Negi *et al.*, 2017). In addition to limiting the species' movement and interfering with reproduction, the man-made infrastructures fragment the habitats (Ginwal *et al.*, 2020). Soil erosion, sedimentation in bodies of water and hydrological changes are all caused by construction (Chatterjee *et al.*, 2014). Comprehensive environmental impact assessments, regulatory compliance and the implementation of sustainable measures like habitat restoration and conservation are essential for reducing these negative effects (Anand *et al.*, 2012). To achieve sustainable development while protecting regional ecosystems, it is essential to include ecological factors into project design and involve local residents (Wang *et al.*, 2019).

9.2.3 Risk assessment

There are three components to risk assessment: susceptibility, danger and exposure. The assessment involves evaluating potential risks to the environment, biodiversity and human health, and provides a better understanding of the sources of vulnerability. Identifying and assessing risks enables the mobilization of stakeholders to take action, to develop effective strategies, and to raise awareness of the situation (Gioli *et al.*, 2019), and for conservationists to minimize the impact (Prasad *et al.*, 2016). Any vulnerability assessment conducted with the intention of lessening the risk under unknown future conditions must take into account both present and future climate change scenarios (Huggel *et al.*, 2020).

Loss of habitat, pollution, invasive species, climate change and human activities are all dangers to ecosystems and the fauna and flora (Tse-Ring *et al.*, 2010). Finding areas of high ecological value, ranking endangered species, and comprehending the effects of diverse pressures on biodiversity are all part of this evaluation (Foden *et al.*, 2013). Ecosystems, flora, wildlife and natural resources in the Himalayas are in danger due to climate change, as temperatures rise, precipitation patterns shift, and severe weather events occur more frequently. It is important to quantify the risks associated with these changes (Ahsan *et al.*, 2022), as it would enable the formulation of adaptation and conservation plans by highlighting potentially at-risk areas, species and ecosystems (Maikhuri *et al.*, 2017).

Human–wildlife conflict is a serious issue in the IHR, especially in overlapping human and wildlife ecosystems (Sharma *et al.*, 2020). Human and animal dangers may be evaluated, as well as potential conflict hotspots, with the use of risk assessment (Klinke and Renn, 2002). Improved livestock management practices, buffer zones and community-based conservation projects need the information in order to be effective (Jackson and Wangchuk, 2001). Habitat loss, fragmentation, water pollution and soil erosion are all examples of risks that may be assessed in the context of infrastructure projects (Pandit and Grumbine, 2012). This information would be useful for decision-makers in planning strategies to lessen any damage (Dubey and Goyal, 2020), and for setting priorities for conservation efforts, allocating resources and formulating policies that take into account the needs of both the environment and the local community (Maikhuri *et al.*, 2017).

9.3 The Link between People's Livelihood and Ecology

Approximately 12% of the people in the world reside in mountains (Everard *et al.*, 2021), with the percentage reaching 20% if people living on cliff faces are included (Everard *et al.*, 2019). There is a high degree of interdependence between the local people and the ecosystem because of the diversity of natural resources and the relative inaccessibility of mountain regions (Bano *et al.*, 2014). This diversity includes locally adapted sustainable food-producing systems (crops, livestock, fishing, hunting and forest produce) (Bisht *et al.*, 2018). Yet poverty and human vulnerability are worse in mountains than in other world habitats due to the poor production and inability of the residents to adapt to change (Fattorini *et al.*, 2021). Because of their dependence on a subsistence economy, the fragile ecosystem, geographic isolation, lack of access to markets and inputs, low resource productivity, and vulnerability to risks from a wide range of natural hazards and disasters, mountain communities face substantial obstacles to their socio-economic well-being (Pandey *et al.*, 2018).

Farming in the forest is the primary means of survival, income generation and rural economic activity in the middle Himalayas in north-central India and Nepal (Silori, 2007). Traditional natural resources, however, have been undergoing shifts recently as a result of the increase in population, fast urbanization, economic globalization and intensification of land use (Tiwari *et al.*, 2018). These alterations have reduced forest cover and biodiversity, and affected the hydrological regimes of Himalayan watersheds (Xu *et al.*, 2009), which has led to a decrease in the recharge of springs and the availability of drinking water, poor sanitation and reduction in crop production, putting people's health and livelihoods at risk (Everard *et al.*, 2019). These changes have emphasized the inter-connection between forest-based ecosystem services and local livelihood dependency in the Himalayan area (Naudiyal and Schmerbeck, 2017; Everard *et al.*, 2019). The community-based conservation initiatives in the IHR are presented in Fig. 9.4.

Ecosystem services and human lives are intertwined, but problems concerning ecological governance and institutional evaluations have received little attention (Xu *et al.*, 2019). One third of India's rural population relies on non-timber forest products (NTFP) commerce for at least half of their yearly income (Pandey *et al.*, 2016), close to US$ 2.7 billion annually. The poor productivity of the land and the isolation of the villages in the Indian Himalaya force locals to rely heavily on environmental commodities and services for their earnings (Upadhyay, 2005). Traditionally, rural communities in the IHR collected medicinal plants in the woods, which they then sold in lowland regions (Pandey, 2002). *Ophiocordyceps sinensis*, a caterpillar fungus originally from China and Nepal, is collected by local residents and it generates a large portion of the income for many residents (Wang *et al.*, 2020). In practice, the revenue and consumption of commodities from outside the area, especially from metropolitan centres, and intensive agriculture in the lowlands are important in sustaining the Indian villages in the Himalayas (Singh, 1995). This poses serious difficulties in striking a balance between the people's livelihood and a healthy ecosystem (Pierce and Jameton, 2004).

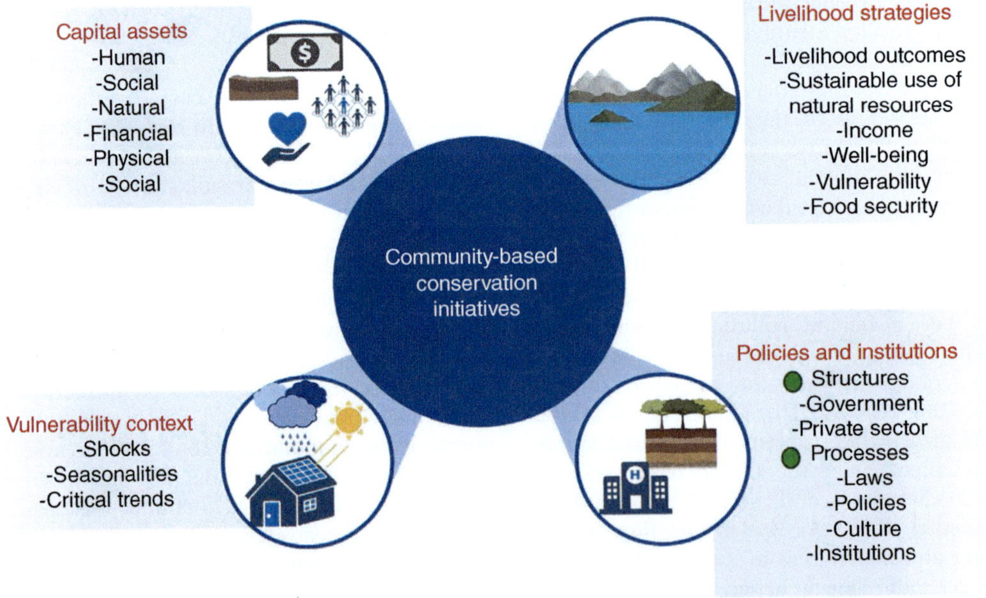

Fig. 9.4. Community-based conservation initiatives in the Indian Himalayan region.

9.4 The Ecological Conservation and Restoration Strategies and Policies of the Government, International Organizations and Local Communities

Land degradation is widespread over much of the Indian subcontinent (Singh *et al.*, 2021). Many animal species, as well as ecosystem services and functions, are at risk because of destruction and fragmentation of the natural habitats brought on by anthropogenic activities (Schaafsma and Bartkowski, 2021). Recent heatwaves, floods, droughts, landslides, and other severe weather occurrences in environmentally vulnerable parts of the nation have been linked to ecosystem deterioration (Rosenzweig *et al.*, 2001; Wheeler, 2013). India's many programmes and missions to revive polluted ecosystems indicate the country's growing commitment to improve the environment (Singh and Tewari, 2022). By 2030, India plans to restore 26 Mha of its degraded lands, an area that includes the 21 Mha it pledged to restore as part of the Bonn Challenge, an international effort to restore 150 Mha of globally degraded land and forests by 2020 and 350 Mha by 2030 (Edrisi *et al.*, 2023). India has proposed in its Nationally Determined Contributions to the Paris Climate Agreement 2015 that it will increase its forest tree cover in urban and rural areas by 2.5–3 billion tonnes of CO_2 equivalent by 2030 through a variety of afforestation and reforestation programmes on lands available through federal and state departments (Amjath-Babu *et al.*, 2019). To better understand the inter-connections among biodiversity, climate change, ecosystem services, health, agriculture and the bioeconomy, the Indian government has recently backed a National Mission on Biodiversity and Human Well-being (Nawn *et al.*, 2020). Knowing the current state of ecological restoration programmes and the practices followed at the local level are crucial for meeting all of these objectives and obligations (Singh and Tewari, 2022; Singh *et al.*, 2022). *Ecological Engineering* journal's timely effort to compile relevant scientific papers and the efforts of ecological restoration practitioners in India during the UN Decade on Ecosystem Restoration (2021–2030) have been beneficial (Abhilash, 2021).

9.4.1 The Annual Spring Revival Workshop

In 2022, the National Institution for Transforming India (NITI) Aayog held a nationwide workshop on spring rejuvenation to further its development efforts in the IHR (Kaur *et al.*, 2022b). Participants included representatives from federal and state governments, as well as academic and non-governmental organizations (Gupta and Koontz, 2019). It enabled people throughout the country to discuss the latest in spring revitalization research, policy and practice. The Indian Himalayan Central University Consortium (IHCUC) initiated research in five distinct areas:

1. estimating the monetary value of women's work in the mountains;
2. agro-ecology in the Himalayan states, with a focus on commercialization;
3. growth of sustainable, low-cost mountain tourism;
4. economic opportunities to prevent people from leaving the hills; and
5. techniques for recycling water.

The Ministry of Education and the University Grants Commission approved the necessary funding for the research. The NRE (Natural Resources and Environment) will assess all topical areas regularly to ensure progress is being made.

9.4.2 Agro-ecology and market growth in the Indian Himalayas: a webinar discussion

A webinar titled Agro-ecology and Market Development in the Indian Himalayan Region was hosted by NITI Aayog and the Indian Himalayan Cooperative Urbanization Consortium (IHCUC) on 25 June 2021. More than 100 people, including Secretaries to the Government of India, Principal Secretaries and Secretaries of the 12 Indian Himalayan states and Union Territories, joined Dr Rajiv Kumar, Vice Chairman NITI Aayog, and Mr Amitabh Kant, CEO NITI Aayog, for this online seminar.

9.4.3 The World Wildlife Fund and the vision for the Himalayas

WWF proposes that biodiversity protection can be improved together with livelihoods and ecosystems if an integrated strategy is employed (Wolmer, 2003). Through the WWF, the government and people of Bhutan, India and Nepal are working closely to restore and safeguard natural systems, lessen the impact of human activities on the environment, and bolster local economies (Conca and Dabelko, 2002). The WWF supports decision-makers and stakeholders from small towns to multi-national corporations to undertake the following game-changing measures in Bhutan, India and Nepal:

1. create an eastern Himalayas conservation and sustainable development strategy;
2. protect the 7 million ha of forest, grassland and wetland in the eastern Himalayas;
3. preserve the natural environments on which endangered animals rely; and
4. protect vital ecosystems so that they can withstand the effects of climate change.

These objectives will enable the eastern Himalayas to fulfil WWF's ultimate vision for the area, that is a mosaic of diverse ecosystems that work together to protect the Himalayas' natural and cultural heritage while also providing livelihoods for the region's inhabitants (Bhusal, 2007).

East Indian Himalayan achievements of WWF

1970s: Project Tiger was launched with the goal of protecting endangered species and creating a reserve for tigers in India (Damodaran, 2007).
2000s: To guarantee the species' continued existence, WWF-India collaborated with the Assam Forest Department and other organizations on the Indian Rhino Vision 2020 programme (Milliken *et al.* 2009). Several protected areas and wildlife in the Indian Himalayan region are presented in detail in Fig. 9.5.

9.5 Suggestions for Future Sustainable Development

The unique, majestic Himalayan region has produced numerous ecosystem products and services in the past and will do so in the future with appropriate planning and management. With environmental degradation there must be attempts at

Fig. 9.5. Indian Himalayan region, several protected areas and wildlife.

restoration. It is much preferable, however, to be able to anticipate the problem and take preventative actions. Improved education, a long-term database, and a comprehensive approach will bring us closer to sustainable development, a better quality of life, an increased standard of living and more protection for the fragile ecosystem. There is also a need to employ methods and means to guarantee that the development does not undermine the traditional lifestyles and social fabric of the region. The Himalayan region should neither be burdened by the past's backwardness, nor be confined by errors made by governments. A movement spreading throughout the Himalayan region is envisioned, bringing in a new era of prosperity and growth in the next century.

References

Aayog, N. (2017) Inventory and revival of springs in the Himalayas for water security. Dept of Science and Technology, Government of India, New Delhi. Available at: https://sdcuk.in/wp-content (accessed 26 September 2023).

Aayog, N. (2018) Sustainable tourism in the Indian Himalayan region. Report of Working Group II, NITI Aayog, Government of India. Available at: https://sdcuk.in/wp-content (accessed 26 September 2023).

Abhilash, P.C. (2021) Restoring the unrestored: strategies for restoring global land during the UN decade on ecosystem restoration (UN-DER). *Land* 10(2), 201. DOI: 10.3390/land10020201

Acharya, A.K. and Kafle, N. (2009) Land degradation issues in Nepal and its management through agroforestry. *Journal of Agriculture and Environment* 10, 133–143. DOI: 10.3126/aej.v10i0.2138

Adger, W.N., Huq, S., Brown, K., Conway, D. and Hulme, M. (2003) Adaptation to climate change in the developing world. *Progress in Development Studies* 3(3), 179–195. DOI: 10.1191/1464993403ps060oa

Ahsan, S., Bhat, M.S., Alam, A., Farooq, H. and Shiekh, H.A. (2022) Evaluating the impact of climate change on extreme temperature and precipitation events over the Kashmir Himalaya. *Climate Dynamics* 58, 1651–1669. DOI: 10.1007/s00382-021-05984-6

Amjath-Babu, T.S., Aggarwal, P.K. and Vermeulen, S. (2019) Climate action for food security in South Asia? Analyzing the role of agriculture in nationally determined contributions to the Paris Agreement. *Climate Policy* 19, 283–298. DOI: 10.1080/14693062.2018.1501329

Anand, A., Chandan, P. and Singh, R.B. (2012) Homestays at Korzok: supplementing rural livelihoods and supporting green tourism in the Indian Himalayas. *Mountain Research and Development* 32(2), 126–136. DOI: 10.1659/MRD-JOURNAL-D-11-00109.1

Aziz, M.A., Lone, S. and Lone, F.A. (2010) An overview of Hangul (*Cervus elaphus hanglu*, Wagner) in Dachigam National Park, Kashmir (India). Forestry Nepal. Available at: https://citeseerx.ist.psu.edu (accessed 26 September 2023).

Bano, A., Ahmad, M., Hadda, T.B., Saboor, A., Sultana, S. *et al.* (2014) Quantitative ethnomedicinal study of plants used in the Skardu valley at high altitude of Karakoram-Himalayan range, Pakistan. *Journal of Ethnobiology and Ethnomedicine* 10, 43. DOI: 10.1186/1746-4269-10-43

Bartarya, S.K. and Valdiya, K.S. (1989) Landslides and erosion in the catchment of the Gaula River, Kumaun Lesser Himalaya, India. *Mountain Research and Development* 9(4), 405–419. DOI: 10.2307/3673588

Bergmann, C. (2016) *The Himalayan Border Region*. Springer, Cham, Switzerland.

Bhalla, R. and Chowdhary, N. (2022) Green workers of Himalayas: evidence of transformation induced regeneration. *Journal of Tourism Futures* 8(3). DOI: 10.1108/JTF-12-2021-0273

Bhandari, R.K. (2006) The Indian landslide scenario, strategic issues and action points. Citeseer, in India disaster management congress, New Delhi, 29–30. Available at: https://citeseerx.ist.psu.edu (accessed 26 September 2023).

Bhattacharyya, R., Ghosh, B.N., Dogra, Mishra, P.K., Santra, P. *et al.* (2016) Soil conservation issues in India. *Sustainability* 8, 565. DOI: 10.3390/su8060565

Bhellum, B.L. (2020) Asteraceae in Jammu and Kashmir Himalaya: a floristic account. In: Dar, G.H. and Khuroo, A.A. (eds) *Biodiversity of the Himalaya: Jammu and Kashmir State.* Springer Singapore, pp. 607–620.

Bhusal, N.P. (2007) Chitwan National Park: a prime destination of eco-tourism in Central Tarai Region, Nepal. *The Third Pole Journal of Geography Education* 5, 70–75. DOI: 10.3126/ttp.v5i0.1956

Bisht, I.S., Mehta, P.S., Negi, K.S., Verma, S.K., *et al.* (2018) Farmers' rights, local food systems, and sustainable household dietary diversification: a case of Uttarakhand Himalaya in north-western India. *Agroecology and Sustainable Food Systems* 42, 77–113. DOI: 10.1080/21683565.2017.1363118

Bizikova, L., Roy, D., Swanson, D., Venema, H.D. and Mccandless, M. (2013) *The Water-Energy-Food Security Nexus: Towards a Practical Planning and Decision-support Framework for Landscape Investment and Risk Management.* IISD report. International Institute for Sustainable Development, Manitoba, Canada.
Blaikie, P.M. (2016) *The Political Economy of Soil Erosion in Developing Countries.* Routledge, London.
Blaikie, P.M. and Muldavin, J.S.S. (2004) Upstream, downstream, China, India: the politics of environment in the Himalayan region. *Annals of the Association of American Geographers* 94, 520–548. DOI: 10.1111/j.1467-8306.2004.00412.x
Chakraborty, S.K. (2021) River pollution and perturbation: perspectives and processes. In: Chakraborty, S.K. (ed.) *Riverine Ecology Volume 2: Biodiversity Conservation, Conflicts and Resolution*. Springer, Cham, Switzerland.
Chandra, S., Singh, A., Mathew, J.R., Singh, C.P., Pandya, M.R. *et al.* (2022) Phenocam observed flowering anomaly of *Rhododendron arboreum* Sm. in Himalaya: a climate change impact perspective. *Environmental Monitoring and Assessment* 194(12), 877. DOI: 10.1007/s10661-022-10466-1
Chatterjee, S., Saikia, A., Dutta, P., Ghosh, D., Pangging, G. *et al.* (2006) Biodiversity significance of north east India. Available at: https://www.researchgate.net/publication/272362960 (accessed 26 September 2023).
Chatterjee, S., Krishna, A.P. and Sharma, A.P. (2014) Geospatial assessment of soil erosion vulnerability at watershed level in some sections of the Upper Subarnarekha river basin, Jharkhand, India. *Environmental Earth Sciences* 71, 357–374. DOI: 10.1007/s12665-013-2439-3
Chen, F., Chen, X., Chen, J., Zhou, A., Wu, D.U.O. *et al.* (2014) Holocene vegetation history, precipitation changes and Indian Summer Monsoon evolution documented from sediments of Xingyun Lake, south-west China. *Journal of Quaternary Science* 29, 661–674. DOI: 10.1002/jqs.2735
Chettri, N., Sharma, E., Shakya, B., Thapa, R., Bajracharya, B. *et al.* (2010) *Biodiversity in the Eastern Himalayas: Status, Trends and Vulnerability to Climate Change.* International Centre for Integrated Mountain Development (ICIMOD), Kathmandu.
Choudhury, A. (2004) Human–elephant conflicts in Northeast India. *Human Dimensions of Wildlife* 9, 261–270. DOI: 10.1080/10871200490505693
Choudhury, A. (2007) The status of endangered species in northeast India. *Bombay Natural History Society* 103(2&3), 157–167.
Choudhury, B.U., Nengzouzam, G., Ansari, M.A. and Islam, A. (2022) Causes and consequences of soil erosion in northeastern Himalaya, India. *Current Science* 122(7), 772–789. DOI: 10.18520/cs/v122/i7/772-789
Conca, K. and Dabelko, G.D. (2002) *Environmental Peacemaking.* Woodrow Wilson Center Press.
Cunfer, G. (2005) *On the Great Plains: Agriculture and Environment.* Texas A&M University Press.
Dahal, R.K. (2016) Lessons learnt from the 2015 Gorkha earthquake in Nepal on the engineering geological perspectives. In: Singh, S.P., Khanal, S. and Joshi, M. (eds) *Lessons from Nepal's Earthquake for the Indian Himalayas and the Gangetic Plains.* Central Himalayan Environment Association (CHEA), India.
Damodaran, A. (2007) The project tiger crisis in India: moving away from the policy and economics of selectivity. *Environmental Values* 16(1), 61–77. DOI: 10.1177/096327190701600105
Dar, G.H. and Khuroo, A.A. (2020) *An Introduction to Biodiversity of the Himalaya: Jammu and Kashmir State.* Springer Singapore.
Dar, S.A., Bhat, S.U. and Rashid, I. (2021) The status of current knowledge, distribution, and conservation challenges of wetland ecosystems in Kashmir Himalaya, India. In: Sharma, S. and Singh, P. (eds) *Wetlands Conservation: Current Challenges and Future Strategies.* John Wiley & Sons, Chichester, UK, pp. 175–200.
de Camino, R.V (2000) *Costa Rica: Forest Strategy and the Evolution of Land Use*. World Bank Publications.
Debnath, J., Sahariah, D., Saikia, A., Meraj, G., Nath, N. *et al.* (2023) Shifting sands: assessing bankline shift using an automated approach in the Jia Bharali river, India. *Land* 12, 703. DOI: 10.3390/land12030703
Dikshit, A., Sarkar, R., Pradhan, B., Segoni, S. and Alamri, A.M. (2020) Rainfall induced landslide studies in Indian Himalayan region: a critical review. *Applied Sciences* 10, 2466. DOI: 10.3390/app10072466
Dubey, S. and Goyal, M.K. (2020) Glacial lake outburst flood hazard, downstream impact, and risk over the Indian Himalayas. *Water Resources Research* 56, e2019WR026533. DOI: 10.1029/2019WR026533
Edrisi, S.A., Bundela, A.K., Verma, V., Dubey, P.K. and Abhilash, P.C. (2023) Assessing the impact of global initiatives on current and future land restoration scenarios in India. *Environmental Research* 216, 114413. DOI: 10.1016/j.envres.2022.114413
Everard, M., Gupta, N., Scott, C.A., Tiwari, P.C., Joshi, B. *et al.* (2019) Assessing livelihood-ecosystem interdependencies and natural resource governance in Indian villages in the Middle Himalayas. *Regional Environmental Change* 19, 165–177. DOI: 10.1007/s10113-018-1391-x

Everard, M., Kataria, G., Kumar, S. and Gupta, N. (2021) Assessing livelihood-ecosystem interdependencies and natural resource governance in a tribally controlled region of India's north-eastern middle Himalayas. *Environment, Development and Sustainability* 23, 7772–7790. DOI: 10.1007/s10668-020-00945-1

Farming, F., Profitable, A. and Farming, D.I. (2014) *Himalayan Ecology*. Available at: https://gbpihedenvis.nic.in (accessed 26 September 2023).

Fattorini, S., Mantoni, C., Di Biase, L. and Pace, L. (2021) Mountain biodiversity and sustainable development. In: Leal Filho, W., Azul, A.M., Brandli, L., Lange Salvia, A. and Wall, T. (eds) *Life on Land*. Encyclopedia of the UN Sustainable Development Goals, Springer, Cham, Switzerland, pp. 640–660.

Fisher, S., Garside, B., Van Epp, M., Dodman, D., D'Errico, S. *et al*. (2016) Planning and implementing climate change responses in the context of uncertainty. Working Paper. IIED.

Foden, W.B., Butchart, S.H.M., Stuart, S.N., Vié, J.-C., Akçakaya, H.R. *et al*. (2013) Identifying the world's most climate change vulnerable species: a systematic trait-based assessment of all birds, amphibians and corals. *PLOS ONE* 8, e65427. DOI: 10.1371/journal.pone.0065427

Ghoshal, A., Bhatnagar, Y.V., Pandav, B., Sharma, K., Mishra, C. *et al*. (2019) Assessing changes in distribution of the endangered snow leopard *Panthera uncia* and its wild prey over 2 decades in the Indian Himalaya through interview-based occupancy surveys. *Oryx* 53, 620–632. DOI: 10.1017/S0030605317001107

Ginwal, H.S., Sharma, R., Chauhan, P., Rai, K.C. and Barthwal, S. (2020) Chloroplast microsatellites reveal genetic diversity and population structure in natural populations of Himalayan cedar (Roxb.) G. Don) in India. *Silvae Genetica* 69, 86–93. DOI: 10.2478/sg-2020-0012

Gioli, G., Thapa, G., Khan, F., Dasgupta, P., Nathan, D. *et al*. (2019) Understanding and tackling poverty and vulnerability in mountain livelihoods in the Hindu Kush Himalaya. In: Wester, P., Mishra, A., Mukherji, A. and Shrestha, A.B. (eds) *The Hindu Kush Himalaya Assessment: Mountains, Climate Change, Sustainability and People*. Springer, pp. 421–455.

Gupta, D. and Koontz, T.M. (2019) Working together? Synergies in government and NGO roles for community forestry in the Indian Himalayas. *World Development* 114, 326–340. DOI: 10.1016/j.worlddev.2018.09.016

Heide, F.Z. (2012) *Feasibility Study for a Lake Tana Biosphere Reserve*. Ethiopia, Deutschland/Bundesamt für Naturschutz.

Hilty, J., Worboys, G.L., Keeley, A., Woodley, S., Lausche, B. *et al*. (2020) *Guidelines for Conserving Connectivity through Ecological Networks and Corridors*. IUCN, Gland, Switzerland.

Huggel, C., Allen, S., von Dach, S.W., Dimri, A.P., Mal, S. *et al*. (2020) An integrative and joint approach to climate impacts, hydrological risks and adaptation in the Indian Himalayan region. In: Dimri, A.P., Bookhagen, B., Stoffel, M. and Yasunari, T. (eds) *Himalayan Weather and Climate and Their Impact on the Environment*. Springer, Cham, Switzerland, pp. 553–573.

Ilyas, O. (2015) Status, habitat use and conservation of Alpine musk deer (*Moschus chrysogaster*) in Uttarakhand Himalayas, India. *Journal of Applied Animal Research* 43, 83–91. DOI: 10.1080/09712119.2014.899495

Ives, J.D. and Messerli, B. (1989) *The Himalayan Dilemma: Reconciling Development and Conservation*. Psychology Press.

Jackson, R. and Wangchuk, R. (2001) Linking snow leopard conservation and people-wildlife conflict resolution: grassroots measures to protect the endangered snow leopard from herder retribution. *Endangered Species Update* 18(4), 138–141.

Jain, C.K. and Singh, S. (2020) Impact of climate change on the hydrological dynamics of River Ganga, India. *Journal of Water and Climate Change* 11, 274–290. DOI: 10.2166/wcc.2018.029

Kala, C.P. (2005) A multifaceted review on the biodiversity conservation of the Valley of Flowers National Park, India. *The International Journal of Biodiversity Science and Management* 1, 25–32. DOI: 10.1080/17451590509618077

Kattel, G.R. (2022) Climate warming in the Himalayas threatens biodiversity, ecosystem functioning and ecosystem services in the 21st century: Is there a better solution? *Biodiversity and Conservation* 31, 2017–2044. DOI: 10.1007/s10531-022-02417-6

Kaur, D., Tiwana, A.S., Kaur, S. and Gupta, S. (2022a) Climate change: concerns and influences on biodiversity of the Indian Himalayas. In: Rani, S. and Kumar, R. (eds) *Climate Change*, Springer International Publishing, 265–281.

Kaur, S., Kaur, D., Tiwana, A.S. and Gupta, S. (2022b) Climate change-related governance and policies in Indian Himalayas. In: Rani, S. and Kumar, R. (eds) *Climate Change*, Springer International Publishing, 309–330.

Klinke, A. and Renn, O. (2002) A new approach to risk evaluation and management: risk-based, precaution-based, and discourse-based strategies 1. *Risk Analysis* 22(6), 1071–1094. DOI: 10.1111/1539-6924.00274

Kotlia, B.S. and Joshi, L.M. (2013) Neotectonic and climatic impressions in the zone of Trans Himadri Fault (THF), Kumaun Tethys Himalaya, India: a case study from palaeolake deposits. *Zeitschrift für Geomorphologie* 57, 289–303. DOI: 10.1127/0372-8854/2013/0057-0101

Kumar, A., Asthana, A.K.L., Priyanka, R.S., Jayangondaperumal, R., Gupta, A.K. *et al.* (2017) Assessment of landslide hazards induced by extreme rainfall event in Jammu and Kashmir Himalaya, northwest India. *Geomorphology* 284, 72–87. DOI: 10.1016/j.geomorph.2017.01.003

Kumar, A., Pinto, M.C., Candeias, C. and Dinis, P.A. (2021) Baseline maps of potentially toxic elements in the soils of Garhwal Himalayas, India: Assessment of their eco-environmental and human health risks. *Land Degradation & Development* 32, 3856–3869. DOI: 10.1002/ldr.3984

Kumar, A., Pramanik, M., Chaudhary, S., Negi, M.S. and Szabo, S. (2022b) Geospatial multi-criteria evaluation to identify groundwater potential in a Himalayan District, Rudraprayag, India. *Environment, Development and Sustainability* 1-42. DOI: 10.1007/s10668-021-02107-3

Kumar, M. (2023) Forestry policies and practices to promote climate change adaptation in the Indian Western Himalayan states. In: Sharma, S., Kuniyal, J.C., Chand, P. and Singh, P. (eds) *Climate Change Adaptation, Risk Management and Sustainable Practices in the Himalaya*. Springer, Cham, Switzerland.

Lakra, W.S., Sarkar, U.K., Dubey, V.K., Sani, R. and Pandey, A. (2011) River inter linking in India: status, issues, prospects and implications on aquatic ecosystems and freshwater fish diversity. *Reviews in Fish Biology and Fisheries* 21, 463–479. DOI: 10.1007/s11160-011-9199-5

Laltlanmawii, R. (2011) An evaluation of use of e-resources in Indian Institute of Technology Library, Guwahati, India. Available at: http://mzuir.inflibnet.ac.in/ (accessed 26 September 2023).

Lorenz, D.F. and Dittmer, C. (2021) Disasters in the 'abode of gods': vulnerabilities and tourism in the Indian Himalaya. *International Journal of Disaster Risk Reduction* 55, 102054. DOI: 10.1016/j.ijdrr.2021.102054

Lynas, M. (2008) *Six Degrees: Our Future on a Hotter Planet*. National Geographic Books.

Lyngdoh, S., Shrotriya, S., Goyal, S.P., Clements, H., Hayward, M.W. *et al.* (2014) Prey preferences of the snow leopard (*Panthera uncia*): regional diet specificity holds global significance for conservation. *PLOS ONE* 9, e88349. DOI: 10.1371/journal.pone.0100071

Maikhuri, R.K., Nautiyal, A., Jha, N.K., Rawat, L.S., Maletha, A. *et al.* (2017) Socio-ecological vulnerability: assessment and coping strategy to environmental disaster in Kedarnath valley, Uttarakhand, Indian Himalayan Region. *International Journal of Disaster Risk Reduction* 25, 111–124. DOI: 10.1016/j.ijdrr.2017.09.002

Maitra, A.K. and Krishan, A. (2005) Agenda 21 and urban India. *Consuming Cities*. Routledge.

Malik, J.N. and Mohanty, C. (2007) Active tectonic influence on the evolution of drainage and landscape: geomorphic signatures from frontal and hinterland areas along the Northwestern Himalaya, India. *Journal of Asian Earth Sciences* 29, 604–618. DOI:10.1016/j.jseaes.2006.03.010

Marsh, J. and Shalvey, J. (2018) When packaged food pollutes our body and Earth: plastic packaged food consumption and waste disposal practices in the Himalayas. Available at: https://digitalcollections.sit.edu/ (accessed 26 September 2023).

Matthew, R. (2012) Environmental change, human security, and regional governance: the case of the Hindu Kush/Himalaya region. *Global Environmental Politics* 12, 100–118. DOI: 10.1162/GLEP_a_00125

McCarthy, T.M. and Chapron, G. (2003) Snow leopard survival strategy. International Snow Leopard Trust and Snow Leopard Network, Seattle, USA, 105. Available at: http://www.catsg.org/fileadmin/filesharing (accessed 26 September 2023).

Meraj, G., Farooq, M., Singh, S.K., Islam, M.N. and Kanga, S. (2022) Modeling the sediment retention and ecosystem provisioning services in the Kashmir valley, India, Western Himalayas. *Modeling Earth Systems and Environment* 8, 3859–3884. DOI: 10.1007/s40808-021-01333-y

Milliken, T., Emslie, R.H. and Talukdar, B. (2009) African and Asian rhinoceroses: status, conservation and trade. Available at: https://citeseerx.ist.psu.edu/ (accessed 26 September 2023).

Mishra, N.B. and Mainali, K.P. (2017) Greening and browning of the Himalaya: spatial patterns and the role of climatic change and human drivers. *Science of the Total Environment* 587, 326–339. DOI: 10.1016/j.scitotenv.2017.02.156

Mishra, P.K., Rai, A., Abdelrahman, K., Rai, S.C. and Tiwari, A. (2022) Land degradation, overland flow, soil erosion, and nutrient loss in the Eastern Himalayas, India. *Land* 11, 179. DOI: 10.3390/land11020179

Nandy, S., Singh, C., Das, K.K., Kingma, N.C. and Kushwaha, S.P.S. (2015) Environmental vulnerability assessment of eco-development zone of Great Himalayan National Park, Himachal Pradesh, India. *Ecological Indicators* 57, 182–195. DOI: 10.1016/j.ecolind.2015.04.024

Narain, V. and Singh, A.K. (2019) Replacement or displacement? Periurbanisation and changing water access in the Kumaon Himalaya, India. *Land Use Policy* 82, 130–137. DOI: 10.1016/j.landusepol.2018.12.004

Nath, A.J., Kumar, R., Devi, N.B., Rocky, P., Giri, K. *et al.* (2021) Agroforestry land suitability analysis in the Eastern Indian Himalayan region. *Environmental Challenges* 4, 100199. DOI: 10.1016/j.envc.2021.100199

National Research, C. (2012) *Himalayan Glaciers: Climate Change, Water Resources, and Water Security*. National Academies Press.

Naudiyal, N. and Schmerbeck, J. (2017) The changing Himalayan landscape: pine-oak forest dynamics and the supply of ecosystem services. *Journal of Forestry Research* 28, 431–443. DOI: 10.1007/s11676-016-0338-7

Nawn, N., Vasan, S. and Bawa, K. (2020) Biodiversity-focused development. *Economic & Political Weekly* 55, 27.

Negi, V.S., Maikhuri, R.K., Pharswan, D., Thakur, S. and Dhyani, P.P. (2017) Climate change impact in the Western Himalaya: people's perception and adaptive strategies. *Journal of Mountain Science* 14, 403–416. DOI: 10.1007/s11629-015-3814-1

Pandey, A.K., Tripathi, Y.C. and Kumar, A. (2016) Non timber forest products (NTFPs) for sustained livelihood: challenges and strategies. *Research Journal of Forestry* 10, 1–7. DOI: 10.3923/rjf.2016

Pandey, D.N. (2002) Traditional knowledge systems for biodiversity conservation. FAO Forestry Paper. Available at: chrome-extension://oemmndcbldboiebfnladdacbdfmadadm/https://www.activeremedy.org/wp-content/uploads/2014/10/pandey_d_n_2004_traditional_knowledge_systems_for_biodiversity_conservation.pdf (accessed 11 February 2024).

Pandey, R. and Jha, S. (2012) Climate vulnerability index-measure of climate change vulnerability to communities: a case of rural Lower Himalaya, India. *Mitigation and Adaptation Strategies for Global Change* 17, 487–506. DOI: 10.1007/s11027-011-9338-2

Pandey, R., Kumar, P., Archie, K.M., Gupta, A.K., Joshi, P.K. *et al.* (2018) Climate change adaptation in the western-Himalayas: household level perspectives on impacts and barriers. *Ecological Indicators* 84, 27–37. DOI: 10.1016/j.ecolind.2017.08.021

Pandit, M.K. and Grumbine, R.E. (2012) Potential effects of ongoing and proposed hydropower development on terrestrial biological diversity in the Indian Himalaya. *Conservation Biology* 26, 1061–1071. DOI: 10.1111/j.1523-1739.2012.01918

Pandit, M.K., Kumar, V. (2013) Land-use change and conservation challenges in the Indian Himalaya: past, present, and future. *Conservation Biology: Voices from the Tropics* 121–133. DOI: 10.1002/9781118679838.ch15

Paudel, D. and Billon, P.L. (2020) Geo-logics of power: disaster capitalism, Himalayan materialities, and the geopolitical economy of reconstruction in post-earthquake Nepal. *Geopolitics* 25, 838–866. DOI: 10.1080/14650045.2018.1533818

Pierce, J. and Jameton, A. (2004) *The Ethics of Environmentally Responsible Health Care*. Oxford University Press, Oxford, UK.

Pradhan, S., Saha, G.K. and Khan, J.A. (2001) Ecology of the red panda *Ailurus fulgens* in the Singhalila National Park, Darjeeling, India. *Biological Conservation* 98, 11–18. DOI: 10.1016/S0006-3207(00)00079-3

Prasad, A.S., Pandey, B.W., Leimgruber, W. and Kunwar, R.M. (2016) Mountain hazard susceptibility and livelihood security in the upper catchment area of the river Beas, Kullu Valley, Himachal Pradesh, India. *Geoenvironmental Disasters* 3, 1–17. DOI: 10.1186/s40677-016-0037-x

Prasad, A.S., Singh, A., Bandooni, S.K. and Negi, V.S. (2021) Water resource management through ecological restoration in Garhwal Himalaya, Uttarakhand, India. In: Pandey, B.W. and Anand, S. (eds) *Water Science and Sustainability.* Springer, Cham, Switzerland, pp. 157–167.

Rai, R., Paudel, B., Changjun, G. and Khanal, N.R. (2020) Change in the distribution of national bird (Himalayan monal) habitat in Gandaki River Basin, central Himalayas. *Journal of Resources and Ecology* 11, 223–231. DOI: 10.5814/j.issn.1674-764x.2020.02.010

Ramesh, K., Sathyakumar, S. and Rawat, G.S. (2009) Methods of capture and radio tracking of Western Tragopan *Tragopan melanocephalus* JE Gray 1829 in the Great Himalayan National Park, India. *Journal of the Bombay Natural History Society* 105(2), 127–132.

Rana, J.C. and Bisht, I.S. (2023) Reviving smallholder hill farming by involving rural youth in food system transformation and promoting community-based agri-ecotourism: a case of Uttarakhand State in North-western India. *Sustainability* 15, 8816. DOI: 10.3390/su15118816

Rasul, G. (2014) Food, water, and energy security in South Asia: a nexus perspective from the Hindu Kush Himalayan region. *Environmental Science & Policy* 39, 35–48. DOI: 10.1016/j.envsci.2014.01.010

Rawal, R., Negi, V.S. and Bhatt, I.D. (2021) Changing outlook on harnessing biodiversity values: a special focus on Indian Himalaya. *Journal of Graphic Era University* 55–82. DOI: 10.13052/jgeu0975-1416.914

Rosenzweig, C., Iglesius, A., Yang, X.B., Epstein, P.R. and Chivian, E. (2001) Climate change and extreme weather events: implications for food production, plant diseases, and pests. *Global Change and Human Health* 2(2), 90–104. DOI: 10.1023/A:1015086831467

Roy, A.B. and Jakhar, S.R. (2002) *Geology of Rajasthan (Northwest India) Precambrian to Recent.* Scientific Publishers (India), Jodhpur, India.

Ruchi, K.C. and Fatima, B. (2021) Emergence of earthquake resistant buildings: review of earthquake resistant tall buildings of India. *Indian Journal of Forensic Medicine and Pathology* 14, 386–394. DOI: 10.21088/ijfmp.0974.3383.14221.54

Sahu, N., Sayama, T., Saini, A., Panda, A. and Takara, K. (2020) Understanding the hydropower and potential climate change impact on the Himalayan river regimes: a study of local perceptions and responses from Himachal Pradesh, India. *Water* 12, 2739. DOI: 10.3390/w12102739

Saikawa, E., Panday, A., Kang, S., Gautam, R., Zusman, E. *et al.* (2019) Air pollution in the Hindu Kush Himalaya. In: Wester, P., Mishra, A., Mukherji, A. and Shrestha, A.B. (eds) *The Hindu Kush Himalaya Assessment: Mountains, Climate Change, Sustainability and People.* Springer, pp. 339–387.

Sati, V.P. (2015) Landscape vulnerability and rehabilitation issues: a study of hydropower projects in Garhwal region, Himalaya. *Natural Hazards* 75, 2265–2278. DOI: 10.1007/s11069-014-1430-y

Schaafsma, M. and Bartkowski, B. (2021) Synergies and trade-offs between ecosystem services. In: Leal Filho, W., Azul, A.M., Brandli, L., Lange Salvia, A. and Wall, T. (eds) *Life on Land. Encyclopedia of the UN Sustainable Development Goals*. Springer, Cham, Switzerland, pp. 1022–1032.

Sharma, A. and Goyal, M.K. (2020) Assessment of the changes in precipitation and temperature in Teesta River basin in Indian Himalayan Region under climate change. *Atmospheric Research* 231, 104670. DOI: 10.1016/j.atmosres.2019.104670

Sharma, M., Areendran, G., Raj, K., Sharma, A. and Joshi, P.K. (2016) Multitemporal analysis of forest fragmentation in Hindu Kush Himalaya: a case study from Khangchendzonga Biosphere Reserve, Sikkim, India. *Environmental Monitoring and Assessment* 188, 596. DOI: 10.1007/s10661-016-5577-8

Sharma, P. and Batish, D.R. (2022) Reasons of biodiversity loss in India. In: Kaur, S., Batish, D.R., Singh, H.P. and Kohli, R. (eds) *Biodiversity in India: Status, Issues and Challenges.* Springer.

Sharma, P., Chettri, N., Uddin, K., Wangchuk, K., Joshi, R. *et al.* (2020) Mapping human–wildlife conflict hotspots in a transboundary landscape, Eastern Himalaya. *Global Ecology and Conservation* 24, e01284. DOI: 10.1016/j.gecco.2020.e01284

Shiva, V. and Bandyopadhyay, J. (1986) The evolution, structure, and impact of the Chipko movement. *Mountain Research and Development* 6(2), 133–142. DOI: 10.2307/3673267

Shrestha, A.B. and Aryal, R. (2011) Climate change in Nepal and its impact on Himalayan glaciers. *Regional Environmental Change* 11, 65–77. DOI: 10.1007/s10113-010-0174-9

Silori, C.S. (2007) Perception of local people towards conservation of forest resources in Nanda Devi Biosphere Reserve, north-western Himalaya, India. *Biodiversity and Conservation* 16, 211–222. DOI: 10.1007/s10531-006-9116-8

Singh, K. and Tewari, S.K. (2022) Does the road to land degradation neutrality in India is paved with restoration science? *Restoration Ecology* 30, e13585. DOI: 10.1111/rec.135

Singh, K., Byun, C. and Bux, F. (2022) Ecological restoration of degraded ecosystems in India: science and practices. *Ecological Engineering* 182. DOI: 10.1016/j.ecoleng.2022.106708

Singh, R.K., Singh, A., Ksherchokpa, L., Rallen, O., Taniang, B. *et al.* (2021) Grassroots approaches for sustaining biocultural diversity and livelihood security: insights from Indian eastern Himalaya. *Environmental Management* 68, 17–37. DOI: 10.1007/s00267-021-01462-1

Singh, S. (1995) *Urbanization in Garhwal Himalaya: A Geographical Interpretation.* MD Publications Pvt. Ltd.

Srivathsa, A., Vasudev, D., Nair, T., Chakrabarti, S., Chanchani, P. *et al.* (2023) Prioritizing India's landscapes for biodiversity, ecosystem services and human well-being. *Nature Sustainability* 1–10. DOI: 10.1038/s41893-023-01063-2

Sur, U., Singh, P., Rai, P.K. and Thakur, J.K. (2021) Landslide probability mapping by considering fuzzy numerical risk factor (FNRF) and landscape change for road corridor of Uttarakhand, India. *Environment, Development and Sustainability* 23, 13526–13554. DOI: 10.1007/s10668-021-01226-1

Telwala, Y., Brook, B.W., Manish, K. and Pandit, M.K. (2013) Climate-induced elevational range shifts and increase in plant species richness in a Himalayan biodiversity epicentre. *PLOS ONE* 8, e57103. DOI: 10.1371/journal.pone.0057103

Thakur, A., Kumari, S., Borker, S.S., Prashant, S.P., Kumar, A. *et al.* (2021) Solid waste management in Indian Himalayan region: current scenario, resource recovery, and way forward for sustainable development. *Frontiers in Energy Research* 9, 609229. DOI: 10.3389/fenrg.2021.609229

Thompson, M. and Warburton, M. (1985) Uncertainty on a Himalayan scale. *Mountain Research and Development* 115–135. DOI: 10.2307/3673250

Tiwari, P.C. (2008) Land use changes in Himalaya and their impacts on environment, society and economy: a study of the Lake Region in Kumaon Himalaya, India. *Advances in Atmospheric Sciences* 25, 1029–1042.

Tse-Ring, K., Sharma, E., Chettri, N. and Shrestha, A.B. (2010) Climate change vulnerability of mountain ecosystems in the Eastern Himalayas. International Centre for Integrated Mountain Development (ICIMOD). Available at: https://www.preventionweb.net/ (accessed 26 September 2023).

Upadhyay, B. (2005) Women and natural resource management: illustrations from India and Nepal. *Wiley Online Library* 224–232. DOI: 10.1111/j.1477-8947.2005.00132.x

Valdiya, K.S. (2002) Emergence and evolution of Himalaya: reconstructing history in the light of recent studies. *Progress in Physical Geography* 26, 360–399. DOI: 10.1191/0309133302pp342ra

Wang, Y., Wu, N., Kunze, C., Long, R. and Perlik, M. (2019) Drivers of change to mountain sustainability in the Hindu Kush Himalaya. In: Wester, P., Mishra, A., Mukherji, A. and Shrestha, A.B. (eds) *The Hindu Kush Himalaya Assessment: Mountains, Climate Change, Sustainability and People.* Springer, pp. 17–56.

Wang, Z., Yang, D., Janak, R. and Pierce, N.E. (2020) Report on the emergence time of a species of *Thitarodes* ghost moth (Lepidoptera: Hepialidae), host of the caterpillar fungus *Ophiocordyceps sinensis* (Ascomycota: Ophiocordycipitaceae) in Uttarakhand, India. *Journal of Economic Entomology* 113, 2031–2034. DOI: 10.1093/jee/toaa096

Wheeler, S.M. (2013) *Planning for Sustainability:Creating Livable, Equitable and Ecological Communities.* Routledge, London.

Wolmer, W. (2003) Transboundary conservation: the politics of ecological integrity in the Great Limpopo Transfrontier Park. *Journal of Southern African Studies* 29, 261–278. DOI: 10.1080/0305707032000060449

Xu, J., Grumbine, R.E., Shrestha, A., Eriksson, M., Yang, X. *et al.* (2009) The melting Himalayas: cascading effects of climate change on water, biodiversity, and livelihoods. *Conservation Biology* 23, 520–530. DOI: 10.1111/j.1523-1739.2009.01237.x

Xu, J., Badola, R., Chettri, N., Chaudhary, R.P., Zomer, R. *et al.* (2019) Sustaining biodiversity and ecosystem services in the Hindu Kush Himalaya. In: Wester, P., Mishra, A., Mukherji, A. and Shrestha, A.B. (eds) *The Hindu Kush Himalaya Assessment: Mountains, Climate Change, Sustainability and People.* Springer, pp. 127–165.

Yu, G.-A., Yao, W., Huang, H.Q. and Liu, Z. (2021) Debris flows originating in the mountain cryosphere under a changing climate: a review. *Progress in Physical Geography: Earth and Environment* 45, 339–374. DOI: 10.1177/0309133320961705

Zahoor, B., Liu, X., Kumar, L., Dai, Y., Tripathy, B.R. *et al.* (2021) Projected shifts in the distribution range of Asiatic black bear (*Ursus thibetanus*) in the Hindu Kush Himalaya due to climate change. *Ecological Informatics* 63, 101312. DOI: 10.1016/j.ecoinf.2021.101312

10 Ecological Protection and Restoration in Myanmar

Qinghui Fang[1,*], Allan Degen[2] and Zhanhuan Shang[1]
[1]*College of Ecology, Lanzhou University, Lanzhou, China;* [2]*Institutes for Desert Research, Ben-Gurion University of the Negev, Israel*

Abstract
Myanmar is rich in natural resources and its ecosystem services provide livelihoods for the local population. However, in recent years, natural resources have been reduced and the ecosystem services have gradually deteriorated, which has impacted the people's livelihoods and the country's economy. This chapter describes the causes of ecological degradation in Myanmar, the relationship between ecosystem services and people's livelihoods, and the relevant ecological protection and restoration strategies, and puts forward some suggestions for reducing the ecological problems in Myanmar.

Key words: biodiversity, ecological protection and restoration, deforestation, land development

10.1 Introduction

Myanmar is the second largest country in southeast Asia with a land area of approximately 676,600 km^2 and a population of about 53.7 million people (The Word Bank, 2019). Elevations range between sea level and more than 5850 m, and intersect with 19 ecological regions worldwide (Dinerstein *et al.*, 2017). Myanmar is known for its species diversity, which includes more than 16,000 plant, 314 mammal, 1131 bird, 293 reptile and 139 amphibian species (Francis, 2019; Middleton *et al.*, 2019). It has large tracts of forests structured by precipitation and temperature gradients that extend across tropical evergreen lowlands and dry subtropical rain shadows to temperate mountain slopes in the eastern Himalayas. The vast central flood plain associated with the Irrawaddy river supports monsoon wetlands, dry forests and savannas, while coastal areas are surrounded by mangroves, tidal mudflats, sandy beaches and other coastal ecosystems. Myanmar has biogeographical links with the Sundaic region, east and south Asia, and the Himalayan region (Ashton and Seidler, 2014), and has nearly 6300 km of coastline bordering the Bay of Bengal and the Andaman Sea (Fig. 10.1).

Myanmar's ecosystem services provide food, water and other basic necessities, many of which have deep cultural and religious importance to millions of people (Barrow, 2019). In recent years, however, Myanmar's ecology has experienced serious degradation (Aung, 2007), but Myanmar's unique socio-political history allows a larger part of its ecology to persist than its neighbours (de Alban *et al.*, 2020). The few researchers who described the extent of environmental change in Myanmar have generally focused on net loss of

*Corresponding author: 120220907251@lzu.edu.cn

DOI: 10.1079/9781800622579.0010

forest area, without identifying risks of ecosystem loss or prioritizing conservation (Webb *et al.*, 2014). Due to decades of inaccessibility, new animal species, such as the recently discovered 12 gecko species in the Salween basin and mountain karst ecosystems, are often identified (Grismer *et al.*, 2017). Myanmar has become a focus of research and conservation aimed at curbing the loss and degradation of ecosystems and species.

In recent years, the creation of natural-history collections has increased (Ito and Barfod, 2014), and remote-sensing studies of forests have outlined changes in a few ecosystem types, such as broadly defined tropical forests and mangroves (Songer *et al.*, 2009). Local descriptive studies of ecosystems have also emerged (Khaing *et al.*, 2019). However, the primary source of Myanmar's national ecosystem inventory is more than 90 years old, and has since been updated with little new content (Turner *et al.*, 2003). The lack of a systematic inventory system in Myanmar impedes the identification of ecosystems that are being lost and degraded, and

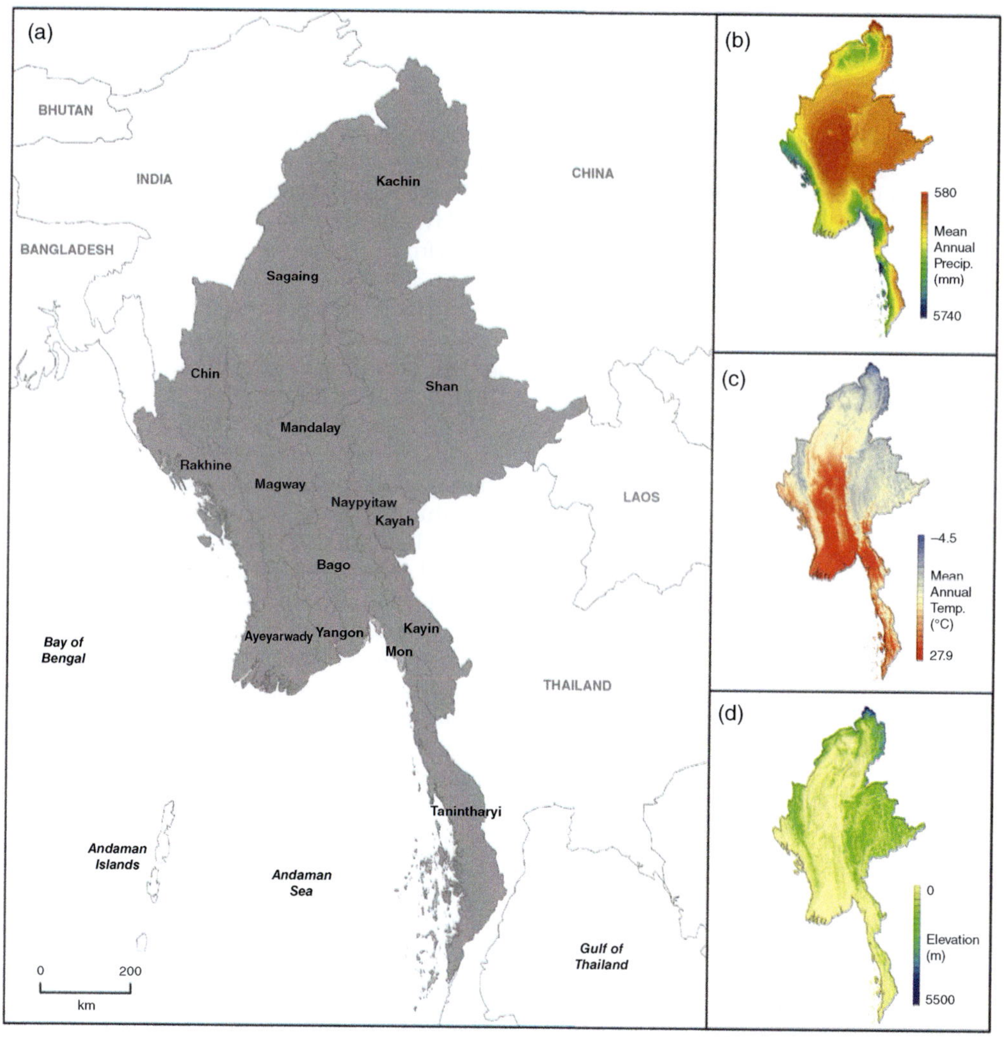

Fig. 10.1. Myanmar and the main climate and topographical drivers of ecological diversity. The panels show the distribution of (a) the states of Myanmar; (b) mean annual precipitation; (c) mean annual temperature; and (d) elevation. (From Murray *et al.*, 2020, used with permission from Elsevier)

also limits the prioritization and co-ordination of conservation actions. Currently, the list of threatened megafauna species remains a major driver of conservation decisions (Ministry of Environmental Conservation and Forestry, 2014); however, these lists do not represent the full range of biodiversity, ecosystem functions and services that need to be protected on a national scale (Keith *et al.*, 2015). Therefore, systematic analysis of the drivers of ecological degradation, assessment of the risks posed by ecological degradation, and understanding the impact of Myanmar's ecology on people's lives are essential to identify broader conservation priorities, establish protected areas, and develop ecological management strategies for sustainable development.

This chapter analyzes the causes of ecological degradation, the relationship between ecosystem services and people's livelihoods, and ecological protection and restoration strategies, and puts forward some feasible suggestions to provide references for ecological protection and restoration in Myanmar in the future.

10.2 Causes of Ecological Degradation and Risk Assessment in Myanmar

Ecosystem degradation is one of the important factors of biodiversity loss. Degradation is generally defined broadly as deviation from the natural variability range of an ecosystem, and can have important adverse effects on the ecosystem's ability to maintain services such as carbon storage, resource supply, water purification, and regulation (Yasuhara *et al.*, 2012). Ecological degradation in Myanmar is mainly due to natural events and anthropogenic activities, such as deforestation, poaching, land development, urbanization and exploitation of mineral resources (Khaing *et al.*, 2019).

10.2.1 Deforestation and poaching

Myanmar is one of the most heavily forested countries in the world. For centuries, it has been known for its forests, ranging from tropical rainforests in the Irrawaddy delta region in the south to alpine forests in the Himalayas in the north. Myanmar's vast forests contribute not only to global carbon sequestration but also to biodiversity conservation (Wang and Mynit, 2016). However, wood has a high market value, so people use illegal means to cut down trees. The problem is exacerbated by the relative ease with which illegal logging and poaching can be carried out in Myanmar due to weak regulations and enforcement.

It is estimated that when Burma declared independence in 1948, forests covered approximately 70% of the territory (Fraser, 2017). Demographic, economic and social changes in the country over the past few decades have exerted considerable pressure on forests and forest resources. A study of changing forest cover patterns using landsat imagery reported that the country's forest cover decreased by approximately 12,000 km^2 over a ten-year period (1990–2000), with an annual deforestation rate of 0.3% (Wang and Mynit, 2016). Forest cover in Myanmar declined from 56% in 1990 to 52% in 2000 (Htun, 2009), and then to 43% in 2015 (FAO, 2015).

Deforestation not only disrupts the balance of ecosystems, but it also causes problems such as soil erosion, loss of soil nutrients, climate change, and biodiversity loss (Ito and Barfod, 2014). Extensive deforestation exposes soil to sunlight, reducing its ability to retain water and causing soil erosion, which, over time, can lead to natural disasters such as river siltation and mudslides (Brandt *et al.*, 2018). With the substantial decrease of clay content, the soil structure of deforested bare land differs from that of natural forest, as bulk density and porosity measurements revealed structural degradation of soils after deforestation. Soil hydrology and the microbial respiration quotient (qCO2) are also adversely affected by deforestation, while soil organic carbon (C), total nitrogen (N), microbial biomass C, N and microfungal biomass all decrease.

The albedo effect cools temperatures due to deforestation, but this cooling effect is much smaller than the warming effect due to greenhouse gas (GHG) emissions (Andréasson *et al.*, 2004). According to the United Nations Intergovernmental Panel on Climate Change (IPCC), deforestation and related activities have become the second largest global source of GHG emissions, accounting for approximately 17–24% of the total (Albanito *et al.*, 2016).

Myanmar is experiencing rapid population growth and strong economic demand. Its population grew from 44.7 million in 1995–1996 (Forest Department, Ministry of Forestry of Myanmar, 1997) to 51.1 million in 2014 (RECOFTC, 2020), which led to overuse of ecosystem services and ecosystem degradation, including reduced carbon stocks (Foley *et al.*, 2011). As a result, deforestation in Myanmar has gradually intensified, and the problem of forest degradation has become urgent (Karuppusamy *et al.*, 2021). The estimated forest area decreased from 71,240 ha to 40,891 ha, while the degraded forest area increased from 88,797 ha to 97,013 ha from 2000 to 2017 (Kyaw *et al.*, 2020). The total forest carbon released was 18.5 million tons, with an annual increase rate of 0.45% (Wang *et al.*, 2019).

Forests serve as habitats for many plant and animal species in Myanmar. Deforestation leads to the loss of their habitats and irreversible damage to the ecological environment (Aukema *et al.*, 2017). These changes can affect the composition, diversity and survival of species, and can lead to their migration and even extinction.

10.2.2 Land development and urbanization

Land development and urbanization have had many negative impacts on ecosystem services in Myanmar. Large-scale timber harvesting, agricultural expansion and urbanization have led to the reduction and fragmentation of forest areas, undermining the integrity and function of forest ecosystems and causing serious damage to ecosystem services, biodiversity and carbon storage (Albanito *et al.*, 2016; Molotoks *et al.*, 2018). Furthermore, urban expansion and infrastructure construction have resulted in large areas of soil being exposed to the external environment and vulnerable to wind and water erosion and loss of soil-water content (Brandt *et al.*, 2018).

Wetlands are an important ecosystem in Myanmar, but due to the pressure of land development and urbanization, many wetlands are being reduced in area, leading to the deterioration of water quality, reduction of water resources and loss of aquatic habitats (Aung *et al.*, 2021). In addition, with land development and urbanization, the demand for water increases, leading to over-exploitation and pollution of water resources, which threaten the health of aquatic populations and ecosystems.

Myanmar's urbanization has been increasing rapidly over the past few decades, from approximately 21.8% in 1990 to 33.0% in 2020 (Central Statistical Organization: http://www.csostat.gov.mm/). The rapid urbanization and infrastructure development has led to the reduction of ecosystem systems and functions (Matelski and Sabrié, 2019), and the fragmentation destruction of wildlife habitats. Urbanization increases the demand for water, which leads to over-exploitation and pollution of water resources, with negative impacts on surrounding water ecosystems and aquatic life.

Soil degradation is one of the main factors causing ecological deterioration. Soil pollution due to industry, waste accumulation, excess pesticides, acidification of air pollutants, and over-fertilization are some of the factors that lead to soil degradation. Essumang *et al.* (2013) reported that, if applied properly, pesticides do not harm the environment and human health. However, pesticides possess long half-lives, and, if applied continuously to control insects, weeds or diseases, residual amounts remain and land degradation can occur (Joko *et al.*, 2017). Furthermore, pesticides applied to soil kill the native microbial communities needed for soil fertility and pesticide degradation itself (Lo, 2010). With its large scale and sensitivity to weather parameters, Myanmar's agriculture sector and economy are the most vulnerable to the effects of soil degradation.

10.2.3 Exploitation of mineral resources

Myanmar is a geologically diverse country with a variety of mineral resources such as gold, silver, copper, tin, tungsten, zinc, jade and gemstones (Gardiner *et al.*, 2014). The country has a long history of mining, but has only recently emerged from decades of military leadership, during which the country's mineral deposits were rarely explored and the mining industry was under-developed (Htun, 2014). Since 2010, Myanmar's political reforms have led to increased international engagement and the easing of long-standing economic sanctions. This process has led to a rapid increase in international trade and investment in the country. Foreign investments in Myanmar are concentrated mainly in natural resource-based industries, with mining being one of the major sectors. However, due to the increasing demand for coal, coal mining in many areas has caused serious damage to the ecological environment (Linn, 2015).

New mining legislation passed in 2015 aimed to encourage foreign investments in Myanmar's mining sector by streamlining the process for further expanding mineral extraction. The law also requires environmental impact assessments for all projects that may cause environmental or social harm (Htun, 2014). Currently, the scale of legal and illegal mining in Myanmar ranges from small-scale artisanal to large open-pit mining (Osawa and Hatsukawa, 2015). Negative impacts of mining include the release of toxic pollutants into the environment, human displacement and public health deterioration (Kristensen *et al.*, 2014). However, Myanmar's records on mining are largely non-digital, do not include illegal mining activities, and are not open to the public. In many parts of the country, natural-resource management remains in conflict between the central government and non-state armed groups (Linn, 2015). As the government, civil society organizations and local communities work to better understand and manage mineral mining in Myanmar, obtaining objective information on the scope and extent of mining activities will be critical.

Mercury (Hg) is listed by the World Health Organization (WHO) as one of the ten most harmful metals, and is considered a public health concern (McNutt, 2013). All common forms, including elemental (metallic), inorganic and organic, are highly toxic. In particular, methyl mercury (MeHg) is the most dangerous form because it can bio-accumulate in microorganisms and bio-magnify or raise nutrient levels in aquatic food webs (Moreno-Brush *et al.*, 2020). Elemental mercury can be converted to methyl mercury in aquatic sediments, and its use in the artisanal and small-scale gold mining (ASGM) sector can be dangerous, as inhalation of mercury vapour can easily penetrate the blood-brain barrier and induce neurotoxicity (Harada, 1995). A survey on atmospheric mercury around ASGM in Myanmar found that the concentration of gaseous mercury, which was measured for the first time, was distributed in strips. The highest concentration, 10,900 ng/m^3, was more than ten times greater than the WHO guideline of 1000 ng/m^3, and was observed near the burning process of ASGM sites (http://www.euro.who.int/__data/assets/ pdf_file/0004/123079/AQG2nd-Ed_6_9Mercury.PDF [accessed 12 February 2024]). Myanmar gold is mined 20–30 m below the surface after digging a vertical tunnel. Large amounts of groundwater emerge from the tunnels and are pumped into the homes of residents in villages near ASGM for daily use, which results in large amounts of mercury in Myanmar's bodies of water.

10.3 The Link between People's Livelihoods and Ecology

10.3.1 The people of Myanmar depend on natural resources for their livelihoods

Myanmar is largely dependent on agriculture, accounting for 37.8% of its gross domestic product (GDP), 25–30% of its total export earnings, and 70% of its employment (FAO, 2014). Ecosystems provide natural resources and ecological services for agriculture, and play a key role in supporting soil fertility, water resources, climate regulation and biodiversity (Emerton and Aung, 2013). In mountainous areas, the slopes are steep, and soil erosion is a serious problem. Slope-management measures, such as terraced farming and the construction of ditches and protective nets, slow water flow and reduce soil erosion. In addition, improving the structure and quality of soil is also key to soil and water conservation. By adding organic matter and maintaining soil cover, water retention and erosion resistance can be improved, and soil erosion can be reduced (Mendelsohn, 2009). Rational and proper land use and ecosystem protection can reduce the occurrence of extreme weather events and the risk of crop damage. For example, forests and wetlands can absorb and store large amounts of water, mitigating the damage of floods and reducing disaster risk (Brandt *et al.*, 2018).

Biodiversity is the genetic resource bank of crops, providing important genetic improvements for agricultural production. Biodiversity also maintains the stability of ecosystems and the ability to resist pests and diseases, reducing losses in agricultural production (Essumang *et al.*, 2013). Protecting and restoring ecosystems, using natural resources rationally, and preserving biodiversity are key measures in ensuring sustainable agricultural development and food security. The government and farmers should strengthen ecological and environmental protection, adopt sustainable agricultural practices, and promote the co-ordinated development of agricultural production and ecological protection.

Myanmar's water resources are critical to people's livelihoods and economic development. People depend on water for irrigation, drinking and fishing. However, poor water management and pollution threaten people's livelihoods and sustainable development. According to the 2016 statistics from the Environment Agency (http://www.mee.gov.mm/), 34.5% of people in Myanmar treated their drinking water at home: 76.2% of these people passed the water through cloth, while 1.4% boiled and 0.6% filtered the water. With the growing industrial development in Myanmar, some enterprises fail to treat sewage properly, resulting in direct discharge of industrial waste into rivers and lakes (Osawa and Hatsukawa, 2015). These contaminants include toxic chemicals, heavy metals and organic material. In addition, fertilizers and pesticides applied on farmland are washed by rainwater into rivers and groundwater, polluting the water. The lack of waste treatment facilities and sewer systems in many places leads to the direct flow of waste into rivers, damaging the ecological environment (Joko *et al.*, 2017).

10.3.2 Importance of ecosystem services to people's livelihoods

Ecotourism goes beyond the popular concept of 'overlap between nature tourism and sustainable tourism' to cover the social dimensions of production and environmental protection. Ecotourism not only creates a range of activities that attract visitors but also offers visitors the opportunity to interact with nature, thus making it possible to preserve or enhance the special qualities of the site and the flora and fauna, while enabling local residents and visitors to enjoy these qualities (Khanal and Babar, 2007). Myanmar, with its rich and diverse natural, cultural and historic assets, is laying the foundation for decades of tourism growth (Aung, 2018). By incorporating local traditional culture and customs into tourism activities, the preservation and inheritance of culture can be promoted. Local residents can display their traditional arts, handicrafts and folk activities, while earning a stable income. Local people are needed to work in tourism-related jobs such as tour guides, accommodation and transportation. To meet the needs of tourists, ecotourism requires the improvement and maintenance of the infrastructure, such as roads, accommodations and restaurants, which provide employment for the residents. In addition, to attract tourists, local residents need to protect and preserve natural resources and minimize environmental damage and pollution, which helps protect the health of local ecosystems (Cutler, 2011).

10.4 Ecological Conservation and Restoration Strategies

Myanmar is committed to protecting and managing forest resources and promoting sustainable forestry management by creating laws, regulations and policies to combat illegal logging and timber trade. For managing forest resources, Myanmar enacted multi-level forest-planning policies. The National Forest Management Plan (NFMP) was initiated in 2001 for a 30-year period (2001–2030), and proposes a series of measures such as forestry development through afforestation and reforestation; effective management of watershed management to extend dam life and increase water storage; strengthening of the legal system against illegal logging and harvesting forest products; increasing the yield of teak and hardwood; broadening forestry scientific research; encouraging the use of other fuels, such as natural gas, instead of wood and charcoal; exporting high-value wood forest and non-wood forest products; and developing ecological tourism. In addition, a master plan for afforestation in dry areas and a ten-year Regional Forest Management Plan (1996–2005) involved 61 administrative regions, each with specific management objectives such as production, afforestation, watershed protection, community forests, non-timber forest products and nature reserves (Webb *et al.*, 2014).

Since the 1990s, Myanmar has passed a number of legal policies on forest protection. The Forest Law of the Union of Myanmar (1992) was the main one, and was supplemented by Wildlife, Plants and Nature Reserves Act (1994), Myanmar Forestry Policy (1995), Union of Myanmar Forestry Charter (1995), National Forest Harvesting Regulations, Myanmar Community Forestry Regulations (1995), Guidelines for Rural Forestry (1996), Criteria and Indicators for Sustainable Forestry Management (1999), and national environmental law framework (http://www.forestdept.gov.mm/). Agenda 21 and Myanmar Environmental Policy were enacted to develop mangrove ecosystems in severely damaged coastal areas and to protect the ecological balance of coastal mangroves.

In addition, the Myanmar government established nature reserves to safeguard the country's biodiversity and protect and manage rare species and ecosystems. These include Hkakaborazi National Park, Nat Ma Taung National Park, Inle Lake Wildlife Sanctuary and Indawgyi Lake Wildlife Sanctuary, Alaungdaw Kathapa National Park, Lenya National Park, Lampi Island Marine National Park, Chatthin Wildlife Sanctuary, Popa Mountain Park and Meinmahla Kyun Wildlife Sanctuary (http://www.wildlifeasia.org/). The Wildlife Protection Law and the Wildlife Trade Law were passed to prohibit illegal hunting, trade and transportation of rare species. In 1994, Government Circular No. 26/94 was issued to strengthen the protection and sustainable use of water resources in Myanmar. International organizations such as the United Nations Environment Programme (UNEP) and the World Bank provide financial and technical support to Myanmar for ecological conservation and restoration projects. Adequate financial assistance was provided by international organizations for the establishment and management of nature reserves, the implementation of ecosystem restoration and wildlife protection and monitoring; and technical instruction and training in ecosystem management, wildlife conservation and sustainable forestry and agricultural practices. Furthermore, international organizations such as the International Union for Conservation of Nature (IUCN) and UNEP advise the Myanmar government on the development and implementation of laws and policies related to biodiversity conservation.

Local communities play an important role in ecological conservation and restoration by participating in decision making, and developing and implementing ecological conservation plans. Through self-management, local ecosystems can be better protected and restored. In Myanmar, local communities set up forestry associations, non-governmental organizations (NGOs) and research institutes to co-operate with the Union Ministry of Forestry on forestry development issues. The Myanmar Forest Products and Timber Consumption Association (MEPTMA) was established in 1993 and is the only commercial organization engaged in the production and sale of timber. Major NGOs include Forest Resources and Environment Development and Conservation Association (FREDA), Biodiversity and Nature Reserve Association (BANCA), Myanmar Rainforest Friendly Organization (FORM), Renewable Energy Association of Myanmar (REAM), Ecological Conservation and Community Development Initiative (ECCDI), and Myanmar Institute of Agriculture, Forestry, Animal Husbandry and Fisheries. These organizations play an important role in forest management, raising public awareness of forest resources and helping people increase their non-agricultural incomes (Oo and Zaw, 2019).

10.5 Suggestions for Future Sustainable Development in Myanmar

Myanmar is a country rich in natural resources and cultural heritage, but it also faces many challenges, such as deforestation, mining, agricultural expansion, urbanization, destruction of natural resources and degradation of ecosystems. These have greatly damaged Myanmar's ecosystem services and functions, as well as livelihoods and the economy (Vaughan, 2017). To ensure sustainable development in the future the following suggestions are recommended:

- **Conservation of natural resources and ecosystems.** To protect the natural resources, and plant and animal species, the Myanmar government should strengthen the implementation of environmental protection laws and enforce more stringent regulations. In addition, more nature and wildlife reserves should be established to protect and restore the ecosystems.

- **Develop sustainable tourism.** Myanmar's natural landscape and many ethnic groups attract a large number of international tourists. To ensure the sustainable development of tourism, the government should formulate policies for tourism development and strengthen the supervision of tourism. In addition, investments should be made in the construction of infrastructure and the upgrading of tourism services to improve the quality and competitiveness of the tourism industry.
- **Promote sustainable agriculture.** Agriculture is Myanmar's main economic livelihood, but it also faces problems such as land degradation and inadequate water. To improve the sustainability of agriculture, organic and sustainable farming practices should be encouraged and farmers should be provided with training and technical support. In addition, the agricultural infrastructure should be strengthened to improve the standard of living of farmers.
- **Promote renewable energy.** Policies should support the development and utilization of renewable energy such as solar, wind and hydro-energy, which would reduce the dependence on traditional energy, carbon emissions and environmental pollution.
- **Improve environmental education and awareness.** Education should be strengthened on environmental protection and sustainable development. Environmental courses should be included in schools, and campaigns should be organized to raise awareness of environmental issues. The public should be encouraged to participate in environmental protection and sustainable development actions, such as clean-energy projects, waste sorting and sustainable agriculture.
- **Promote social justice.** Sustainable development should ensure that the benefits of economic growth are distributed equally to all, including farmers, indigenous people and vulnerable groups. Social protection and justice and public services should be strengthened, and education and health services should be improved.
- **Strengthen international co-operation.** Sustainable development is a global challenge that requires the co-operation and support of the international community. Myanmar should strengthen co-operation with the international community in technology transfer and knowledge sharing.
- **Support sustainable urban development.** With the acceleration of urbanization, sustainable urban development has become critical. Myanmar should develop urban infrastructure, public transportation, clean energy and green buildings. In addition, urban landscaping and urban ecosystems should be included in the planning.
- **Strengthen environmental governance and supervision.** Environmental governance and supervision of environmental regulations should be enforced. Environmental law enforcement departments should be established, and environmental violations should be punished. In addition, environmental impact assessments should be implemented regularly to ensure the sustainability of environmental projects.

References

Albanito, F., Beringer, T., Corstanje, R., Poulter, B., Stephenson, A. *et al.* (2016) Carbon implications of converting agricultural cropland to bioenergy crops or forest for climate mitigation: a global assessment. *Global Change Biology Bioenergy* 8, 81–95. DOI: 10.1111/gcbb.12242

Andréasson, J., Bergström, S., Carlsson, B., Graham, L.P. and Lindström, G. (2004) Hydrological change: climate change impact simulations for Sweden. *Ambio* 33(4/5), 228–234.

Ashton, P.S. and Seidler, R. (2014) *On the Forests of Tropical Asia: Lest the Memory Fade.* Kew Publishing, London.

Aukema, J.E., Pricope, N.G., Husak, G.J. and Lopez-Carr, D. (2017) Biodiversity areas under threat: overlap of climate change and population pressures on the world's biodiversity priorities. *PLOS ONE* 12(1), e0170615. DOI: 10.1371/journal.pone.0170615

Aung, M. (2018) A way to achieve success for Myanmar ecotourism. *International Journal on Recent Trends in Business and Tourism (IJRTBT)* 2(3), 92–98.

Aung, T.D.W., Kyi, S.W., Suzue, K., Mon, S.Y., Tsujita, K. *et al.* (2021) Rapid ecosystem service assessment of a protected wetland in Myanmar, and implications for policy development and management. *Ecosystem Services* 50, 101336. DOI: 10.1016/j.ecoser.2021.101336

Aung, U.M. (2007) Policy and practice in Myanmar's protected area system. *Journal of Environmental Management* 84, 188–203. DOI: 10.1016/j.jenvman.2006.05.016

Barrow, E. (2019) *Our Future in Nature: Trees.* Balboa Press, Bloomington, Indiana.

Brandt, M., Yue, Y., Wigneron, J.P., Tong, X., Tian, F. *et al.* (2018) Satellite observed major greening and biomass increase in South China karst during recent decade. *Earths Future* 6, 1017–1028. DOI: 10.1029/2018EF000890

Cutler, S.Q. (2011) Geotourism: the tourism of geology and landscape. *Tourism Management* 32(1), 205–206. DOI: 10.1016/j.tourman.2010.06.021

de Alban, J.D.T., Jamaludin, J., de Wen, D.W., Than, M.M. and Webb, E.L. (2020) Improved estimates of mangrove cover and change reveal catastrophic deforestation in Myanmar. *Environmental Research Letters* 15, 034034. DOI: 10.1088/1748-9326/ab666d

Dinerstein, E., Olson, D., Joshi, A., Vynne, C., Burgess, N.D. *et al.* (2017) An ecoregion based approach to protecting half the terrestrial realm. *BioScience* 67, 534–545. DOI: 10.1093/biosci/bix014

Emerton, L. and Aung, Y.M. (2013) *The Economic Value of Forest Ecosystem Services in Myanmar and Options for Sustainable Financing.* International Management Group, Yangon, Myanmar.

Essumang, D., Asare, E. and Dodoo, D. (2013) Pesticides residues in okra (non-target crop) grown close to a watermelon farm in Ghana. *Environmental Monitoring and Assessment* 185, 7617–7625. DOI: 10.1007/s10661-013-3123-5

FAO (2014) Myanmar at a glance: Myanmar economy and the role agriculture. Available at: http://www.fao.org/myanmar/fao-in-myanmar/myanmar/en/ (accessed 27 September 2023).

FAO (2015) *Global Forest Resources Assessment 2015. How Are the World's Forests Changing?* Food and Agriculture Organization of the United Nations, Rome.

Foley, J.A., Ramankutty, N., Brauman, K.A., Cassidy, E.S., Gerber, J.S. *et al.* (2011) Solutions for a cultivated planet. *Nature* 478, 337–342. DOI: 10.1038/nature10452

Forest Department, Ministry of Forestry of Myanmar (1997) *Asia-Pacific Forestry Sector Outlook Study: Country Report – Union of Myanmar.* Forestry Policy and Planning Division, Food and Agricultural Organization of the United Nations, Rome, pp. 1–43.

Francis, C. (2019) *Field Guide to the Mammals of South-East Asia.* Bloomsbury Publishing, London.

Fraser, K.C. (2017) *Historical Dictionary of Burma (Myanmar)*, 2nd edition. Emerald Publishing Limited, Leeds, UK.

Gardiner, N.J., Robb, L.J. and Searle, M.P. (2014) The metallogenic provinces of Myanmar. *Applied Earth Science* 123, 25–38. DOI: 10.1179/1743275814Y.0000000049

Grismer, L.L., Wood Jr, P.L., Thura, M.K., Thaw, Z.I.N., Quah, E. *et al.* (2017) Twelve new species of *Cyrtodactylus gray* (Squamata: Gekkonidae) from isolated limestone habitats in east-central and southern Myanmar demonstrate high localized diversity and unprecedented microendemism. *Zoological Journal of the Linnean Society* 182, 862–959. DOI: 10.1093/zoolinnean/zlx057

Harada, M. (1995) Minamata disease: methylmercury poisoning in Japan caused by environmental pollution. *Critical Reviews in Toxicology* 25, 1–24. DOI: 10.3109/10408449509089885

Htun, K. (2009) *Asia-Pacific Forestry Sector Outlook Study II: Myanmar Forestry Outlook Study.* Working Paper No. ASFSOS II/WP/2009/07. Food and Agriculture Organization of the United Nations, Regional Office for Asia and the Pacific, Bangkok.

Htun, K. (2014) Sustainable mining in Myanmar. *Applied Environmental Research* 36, 25–35. DOI: 10.35762/AER.2014.36.1.5

Ito, M.Y. and Barfod, M.A.S. (2014) An updated checklist of aquatic plants of Myanmar and Thailand. *Biodiversity Data Journal.* DOI: 10.15468/jzjzee

Joko, T., Anggoro, S., Sunoko, H.R. and Rachmawati, S. (2017) Pesticides usage in the soil quality degradation potential in Wanasari subdistrict, Brebes, Indonesia. *Applied and Environmental Soil Science.* DOI: 10.1155/2017/5896191

Karuppusamy, B., Sarma, D.K., Lalmalsawma, P., Pautu, L., Karmodiya, K. and Nina, P.B. (2021) Effect of climate change and deforestation on vector borne diseases in the North-Eastern Indian state of Mizoram bordering Myanmar. *The Journal of Climate Change and Health* 2, 100015. DOI: 10.21203/rs.3.rs-126530/v1

Kawakami, T., Konishi, M., Imai, Y. and Soe, P.S. (2019) Diffusion of mercury from artisanal small-scale gold mining (ASGM) sites in Myanmar. *Geomate Journal* 17(61), 228–235. DOI: 10.21660/2019.61.4823

Keith, D.A., Rodríguez, J.P., Brooks, T.M., Burgman, M.A., Barrow, E.D. *et al.* (2015) The IUCN red list of ecosystems: motivations, challenges, and applications. *Conservation Letters* 8, 214–226. DOI: 10.1111/conl.12167

Khaing, T.T., Pasion, B.O., Lapuz, R.S. and Tomlinson, K. (2019) Determinants of composition, diversity and structure in a seasonally dry forest in Myanmar. *Global Ecology and Conservation* 19, e00669. DOI: 10.1016/j.gecco.2019.e00669

Khanal, B.R. and Babar, J.T. (2007) Community based ecotourism for sustainable tourism development in the Mekong region. *Policy Brief* 1, 2007.

Kristensen, A.K.B., Thomsen, J.F. and Mikkelsen, S. (2014) A review of mercury exposure among artisanal small-scalegold miners in developing countries. *International Archives of Occupational and Environmental Health* 87, 579–590. DOI: 10.1007/s00420-013-0902-9

Kyaw, T.Y., Germain, R.H., Stehman, S.V. and Quackenbush, L.J. (2020) Quantifying forest loss and forest degradation in Myanmar's 'home of teak'. *Canadian Journal of Forest Research* 50(2), 89–101. DOI: 10.1139/cjfr-2018-0508

Linn, S.N. (2015) Myanmar's mining investment and its discontents. Available at: http://www.eastasiaforum.org/2015/09/23/myanmars-mining-investment-and-its-discontents/ (accessed 27 September 2023).

Lo, C.C. (2010) Effect of pesticides on soil microbial community. *Journal of Environmental Science and Health B* 45, 348–359. DOI: 10.1080/03601231003799804

Matelski, M. and Sabrié, M. (2019) Challenges and resilience in Myanmar's urbanization: a special issue on Yangon. *Moussons* 33, 11–31. DOI: 10.4000/moussons.4844

McNutt, M. (2013) Mercury and health. *Science* 341, 14–30. DOI: 10.1126/science.1245924

Mendelsohn, R. (2009) The impact of climate change on agriculture in developing countries. *Journal of Natural Resources Policy Research* 1, 5–19. DOI: 10.1080/19390450802495882

Middleton, D., Armstrong, K., Baba, Y., Balslev, H., Chayamarit, K. *et al.* (2019) Progress on Southeast Asia's flora projects. *Garden's Bulletin Singapore* 71(2), 267–319. DOI: 10.26492/gbs71(2).2019-02

Ministry of Environmental Conservation and Forestry (2014) *The Republic of the Union of Myanmar: Fifth National Report to the Convention on Biological Diversity.* Ministry of Environmental Conservation and Forestry, Nay Pyi Taw, Myanmar.

Molotoks, A., Stehfest, E., Doelman, J., Albanito, F., Fitton, N. *et al.* (2018) Global projections of future cropland expansion to 2050 and direct impacts on biodiversity and carbon storage. *Global Change Biology* 24(12), 5895–5908. DOI: 10.1111/gcb.14459

Moreno-Brush, M., McLagan, D.S. and Biester, H. (2020) Fate of mercury from artisanal and small-scale gold mining in tropical rivers: hydrological and biogeochemical controls: a critical review. *Critical Reviews in Environmental Science and Technology* 50, 437–475. DOI: 10.1080/10643389.2019.1629793

Murray, N.J., Keith, D.A., Duncan, A., Tizard, R., Ferrer-Paris, J.R. *et al.* (2020) Myanmar's terrestrial ecosystems: status, threats and conservation opportunities. *Biological Conservation* 252, 108834. DOI: 10.1016/j.biocon.2020.108834

Oo, T.M. and Zaw, A. (2019) The role of civil society in Myanmar's political transition. *Asian Affairs: An American Review* 46(2), 55–75.

Osawa, T. and Hatsukawa, Y. (2015) Artisanal and small-scale gold mining in Myanmar. *International Journal of Human Culture Studies* 2015(25), 221–230. DOI: 10.9748/hcs.2015.221

RECOFTC (2020) *Social Forestry and Climate Change in the ASEAN Region: Situational Analysis 2020.* Regional Community Forestry Training Center for Asia and the Pacific, Bangkok.

Songer, M., Aung, M., Senior, B., DeFries, R. and Leimgruber, P. (2009) Spatial and temporal deforestation dynamics in protected and unprotected dry forests: a case study from Myanmar (Burma). *Biodiversity and Conservation* 18, 1001–1018. DOI: 10.1007/s10531-008-9490-5

Turner, I., Kress, W., DeFilipps, R., Farr, E., Kyi, D.Y.Y. *et al.* (2003) A checklist of the trees, shrubs, herbs, and climbers of Myanmar. *Systematic Botany* 29(2), 462–463. DOI: 10.1600/036364404774195674

UNEP (2019) *Global Mercury Assessment 2018*. United Nations Environment Programme, New York.

Vaughan, E. (2017) *Myanmar Country Report*. ESPA (Ecosystems Services for Poverty Alleviation Directorate) and ACCCRN (Asian Cities, Climate Change Resilience Network), Yangon, Myanmar.

Wang, C. and Myint, S.W. (2016) Environmental concerns of deforestation in Myanmar 2001–2010. *Remote Sensing* 8(9), 728. DOI: 10.3390/rs8090728

Wang, C., Myint, S.W. and Hutchins, M. (2019) The assessment of deforestation, forest degradation, and carbon release in Myanmar 2000–2010. In: Saikia, A. and Thapa, P. (eds) *Environmental Change in the Himalayan Region*. Springer, Cham, Switzerland, pp. 47–64.

Webb, E.L., Jachowski, N.R., Phelps, J., Friess, D.A., Than, M.M. *et al.* (2014) Deforestation in the Ayeyarwady Delta and the conservation implications of an internationally-engaged Myanmar. *Global Environmental Change* 24, 321–333. DOI: 10.1016/j.gloenvcha.2013.10.007

World Bank Data (2019) Country profile: Myanmar. Available at: https://databank.worldbank.org/views/reports/reportwidget.aspx?Report_Name=CountryProfile&Id=b450fd57&tbar=y&dd=y&inf=n&zm=n&country=MMR (accessed 27 September 2023).

Yasuhara, M., Hunt, G., Breitburg, D., Tsujimoto, A. and Katsuki, K. (2012) Human-induced marine ecological degradation: micropaleontological perspectives. *Ecology and Evolution* 2(12), 3242–3268. DOI: 10.1002/ece3.425

11 Ecological Restoration in Nepal: A Comprehensive Policy Overview

Ramesh Raj Pant[1,*], Madan Koirala[1], Diwakar Adhikari[1] and Narayan Prasad Gaire[2]

[1]Central Department of Environmental Science, Institute of Science and Technology, Tribhuvan University, Nepal; [2]Department of Environmental Science, Patan Multiple Campus, Tribhuvan University, Nepal

Abstract

This chapter presents a comprehensive description of Nepal's conservation and restoration strategies and policies. The goal of the policies is to foster sustainable development and safeguard the unique and diverse ecosystems. The escalating population has increased the demand for natural resources like forestry products, water, land and minerals, which could lead to a degradation of ecosystem services. Some restoration projects are improving the ecosystem functions and services, but intensive efforts are still needed. By adopting ecosystem restoration and resource conservation strategies, Nepal improves ecosystem functions and services, and livelihoods. An integrated approach to restoration offers additional benefits, creating employment opportunities in the country's post-pandemic economic recovery.

11.1 Introduction

An ecosystem is a complex integration of interacting organisms and the physical environment, encompassing matter and energy assimilated in specific directions (Parker and Pickett, 1997). These systems and their services play a crucial role in human well-being and survival, acting as life-supporting mechanisms for the planet (Costanza *et al.*, 2014; TEBB, 2010). Unfortunately, many of the world's ecosystems have undergone degradation, leading to adverse effects on biological diversity, ecological function and people's ability to meet basic needs like food, water, shelter and clothing (Millennium Ecosystem Assessment, 2005; Ruckelshaus *et al.*, 2020). Ecosystem restoration, a human-initiated or accelerated recovery process, is crucial in restoring the health, integrity and sustainability of ecosystems (SER, 2004). Given the accumulating negative impacts on ecosystems in recent years (Davis and Slobodkin, 2004; Choi, 2007), it becomes imperative to return degraded ecosystems to their pre-disturbance states.

Global ecosystem degradation and loss is well documented (Vitousek *et al.*, 1997; Hoekstra *et al.*, 2005; Kok *et al.*, 2014; IPBES, 2019), resulting in reduced ecosystem functions, services and buffering capacity (Costanza *et al.*, 1997; Millennium Ecosystem Assessment, 2005; WHO, 2005). Human activities, such as exploiting natural resources for food, water, wood, fibre and energy, combined with technological advancements, have caused an irreversible decline in Earth's biodiversity (Millennium Ecosystem Assessment, 2005). Therefore, restoring degraded

*Corresponding author: rpant@cdes.edu.np

DOI: 10.1079/9781800622579.0011

ecosystems has become essential in the face of growing ecological challenges (Davis and Slobodkin, 2004; Choi, 2007). In the past 50 years, human-driven changes to ecosystems have outpaced any previous period due to the escalating demand for resources. Consequently, it is now crucial to prioritize the preservation and restoration of ecosystems. The advancement of research in ecology, conservation biology and sustainable management practices has underscored the importance of maintaining ecosystem health for both human and non-human life (Vihervaara *et al.*, 2010). As such, safeguarding and restoring ecosystems have become urgent requirements for achieving sustainable development and preserving human well-being.

Nepal, with its inherently fragile ecosystems that are vulnerable to anthropogenic activities (Maskey *et al.*, 2003), is not immune to this global issue. Ecosystems in the country have suffered degradation, damage, transformation and destruction due to human-induced actions like deforestation, pollution, over-fishing, urbanization, invasive species and climate change, as well as natural events such as landslides, erosion, wildfires, floods, storms and volcanic eruptions (SER, 2004).

Restoration options vary from passive to active, depending on the severity of ecosystem damage. Efforts are critical to address habitat loss, declining biodiversity and environmental imbalances, while also promoting sustainable practices and raising public awareness to conserve and revive degraded ecosystems. Passive restoration involves allowing ecosystems to recover through natural processes like ecological succession, soil building and re-colonization of extirpated species, achieved by removing stressors causing ecosystem damage, degradation and destruction. This approach may be suitable for ecosystems that have not undergone significant changes. However, for cases with substantial physical, chemical or biotic alterations, active restoration may necessitate human intervention, such as reintroducing native species, managing invasive species and reconstructing physical structures like soil, waterways and habitats. These interventions can help the ecosystem recover and restore its ecological processes and functions (Allen *et al.*, 2002).

Overall, ecosystem restoration is crucial in addressing the environmental challenges stemming from human activities and natural disasters. The approach to restoration depends on the extent of ecosystem degradation and the level of human intervention required to return it to its pre-disturbance state or historical developmental trajectory.

11.1.1 The context of Nepal

Nepal, nestled in the central Himalayas, boasts a landscape dominated by hills and majestic mountains covering around 83% of its terrain, while the remaining 17% constitutes the flatlands known as the Terai. Nepal's altitude varies from 60 m above sea level (masl) in the south-east Terai to 8849 masl at Mount Everest (MoFSC, 2009). The remarkable geographical position and diverse climate have nurtured a flourishing ecosystem abundant with various species and genetic biodiversity. The country's diverse landscape spans multiple biogeographical regions, contributing to a unique and diverse range of terrestrial biodiversity (MoFSC, 2009). Table 11.1 exemplifies the legal bindings, policies, implementation and commitments based on Nepal's constitutional history. The social, economic and environmental perspectives underscores the profound impacts on nation and people.

Causes of ecological degradation in Nepal

The causes of ecological degradation can be divided broadly into two categories: natural and anthropogenic causes.

Natural causes

Natural disasters like landslides, floods, earthquakes and changing climate patterns are inherent contributors to ecological degradation in Nepal. Positioned in the seismically active Himalayan region, Nepal is particularly susceptible to such events. Climate-change impact (UNDP, 2020) and human activities have escalated the frequency and severity of these disasters. The country's rugged topography, unstable landforms and active geological processes have caused irreversible biodiversity loss. Extreme climate and resource degradation also trigger soil erosion, land degradation and mass wasting, undermining the integrity of associated ecosystems (Bhattarai *et al.*, 2009). Nepal's fragile geological conditions necessitate urgent conservation efforts.

Table 11.1. Legal bindings, policies, implementation and commitments based on Nepal's constitutional history.

1950–1975 (2007–2031 BS)	1976–1990 (2032–2046 BS)
1951: The interim constitution of Nepal 1956: The government launched the first Five-year Plan, which focused on agriculture and rural development. 1957: Private Forest Nationalization Act 1964: Nepal's membership of the United Nations 1967: Forest Protection Act (special arrangement) 1970: Forest Product Rules (sales and distribution) 1972: Signatory to the Convention on Wetlands of International Importance 1973: The National Parks and Wildlife Conservation Act 1973: Establishment of Chitwan National Park 1974: Watershed management in Nepal began with the establishment of the Department of Soil Conservation and Watershed Management	1976: National Forest Policy 1977: Signatory to the Convention Concerning the Protection of World Cultural and Natural Heritage 1982: Soil and Watershed Conservation Act 1985: Seventh Five-year Plan 1988: National Conservation Strategy 1990: The Constitution of the Kingdom of Nepal
1990–2007 (2046–2063 BS)	**(2006–2016) 2063–2072 BS**
1992: New industrial policy 1992: Trade and transition arrangement for medicinal plants 1993: Revision of National Parks and Wildlife Conservation Act 1993: Forest Act 1997: Environmental Protection Act 1998: Biodiversity Conservation Strategy 2002: National Biodiversity Action Plan 2003: National Wetland Policy 2005: Millenium Development Goals 2006: Nepal Biodiversity Strategy Implementation Plan 2007: Interim constitution of Nepal	2010: National Adaptation Program of Action 2011: LAPA 2011: Climate Change Policy 2011: Solid Waste Management Act 2014: National Biodiversity Strategy and Action Plan (NBSAP) 2015: Paris Agreement 2015: Sustainable Development Goals 2015: National Adaptation Plan
2016–present (2072 BS–present)	
2017: National REDD+ strategy 2019: Climate Change Policy 2019: Forest Act 2019: Environmental Protection Act 2019: Zero Waste Himalaya Initiative 2020: Environmental Protection Regulation	

Anthropogenic causes

Human activities in Nepal have become the leading force driving ecological degradation, resulting in the destruction of forests, soil erosion, loss of biodiversity and pollution in rivers (Paudel, 2018). The escalating demand for resources, coupled with the expansion of agriculture, settlements and infrastructure, has had alarming consequences for the environment. The severity, duration and scale of the degradation have resulted in the depletion and degradation of ecosystems (Ives, 1987; UNEP, 2001; Karkee, 2004; MoEST, 2006). This depletion has occurred due to excessive utilization of natural resources such as forests, water and minerals, while the use of environmentally harmful technologies has contributed to air and water pollution. The excessive use of chemical fertilizers and pesticides in Nepal's agricultural lands has led to contamination of soil and water, posing risks to ecological integrity and biodiversity. The growing agricultural production, driven by the increasing population, urbanization and drought, has put immense pressure on the limited remaining land, exacerbating the situation.

Traditionally, Nepal's people have followed sustainable agricultural and land management practices like contour farming, agroforestry and terracing. However, modern practices involving the excessive use of chemical substances have caused further land degradation through soil erosion, fertility loss and landslides, adversely affecting the entire ecosystem. An example of this degradation is seen in Nepal's forests and rangelands, particularly in the Siwaliks and Terai regions, where increased resource demand and utilization to support the growing population have led to detrimental effects. Over the past few decades, these forests have faced intensified loss and degradation, resulting in adverse environmental and social implications. Structural decline, changes in composition, reduced productivity, invasion by non-native species, and loss of biodiversity have been some of the consequences of this degradation (Acharya *et al.*, 2011).

River pollution is another consequence of human activities in Nepal, with the Bagmati river and the Sirsiya river being severely affected by untreated industrial effluents, sewage and solid waste. Pollution in the Bagmati river has reached critical levels, jeopardizing the aquatic ecosystem and the communities relying on it for drinking water and other needs (Pant *et al.*, 2021). Addressing these pressing issues requires concerted efforts to promote sustainable practices, implement effective waste management systems, and raise awareness about the importance of preserving and conserving Nepal's natural resources and ecosystems.

Problems associated with ecological degradation in Nepal

The above-mentioned causes have sparked a growing concern regarding the restoration of degraded ecosystems in Nepal, particularly in light of shifting ecosystem structures and climate change. However, the efforts and work directed towards addressing these concerns have been nearly negligible. The consequences of ecological degradation pose significant challenges to both the ecosystems and the well-being of the people, as outlined below.

LOSS OF FORESTS. Deforestation in Nepal poses a pressing challenge, leading to diminished forest areas and wildlife habitats. The country's forest cover has declined rapidly due to activities like commercial logging, agricultural expansion and fuelwood extraction. This has direct repercussions, causing soil erosion, landslides and reduced water availability. Indirectly, it lowers biodiversity and increases carbon emissions. Nepal's forest degradation severely impacts crucial ecosystem services like carbon sequestration, water regulation and biodiversity preservation. Over a broader span, from 1990 to 2010, approximately 1.5 million ha of forest cover were lost (MoFSC, 2014). These statistics demonstrate the alarming scale of deforestation in the country, resulting in significant environmental challenges that adversely affect both the ecosystem and the livelihoods of local communities.

SOIL EROSION AND LAND DEGRADATION. Nepal faces urgent challenges from soil erosion and land degradation. A study by Pandey *et al.* (2016) found that nearly 68% of the country's land is at risk of erosion, with 35% experiencing moderate to severe soil erosion. This alarming situation requires immediate action. Deforestation and improper land-use practices are major contributors to soil erosion. Natural disasters like landslides and floods worsen the problem, making the land even more vulnerable. The consequences of soil erosion are far-reaching. Reduced soil fertility and compromised crop yields directly affect local livelihoods and the nation's agricultural productivity. Moreover, the increased sedimentation in rivers and lakes due to erosion poses challenges to water management and aquatic ecosystems. Access to clean drinking water for Nepalese communities is also affected due to the negative impact on water quality. Additionally, soil erosion heightens the risk of natural disasters in affected regions. A study by the International Centre for Integrated Mountain Development (Uddin *et al.*, 2018) revealed that Nepal's soil erosion rates were between ten and 100 times higher than the natural rate, emphasizing the need for comprehensive and sustainable measures. Hill and mountain regions, with steep slopes and inadequate soil conservation practices, are particularly vulnerable and require targeted interventions.

To protect its rich cultural and ecological heritage, Nepal must prioritize conservation efforts and adopt environmentally friendly practices. Raising awareness among local communities and stakeholders about sustainable land management practices and afforestation initiatives could play a crucial role in curbing soil erosion. Immediate and collective action is needed to address this pressing challenge. By implementing well-designed conservation measures, Nepal can safeguard its valuable land resources and ensure a more sustainable and prosperous future.

WATER SCARCITY AND POLLUTION. Nepal faces ecological degradation, resulting in severe water pollution and scarcity. Deforestation and land mismanagement deplete water resources, leading to drying rivers, declining groundwater and unsafe drinking water for approximately 3 million Nepalese. This poses public health and economic concerns, especially in rural areas relying on agriculture and livestock. Pollution in water bodies like the Bagmati and Sirsiya rivers, and Phewa lake worsens the situation due to unprocessed wastewater, agricultural run-off and improper waste disposal, endangering biodiversity and public health. Water-borne diseases burden the healthcare system, disrupt ecological balance, and impact fishing, tourism, food security and agriculture. To tackle these challenges, Nepal must prioritize environmental conservation by effective waste management, wastewater treatment, and eco-friendly agriculture. Preserving natural habitats along rivers and lakeshores could safeguard water quality. Optimizing water allocation for upstream and downstream regions is vital, considering climate change and human activity impacts. Policy initiatives focusing on comprehensive water resource allocation and sustainability are crucial for local communities.

Restoration plans should integrate scientific data and research on climate change and water availability. Collaboration among government, communities, industries and environmental organizations is essential for reducing pollution in urban river systems and Ramsar-listed lakes. Data-driven insights, robust policies and collaborative efforts aim to secure the sustainability of water resources and aquatic ecosystems in Nepal.

BIODIVERSITY LOSS. Nepal, renowned for its breathtaking landscapes and diverse wildlife, faces ecological degradation endangering precious habitats and unique biodiversity. Nearly one third of Nepal's mammals, birds and amphibians are on the verge of extinction, necessitating urgent action on multiple fronts. Preserving this natural heritage requires enforcing conservation laws and establishing new ones. Collaboration between governmental agencies, local communities and international organizations is vital to safeguard endangered species and their habitats. Education and awareness campaigns can foster environmental responsibility, while investment in research and monitoring changes can improve conservation strategies.

Beyond ecological implications, biodiversity holds cultural significance, as Nepalese communities revere their natural surroundings. Preserving biodiversity ensures that the cultural identity remains intact for future generations. Moreover, protecting biodiversity can drive economic growth through ecotourism and sustainable industries. Nepal must act urgently and collaboratively to become a global leader in biodiversity preservation. Robust conservation, public awareness and acknowledging biodiversity's cultural and economic value will secure a vibrant natural legacy for generations to come.

CLIMATE CHANGE. Nepal faces high vulnerability to climate change, with glacial melt, floods and landslides posing risks. Melting glaciers and increased mountain precipitation lead to flooding and landslides, threatening human settlements and infrastructure. Urgent attention is required to address the far-reaching consequences of climate change in this diverse landscape. Moreover, climate warming cause rivers and streams to dry up, intensifying water scarcity and impacting urban and rural communities' water needs. As an agrarian society, Nepal relies heavily on water for agriculture, making water scarcity a pressing concern that directly affects food security and livelihoods. Shifting growing seasons, altered precipitation, and new pests and diseases reduce crop yields and income for farmers, while tourism faces risks due to ecosystem and biodiversity instability.

Nepal's vulnerability to climate change is not solely due to its emissions but also by emissions from industrialized nations. Thus, international co-operation and support for climate adaptation and mitigation efforts are crucial for effectively addressing this global challenge. To combat climate change, a multi-pronged approach is necessary. Mitigation efforts should focus on reducing greenhouse gas emissions by promoting renewable energy sources, energy-efficient practices and sustainable land use planning. Nepal's commitment to the Paris Agreement highlights its dedication to global climate change efforts. Adaptation strategies are equally essential, involving climate-resilient infrastructure, enhanced early-warning systems, and equitable water management practices. Furthermore, integrating climate change education into school curricula and public awareness campaigns can foster a culture of sustainability and empower citizens to take individual actions.

Preserving Nepal's unique biodiversity and protecting communities from climate-related disasters demand immediate and comprehensive actions. Combining conservation with robust climate mitigation and adaptation strategies is vital for a sustainable and resilient future. By engaging in collective action at national and global levels, Nepal can pave the way for a greener and climate-resilient future.

POVERTY. The intricate relationship between poverty and ecological degradation in Nepal underscores the challenges faced by its people and environment. The majority of Nepalese rely directly on natural resources like forests and rivers for their livelihoods. This balance is crucial yet vulnerable; unsustainable practices can lead to over-exploitation and degradation, placing immense pressure on the ecosystem's resilience. This cycle begins with economic pressures leading to resource over-exploitation, which, in turn, exacerbates poverty. According to the UNDP, Nepal's multi-dimensional poverty rate stands at 28.6%, highlighting the complexity of the issue. Addressing poverty requires a holistic approach, which includes education, healthcare, sanitation and environmental sustainability.

To break this cycle, integrated strategies are needed. Promoting sustainable resource management, eco-friendly alternatives and afforestation efforts can help restore degraded areas. Investments in education and skill development can empower communities to diversify their livelihoods, reducing reliance on natural resources. Equitable access to healthcare and family planning can curb population growth and resource strain. Collaborative efforts between the government and international organizations can promote sustainable economic growth while safeguarding the environment and empowering local communities.

Nepal's web of poverty and ecological degradation calls for a comprehensive and integrated approach. By promoting sustainability, investing in education and healthcare, and empowering communities, Nepal can forge a path towards an equitable and environmentally resilient future, preserving its invaluable natural heritage.

WEAK GOVERNANCE. Nepal's weak governance has been a contributing factor to environmental degradation. The lack of effective policies, enforcement and co-ordination has led to challenges in managing and preserving natural resources. As a result, various environmental issues have emerged, affecting the well-being of both the environment and the people. One of the major consequences of weak governance is the improper management of waste and pollution. Inadequate waste disposal systems and the absence of strict regulations have led to the contamination of water bodies and soil, posing serious health risks to the population.

Furthermore, deforestation and illegal logging are rampant due to weak enforcement of forest conservation laws. The exploitation of natural resources without proper control has led to the loss of precious forests, exacerbating climate change and threatening biodiversity. Another pressing concern is the unchecked expansion of industries and infrastructure without proper environmental impact assessments. This has resulted in habitat destruction, loss of agricultural land and disruption of ecosystems.

Moreover, weak governance has affected the enforcement of wildlife protection laws, leading to increased poaching and illegal wildlife trade. Iconic species like the Bengal tiger and one-horned rhinoceros are facing severe threats. Addressing the issue of weak governance is crucial to mitigate environmental degradation in Nepal. Strengthening institutional structures, enhancing co-ordination among government agencies, and enforcing strict environmental regulations are essential steps to tackle these challenges effectively. Additionally, empowering local communities and fostering public participation in environmental decision making can promote sustainable practices and raise awareness about the importance of protecting natural resources.

By prioritizing strong governance and adopting proactive measures, Nepal can strive towards a more sustainable and ecologically responsible future, safeguarding its environment for generations to come.

The need for ecological restoration

RESTORATION OF DEGRADED LANDSCAPES. Nepal's delicate mountain ecosystems are under significant threat from deforestation, over-grazing and soil erosion. Restoring and safeguarding these degraded landscapes are vital, necessitating actions like reforestation and soil conservation. The government's Community-based Rangeland Management Project has shown success in restoring grasslands in the Karnali region by involving local communities in sustainable land management practices (Gautam, 2020). Community engagement and awareness campaigns are essential to instill a sense of responsibility in citizens to protect their water resources. Co-operative efforts among government agencies, communities and NGOs are crucial for comprehensive and sustainable water resource management.

Nepal's water scarcity and pollution require immediate action to preserve its valuable water resources and ensure a healthier and more sustainable future. By addressing the underlying causes of ecological degradation, embracing eco-friendly practices, and promoting responsible stewardship of natural resources, Nepal can create a brighter and cleaner future.

CONSERVATION OF BIODIVERSITY. Nepal is a biodiversity hotspot, harbouring endangered species like the Bengal tiger, one-horned rhinoceros and snow leopard. Human activities have led to habitat loss and fragmentation, endangering the country's biodiversity. Ecological restoration, exemplified by Nepal's Annapurna Conservation Area Project, has restored forests, grasslands and wetlands, enhancing habitat connectivity and promoting biodiversity conservation.

MITIGATION OF CLIMATE CHANGE. Ecological restoration, like reforestation and wetland restoration, plays a vital role in combating climate change by capturing carbon and curbing greenhouse gas emissions. The Nepal REDD+ programme successfully restored degraded forests and encouraged sustainable forest management, resulting in carbon sequestration and emission reduction.

SOCIO-ECONOMIC BENEFITS. Ecological restoration offers socio-economic benefits, including job opportunities, improved water resources, and enhanced food security. For instance, the Nepal Watershed Management Project successfully restored degraded watersheds, benefitting locals with better irrigation, drinking water and sanitation facilities. Nepal has made notable strides in ecological restoration through government and community-led efforts. Nonetheless, further scaling up of these initiatives is essential to ensure long-term sustainability. Making ecological restoration a central component of Nepal's sustainable development agenda is crucial to safeguard its diverse biodiversity and vulnerable mountain ecosystems for future generations.

11.2 People, Livelihood and Restoration

The close connection between people's livelihoods and ecosystem health is widely acknowledged, and ecosystem degradation can affect those who depend on them. An example is the concept of ecosystem services, which refers to the benefits people receive from the environment, such as clean water, food and climate regulation. Another relevant example is the social-ecological system, which recognizes how social and ecological systems intertwine and influence each other. Moreover, the sustainable livelihoods approach and political ecology theories explain this inter-connection further. The former emphasizes the importance of diverse and resilient livelihood strategies, while the latter explores power dynamics in resource access and decision making.

Debates persist concerning the relationship between livelihoods and ecology, particularly in environmental conservation and development. Some argue that conservation efforts might impact local livelihoods negatively, while others stress the need to protect ecological systems for sustainable development. These debates underscore the intricate trade-offs and challenges involved in balancing ecological conservation and livelihoods.

11.2.1 Ecology and livelihoods

The interdependence of people's livelihoods and ecology is crucial. Natural resources and ecosystem services support livelihoods like agriculture, fishing, forestry and livestock production. However, these livelihoods can also impact ecology negatively, leading to environmental degradation and resource depletion. Striking a balance between livelihoods and ecology through sustainable practices is essential.

In Nepal, sustainable agriculture practices such as crop rotation and organic fertilizers maintain soil fertility and boost productivity. Similarly, sustainable forestry practices like selective harvesting and reforestation ensure a steady supply of forest products. Implementing fishing quotas and protecting breeding grounds promote the long-term sustainability of fish stocks. Sustainable livelihoods are those that withstand shocks while preserving assets and the natural resource base. Addressing poverty and promoting sustainable development are widely discussed. Ecosystem restoration is crucial for sustainable livelihoods as it provides essential natural resources and services.

Studies revealed the positive impacts of ecosystem restoration, such as increased water resources, improved soil quality, and ecotourism opportunities. Community participation is crucial for successful ecosystem restoration, enhancing local capacity and long-term sustainability. Further research is needed to explore connections between sustainable livelihoods and ecosystem restoration to develop effective future strategies.

11.2.2 Indigenous knowledge and ecosystem restoration

In Nepal, indigenous communities depend heavily on natural resources like forests, water sources and grazing lands for their livelihoods. They manage these resources using traditional practices rooted in local knowledge and customs.

Indigenous knowledge is crucial for natural resource management, passed down through generations and refined over centuries through trial and error. It relies on a deep understanding of the local environment, including flora and fauna behaviour, weather patterns and water sources. The local communities and indigenous people of Nepal have adopted various ecological conservation and restoration approaches. Examples are provided below.

Community forest management (CFM) is an effective approach wherein local communities collectively manage forests. In Nepal, CFM has led to successful forest restoration and biodiversity conservation. This system grants communities the rights to manage and use forest resources while ensuring protection from overuse and degradation. CFM has also brought economic benefits to local communities through sustainable forest management.

Agroforestry, a farming system commonly adopted by local communities and indigenous people in Nepal, involves integrating trees, crops and livestock. It fosters ecological conservation and restoration by enhancing soil fertility, reducing erosion and boosting biodiversity.

Traditional water management systems (Box 11.1) are employed by local communities to preserve water resources and counter soil erosion. Centuries-old techniques like Dhunge Dhara, or stone spouts, redirect water from natural sources to fields. Other methods involve community-run irrigation canals, ponds and water-harvesting structures.

Biodiversity conservation areas: community-based conservation areas, such as the Annapurna conservation area and the Kanchenjunga conservation area, have been established in Nepal. These areas are managed using traditional knowledge and practices and have conserved biodiversity and restored ecosystems.

Sacred sites: Nepal's local communities cherish sacred sites, such as forests, lakes and mountains. These spots employ traditional wisdom to effectively preserve biodiversity and revive ecosystems. Recently, there has been a growing recognition of indigenous knowledge in Nepal's natural resource management. The government now integrates local practices into policies, promoting indigenous community involvement in decision making. However, preserving and

Box 11.1. *Pani Ghatta*, an indigenous knowledge water resource management in Nepal.

Traditional water mills, known as *ghatta*, have been used for water resource management for centuries in Nepal. They are powered by natural water flow in streams and rivers and perform tasks like grinding grains, husking rice and milling oilseeds. The ghatta system, an indigenous practice passed down through generations, provides a sustainable and eco-friendly way to utilize water resources for local livelihoods. These mills, usually made of wood and stone, are situated on stream and riverbanks. Water is directed into a channel, powering the millstones through a turbine. A key advantage is that it requires no external energy source like electricity or fossil fuels, making it a cost-effective and sustainable method.

The ghatta system has contributed significantly to rural development in Nepal, providing income and food security to local communities while preserving traditional agricultural practices and knowledge essential to Nepal's cultural heritage.

However, modernization and the rise of motorized mills pose a threat to the ghatta system. Efforts are being made to promote and preserve these traditional water mills for sustainable water resource management and rural development in Nepal.

promoting indigenous knowledge in Nepal faces challenges, like the erosion of traditional practices due to modernization and the lack of recognition and support for indigenous communities.

11.3 Legal Bindings, Policies and Commitments

11.3.1 Initiatives to ecological restoration in Nepal

Nepal has made major strides in implementing ecological restoration policies and plans in recent years. The country has recognized the importance of preserving its rich biodiversity and natural resources for sustainable development, poverty reduction and climate resilience.

The government of Nepal has developed several policies and programmes aimed at restoring degraded ecosystems, conserving biodiversity, and promoting sustainable land-use practices. National Biodiversity Strategy and Action Plan (NBSAP) (2014) is a national plan for conserving Nepal's biodiversity, including its ecosystems. The plan focuses on restoring degraded ecosystems, conserving endangered species and promoting sustainable development. National Adaptation Program of Action (NAPA) (2010) is a national plan that identifies and addresses the most urgent adaptation needs related to climate change. One of its main objectives is to enhance the resilience of ecosystems and biodiversity to climate change. National REDD+ Strategy (2017) aims to reduce deforestation and forest degradation emissions and promote forest conservation and sustainable management. The strategy includes provisions for restoring degraded forests and other ecosystems. In addition, the government has launched several programmes and initiatives to restore degraded forests, wetlands and other ecosystems. The community forestry programme involves local communities in forest management and conservation, has successfully restored degraded forests, and promoted sustainable livelihoods. Similarly, the wetland conservation and management policy has been instrumental in restoring and conserving wetlands in the country.

The government has also implemented programmes to promote sustainable agriculture and land-use practices. The prime minister's Agriculture Modernization Project, which aims to increase agricultural productivity while reducing environmental degradation, is one such programme. The government has also launched initiatives to promote agroforestry, organic farming and sustainable land-use practices. Nepal's ecological restoration policies and plans demonstrate a solid commitment to preserving its natural resources and promoting sustainable development. While significant progress has been made, there is still much work to be done to address the challenges of biodiversity loss, climate change and environmental degradation in the country.

National Parks and Wildlife Conservation Act (1973) provides the legal framework for establishing and managing national parks, wildlife reserves and conservation areas in Nepal. These protected areas are vital in conserving Nepal's ecosystems and biodiversity. The National

Wetland Policy (2003) aims to promote the conservation and sustainable management of Nepal's wetlands, which are critical ecosystems that provide a range of ecosystem services. The policy includes provisions for restoring degraded wetlands. The Agriculture Development Strategy (2015) aims to promote sustainable agriculture practices, including restoring degraded agricultural land. The strategy includes provisions for promoting agroforestry and other sustainable land-use practices.

The Local Government Operation Act (2017) decentralizes power to local governments and empowers them to manage natural resources, including forests and other ecosystems. These legal policies demonstrate Nepal's commitment to conservation and ecological restoration efforts; however, implementing and enforcing these policies and laws remain crucial to achieving the desired outcomes. The Forest Act (2019) provides the legal framework for the sustainable management of Nepal's forests. The act includes community forestry provisions, encouraging local communities to manage and restore degraded forests. The Climate Change Policy (2019) emphasizes integrating climate change adaptation and mitigation measures into development planning and promotes ecosystem-based approaches to climate change mitigation and adaptation.

In addition to these policies, Nepal has also implemented a range of ecosystem restoration and conservation practices. These include community forestry, watershed management, soil conservation and biodiversity conservation initiatives, which involve local communities and stakeholders and are essential for the long-term sustainability of the country's natural resources. The entire legal development history of Nepal is presented in Table 11.1.

11.3.2 International commitments

The government of Nepal has been an active member and has committed to several international ecological conservation and restoration strategies and policies to protect natural resources. Several, which have been implemented to protect the world's natural resources, are described below.

Convention on Biological Diversity (CBD) is an international treaty that aims to conserve biodiversity, promote the sustainable use of its components, and ensure the fair and equitable sharing of the benefits derived from genetic resources. In total, 196 countries, including Nepal, have ratified the CBD.

Nepal has committed to supporting Aichi Targets 5, 11 and 15 to implement ecosystem restoration programmes as a signatory of the Convention on Biological Diversity and the Aichi Declaration. The Food and Agriculture Organization has proposed establishing the Forest and Landscape Restoration (FLR) mechanism to help countries achieve their commitments towards the Bonn Challenge and the Aichi Biodiversity Targets. This mechanism will operate globally and aims to support the implementation, monitoring and reporting of forest and landscape restoration by facilitating a multi-stakeholder process, developing and disseminating tools and best practices related to FLR, supporting the establishment of pilot projects, helping broker new large-scale projects and programmes, and ensuring compliance with accepted guidelines, norms and standards.

Other commitments of Nepal are:

Paris Agreement: an international treaty that aims to combat climate change by limiting global warming to below 2°C above pre-industrial levels and pursuing efforts to limit the temperature increase to 1.5°C. The agreement also promotes the adaptation of communities to the impacts of climate change.

United Nations Framework Convention on Climate Change (UNFCCC): an international treaty that aims to stabilize greenhouse gas concentrations in the atmosphere. The treaty provides a framework for international co-operation to mitigate and adapt to climate change.

Sustainable Development Goals (SDGs): 17 goals adopted by the United Nations General Assembly in 2015, which aim to end poverty, protect the planet and ensure that everyone enjoys peace and prosperity. Several goals are related directly to ecological conservation and restoration, including Goal 13 on climate action,

Goal 14 on life below water and Goal 15 on land.

Ramsar Convention: an international treaty that provides the framework for the conservation and wise use of wetlands and their resources; 171 countries have ratified the convention and, currently, there are 2341 wetlands designated as Ramsar sites.

The New York Declaration on Forests (2014) is a voluntary commitment by governments, companies, indigenous peoples and civil-society organizations to halt global deforestation by 2030 and to restore 150 million ha of degraded forests and lands by 2020, and an additional 200 million ha by 2030. The declaration is supported by a coalition of partners, including the UNFF, the CBD, the UNFCCC and the UNCCD.

The United Nations Convention to Combat Desertification (UNCCD) (1994) aims to combat desertification, land degradation and drought and to promote sustainable land management. The UNCCD recognizes the importance of restoring degraded lands and landscapes to achieve its objectives.

The Bonn Challenge (2011) aims to restore 150 million ha of degraded and deforested land by 2020 and 350 million ha by 2030. The initiative is led by the government of Germany and the International Union for Conservation of Nature (IUCN). Various governments, international organizations and civil-society groups support it.

REDD+ (Reducing Emissions from Deforestation and Forest Degradation) is an international programme that aims to incentivize the reduction of greenhouse gas emissions from deforestation and forest degradation in developing countries. The programme provides financial and technical support to countries committed to reducing deforestation and forest degradation emissions.

11.3.3 Ongoing efforts to ecological restoration in Nepal

Several ongoing ecological restoration projects in Nepal aim to restore degraded ecosystems and enhance biodiversity conservation. Some examples are:

Community-based Forest Management (1990s) is a successful model of community-based forest management in Nepal. The programme involves local communities in the management of forests, including restoration activities such as tree planting and soil conservation measures.

Rangeland management (2015) aims to restore degraded grasslands by promoting sustainable grazing practices and habitat restoration activities.

Wetland restoration: the Koshi Tappu Wildlife Reserve Wetland Restoration Project (2014) aims to restore degraded wetlands in reserves through habitat restoration, water management and community engagement.

Watershed management (2016) aims to restore degraded watersheds through soil conservation, reforestation and sustainable land-use practices.

Urban green infrastructure: the Kathmandu Metropolitan City Green Infrastructure Project (2019) aims to enhance urban biodiversity and ecosystem services by promoting green roofs, walls and urban parks.

Terai Arc Landscape (TAL) programme is a landscape-level conservation programme that aims to restore degraded ecosystems and enhance the provision of ecosystem services in the Terai region. The programme includes community forestry, habitat restoration and wildlife conservation measures.

National Biodiversity Strategy and Action Plan (NBSAP) provides a framework for the conservation and sustainable use of Nepal's biodiversity. The plan includes measures for the restoration of degraded ecosystems and the protection of critical habitats.

National Adaptation Programme of Action (NAPA) outlines Nepal's response to climate change and includes measures for restoring degraded ecosystems and adapting communities to changing environmental conditions.

The Community Forestry Programme is considered one of Nepal's most successful forest resource management programmes, particularly in the mid-hill regions. This programme has improved the forest structure and dynamics in many areas, and, to date, 1.45 million households, or 35% of the Nepalese population, are involved in community forestry management programmes. In recent years, the government has also started managing some production forests scientifically in the Terai under a collaborative arrangement to slow down deforestation and forest degradation. Trees grown on farmland and community-based forest management programmes, including leasehold forestry and buffer-zone community forestry, have also helped to mitigate deforestation and forest degradation.

These restoration projects involve various stakeholders, including government agencies, local communities, civil-society organizations and international donors. They demonstrate the potential of ecological restoration to enhance ecosystem services, promote biodiversity conservation and support sustainable livelihoods. Ecosystem restoration programmes and activities in Nepal are still in their infancy, and there is a pressing need to strengthen them.

11.3.4 Institutions involved in ecological restoration

In Nepal, the legal provisions for ecological restoration can traced to the federal, provincial and local levels. The Constitution of Nepal (2015), in Article 30, mandates protecting and conserving the environment and biodiversity. Additionally, the National Parks and Wildlife Conservation Act, 2029 BS (1973) and the Forest Act, 2049 BS (1993) provide the legal framework for protecting and managing forest and wildlife resources in the country.

At the provincial and local levels, each province and local government has laws and regulations related to environmental protection and management which vary in scope, depth and enforcement mechanisms. For instance, the Bagmati province has enacted the Bagmati Province Environmental Protection Act, 2077 BS (2021), which lays down provisions related to the protection, conservation and restoration of the environment and natural resources in the province.

For specific information on the legal provisions for ecological restoration in Nepal, the following sources can be consulted:

The Ministry of Forests and Environment is responsible for formulating and implementing policies, plans and programmes related to forest and environment conservation and management in Nepal.

The Department of National Parks and Wildlife Conservation protects, manages and conserves wildlife and their habitats in Nepal.

The Environmental Protection and Study Center is a non-governmental organization that handles environmental issues and provides research and consultancy services related to environmental management and protection.

The Center for Environmental and Agricultural Policy Research, Extension and Development (CEAPRED) is a research-based organization that handles environmental and agricultural issues and provides consultancy and technical support to various stakeholders.

The Ministry of Forests and Soil Conservation has implemented special landscape management programmes like the Terai Arc Landscape, Sacred Himalayan Landscape, Kailash Sacred Landscape, and other regional watershed management programmes for ecosystem restoration.

Several NGOs and INGOs have initiated a landscape management programme in the Chitwan-Annapurna Landscape since 2011, to enhance ecological processes and conservation of threatened species.

Other governmental organizations in Nepal, including the National Planning Commission, Ministry of Science, Technology and Environment, Ministry of Federal Affairs and Local Development, and Ministry of Forests and Soil Conservation are working in the field of ecosystem restoration and rehabilitation. Several national and international organizations have also initiated restoration activities through restoration projects, contributing to Nepal's ecosystem restoration. It is important to note that laws and regulations related to ecological restoration are subject to change and updating. Therefore, it is advisable to consult the latest information and resources to obtain accurate and up-to-date information for Nepal.

11.3.5 Environmental Protection Act (2019)

The Environmental Protection Act (2019) in Nepal is a comprehensive legislation that aims to protect the environment and promote sustainable development. Here are some of the key features of the act:

The institutional framework establishes a new institutional framework for environmental protection in Nepal. It creates the National Environment Commission as the apex body for environmental governance and establishes the Environment Protection Council as an advisory body to the commission.

Environmental impact assessment (EIA) is required by developers for projects that may have significant environmental impacts. The EIA process includes public consultation and requires the developer to address any concerns raised by stakeholders.

Pollution control measures includes provisions for controlling pollution, such as air, water and soil pollution. It requires industries to obtain permits to discharge pollutants and establishes penalties for non-compliance.

Biodiversity conservation includes provisions for biodiversity conservation, including establishing protected areas and regulating activities that may impact biodiversity.

Waste management includes provisions for the management of waste, including the regulation of hazardous waste and the establishment of waste management facilities.

Climate change includes provisions for mitigating and adapting to climate change. It requires the government to develop and implement climate change policies and strategies and establish a fund for climate change mitigation and adaptation activities.

Public participation emphasizes the importance of public participation in environmental decision making. It requires the government to provide access to environmental information and to consult with stakeholders on environmental matters.

Nepal's Environmental Protection Act 2019 provides a thorough legal framework for environmental protection and sustainable development. Its key features aim to promote responsible development that consider the environmental impacts of activities and promotes the conservation of natural resources.

11.3.6 Constitution of Nepal (2015) and environmental and natural resources

The Constitution of Nepal (2015) includes several key provisions related to the environment and natural resources. These provisions recognize the importance of environmental protection and sustainable development, the role of local communities in natural resource management, and the need for government policies and programmes to support conservation and environmental management. Article 30 in the constitution grants every citizen the fundamental right to live in a healthy environment, emphasizing the intrinsic link between environmental health and people's well-being. To effectively address environmental issues, responsibilities are divided among different levels of government in Nepal.

The federal government is responsible for developing national policies, acts, rules, guidelines, norms and standards related to environmental aspects such as wetlands of international importance, carbon services, environmental management, national forests, international water agreements and national parks.

The provincial government prepares policies and regulations specific to environmental issues within the province. The provincial government is also responsible for monitoring and co-ordinating activities within its jurisdiction, addressing matters like resource profiles, disaster risk reduction and management (DRRM), water use and environmental resources.

The local government plays a crucial role in implementing federal and provincial policies related to environmental matters. It co-ordinates activities within its jurisdiction and addresses issues such as DRRM, water management, conservation of mines and minerals, protection of watersheds, wildlife conservation and local development.

Additionally, the Nepal Fifteenth Plan (2019–2024) focuses on pollution control, environmental impact assessment monitoring and evaluation in development processes, public–private partnerships in waste management, green urban spaces, integrated river management, and collaboration with universities for research and development on pollution control. This plan also emphasizes managing climate change, promoting green development and encouraging the use of clean energy.

Nepal has adopted various policies and acts to effectively conserve and manage its natural resources and biodiversity. Notable initiatives include the National Wetland Policy, community-based forest management programmes, and the

National Parks and Wildlife Conservation Act. Through comprehensive plans, policies and governance mechanisms Nepal is working towards achieving environmental sustainability and a healthier future for its citizens and natural environment:

Article 27: Right to information related to environment and natural resources
Article 30: Right to live in a clean and healthy environment
Article 35: Protection and promotion of traditional knowledge and practices related to the environment and natural resources
Article 50: Federalism and decentralization in environmental and natural resource management
Article 51: Mitigation of climate change and promotion of sustainable development
Article 51: Protection of water resources
Article 56: Conservation of biodiversity
Article 59: Community forestry

These articles are just a few examples of the various environmental and natural resources in the Constitution of Nepal 2015. Here are some of the key features:

Fundamental rights: the constitution recognizes the right of every person to live in a clean and healthy environment. It also recognizes the right of indigenous peoples and local communities to participate in the management of natural resources.
Environmental protection: the constitution includes provisions for protecting the environment and conserving natural resources. It requires the government to develop and implement policies and programmes for conserving biodiversity, protecting water resources and mitigating climate change.
Community forestry: the constitution recognizes the importance of community forestry and includes provisions for local communities' management of community forests.
Right to information: the constitution guarantees the right of every citizen to access information related to the environment and natural resources.
Federalism and decentralization: the constitution establishes a federal system of government, with powers and responsibilities related to environmental management and natural resource management shared among the federal, provincial and local governments.
Traditional knowledge and practices: the constitution recognizes the importance of traditional knowledge and practices related to the environment and natural resources and requires the government to protect and promote them.
Sustainable development: the constitution emphasizes the importance of sustainable development and requires the government to promote the sustainable use of natural resources and biodiversity conservation (Table 11.2).

11.4 Ecological Restoration and Sustainable Development Goals

Ecological restoration can play a significant role in achieving the UN Sustainable Development Goals by improving the health and resilience of ecosystems, which in turn can enhance human well-being and promote sustainable development. Table 11.3 gives some examples of how ecological restoration and sustainable development goals are linked.

Goal 1: No poverty. Ecological restoration can reduce poverty by providing employment and generating income for local communities through reforestation, watershed restoration and sustainable agriculture.
Goal 2: Zero hunger. Ecological restoration can help achieve food security and nutrition goals by promoting sustainable agriculture and agroforestry practices that improve soil health, increase crop yields and support biodiversity.
Goal 3: Good health and well-being. Ecological restoration can improve human health by providing clean air and water, reducing the risk of natural disasters, and promoting recreational activities that support physical and mental well-being.
Goal 6: Clean water and sanitation. Ecological restoration can improve water quality and quantity by restoring wetlands, protecting watersheds and promoting sustainable water management practices.
Goal 7: Affordable and clean energy. Ecological restoration can contribute to renewable energy production through activities such as reforestation and the use of bioenergy crops.
Goal 13: Climate action. Ecological restoration can help mitigate climate change by sequestering carbon in soils and vegetation, restoring degraded

Table 11.2. Judicial regulations related to recovery and environmental sustainability.

Schedule	Jurisdiction	Restoration and environmental sustainability
5	Federal	7. International boundary rivers 11. Policies relating to conservation and multiple use of water resources 27. National and international environment management, national parks, wildlife reserves, wetland, national forest, carbon services 29. Land use, human settlement, development, tourism, environment adaptation
6	State	7. State-level electricity, irrigation, water supply services, navigation 19. Use of forests and water, management of environment within the state
7	Federal and state	13. State boundary river, waterways, environment protection, biodiversity 23. Utilization of forests and mountains, forest conservation
8	Local	10. Local market management, environment protection 11. Irrigation 19. Water supply, small hydro-power projects, alternative energy 21. Protection of watersheds, wildlife, mines and minerals
9	Federal, state and local	5. Services such as electricity, water supply and irrigation 6. Service fees, charges, penalties and royalties from natural resources 7. Forests, wildlife, birds, water uses, environment, ecology, biodiversity 14. Royalties from natural resources

Table 11.3. Ecological restoration that helps to meet SDGs.

SDG	Links to restoration
1: No poverty	Restored ecosystems provide increased and sustained food supply helping to minimize poverty or, if no poverty, then degraded ecosystems due to poverty will be restored.
4: Quality education	Well-educated people can use resources and ecosystem services rationally and also can contribute to restoration of the degraded ecosystem through their knowledge and skills.
6: Clean water and sanitation	This goal is closely tied to restoration ecology as it emphasizes the protection and restoration of water-related ecosystems and their services.
11: Sustainable cities and communities	Sustainable cities will restore degraded urban ecosystems or will minimize damage to existing ecosystems.
13: Climate action	Climate action includes efforts to restore degraded ecosystems, such as reforestation and afforestation, to enhance carbon sequestration and mitigate climate change.
14: Life below water	This goal focuses on the conservation and restoration of marine and coastal ecosystems, which are crucial for biodiversity and sustainable fisheries.
15: Life on land	Specifically addresses the protection, restoration and sustainable management of terrestrial ecosystems, including forests, wetlands and other habitats.

forests and grasslands, and promoting sustainable land-use practices.

Goal 15: Life on land. Ecological restoration can help protect biodiversity and support the conservation of threatened species by restoring degraded ecosystems and promoting sustainable land-use practices.

Ecological restoration can play an important role in achieving several SDGs by promoting biodiversity, improving water and soil quality, mitigating climate change, and enhancing ecosystem services. It can contribute to achieving several Sustainable Development Goals by providing ecological, social and economic benefits that support sustainable development.

Sustainable development and environmental protection are inextricably linked, and any meaningful effort towards achieving sustainable development must prioritize protecting and conserving the planet's natural resources. As the world's population continues to grow and the demand for resources increases, it is crucial that we adopt sustainable practices that preserve our environment for future generations. This can be achieved through policies that promote the responsible use of natural resources, the development of clean and renewable energy sources, and the adoption of environmentally friendly technologies.

11.5 Conclusions and Future Perspectives

To summarize, successful ecosystem restoration requires dedicated efforts, appropriate technology and optimal resources. Addressing both autogenic and allogeneic ecosystem processes is essential. Sustainable restoration and conservation play a vital role in safeguarding biodiversity and ecosystem services, positively impacting the livelihoods and well-being of millions. Although the government of Nepal has committed to ecological restoration, there is a necessity to delve deeper into the root causes of degradation. Utilizing integrated restoration approaches can enhance cost-effectiveness and feasibility. To achieve sustainable ecological restoration and conservation in Nepal, several key actions are recommended:

- Focus on restoring degraded ecosystems while considering the impact of human activities. This involves identifying areas that have been adversely affected by human actions and implementing restoration strategies to reverse the damage.
- Implement restoration measures both within and outside protected areas. It is essential to address environmental degradation and biodiversity loss not only within protected zones but also in areas outside these designated regions.
- Enhance public awareness about ecosystem services. Educating the public about the value and importance of ecosystem services can foster greater support and participation in biodiversity conservation efforts.
- Identify areas in urgent need of restoration and promptly develop suitable plans for restoration. Identifying priority areas that require immediate attention and implementing well-designed restoration plans are crucial for effective results.
- Place importance on post-treatment considerations, research and capacity building. After restoration, it is vital to assess the outcomes, conduct further research to improve future restoration practices, and build the capacity of local communities and stakeholders to engage in ongoing conservation.
- Foster sustainable development through various initiatives. Encourage the adoption of renewable energy sources, promote eco-friendly tourism practices, support sustainable agriculture, implement waste management strategies, conserve ecosystems, ensure water efficiency, invest in infrastructure development and empower local communities. These efforts collectively contribute to a more sustainable future for Nepal and its ecosystems.

By implementing these recommendations, Nepal can progress towards successful ecological restoration and contribute to the well-being of future generations.

References

Acharya, K.P., Dangi, R.B. and Acharya, M. (2011) Understanding forest degradation in Nepal. *Unasylva* 62(238), 31–38.

Allen, C.D., Savage, M., Falk, D.A., Suckling, K.F., Swetnam, T.W. *et al.* (2002) Ecological restoration of southwestern ponderosa pine ecosystems: a broad perspective. *Ecological Applications* 12(5), 1418–1433. DOI: 10.2307/3099981

Bhattarai, K., Conway, D. and Yousef, M. (2009) Determinants of deforestation in Nepal's Central Development Region. *Journal of Environmental Management* 91(2), 471–488. DOI: 10.1016/j.jenvman.2009.09.016

Choi, Y.D. (2007) Restoration ecology to the future: a call for new paradigm. *Restoration Ecology* 15(2), 351–353. DOI: 10.1111/j.1526-100X.2007.00224.x

Costanza, R., d'Arge, R., de Groot, R., Farber, S., Grasso, M. *et al.* (1997) The value of the world's ecosystem services and natural capital. *Nature* 387, 253–260. DOI: 10.1038/387253a0

Costanza, R., de Groot, R., Sutton, P., van der Ploeg, S., Anderson, S.J. *et al.* (2014) Changes in the global value of ecosystem services. *Global Environmental Change* 26, 152–158. DOI: 10.1016/j.gloenvcha.2014.04.002

Davis, M.A. and Slobodkin, L.B. (2004) The science and values of restoration ecology. *Restoration Ecology* 12, 1–3.

Foundations, M. (2010) The economics of ecosystems and biodiversity. In: *The Wetland Book.* Ecological and Economic Foundations, London and Washington, DC.

Gautam, D. (2020) *A Refresh Book of Rangeland Management*. Institute of Forestry, Tribhuvan University, Nepal. Available at: https://www.researchgate.net/publication/338719333_RANGELAND_MANAGEMENT (accessed 20 September 2023).

Hoekstra, J.M., Boucher, T.M., Ricketts, T.H. and Roberts, C. (2005) Confronting a biome crisis: global disparities of habitat loss and protection. *Ecology Letters* 8(1), 23–29. DOI: 10.1111/j.1461-0248.2004.00686.x

Intergovernmental Panel on Biodiversity and Ecosystem Services (IPBES) (2019) Global Assessment Report on Biodiversity and Ecosystem Services. Available at: https://ipbes.net/global-assessment (accessed 20 September 2023).

Ives, J.D. (1987) The theory of Himalayan environmental degradation: its validity and application challenged by recent research. *Mountain Research and Development* 7, 189–199.

Karkee, K. (2004) Land degradation in Nepal: a menace to economy and ecosystems. Master's programme in Environmental Science (LUMES), Lund University, Lund, Sweden. Available at: http://info.thinkgeography.org.uk/ Year%2010% 20Geog/Year%2010/Soils/15KrishnaKarkee.pdf (accessed 20 September 2023).

Kok, M., Alkemade, R., Bakkenes, M., Boelee, E., Christensen, V. *et al.* (2014) How sectors can contribute to sustainable use and conservation of biodiversity. PBL Netherlands, Environmental Assessment Agency, Secretariat of the Convention on Biological Diversity: Montreal, QC, Canada, 2014. Technical Series No. 79. Available at: https://www.cbd.int/doc/publications/cbd-ts-79-en.pdf (accessed 20 September 2023).

Maskey, R.B., Sharma, B.P. and Joshi, M. (2003) Human dimensions in sustainable land use management in degraded land areas of Nepal. Paper prepared for presentation at the open meeting of the Global Environmental Change Research Community, Montreal, Canada, 16–18 October.

Millennium Ecosystem Assessment (2005) *Ecosystems and Human Well-being* (Vol. 5). Island Press, Washington, DC.

MoEST (2006) *Implementation of UN Convention to Combat Desertification: Third National Report Nepal.* Ministry of Environment, Science and Technology, Kathmandu.

MoFSC (2009) *Nepal's Fourth Assessment Report to Convention on Biological Diversity*. Ministry of Forests and Soil Conservation, Babarmahal, Kathmandu.

MoFSC (2014) *Nepal's Fifth Assessment Report to Convention on Biological Diversity*. Ministry of Forests and Soil Conservation, Babarmahal, Kathmandu.

Pandey, A., Himanshu, S.K., Mishra, S. and Singh, V.P. (2016) Physically based soil erosion and sediment yield models revisited. *Catena* 147, 595–620. DOI: 10.1016/j.catena.2016.08.002

Pant, R.R., Bishwakarma, K., Qaiser, F.U.R., Pathak, L., Jayaswal, G. *et al.* (2021) Imprints of COVID-19 lockdown on the surface water quality of Bagmati river basin, Nepal. *Journal of Environmental Management* 289, 112522. DOI: 10.1016/j.jenvman.2021.112522

Parker, V.T. and Pickett, S.T.A. (1997) Restoration as an ecosystem process: implications of the modern ecological paradigm. *Restoration Ecology and Sustainable Development* 17.

Ruckelshaus, M.H., Jackson, S.T., Mooney, H.A., Jacobs, K.L., Kassam, K.A.S. *et al.* (2020) The IPBES global assessment: pathways to action. *Trends in Ecology & Evolution* 35(5), 407–414. DOI: 10.1016/j.tree.2020.01.009

SER (2004) *The SER International Primer on Ecological Restoration*. Society for Ecological Restoration International, Science and Policy Working Group, Tucson, Arizona.

Uddin, K., Abdul Matin, M. and Maharjan, S. (2018) Assessment of Land Cover Change and Its Impact on Changes in Soil Erosion Risk in Nepal. *Sustainability* 10(12), 4715.

UNEP (2014) *Convention on Biological Diversity (Ecosystem Conservation and Restoration).* United Nations Environment Programme (UNEP), Pathumthani, Thailand.

Vihervaara, P., Rönkä, M. and Walls, M. (2010) Trends in ecosystem service research: early steps and current drivers. *Ambio* 39(4), 314–324. DOI: 10.1007/s13280-010-0048-x

Vitousek, P.M., Mooney, H.A., Lubchenco, J. and Melillo, J.M. (1997) Human domination of Earth's ecosystems. *Science* 277(5325), 494–499. DOI: 10.1126/science.277.5325.494

WHO (2005) *Ecosystems and Human Well-Being (Health Synthesis).* World Health Organization, Geneva, Switzerland.

12 Ecological Conservation and Restoration in Pakistan

Muhammad Khalid Rafiq[1,*], Rukhsanda Aziz[2], Jalal Hayat Khan[1], Usman Ali[1] and Zhanhuan Shang[3]

[1]*Rangeland Research Institute, National Agricultural Research Center, Islamabad, Pakistan;* [2]*Center for Interdisciplinary Research in Basic Sciences, International Islamic University, Islamabad, Pakistan;* [3]*College of Ecology, Lanzhou University, Lanzhou, China*

Abstract

Pakistan is the world's fifth most populous country, with 241 million people, and boasts one of the largest irrigation systems in the world; but today, Pakistan is a water-scarce country. Rangelands, forests and agriculture are the major land uses in Pakistan. Over-exploitation and mismanagement of resources, and natural hazards such as floods, have led to ecological degradation and desertification resulting in food insecurity and reduced economy. Consequently, the current status of ecological degradation should be assessed and measures for ecological conservation should be developed. This chapter focuses on: (i) the causes, problems and risk assessment of ecological degradation in Pakistan; (ii) the link between people's livelihoods and ecology; (iii) the ecological conservation and restoration strategies and policies of local communities, the government and international organizations; and (iv) suggestions and recommendations for sustainable development in the future.

Keywords: ecological degradation, ecosystem restoration, sustainable development, natural hazards

12.1 Introduction

Pakistan occupies an area of 88 million ha (Mha) lying between 23° and 37° north latitude and 61° and 76° east longitude, and has a number of agroclimatic zones (Fig. 12.1). Mountainous areas on the northern and western boundaries cover more than two-thirds of the country, the Indus plains (I & II) extend over 20.7 Mha, and the Thar desert (III), a vast region of sand ridges, covers 6.5 Mha. The climate is generally arid subtropical, but intensively modified by altitude. The annual rainfall ranges from 125 mm in the extreme southern plains to between 500 mm and 875 mm in the country's mountainous and northern plains, with about 70% of the total precipitation occuring as heavy storms between July and September. Summer months in the Indus plains and the Thar desert are very hot with monthly mean maximum temperatures exceeding 40°C, and temperatures reaching up to 53°C. By contrast, in winter months, mean monthly temperature in the plains can be as low as 2–5°C, and often falls below zero in mountainous areas.

Agro-ecosystems in Pakistan are straining under increasing demands for food and forest resources. Currently, Pakistan is the world's fifth

*Corresponding author: khalid_parc@hotmail.com

DOI: 10.1079/9781800622579.0012

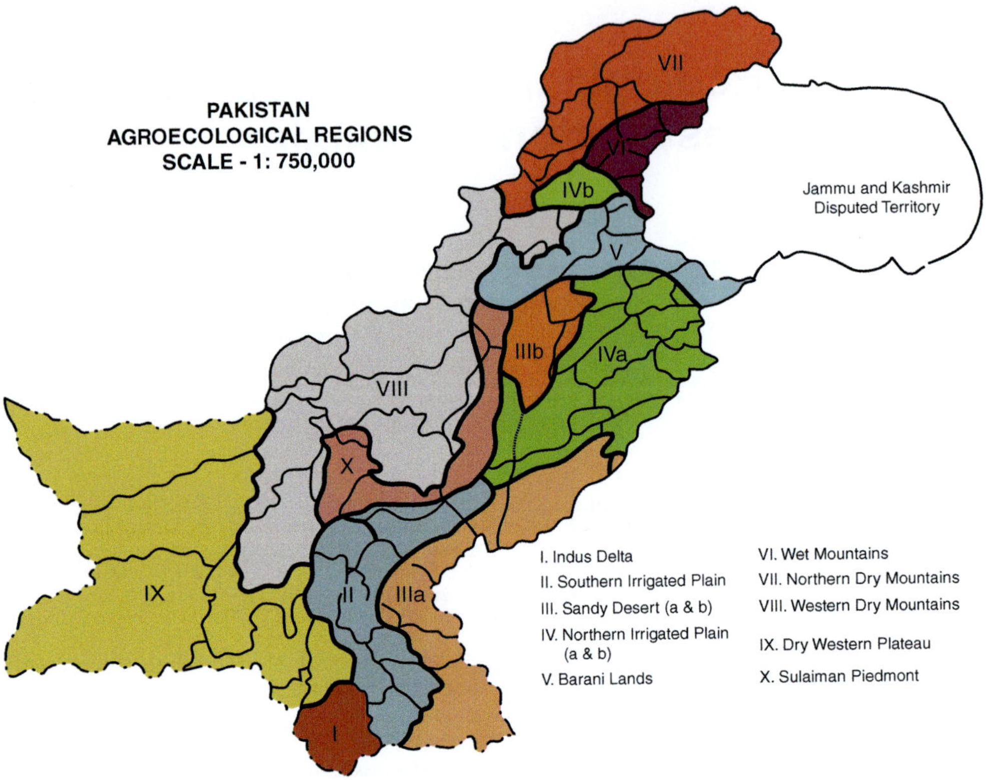

Fig. 12.1. Agro-ecological regions in Pakistan. (From PARC, 2019; Baig *et al.*, 2020; courtesy of Springer Nature BV. Permission conveyed through Copyright Clearance Center, Inc.)

most populated country, with 241 million people and a mean density of 2.6 people per ha (Finance Division Government of Pakistan, 2020). The country's population increased from 32.5 million at the time of independence in 1947 to 208 million by 2017 (Pakistan Bureau of Statistics, 2017a; Finance Division Government of Pakistan, 2020), an average growth rate of 2.7% per annum. With such growth rates, the country will have to double its annual food production every 15 years (Ali *et al.*, 2017), but this is challenged by limited resources due to overexploitation and mismanagement. Slightly over 20 Mha are cultivated and 15.6 Mha are irrigated (Ministry of Finance, 2007) and, of the cultivated area, about 21% is double-cropped, while 25% remains fallow at any time. More than 80% of the arable land is irrigated producing more than 90% of the total food for people and feed for livestock (Government of Pakistan, 2008). However, due to competition among the farming communities for irrigation water, huge volumes of water are lost in water transportation through the canals (Ma *et al.*, 2008; Ali *et al.*, 2017). Rangelands and forestry resources are being over-exploited leading to ecological instability and acceleration of land desertification. At least ten ecosystems of importance for their species richness and/or unique communities of flora and fauna are threatened with habitat loss and degradation. Determining the current status of ecological degradation and making suggestions for measures for ecological conservation are urgently needed. This chapter focuses on: (i) the causes, problems and risk assessment of ecological degradation; (ii) the link between people's livelihoods and ecology; (iii) the ecological conservation and restoration strategies and policies of the government, international organizations and local communities; and (iv) suggestions for sustainable development for Pakistan in the future.

12.2 Causes, Problems and Risk Assessment of Ecological Degradation in Pakistan

There is substantial evidence of widespread environmental degradation throughout the terrestrial ecosystems in Pakistan. The northern mountains are the major source of water for the Tarbela and Mangla dams. However, due to heavy soil erosion caused by deforestation in the catchments, these reservoirs are silting up, thus reducing the capacity of power generation and availability of irrigation water. Barani lands are subjected to heavy soil erosion primarily due to improper land use by crop cultivation, livestock grazing and illegal removal of vegetation cover. Deserts have acute problems of shifting sand dunes and salinity, whereas the productivity of rangelands is hampered by heavy livestock pressure. The details of the contributing factors towards environmental degradation and associated problems faced by Pakistan are described in the following sections.

12.2.1 Deforestation

Forests cover only 5.2% of the area, which is insufficient to meet the needs as well as socio-economic stability of the country. Due to deforestation, forest cover is shrinking by 3.1% and woody biomass by 5% annually. Natural vegetation reduces sedimentation in water reservoirs and stabilizes sand dunes, while mangrove forests protect the port of Karachi against wave action and act as a nursery for shrimp. Every major type of forest or protective cover in Pakistan has suffered heavily in the recent past from indiscriminate cutting, over-grazing, poor management and anthropogenic activities, which have increased desertification, erosion and silting of reservoirs and reduced biological defences against water-logging and salinity. The ability to resist soil erosion caused by monsoon rains is largely dependent on vegetation and vigour of vegetation at ground level and, therefore, degraded land with sparse vegetation cover increases soil erosion and siltation of rivers and channels. The removal of topsoil reduces soil fertility resulting in low production of forage, fodder, fuelwood, timber and grains. Soil erosion in the watershed areas of rivers increases sedimentation load, which reduces the storage capacity of dams.

12.2.2 Rangeland degradation

Rangelands cover more than 60% of the total area of Pakistan, extending from the northern part of the country to the temperate and Mediterranean ranges in the western mountains. They also cover a considerable part of the Indus plain. With the increase in livestock, there is heavy pressure on natural vegetation. During the period of feed scarcity, thousands of cattle, sheep and goats depend on fodder plants, which are harvested to the main stem. Unsystematic livestock grazing reduces the productivity of rangelands due to soil compaction, over-grazing of fragile slopes and damage to terraces, growing trees and shrubs.

12.2.3 Water-logging and salinity

The major factor contributing to water-logging in cultivated areas is excessive percolation from the canals, which builds up the groundwater level. Human activities such as cultivation of high delta crops on highly or moderately permeable soils, obstruction of natural drainage channels by the construction of buildings and roads, improper alignment and poor maintenance of artificial open-drainage systems, and inefficient disposal of excess rainwater also contribute to the water-logging problem. A total water-logged area with a water table depth of 5–10 feet covers approximately 11 million ha. Salinity and sodicity are associated with irrigation but these also occur as a consequence of soil formation processes over the centuries. At present, over 2.8 million ha are affected by salinity/sodicity in Pakistan. Water-logging, salinity and sodicity have reduced the drainage capacity of the soils resulting in lower soil fertility, decline in crop yields and loss of biodiversity.

12.2.4 Droughts and flooding

Periodic droughts are seen as a major cause of desertification, and adverse effects of droughts can last for many years. Recently, the provinces of Balochistan, Sindh and southern Punjab were affected severely by a prolonged dry spell. It is estimated that 2.2 million people and 7.2 million head of livestock were affected. Flood occurrence could be an occasional or regular feature of the region. Being an arid to semi-arid country, floods are usually caused by excessive precipitation, particularly during the monsoon season. Floods in 1973 and 1992 caused land degradation and loss of biodiversity and severe damage to the national economy. Floods prohibit cultivation in the summer season, destroy farmhouses and other facilities and bury fertile topsoil under relatively infertile sediments.

12.2.5 Socio-economic constraints

The implications of rapid population growth are grave and multi-faceted. It adversely fosters the process of subdivision and fragmentation of farmlands, adds to the fragile and marginal lands by denudation of forests and rangelands, and can lead to massive migration from rural to urban areas. It is the poor who suffer from desertification and drought. Unless properly managed, the already degraded resources come under heavy pressure. Agriculture, forestry, hunting and fishing account for 66% of the rural workforce, with over half the labour force engaged in agriculture.

12.3 Ecological Degradation in Pakistan and Risk Assessment

Ecological degradation and desertification is a serious problem globally, but it is more acute in Pakistan where almost 75% of the land is either already affected or likely to be affected. Population pressure is the fundamental underlying cause for degradation and desertification in Pakistan, while major causes include (i) excessive felling of trees; (ii) uncontrolled and excessive livestock grazing and browsing in forests and rangelands; (iii) inappropriate cultivation practices in hilly and rain-fed farming areas contributing to wind and water erosion; (iv) inadequate drainage in heavily irrigated areas causing water-logging and salinity; and (v) inadequate efforts with regard to watershed protection and management in the catchment areas of reservoirs.

12.3.1 Ecological degradation and natural hazard risk in Pakistan

Natural resources are being used continuously, which is leading to habitat deterioration, ecosystem disturbances, reduction of species and depletion of water and soil. Depletion and degradation of natural resources, fuelwood deficiency, air and water contamination, soil erosion and extinction of species are among the consequences of many human activities to fulfil their needs and demands. Pakistan is a developing country with most of the population living in rural areas.

Pakistan ranked 18 out of 191 countries in the world in the 2020 Inform Risk Index. The country is particularly vulnerable to earthquakes and internal conflicts but is also at risk of flooding (ranked jointly eighth), including riverine, flash, coastal, tropical cyclones and their associated hazards (ranked jointly 40th) and drought (ranked jointly 43rd). Social vulnerability (ranks 37th) is driven by its high rate of poverty.

12.4 Link between People's Livelihoods and Ecology

12.4.1 Socio-economic status of Pakistan

About 64% of Pakistan's population lives in rural areas and over 50% is engaged in agriculture using primitive methods. Agriculture accounts for 18.9% of the GDP and 42.3% of total employment (Finance Division Government of Pakistan, 2020). Therefore, a successful agriculture policy is critical to national economic growth and poverty reduction (Ministry of Finance, 2007). The average household in rural areas is composed of six residents, with two earners per household (Pakistan Bureau of Statistics, 2017a,b; Farooq *et al.*, 2018), and, thus, a dependency ratio of 3:1. In addition, household incomes are much lower in rural areas than in urban areas. Although poverty is declining, 24.3% of people in Pakistan live below the poverty line (World Bank, 2006). Country-wide literacy rates are about 62%, with rates for males (72.5%) substantially greater than for females (51.8%) (Ministry of Finance, 2019). The socio-economic arrangement of rural Pakistani villages can be categorized into three broad patterns based on ownership and control (Baig *et al.*, 1999). In the first scenario, a single family owns the entire village, thereby controlling the whole local economy. In the second scenario, more than one family owns and controls the community but control in these villages is similar to that of single-family. In the third scenario, several small landowners own the village and they work together with tenants to provide food and shelter (Baig *et al.*, 1999). About 80% of Pakistan's rural population consists of smallholders or landless people (Shahbaz and Ata, 2014; Finance Division Government of Pakistan, 2020). These people typically suffer from poverty, malnutrition, disease and ignorance and have no economic security. Most rural areas contain no large-scale industries and the few craft and cottage industries that exist depend on the availability of raw material (Baig *et al.*, 1999). Most goods and commodities consumed by the villages are produced within the villages. Among village artisans are carpenters and blacksmiths who can construct and repair agricultural implements and tools. Historically, such artisans were paid with crops at harvest time, but this has declined as skilled professionals have migrated to urban areas and factories, and now specialize in manufacturing agricultural tools and implements (Baig and Ehrenreich, 1997; Baig *et al.*, 1999).

Economic growth in Pakistan has generally been achieved at the expense of the environment, since most production and manufacturing processes exploit healthy but scarce ecosystems. The heavy dependence on natural resources and resulting environmental degradation has generated a cycle of 'ecological poverty' with implications for the sustainability of economic growth.

12.4.2 Dependence on natural resources and increased pressure on the ecosystems

The country's large and growing population is sustained by, and critically dependent on, the Indus river and its tributaries. Irrigated agriculture accounts for more than 90% of food and fibre, and for most fodder production. Population growth has reduced per capita water availability to less than 1100 m^3 in 2007. Furthermore, irrigation has serious environmental implications such as water-logging and salinity, both of which affect crop yields. Around 25–30% of the canal-irrigated area becomes water-logged after the monsoon season. Irrigation leads to an additional 1.7 tons of salt deposited per ha every year, and, consequently, 8% of land suffers from severe salinity. Although the use of tube wells has reduced the extent of water-logging, the use of groundwater containing high levels of dissolved salts has resulted in sodicity, which also affects crop yields, especially in clayey soils. The drawdown of fresh groundwater has led to the intrusion of brackish water from surrounding zones, and an overall deterioration in groundwater quality. There is also an increased risk of flooding owing to rising riverbeds in the lower Indus basin. Diversions and inadequate environmental flows have resulted in seawater intrusion and the degradation of the Indus delta and adjacent coastal areas.

12.4.3 Dependence of poor communities on natural resources

The vast majority of Pakistan's population lives in rural areas. Approximately 34% of the rural population is poor and includes mainly landless households or small tenant farmers. In the absence of any formal sources of employment, their livelihoods are linked to agriculture and the use of natural resources. Many provide services to the agriculture sector and graze their small stock on marginal lands around farm fields. Thus, the degradation of various ecosystems has an immediate and direct impact on the livelihoods of the rural poor. Similarly, the pollution of land, water and air affects the health and livelihoods of poor households who do not have coping strategies against unexpected stress factors. There has been an impressive decline in national poverty in the four years between 2001 and 2005; however, rural poverty declined at a slower rate than the national average despite satisfactory growth in the agricultural GDP in three of the four years. One implication of this is that poverty in Pakistan is concentrated in rural areas among households who have neither access to adequate holdings of cultivable land nor security of tenure.

12.4.4 Cost of environmental degradation in Pakistan

Ecological and environmental degradation causes huge losses to the economy in Pakistan. The World Bank estimated that pollution and natural resources degradation cost billions of rupees in damages and had negative impacts on the livelihoods of the people. Further research must be conducted on a number of important ecological issues, e.g. the offsite costs of deforestation, ecosystem services valuation, the long-term health-related costs of toxic waste generation by the industry sector, and the impact of environmental degradation on the storm-surge protection function of Pakistan's mangroves.

12.5 Pakistan's Commitment to Ecological Restoration

Ecosystem restoration not only protects but also improves the livelihoods of people who depend on them. It also aids in disease control and reduces the risk of natural disasters. Pakistan has long struggled with rising levels of environmental degradation, which costs the national economy billions of rupees each year. However, recently, a number of initiatives, at an investment of billions of rupees, have been undertaken to restore ecosystems through nature-based solutions as part of Clean Green Pakistan vision. This form of restoration would also aid in the achievement of many of the UN's Sustainable Development Goals (Khan *et al.*, 2019).

12.5.1 Multi-lateral environmental agreements

Multi-lateral environmental agreements (MEAs) help address the environmental problems being faced by the international community. The ultimate goal of reaching MEAs is to realize sustainable development across the globe. According to the World Bank (2006), Pakistan's economy loses Rs 365 billion annually to environmental hazards. Pakistan is signatory to several MEAs and has acceded to other non-legally binding instruments, such as Agenda-21, Rio Principles and Johannesburg Plan of Implementation, aiming at sustainable development of natural resources. As a party to various conventions and protocols, Pakistan has been participating in different Conferences of Parties (CoPs), Meetings of Parties (MoPs), and international consultative and technical meetings organized by the UN and other international bodies.

12.5.2 Current project initiatives for ecological restoration in Pakistan

Pakistan is maintaining approximately 4.786 million ha (5.45%) under forest cover. By forest type, dry temperate forests have the largest proportional coverage (36%) followed by subtropical broad-leaved shrub (19%), moist temperate (15%), chir pine (13%), riverine (4%), irrigated plantation (4%), thorn (3%), mangrove (3%) and sub-alpine forests (2%). Climatic conditions, rural poverty, dependence on natural resources, insufficient forest cover and rapid deforestation rate have rendered the country one of the most vulnerable to ecological degradation. Forest, biodiversity and wildlife resources have also suffered from the adverse effects of climate change. Furthermore, these resources are under tremendous pressure because of land-use changes and habitat destruction. Due to population increase, the harvesting of fuelwood and timber has increased. Such pressures have rendered most forests of poor and medium density in need of urgent restoration. The overall improvement of the sector requires long-term programmes to restore ecological stability. The following sections describe key projects initiated by the government for the ecological restoration in Pakistan.

Ten Billion Tree Tsunami Programme

The Ten Billion Tree Tsunami Programme (TBTTP) was initiated in 2019 with a total cost of Rs 125.1843 billion on a cost-sharing basis for four years (2019–2023) to plant/regenerate 3.29 billion plants. A total of 528.34 million plants were planted/regenerated/distributed and the provincial Forest and Wildlife Departments, Azad Jammu and Kashmir (AJK) and Gilgit Baltistan (GB) provided 85,000 green jobs during 2020–2021, while a cumulative total of 1007 million plants were planted/regenerated/distributed to increase tree cover on 0.488 million ha over the past two years. An independent third-party consortium of IUCN, WWF and FAO conducted preliminary assessment of the achievements of this programme, while this initiative of the government was registered under the Bonn Challenge.

Protected Areas Initiative

The programme was launched by the prime minister to improve management and governance of 23 protected areas to preserve rare fauna and flora and promote ecotourism. The total protected area will increase from 12% to 15% by 2023. Nanga Parbat National Park and Himalayan National Park were inaugurated in Gilgit Baltistan, while Tilla Joggian Park and Salt Range National Park were inaugurated in Punjab.

REDD+ Readiness and Preparation Project

'Reducing emissions from deforestation and forest degradation, conservation of existing forest carbon stocks, sustainable forest management and enhancement of forest carbon stocks' (REDD+) were concepts adopted in 2010 by countries under the UN Framework Convention on Climate Change (UNFCCC). The concepts relate to the absorption of atmospheric carbon through forest resources. Due to the accumulation of carbon in trees, their financial value increases, as carbon stocked in forests can be traded in carbon markets.

The Ministry of Climate Change, with a grant of US$ 7.81 million from the Forest Carbon Partnership Facility (FCPF) of the World Bank initiated the REDD+ Readiness Preparation Project. Four essential elements of REDD+ were required to fulfil the criteria of receiving results-based payments under REDD+ mechanism as follows:

i. development of National REDD+ strategy and implementation framework;
ii. national forest monitoring system;
iii. forests reference emission levels and/or forest reference levels; and
iv. safeguard information system.

Sustainable land management to combat desertification in Pakistan (SLMP-II)

The SLMP-II project, at a total cost of Rs 1666.695 million, was approved by the Central Development Working Party (CDWP) in March 2015 and started in September 2015. It is a modified version of the SLMP pilot project to be implemented in 14 dry land districts in four provinces 'to combat land degradation and desertification in Pakistan' with the primary objective 'to promote sustainable management of land and natural resources in the arid and semi-arid regions of Pakistan in order to restore degraded ecosystems and their essential services, reduce poverty, and increase resilience to climate change'. Integrated activities were performed in agriculture, forests, irrigation, livestock, rangelands, and soil conservation and stabilization sectors.

12.6 Research and Development Recommendations for Ecological Restoration in Pakistan

1. Natural resources inventory procedures should be standardized through the involvement of stakeholders engaged in rangelands, namely IUCN, WWF, FAO and government institutions so that records on the extent and status of resource bases are accurate. Land suitability classification should be developed at a national scale.
2. Projects on ecological restoration and climate change should be reviewed by public sector institutes. The National Climate Change Policy of 2012 and other relevant policies provide the centre with the authority to unite the provinces; however, little to no action has been taken. The Pakistan Climate Change Authority, Climate Change Council and Climate Change Fund were all to be built into the 2017 Climate Change Act, but they have yet to be fully implemented. A National Plan for Ecosystem Restoration and subnational plans at every provincial level focused on the UN Decade Strategy must be generated quickly to demonstrate responsiveness.
3. Public–private partnerships should be promoted to help fund restoration. Technical and technological capabilities of public sector institutes, research and development need improvement.
4. Environmental studies and ecosystem restoration should be added to the national curriculum so that the new generation understands the damages caused by degrading the environment. The public should be made aware of the dangers of over-exploitation of natural resources and the need for resource conservation.

5. Deployment of low carbon-emitting and polluting transportation and the promotion of renewable resources (biofuels) to assist Pakistan in conserving its biodiversity through resource management and ecosystem restoration should be encouraged.

6. Policies should be followed that ensure the protection of human settlements, including hazardous-waste management, and healthy and sustainable agricultural and rural growth.

7. Biological/green corridors should be built alongside economic corridors (such as in case of China Pakistan Economic Corridor (CPEC)) to create coherence between socio-economic growth and environmental sustainability. Pakistan can learn from the precedents of projects initiated by the Asian Development Bank such as Biodiversity Conservation Corridor Initiative (BCI) and the Core Environment Program (CEP), which were successful in improving environmental quality. The CEP and BCI were planned to balance growth and the environment (e.g. tourism, climate change, institutional support and an integrated spatial and strategic approach) and are successful in resolving environmental degradation issues (Ancha, 2011) and establishing green corridors (SADC, 1997; Khan *et al.*, 2019). The design of the biological conservation complex in Bhutan could aid in protecting national parks and wildlife sanctuaries across Pakistan. Furthermore, the Mesoamerican Biological Corridor (MBC), which protects biodiversity sites while also promoting socio-economic development (Miller *et al.*, 2001), could aid Pakistan in reducing natural resources exploitation and poverty. Projects such as the Nicaragua Atlantic Biological Corridor Project (1997–2005), the Honduras Biodiversity in Priority Areas Project (1997–2005), the Panama Atlantic Mesoamerican Biological Corridor Project (1998–2005), Costa Rica's Ecomarkets Project (2000–2006) and Mexico's Mesoamerican Biological Corridor Project (2000–2009), all sponsored by the World Bank, show promise in mitigating the negative effects of growth (Brown *et al.*, 2014). These projects have the potential to be beneficial for Pakistan, as they allow for both industrial expansion and biodiversity conservation, and they are focused on internationally recognized green growth initiatives.

8. Continuous evaluation of previous and ongoing restoration projects and the monitoring of climate variables are needed to generate plans so that Pakistan could deliver efficiently and productively on the agenda of sustainable development by the year 2030 (Abhilash, 2021).

References

Abhilash P.C. (2021) Restoring the unrestored: strategies for restoring global land during the UN Decade on Ecosystem Restoration (UN-DER). *Land* 10(2), 201. DOI: 10.3390/land10020201

Ali, A., Rahut, D.B. and Imtiaz, M. (2017) Implications of less tail end water on livelihoods of small farmers in Pakistan. *Outlook on Agriculture* 46(1), 36–43. DOI: 10.1177/0030727016689735

Ancha, S. (2011) *Regional: Core Environment Program and Biodiversity Conservation Corridors Initiative in the Greater Mekong Subregion, Phase 2.* Philippines. Available at: https://policycommons.net/artifacts/388716/regional/1353235/ (accessed 20 September 2023) CID: 20.500.12592/5hsw9q.

Baig, M.B. and Ehrenreich, J.H. (1997) The potential prospects and problems of agroforestry in Pakistan-1: research and development. *Pakistan Journal Agriculture Science* 34, 22–28.

Baig, M.B., Akbar, G., Straquadine, G.S. and Razzaq, A. (1999) Agroforestry extension and technology transfer to farmers in Pakistan. *Journal Science Vision* 5(1), 42–50.

Baig, S.U., Khan, H., Muneeb, F. and Dad, K. (2020) Formation of a hazardous ice-dammed glacier lake: a case study of anomalous behavior of Hassanabad glacier system in the Karakoram. *SN Applied Sciences* 2(1285).

Brown, E., Cloke, J., Gent, D., Johnson, P.H. and Hill, C. (2014) Green growth or ecological commodification: Debating the green economy in the global south. *Geografiska Annaler: Series B, Human Geography* 96(3), 245–259. DOI: 10.1111/geob.12049

Farooq, T.H., Gautam, N.P., Rashid, M.H.U., Gilani, M.M., Nemin, W., Nawaz, M.F., Islam, W., Zainab, M. and Wu, P. (2018) Contributions of agroforestry on socioeconomic conditions of farmers in central Punjab, Pakistan—A case study. *Cercetari Agronomice In Moldova (Agronomic Research in Moldavia)* 51, 91–101. DOI: 10. 2478/cerce-2018-0020

Finance Division Government of Pakistan (2020) Pakistan Economic Survey. Islamabad, Pakistan. Available at: https://www.pc.gov.pk/uploads/cpec/PES_2020_21.pdf (accessed 20 September 2023)

Government of Pakistan (2008) *Agricultural statistics of Pakistan 2008*. Economics Division, Ministry of Food, Agriculture and Livestock, Government of Pakistan, Islamabad.

Khan, N., Shah, S.J., Rauf, T., Zada, M., Yukun, C., *et al.* (2019) Socioeconomic impacts of the Billion Trees Afforestation Program in Khyber Pakhtunkhwa Province (KPK), Pakistan. *Forests* 10, 703. DOI: 10.3390/f10080703

Ma, W.J., Mao, Z.Q., Yu, Z.R., Van Mensvoort, M.E.F. and Driessen, P.M. (2008) Effects of saline water irrigation on soil salinity and yield of winter wheat–maize in North China Plain. *Irrigation and Drainage Systems* 22, 3–18. DOI: 10.1007/s10795-007-9027-1

Miller, K., Chang, E. and Johnson, N. (2001) *Defining common ground for the Mesoamerican Biological Corridor*. World Resources Institute, Washington, DC, pp: 45–45.

Ministry of Finance (2007) *Economic Survey of Pakistan (2006- 2007)*. Ministry of Finance, Government of Pakistan, Islamabad, Pakistan. Available at: http://www.finance.gov.pk/survey/ chapters/02-Agriculture.PDF (accessed 20 September 2023)

Ministry of Finance (2019) *Pakistan Economic Survey*. Ministry of Finance, Government of Pakistan, Islamabad, Pakistan. Available at: http://www.finance.gov.pk/survey/chapters_19/10-Education.pdf (accessed 20 September 2023)

Pakistan Bureau of Statistics (2017a) *Provisional Summary Results of 6th Population and Housing Census—2017*. Available at: http:// www.pbs.gov.pk/content/provisional-summaryresults6th-population-and-housing-census-2017-0 (accessed 20 September 2023)

Pakistan Bureau of Statistics (2017b) *Population Census*. Government of Pakistan, Islamabad, Pakistan. Available at: http://www.pbs.gov.pk/content/ population-census (accessed 20 September 2023)

PARC (2019) *Agro-ecological zones of Pakistan*. Agricultural maps of Pakistan. Available at: http://www.parc.gov.pk/index.php/en/ component/content/article/43-maps/19-agrimaps. (accessed 20 September 2023)

SADC (1997) *Southern African Development Community*. United Nations Commission on Sustainable Development.

Shahbaz, B. and Ata, S. (2014) *Agricultural extension services in Pakistan: challenges, constraints and ways forward*. Background Paper No. 2014/1, Enabling agricultural policies for benefiting smallholders in dairy, citrus and mango industries of Pakistan—Project No. ADP/2010/091. Victoria Institute of Strategic Economic Studies, Victoria University, Australia. Available at: www.vises.org.au/documents/2014_Sept_Background_Paper_1_Agriculture_Extension_Services_Pakistan.pdf (accessed 24 April 2024).

World Bank (2006) *Pakistan: Strategic Country Environment Assessment*. Report No. 36946-PK, South Asia Environment and Social Development Unit, South Asia Region. Available at: http://siteresources.worldbank.org/SOUTHASIA EXT/Resources/Publications/448813- 1188777211460/pakceavolume1.pdf (accessed 20 September 2023)

13 Forests: Conservation, Restoration and Sustainable Utilization in the Hindu Kush Himalayan Region

Wenyin Wang[1,*], Allan Degen[2], Zhen Peng[1], Binyu Luo[1] and Zhanhuan Shang[1]
[1]College of Ecology, Lanzhou University, Lanzhou, China; [2]Institutes for Desert Research, Ben Gurion University of the Negev, Israel

Abstract

Forest ecosystems in the Hindu Kush-Himalayan (HKH) region are rich in biodiversity and provide a range of ecosystem services; however, forest degradation is spreading and is threatening the forest resources. Consequently, rational management strategies are urgently needed to mitigate forest degradation and to promote sustainable development of the forests in the HKH region. Therefore, this chapter describes, systematically, the status of forests in the HKH region, including: (i) threats to forest resources and their impacts; (ii) conservation of forest resources; (iii) livelihoods from forest resources; and (iv) management strategies and restoration of forests. The chapter provides valuable information for policy-makers and practitioners engaged in forest ecosystem restoration in the HKH region.

Keywords: forest ecosystem, forest degradation, forest carbon stock, management strategy

13.1 Introduction

The HKH region is a diverse ecological buffer zone, covering 4,200,000 km^2, while, in 2021, forests in the region covered approximately 838,000 km^2 (ICIMOD, 2022). They are a major source of wood for construction, furniture and fuel, act as carbon sinks and are rich in biodiversity (Fig. 13.1). In addition, forests provide recreational and cultural services, and are a source of livelihoods for the local residents. Mountain communities are heavily dependent on forests; however, tourism, global climate change, changes in land use, ineffective management, over-development, pollution and mountain disasters are threatening mountain forests. Conservation strategies such as afforestation and reforestation, well-planned tourism, sustainable agriculture, alternative sources of energy and income, climate change mitigation, adaptation measures and community involvement in conservation are urgently needed to sustain forests and ecosystem services.

Although the region is rich in biodiversity, it is highly vulnerable to climate change and socio-demographic interactions (Kumar *et al.*, 2019). Under climate change impacts, there have been increases in the frequency, duration and severity of drought and heat stress and in pest and fire outbreaks, changes in vegetation phenology and soil nutrient dynamics, and a decrease in vegetation productivity (Wester *et al.*, 2019), which are altering the composition of the

*Corresponding author: wangwy@lzu.edu.cn

DOI: 10.1079/9781800622579.0013

Fig. 13.1. Forest landscape of Medog county, south-east Tibet. (Photo: Binyu Luo)

forest community structure and productivity, and affecting ecosystem services and forest-based livelihoods. Furthermore, over the past few decades, the HKH region has experienced rapid economic, social and environmental changes due to unsustainable development (Sharma *et al.*, 2019). According to GlobeLand30, the Tibetan Plateau lost 3345 km^2 of forest from 2000 to 2010. After the state issued the Plan for the Protection and Construction of the Ecological Safety Barrier in Tibet for 2008–2030 and the Plan for Regional Ecological Construction and Environmental Protection on the Tibetan Plateau for 2011–2030, the situation of forest loss reversed, and the forest area on the Tibetan Plateau increased by 2763 km^2 from 2010 to 2020.

13.2 Threats and Impacts to Forest Resources

Forests are important for the livelihoods of the majority of the population in the HKH region. For example, nearly 35% of the population in Nepal depends on forest resources for their livelihoods; the country has 6.61 million ha of forest land, accounting for 45% of the total area (FAO, 2018). Consequently, forest degradation and deforestation are major environmental concerns in Nepal (Chaudhary *et al.*, 2017). Climate warming and road accessibility are the main drivers of forest cover change (Murthy *et al.*, 2016). Furthermore, as with other mountain landscapes, deforestation, over-grazing, human activities, infrastructure development, and steep slope planting have led to the fragmentation of forests in the HKH region (Rueff, 2014; Chaudhary *et al.*, 2017). Excessive exploitation of natural resources for local use and trade, such as illegal wood harvesting, hunting of wildlife and habitat destruction are the root causes of biodiversity loss in the HKH region, while climate change accelerates the process (Khan *et al.*, 2022b).

13.2.1 Climate change and forest diversity in the HKH region

Anthropogenic activities have been identified as a major cause of global habitat fragmentation and biodiversity loss. The HKH region is one of the least disturbed habitats by human activity, in the world but climate warming is a serious threat (Kattel, 2022). Since 1960, the temperature has increased by 0.2°C per decade in the region, which is a much faster rate than the global average (Chen *et al.*, 2013). Climate warming is likely to be one of the most important drivers of biodiversity loss in the HKH region in the 21st century, as changes in species richness and β diversity respond strongly to climate warming (Bhattacharjee *et al.*, 2017), especially in tree-line ecotone zones. For example, driven by climate change, *Betula utilis* (Himalayan birch) shifted from the western to the eastern Himalayas (Hamid *et al.*, 2019). The interaction of photosynthetically active radiation and increased CO_2 in the tree ecotone led to species redundancy, resulting in the transition of *Betula utilis* as the dominant species to evergreen *Rhododendron* (Behera *et al.*, 2019). Furthermore, the dependence on fossil fuels will exacerbate the effect of climate change in HKH after the end of the COVID-19 pandemic.

Rapid warming at high altitudes could increase the sensitivity of trees to temperature. Tree growth was correlated positively with winter and spring precipitation in the arid western Himalayas, but with temperature and spring precipitation in humid south-east Tibet. This suggests that if temperature continues to rise in the HKH region, the forest area may decline in the western Himalayas due to water deficit caused by warming, but may expand in south-east Tibet due to trees benefitting from warming (Zheng *et al.*, 2021).

13.2.2 Land-use change and forest degradation

Soil erosion

Forests have a complex relationship in the 'percolation-evapotranspiration trade-off' chain. Natural forests have a greater capacity to store and release water, a slower water run-off rate, and a much longer duration of low flow than degraded forests (Roa-Garcia *et al.*, 2011; Krishnaswamy *et al.*, 2013). In the hilly regions of central Nepal, replacing degraded pastures with mature broad-leaved forests has a near-neutral effect on dry-season water flows, as the gains from infiltration are roughly the same as the losses from evaporation (Ghimire *et al.*, 2014). Vegetation plays an important role in reducing soil erosion. With forest degradation or deforestation, soil cover is reduced, which accelerates soil erosion due to the absence of roots to bind the soil. During monsoons, the Himalayan mountain forest transition zone is particularly vulnerable to intense and persistent rainfall.

Deforestation and forest degradation

Exploitation of natural resources is one of the main factors preventing forest area expansion. Over-exploitation includes clearing forests (Fig. 13.2) or shrubs for fuelwood and commercial timber, over-harvesting of non-timber products such as medicinal products, over-grazing, and unsustainable or illegal mining. Other factors include agricultural expansion, urbanization and infrastructure development (Chazdon, 2008;

Fig. 13.2. Forest status in south-eastern Tibet after two years (*left*) and 20 years (*right*) of logging. (Photo: Zhen Peng)

Lewis *et al.*, 2015). Reduction in forest resources threatens the livelihoods of millions of people (Erbaugh and Oldekop, 2018).

Increased timber demand due to population growth is one of the main reasons for large-scale logging across the HKH region, as locals are completely dependent on wood for house construction (Zeb, 2019). Warm temperate coniferous forests in the south-eastern Tibetan Plateau have almost disappeared, mainly due to commercial logging before the late 1990s (Cui and Graf, 2009). In the high mountains of Pakistan, the forest area decreased by 30% from 1968 to 2007, a third of which was due to agricultural expansion (Qasim and Mane, 2013). In the same region, agricultural land at middle and high altitudes expanded by 70% and forest area decreased by 50%, while from 1968 to 2007, agricultural land at low altitude increased by 130%, and forest area decreased by more than 30% (Qasim and Mane, 2013). The forest area in the mountains of India decreased by 0.2% per year (Reddy *et al.*, 2013), and that of Myanmar by 0.3% per year (Leimgruber *et al.*, 2005), but in Bhutan it increased by 0.22% per year (Gilani *et al.*, 2015).

According to data provided by Global Forest Watch, loss of tree cover from 2001 to 2022 ranged between a low of 18.9 km^2 in Afghanistan to a high of 45,600 km^2 in Myanmar. The loss of forest area through deforestation, other types of land use and forest degradation is a major indicator of carbon loss in a country or region. CO_2 emissions from tree-cover reduction from 2001 to 2022 was estimated to be as high as 2.48 Gt in Myanmar and 1.11 Gt in India (Table 13.1).

Animal hunting, plant harvesting and fires

Forest fires are one of greatest threats in the HKH region, causing long-lasting damage to forest biodiversity, forest carbon stocks, nutrient cycling and other ecosystem services (Fig. 13.3). In summer, forest fires are common in the western Himalayas due to reduced soil-moisture content and the accumulation of dead leaves on the forest surface (Zhang-Turpeinen *et al.*, 2020).

Table 13.1. Changes of forest area in the HKH region.

	Tree cover loss 2001–2022 (km^2)	Loss of moist primary forest 2002–2022 (km^2)	Carbon dioxide emissions
Afghanistan	18.9	—	—
Bangladesh	2280	76.4	124 Mt
Bhutan	244	81.0	13.8 Mt
China (Tibet)	337	24.9	11.8 Mt
India	21900	3930	1.11 Gt
Myanmar	45600	7030	2.48 Gt
Nepal	534	43.6	29.3 Mt
Pakistan	98.0	—	3.62 Mt

Data from the Global Forest Watch, Food and Agriculture Organization of the United Nations (FRA, https://fra-data.fao.org/assessments/fra/2020 [accessed 3 November 2023])

Fig. 13.3. Forests after different years of forest fires in south-east Tibet: (a) 1 year; (b) 3 years. (Photo: Zhen Peng)

From 2001 to 2020, there were more than 920,000 forest fires in the HKH region, with the highest number, 73,305, occurring in 2009 (Bhattarai *et al.*, 2022). From 2003 to 2017, there were 520,861 forest fires in India alone, mainly concentrated in dense evergreen and deciduous forests in the eastern Himalayas (Sannigrahi *et al.*, 2020). The year 2021 was one of the worst years for forest fires, with approximately 130 km^2 of forest area destroyed (DownToEarth, 2021). Forest fires in Nepal in 2021 were considered the worst in a decade (Azarkan, 2021).

Most fires in Himalayan forests are caused by humans (Giriraj *et al.*, 2010) and not by natural causes, as reported in Australia and California (Helvarg, 2019; Tran *et al.*, 2020). The number of forest fires is much higher in Myanmar and Mizoram (state in north-east India) than elsewhere because of the slash-and-burn agricultural practices (also known as *jhum* cultivation) (Springate-Baginski, 2018). In Kudupakia, India, villagers collect mahua flowers to prepare a local drink. Mahua flowers are found among dry leaves, and to pick the flowers, villagers set the dried leaves on fire, which, if left unchecked, spreads and causes forest fires. In addition, poachers set fire to forests to flush out animals (DownToEarth, 2023).

To reduce the risk of fire in the HKH region, the locals of Kudupaika village wrote songs in the local Kui dialect about why and how to protect the forest from fire (DownToEarth, 2023).

Forest pests and diseases

Plant pathogens and pests seriously affect trees and forest productivity, reduce tree cover and pose a serious threat to forest ecosystems. Many soil-borne fungi damage economically important trees in both planted and natural forests. According to the Indian Board of Forestry Research and Education, pathogens cause disease in leaves, stems and roots of trees. There are five major leaf diseases, namely leaf spot, leaf blight, pink disease, leaf rust, and powdery mildew; stem diseases include stem canker and dieback of twigs and branches; while root diseases include root rot and dieback (ICFRE, 2020).

In Nepal, 44.7% of the land is covered by forests (DFRS, 2015) and due to the large changes in altitude, Nepal has various types of forests. *Shorea robusta*, especially in the Sindhupalchowk area, is susceptible to the rosy gypsy moth, *Lymantria mathura*, a major defoliator of deciduous trees. *Pterocarpus marsupium* and *Tectona grandis* are trees in natural forests with relatively high commercial value, especially in low-altitude areas of Nepal. Many of the leaves are infected by pathogenic bacteria, resulting in leaf folding and leaf surface loss (Malla and Pokharel, 2018).

13.2.3 Carbon sequestration

Forests store the most carbon of the terrestrial ecosystems; the amount varies among countries in the HKH region and may be related to the species of trees. Under the provisions of the Paris Agreement, India plans to store additional carbon of 2.5–3 billion tonnes of CO_2 equivalent by 2030 by increasing forests and tree cover (Government of India, 2022). In 2016, Pakistan passed the Ten Billion Tree Tsunami Program (TBTTP) to sequester 149 MtCO2e emissions over ten years (Government of Pakistan, 2021). China set a target of reaching carbon storage peak by 2030 and reaching carbon neutrality by 2060. The forest-growing stock and carbon stock of HKH in 2020 is presented in Table 13.2.

Forest carbon storage is divided into five types: above-ground biomass, below-ground biomass, dead wood, litter and soil. The above- and below-ground biomasses of carbon reserves in Bhutan in 2020 amounted to 127 tonnes/ha and 54.5 tonnes/ha, respectively. In China, the

Table 13.2. Forest-growing stock (tree trunk volume) and carbon stock of HKH in 2020.

	Bangladesh	Bhutan	China	India	Myanmar	Nepal	Pakistan
Growth stock (million m^3 over bark)	91.8	943	19191	5142	901	982	212
Carbon stock in biomass (tonnes/ha)	43.5	182	39.7	41.5	75.5	108	46.2
Total carbon stock (tonnes/ha)	126	255	45.3	100	77.6	187	46.2

Data obtained from Global Forest Resources Assessments, Food and Agriculture Organization of the United Nations (FRA, https://fra-data.fao.org/assessments/fra/2020, accessed 3 November 2023)

carbon storage of dead wood was 5.67 tonnes/ha, while Nepal had the highest carbon storage in litter at 11.9 tonnes/ha. In addition to the above- and below-ground carbon storage, the carbon fixed by forest trees is also stored in the soil. The soil carbon stock was 80.4 tonnes/ha in Bangladesh, 66.9 tonnes/ha in Nepal, 64.1 tonnes/ha in Bhutan and 56.2 tonnes/ha in India (Table 13.3).

13.2.4 Forest ecosystem service functions

Forest ecosystems in the HKH region function twofold: directly, by providing resources such as fuelwood (Fig. 13.4), feed for livestock (Fig. 13.5) and wild herbivores; and food for people such as mushrooms and wild berries; and, indirectly, by intangibles such as clean air, water purification, soil nutrients, outdoor recreation and culture, and tourism (ICIMOD, 2022). Local residents in the HKH region rely on water from natural springs or streams for domestic and agricultural water as a direct benefit (Sharma *et al.*, 2019).

The ecological, social, cultural and economic contributions of the HKH region are substantial, and the forest ecosystem services of the region contribute to the life-sustaining capacity not only in the mountain areas but also in the adjoining plains and downstream landscapes (Negi and Agarwal, 2006; Singh, 2007; Joshi and Negi, 2011). Agriculture in the HKH region is highly dependent on forest resources for energy flow, fertilizer, soil nutrients and food pro-

Table 13.3. Forest carbon stocks in HKH region in 2020 (tonnes/ha).

Categories	Bangladesh	Bhutan	China	India	Myanmar	Nepal	Pakistan
Carbon in above-ground biomass	31.4	127	31.7	31.6	59.0	81.0	35.4
Carbon in below-ground biomass	12.1	54.5	7.94	9.90	16.5	27.0	10.8
Carbon in dead wood	1.70	3.03	5.67	0.40	—	—	—
Carbon in litter	0.27	6.23	—	1.90	2.10	11.9	—
Soil carbon	80.4	64.1	—	56.2	—	66.9	—

Data obtained from Global Forest Resources Assessments, Food and Agriculture Organization of the United Nations (FRA, https://fra-data.fao.org/assessments/fra/2020, accessed 3 November 2023)

Fig. 13.4. Wood collected in the forest in the mid-hills of Nepal. (Photo: Allan Degen)

duction, as fertile soil can be collected from forests and mixed with agricultural soils. Among the immediate benefits, food production and distribution at the regional level are important and critical. For example, vegetable production has found a good market in nearby towns and there is also an out-of-season supply of vegetables in the plains.

The forest ecosystem in the HKH region plays a role in regulating water cycles, protecting watersheds, maintaining soil and soil moisture, and influencing rainfall (Joshi and Negi, 2011). In addition to water retention, it is also vital in mediating the flow of rainwater and recharge of groundwater, as well as the availability of fresh water (Fig. 13.5). In low-altitude areas, the distance to access drinking water can reach an average of 3–4 km; however, in high-altitude areas it averages about 1 km, because high-altitude forests are rich in water. Water purification and waste disposal are important ecological services provided by forest ecosystems. Forest ecosystems filter and break down organic waste in water and absorb/detoxify compounds through soil and subsoil processes. Furthermore, forests play an important role in filtering dust particles from the air (Singh, 2007; Negi and Dhyani, 2012).

Forest ecosystem services can provide important leverage to positively adjust the relationship between ecosystems and society (Colloff *et al.*, 2021). Forests in the HKH region play a key role in controlling soil erosion and promoting slope stability. Soil erosion is evident in degraded forests, especially during the rainy season, whereas areas without vegetation are prone to landslides during the rainy season. Roads at high altitudes have suffered massive landslides on patches of bare hills (Joshi and Joshi, 2019). In addition, forests play a crucial role in soil formation and maintenance of soil fertility. Succession of forest ecosystems increases the ability of degraded landscapes to provide better-quality goods and services, and trees on hillsides are able to stabilize landslide sites (Singh, 2006).

Forest ecosystems in the HKH region promote tourism. The region has dense forests and many small lakes such as Lake Kulpatar and Lake Sariyatar, which are interesting tourist destinations. Activities such as hiking and birdwatching are popular for tourists in the HKH region forests, while many cultural events are held at high-altitude areas. For example, in some places, the original forest is preserved as a sacred grove between small temples. Conservation practices of the local forests include prayers to

Fig. 13.5. Forest ecosystems regulate the water cycle. (Photo: Binyu Luo)

demigods and goddesses through religious rituals. Another traditional practice is the driving away of evil spirits by witch doctors using wooden branches and leaves of bodhi and sycamore trees. Many other local deities, such as Goluji, Gangnath and Bhumiya, are also said to appear on forest patches and use forest resources (Joshi and Joshi, 2019).

Pollination and pollinators are important services provided by forest ecosystems to maintain a healthy gene pool and transfer genetic material. A reduction in pollinators can affect the total yield and quality of the harvest (Hassan *et al.*, 2005). Beekeeping, bee populations and honey production decreased in the study area due to the increased use of agricultural chemicals, changes in crop patterns and forest degradation.

13.3 Conservation and Counter-measures of Forest Resources in the HKH Region

The disruption to daily life caused by COVID-19 has forced people to recognize that the loss of biodiversity could cost society billions of dollars (OECD, 2021). Understanding the vulnerability of forests and the degradation of ecosystems is the first step in defining adaptation and management strategies. Mitigation has traditionally been the main focus of studying climate change from a scientific and policy perspective; however, researchers have been emphasizing the importance of considering adaptation as a response to climate change. Climate change is unavoidable and adaptive measures can be undertaken at both local and regional levels (Millar *et al.*, 2007).

Beginning in the 17th century, selective and rotating ownership models provided limited incentives for resource conservation. It also created tension between *de jure* and *de facto* owners, which persists and is a source of forest degradation today. Mismanagement of resource rights, corruption and perverse incentives exacerbate forest degradation (Khan, 2009). Pakistan should increase forest cover from the current 4–5% to 25%, and afforestation and reforestation initiatives should be accelerated to achieve the goals. Deforestation should be banned in the region and alternatives should be provided to reduce human pressure on natural resources. Governments and other agencies should plan urbanization, industrialization and ecotourism to minimize deforestation and pollution (Khan *et al.*, 2022a).

The World Resources Institute (WRI), through its Global Forest Watch (GFW) initiative, developed a new online forest monitoring system that combines satellite technology and human networks to mobilize faster and more effective forest protection and sustainable forest management. The GFW combines a near-real-time deforestation alert system, complementary satellite data and GIS layers to create transparency on a global scale. ICIMOD and WRI are collaborating on activities related to the GFW initiative to increase the sharing of technology and data to improve forest-cover information in the HKH region. They are jointly hosting a stakeholder workshop to present new advances in forest-cover mapping and discuss the development of a regional forest-cover monitoring system using GFW data and technology for the HKH region, with an initial focus on Nepal.

Forest ecosystems in the HKH region are facing urgent climate change challenges, but the government's climate change adaptation measures are largely incremental and not well integrated with development plans and projects (Bennett, 2003; Hansen and DeFries, 2007). Despite these challenges, the region still has the opportunity to adopt a higher-level, inclusive and more comprehensive climate change adaptation response. However, strengthening the resilience of the HKH region's forest ecosystems to climate change will require a substantial increase in existing financial support.

National policies in the HKH region do not adequately take into account the oppression and exclusion that women face, nor do they adequately recognize the roles and problems that women face in family, community and market settings (Biland *et al.*, 2021). Despite the dual responsibilities of production and reproduction, women throughout the region have little decision-making power or control over resources. Migration has driven broad economic, social and political changes throughout the HKH region. While remittances from workers that left the area can boost resilience to climate change, investment in agriculture or climate adaptation are rarely a priority for mountain migrant families. Most of the policies and processes for environmental governance occur at the national and local levels and lack regional co-ordination. Countries in the HKH region lack mechanisms to connect upstream and downstream communities. Cross-border co-operation is crucial to improving environmental governance in the region.

The poverty rate in mountain areas (31%) exceeds the national average for all HKH countries (26%) (Gerlitz *et al.*, 2012), and therefore poverty statistics at the national level mask serious differences between mountainous and non-mountainous areas. Mountainous areas are characterized by remoteness, poor accessibility and high dependence on natural resources. More than 80% of the rural population of the countries in the HKH region – a large proportion of whom live in mountains and hills – rely on traditional biomass fuels for cooking (Fig. 13.6), and about 400 million people still lack basic access to electricity (Hussain *et al.*, 2019). Although hydropower has great potential as a source of renewable energy in the region, it is still untapped.

13.4 Local Livelihoods and Ecological Restoration of Forest Resources

Fuelwood contributes the highest proportion of income to the household economy in the HKH region and households with more people are more likely to rely on forests for income (Biland *et al.*, 2021). A reasonable explanation might be that the amount of forest products actually collected depends on the family size, and the decision to allocate part of their workforce to forest-resource harvesting may depend on the availability of labour. For families living near forests, the most accessible work for children and women is collecting fuelwood, while adult men work mostly in non-agricultural jobs. Interestingly, however, large families may be more inclined to harvest forest resources to meet their household needs rather than sell them (Mendako *et al.*, 2022). From the perspective of environmental protection, biodiversity and economic benefits, the best option to improve forest livelihoods is to restore the degraded forests in the HKH region to achieve some level of productivity.

The broad classification of degraded forests includes: secondary forests after natural disasters or tree harvesting, abandoned non-forest secondary forests, and degraded land that is deserted or covered by weeds (*imperata*). A number of techniques can be used for forest restoration, including natural regeneration, and planting of monoculture and mixed tree species. The last category includes agroforestry, non-uniformly aged plantations, and uniformly aged plantations with short- or long-rotation tree species (Eliott

Fig. 13.6. Dried yak dung on the Tibetan Plateau to be used for heating and cooking. (Photo: Allan Degen)

et al., 2013). To improve wood production in degraded forests, the most common method is enrichment planting (gap planting or row planting), in which seedlings of desired species are planted as part of a commercial plantation (Lamb, 2011).

A simple and cost-effective restoration method that employs the natural restoration potential of forest ecosystems is the assisted natural regeneration (ANR) method, pioneered in the Philippines in the 1980s (Lamb, 2011; Appanah, 2016). ANR accelerates succession, transforming areas of degraded vegetation into more productive forests by removing or reducing barriers to natural forest regeneration, such as weeds and fires. The cost of restoring forests through ANR is usually less than half that of traditional reforestation methods (Bagong Pagasa Foundation, 2009). Agroforestry is also becoming increasingly popular as a restoration option and climate change adaptation strategy and is especially relevant to farmers with small plots of land (Landsberg *et al.*, 2014).

HKH countries have different strategies in forest restoration, depending on economic development paths, governance systems, social contexts, biophysical conditions and other factors that influence patterns of forest resource use. Almost all countries have experienced severe deforestation, and some countries are still suffering, while others have succeeded in curbing or even reversing the loss of forests. In the past, harvesting wood relied mainly on the establishment of monocultural plantations of exotic species. However, many countries are recognizing the importance of adapting the broader objectives of forest restoration at the landscape level.

While Myanmar has a rich history in forest management, it focuses primarily on natural forests. There are few planted forests and community forests have not been developed. The main methods used to restore forest cover in Myanmar include the establishment of commercial plantations of native tree species such as teak, rosewood and pine, industrial plantations of fast-growing exotic tree species, watershed restoration, and plantations that meet local timber needs, mainly wood for fuel. The Myanmar Reforestation and Restoration Plan (MRRP) 2017–2027 aims to protect 5900 km^2 of reserved forest (RF), manage 2500 km^2 of plantations on public and private lands, develop community forestry management and agroforestry on 3000 km^2, and establish protected and public protected forests on 41,000 km^2 (Maw, 2020).

The Bangladesh Department of Forestry (BFD) is currently implementing the World Bank-supported Sustainable Forests and Livelihoods Project (SUFAL) by funding nearly 790 km^2 of forest on public and private land, including 220 km^2 of coastal green belts in 147 upazilas (sub-districts) (Bangladesh Forestry Department, 2018). The project aims to increase sustainable livelihood options for forest-dependent communities by improving forest management in target areas and also to ensure the sustainability of forest resources. Completed sustainable management projects include: Climate Resilient Ecosystems and Livelihoods (CREL), integrating community adaptation into the Bangladesh Afforestation and Reforestation Programme, and Climate Resilient Participatory Afforestation and Reforestation Project (CRPARP). These projects were designed to reduce forest degradation and to build long-term resilience to climate change. Bangladesh's mitigation activities for restoration include: increasing tree cover from 22.4% (2014) to 24.0% by 2023; afforestation and reforestation of 1500 km^2 in coastal areas, islands and degraded areas; and restoration of 1378 km^2 of deforested forests and 2000 km^2 of degraded forests. Possible mitigation activities include: maintenance of forest and tree cover through co-operative forest management, social forestry and other projects; protecting forests by expanding other income-generating activities for forest-dependent communities; joint management in forest protected areas (720 km^2); increasing coastal afforestation activities; and maintaining the restoration of degraded forests or deforested areas (Ministry of Environment, Forest, and Climate Change, 2021).

13.5 Forest Conservation Management Strategies, Adaptation and Path Changes

To ensure that forests continue playing a role in offsetting the effects of global warming, ICIMOD and the Forest Institute are promoting research on human-centred sustainable management practices (Wester *et al.*, 2019). Measures are being examined to promote HKH cross-border biodiversity conservation, including co-ordinated planning of border isolation facilities according to species distribution, establishment of peace parks in border areas, development of cross-border ecotourism, and implementation of cross-border ecological compensation (Chen and Hull, 2023).

Governance and institutions are at the heart of the paradigm shift in managing forests in the HKH region and can initiate policies that facilitate transboundary co-operation (Basnet and Karki, 2020). This can be accomplished by: (i) appointing long-term landscape-level governance institutions, such as the Mekong Basin Commission; (ii) instituting policy development projects in the eight HKH countries to promote cross-border co-operation; and (iii) ensuring that women and gender issues are an integral part of conservation and development, from conception to implementation and monitoring.

A universal basic income policy should be implemented, mainly for women, and social protection policies and programmes should be strengthened to help the most vulnerable groups. A gender-sensitive perspective in decision making should be formulated to ensure that women are not marginalized but included in decision-making processes and policies. Furthermore, women workers should be provided with paid family leave and paid sick leave, as well as flexible working hours (Smith, 2022).

13.5.1 Forest dynamics and management

There is a need to link science with policy and practice in order to better manage forests (Nesshöver *et al.*, 2017). Innovative ways to exchange knowledge, bridge information gaps and create linkages among stakeholders have been suggested. ICIMOD and the Forest Institute recommend the following actions on forest management in the HKH region: (i) maintaining and strengthening the 'carbon pool' of forests; prioritize mixed forests and make them resilient to any form of disturbance and to improve the habitat they provide for wildlife; (ii) promoting sustainable uses of the area, which implies actively planned thinning of trees that may result in a higher amount of carbon being fixed; (iii) harvesting timber to develop different regenerative niches; (iv) monitoring forest areas through remote sensing; (v) establishing permanent land and small wildlife enclosures in forests; (vi) developing wildlife management programmes that are compatible with sustainable mountain forestry and human welfare; and (vii) implementing land-use policies to guide development investments. Subsidized or free clean fuels should be provided to mountain people to reduce pressure on forests for firewood and promote the use of clean fuels for cooking, such as induction stoves, solar energy (Fig. 13.7) and liquefied petroleum gas.

13.5.2 From survival to regional markets

The economic base of livelihoods in the HKH region could be improved for the local inhabitants by: (i) introducing technical interventions that benefit local communities; (ii) developing a market network with minimal regulatory barriers through cross-border co-operation between policy-makers and practitioners (Han and Li, 2018); (iii) expanding the value chain to the 'minimum quantity' level; (iv) developing regional health standards with a clear focus on sustainable management, products and processes related to poverty and inclusion; and (v) producing niche or specialty products to ensure the viability of community businesses (Sharma *et al.*, 2023).

13.5.3 Forestry knowledge forum and regional co-operation

There should be support for formal and informal networks and cross-border co-operation, and encouragement for legal trade in mountain products. In science, there should be attempts to: (i) identify and prioritize research topics; (ii) encourage private sector involvement; and (iii) link research to business and patents (Vieira Nunhes *et al.*, 2021). In policies, it is important to: (i) establish a network of academic institutions; and (ii) strengthen the Himalayan University Alliance and encourage inter-university co-operation (Arnaldo Valdés, 2022). In practice, there should be a focus on promoting exchanges of knowledge

Fig. 13.7. Solar panels heating a kettle of water on the Tibetan Plateau. (Photo: Allan Degen)

and information at all levels. Mechanisms should be established to allow for communication among scientists, practitioners and end-users of knowledge (Mazur *et al.*, 2021).

13.6 Conclusions

In the past two decades, the forest cover in HKH has decreased, while the degradation of forest has increased substantially. Degradation, due mainly to climate warming, but also land-use change, soil erosion, deforestation, forest fires, forest pests and diseases, has affected carbon sequestration of forests and the livelihoods of local residents, and reduced the service functions of the forest ecosystem. To mitigate the effects of global warming, government and institutions should be involved, incentives and community participation should be integrated, and academic networks should be established to manage forest ecosystems in the HKH.

References

Appanah, S. (2016) The Food and Agriculture Organization of the United Nations and RECOFTC – The Center for People and Forest Bangkok. Available at: https://portal.gms-eoc.org/uploads/resources/1347/attachment/forest_landscape_restoration_for_asia-pacific_forests_2016_04_eng.pdf (accessed 26 September 2023).

Arnaldo Valdés, R.M. and Gómez Comendador, V.F. (2022) European universities initiative: how universities may contribute to a more sustainable society. *Sustainability* 14(1), 471. DOI: 10.3390/su14010471

Azarkan, M. (2021) Nepal faces its worst fire season in years. *Atalayar*, 12 April. Available at: https://www.atalayar.com/en/articulo/society/nepal-faces-its-worst-fire-season-years/20210412090315150734.amp.html (accessed 26 September 2023).

Bagong Pagasa Foundation (2009) Cost comparison analysis ANR vs. conventional reforestation. Paper presented at the concluding seminar of FAO-assisted project TCP/PHI/3010 (A), Advancing the Application of Assisted Natural Regeneration (ANR) for Effective, Low-Cost Forest Restoration.

Bangladesh Forestry Department (2018) Resettlement policy framework for the sustainable forests & livelihood (SUFAL) project. Available at: https://documents1.worldbank.org/curated/en/969011533550705191/pdf/Resettlement-Policy-Framework-BD-SUFAL-final.pdf (accessed 16 February 2024).

Basnet, S. and Karki, B.S. (2020) REDD+ across transboundary landscapes: a look into the opportunities and challenges of participatory forest management systems in receiving results based payments in the Hindu Kush Himalayan Region. *Small-scale Forestry* 19, 399–418. DOI: 10.1007/s11842-020-09448-3

Behera, M.D., Behera, S.K. and Sharma, S. (2019) Recent advances in biodiversity and climate change studies in India. *Biodiversity and Conservation* 28, 1943–1951. DOI: 10.1007/s10531-019-01781-0

Bennett, A.F. (2003) *Linkages in the Landscapes: The Role of Corridors and Connectivity in Wildlife Conservation.* IUCN, Gland, Switzerland.

Bhattacharjee, A., Anadón, J.D., Lohman, D.J., Doleck, T., Lakhankar, T. *et al.* (2017) The impact of climate change on biodiversity in Nepal: current knowledge, lacunae, and opportunities. *Climate* 5(4), 80. DOI: 10.3390/cli5040080

Bhattarai, N., Dahal, S., Thapa, S., Pradhananga, S., Karky, B.S. *et al.* (2022) Forest fire in the Hindu Kush Himalayas: a major challenge for climate action. *Journal of Forest and Livelihood* 21(1), 14–31.

Biland, M., Zeb, A., Ullah, A. and Kaechele, H. (2021) Why do households depend on the forest for income? Analysis of factors influencing households' decision-making behaviors. *Sustainability* 13(16), 9419. DOI: 10.3390/su13169419

Chaudhary, S., Tshering, D., Phuntsho, T., Uddin, K., Shakya, B. *et al.* (2017) Impact of land cover change on a mountain ecosystem and its services: case study from the Phobjikha valley, Bhutan. *Ecosystem Health and Sustainability* 3(9), 1393314. DOI: 10.1080/20964129.2017.1393314

Chazdon, R.L. (2008) Beyond deforestation: restoring forests and ecosystem services on degraded lands. *Science* 320(5882), 1458–1460. DOI: 10.1126/science.1155365

Chen, H., Zhu, Q., Peng, C., Wu, N. and Wang, Y. (2013) The impacts of climate change and human activities on biogeochemical cycles on the Qinghai-Tibetan Plateau. *Global Change Biology* 19(10), 2940–2955. DOI: 10.1111/gcb.12277

Chen, X. and Hull, V. (2023) Border conservation in Hindu Kush-Himalaya. *Science* 380(6652), 1330–1332. DOI: 10.1126/science.adi7958

Colloff, M.J., Gorddard, R., Abel, N., Locatelli, B., Wyborn, C. *et al.* (2021) Adapting transformation and transforming adaptation to climate change using a pathways approach. *Environmental Science & Policy* 124, 163–174. DOI: 10.1016/j.envsci.2021.06.014

Cui, X. and Graf, H. (2009) Recent land cover changes on the Tibetan Plateau: a review. *Climatic Change* 94, 47–61. DOI: 10.1007/s10584-009-9556-8

DFRS (2015) *State of Nepal's Forests.* Forest Resource Assessment (FRA) Nepal, Department of Forest Research and Survey (DFRS), Kathmandu.

DownToEarth (2021) Still burning: forest fires continue to rage in Uttarakhand. 14 April. Available at: https://www.downtoearth.org.in/video/climate-change/still-burning-forest-fires-continue-to-rage-in-uttarakhand-76487 (accessed 26 September 2023).

DownToEarth (2023) Youth from Odisha's Kandhamal show the way to combat forest fires. 4 May. Available at: https://www.downtoearth.org.in/blog/natural-disasters/youth-from-odisha-s-kandhamal-show-the-way-to-combat-forest-fires-89087 (accessed 26 September 2023).

Eliott, R.M., Nogueira, M.F., Silva Sobrinho, A.S., Couto, B.A., Maciel, H.S. *et al.* (2013). Tar reforming under a microwave plasma torch. *Energy & Fuels* 27(2), 1174–1181. DOI: 10.1021/ef301399q

Erbaugh, J.T. and Oldekop, J.A. (2018) Forest landscape restoration for livelihoods and well-being. *Current Opinion in Environmental Sustainability* 32, 76–83. DOI: 10.1016/j.cosust.2018.05.007

FAO (2018) The state of the world's forests 2018 – forest pathways to sustainable development. Available at: https://fra-data.fao.org/assessments/fra/2020 (accessed 31 November 2023) (reproduced with permission from IGO, Rome)

Gerlitz, J.Y., Hunzai, K. and Hoermann, B. (2012) Mountain poverty in the Hindu-Kush Himalayas. *Canadian Journal of Development Studies [Revue canadienne d'études du développement]* 33(2), 250–265. DOI: 10.1080/02255189.2012.689613

Ghimire, C.P., Bruijnzeel, L.A., Bonell, M., Coles, N., Lubczynski, M.W. *et al.* (2014) The effects of sustained forest use on hillslope soil hydraulic conductivity in the Middle Mountains of Central Nepal. *Ecohydrology* 7(2), 478–495. DOI: 10.1002/eco.1367

Gilani, H., Shrestha, H.L., Murthy, M.S.R., Phuntso, P., Pradhan, S. *et al.* (2015) Decadal land cover change dynamics in Bhutan. *Journal of Environmental Management* 148, 91–100. DOI: 10.1016/j.jenvman.2014.02.014

Giriraj, A., Babar, S., Jentsch, A., Sudhakar, S. and Murthy, M.S.R. (2010) Tracking fires in India using advanced along track scanning radiometer (A) ATSR data. *Remote Sensing* 2(2), 591–610. DOI: 10.3390/rs2020591

Government of India (2022) India's updated first nationally determined contribution under Paris Agreement (2021–2030). Available at: https://unfccc.int/sites/default/files/NDC/2022-08/India%20Updated%20First%20Nationally%20Determined%20Contrib.pdf (accessed 17 April 2024).

Government of Pakistan (2021) Pakistan: Updated Nationally Determined Contributions 2021. Available at: https://unfccc.int/sites/default/files/NDC/2022-06/Pakistan%20Updated%20NDC%202021.pdf (accessed 17 April 2024).

Hamid, M., Khuroo, A.A., Charles, B., Ahmad, R., Singh, C.P. *et al.* (2019) Impact of climate change on the distribution range and niche dynamics of Himalayan birch, a typical treeline species in Himalayas. *Biodiversity and Conservation* 28, 2345–2370. DOI: 10.1007/s10531-018-1641-8

Han, Y. and Li, Y. (2018) Promotion of resource utilization and livelihood improvement in the Hindu Kush-Himalaya region through international agricultural cooperation. In: 2nd International Conference on Management, Education and Social Science (ICMESS, 2018). Atlantis Press, The Netherlands.

Hansen, A.J. and DeFries, R. (2007) Ecological mechanisms linking protected areas to surrounding lands. *Ecological Applications* 17(4), 974–988. DOI: 10.1890/05-1098

Hassan, R.M., Scholes, R. and Ash, N. (2005) *Ecosystems and Human Well-being: Current State and Trends*. Island Press, Washington, DC.

Helvarg, D. (2019) How will California prevent more mega-wildfire disasters? Available at: https://www.nationalgeographic.com/science/article/how-will-california-prevent-more-mega-wildfire-disasters (accessed 26 September 2023).

Hussain, A., Sarangi, G.K., Pandit, A., Ishaq, S., Mamnun, N. *et al.* (2019) Hydropower development in the Hindu Kush Himalayan region: issues, policies and opportunities. *Renewable and Sustainable Energy Reviews* 107, 446–461. DOI: 10.1016/j.rser.2019.03.010

ICFRE (2020) *A User Manual on Forest Insect Pests and Diseases*. Indian Council of Forestry Research and Education, Dehradun, India.

ICIMOD (2022) Land cover of HKH region [data set]. Available at: ICIMOD. https://doi.org/10.26066/RDS.1972511 (accessed 26 September 2023).

Joshi, A.K. and Joshi, P.K. (2019) Forest ecosystem services in the central Himalaya: local benefits and global relevance. *Proceedings of the National Academy of Sciences, India Section B: Biological Sciences* 89, 785–792. DOI: 10.1007/s40011-018-0969-x

Joshi, G. and Negi, G.C. (2011) Quantification and valuation of forest ecosystem services in the western Himalayan region of India. *International Journal of Biodiversity Science, Ecosystem Services & Management* 7(1), 2–11. DOI: 10.1080/21513732.2011.598134

Kattel, G.R. (2022) Climate warming in the Himalayas threatens biodiversity, ecosystem functioning and ecosystem services in the 21st century: Is there a better solution? *Biodiversity and Conservation* 31(8–9), 2017–2044. DOI: 10.1007/s10531-022-02417-6

Khan, B., Kesang, W., Nakul, C. and Dongal, G. (2022a) *Understanding the Scale, Trend, and Patterns of Illegal Wildlife Trade in the Hindu Kush Himalaya*. ICIMOD, Nepal.

Khan, M.Z., Baig, T., Ali, S. and Shah, A. (2022b) *Mountain Studies: Understanding and Managing Mountains for People and Nature*. Karakoram International University, Gilgit, Pakistan.

Khan, S.R. (2009) Assessing poverty-deforestation links: evidence from Swat, Pakistan. *Ecological Economics* 68(10), 2607–2618.

Krishnaswamy, J., Bonell, M., Venkatesh, B., Purandara, B.K., Rakesh, K.N. *et al.* (2013) The groundwater recharge response and hydrologic services of tropical humid forest ecosystems to use and reforestation: support for the 'infiltration-evapotranspiration trade-off hypothesis'. *Journal of Hydrology* 498, 191–209. DOI: 10.1016/j.jhydrol.2013.06.034

Kumar, M., Savita, Singh, H., Pandey, R., Singh, M.P., Ravindranath, N.H. *et al.* (2019) Assessing vulnerability of forest ecosystem in the Indian Western Himalayan region using trends of net primary productivity. *Biodiversity and Conservation* 28, 2163–2182.

Lamb, D. (2011) *Regreening the Bare Hills: Tropical Forest Restoration in the Asia-Pacific Region* (Vol. 10, 978–990). Springer, Dordrecht, Germany.

Landsberg, F., Reij, C., Winterbottom, R., Ouédraogo, M. and Belemviré, A. (2014) Potential contributions to food security from scaling up agroforestry and improved soil and water management practices in Burkina Faso. Available at: https://pdf.usaid.gov/pdf_docs/PA00K42J.pdf (accessed 26 September 2023).

Leimgruber, P., Kelly, D.S. and Steininger, M.K. (2005) Forest cover change patterns in Myanmar (Burma) 1990–2000. *Environmental Conservation* 32(4), 356–364. DOI: 10.1017/S0376892905002493

Lewis, S.L., Edwards, D.P. and Galbraith, D. (2015) Increasing human dominance of tropical forests. *Science* 349(6250), 827–832. DOI: 10.1126/science.aaa9932

Malla, R. and Pokharel, K.K. (2018) *Forest Pests and Pathogens Problem in Different Forest Types of Nepal*. Department of Forest Research and Survey, Kathmandu.

Maw, M.C. (2020) The socioeconomic impact of community forest management in Meiktila District (case study in Yoe Sone Community Forest, Wun Dwin Township). Available at: https://meral.edu.mm/records/7486 (accessed 26 September 2023).

Mazur, E.N., Walker, S.E., Don, C. and Conway, F.D. (2021) Designing and delivering webinars to improve science communication and engagement between environmental researchers and natural resource practitioners. *Coastal Management* 49(4), DOI: 392–412. DOI: 10.1080/08920753.2021.1928458

Mendako, R.K., Tian, G. and Matata, P.M. (2022) Identifying socioeconomic determinants of households' forest dependence in the Rubi-Tele hunting domain, DR Congo: a logistic regression analysis. *Forests* 13(10), 1706. DOI: 10.3390/f13101706

Millar, C.I., Stephenson, N.L. and Stephens, S.L. (2007) Climate change and forests of the future: managing in the face of uncertainty. *Ecological Applications* 17(8), 2145–2151. DOI: 10.1890/06-1715.1

Ministry of Environment, Forest, and Climate Change (2021) Nationally determined contributions (NDCs) 2021 Bangladesh (updated). Available at: http://nda.erd.gov.bd/en/c/publication/nationally-determined-contributions-ndcs-2021-bangladesh-updated (accessed 16 February 2024).

Murthy, M.S.R., Pulakesh, D. and Behera, M.D. (2016) Road accessibility, population proximity and temperature increase are major drivers of forest cover change in the Hindu Kush Himalayan Region. *Current Science* 111(10), 1599–1602.

Negi, G.C.S. and Agrawal, D.K. (2006) Meeting report: measuring and valuing ecosystem services: Himalayan mountain context. *Current Science* 91(5), 573.

Negi, G.C.S. and Dhyani, P.P. (2012) *Glimpses of Forestry Research in the Indian Himalayan Region: Special Issue in the International Year of Forest – 2011*. ENVIS Centre on Himalayan Ecology, GB Pant Institute of Himalayan Environment and Development.

Nesshöver, C., Assmuth, T., Irvine, K.N., Rusch, G.M., Waylen, K.A. *et al.* (2017) The science, policy and practice of nature-based solutions: an interdisciplinary perspective. *Science of the Total Environment* 579, 1215–1227. DOI: 10.1016/j.scitotenv.2016.11.106

OECD (2021) *Biodiversity, Natural Capital and the Economy: A Policy Guide for Finance, Economic and Environment Ministers.* OECD Environment Policy Papers 26. OECD Publishing, Paris.

Qasim, W. and Mane, A.V. (2013) Characterization and treatment of selected food industrial effluents by coagulation and adsorption techniques. *Water Resources and Industry* 4, 1–12. DOI: 10.1016/j.wri.2013.09.005

Reddy, M.M., Vivekanandhan, S., Misra, M., Bhatia, S.K. and Mohanty, A.K. (2013) Biobased plastics and bionanocomposites: current status and future opportunities. *Progress in Polymer Science* 38(10–11), 1653–1689. DOI: 10.1016/j.progpolymsci.2013.05.006

Roa-García, M.C., Brown, S., Schreier, H. and Lavkulich, L.M. (2011) The role of land use and soils in regulating water flow in small headwater catchments of the Andes. *Water Resources Research* 47(5). DOI: 10.1029/2010WR009582

Rueff, H. (2014) High-altitude rangelands and their interfaces in the Hindu Kush Himalayas. *Mountain Research and Development* 34(3), 305–306. DOI: 10.1659/mrd.mm141

Sannigrahi, S., Pilla, F., Basu, B., Basu, A.S., Sarkar, K. *et al.* (2020) Examining the effects of forest fire on terrestrial carbon emission and ecosystem production in India using remote sensing approaches. *Science of the Total Environment* 725, 138331. DOI: 10.1016/j.scitotenv.2020.138331

Sharma, E., Molden, D., Rahman, A., Khatiwada, Y.R., Zhang, L. *et al.* (2019) Introduction to the Hindu Kush Himalaya assessment. In: Wester, P., Mishra, A., Mukherji, A. and Shrestha, A. (eds) *The Hindu Kush Himalaya Assessment: Mountains, Climate Change, Sustainability and People.* Springer Nature, pp. 1–16.

Sharma, U.C., Datta, M. and Sharma, V. (2023) Land use and management. In: Sharma, U.C., Datta, M. and Sharma, V. (eds) *Soils in the Hindu Kush Himalayas: Management for Agricultural Land Use*. Springer International Publishing, Cham, Switzerland, pp. 295–462.

Singh, J.S. (2006) Sustainable development of the Indian Himalayan region: linking ecological and economic concerns. *Current Science* 90, 784–788.

Singh, S.P. (2007) *Himalayan Forest Ecosystem Services: Incorporating in National Accounting.* Central Himalayan Environment Association (CHEA), Nainital.

Smith, J.C. (2022) Representation in times of crisis: women's executive presence and gender-sensitive policy responses to crises. *Journal of European Public Policy* 13, 1–26. DOI: 10.1080/13501763.2022.2110142

Springate-Baginski, O. (2018) Decriminalise agro-forestry: a primer on shifting cultivation in Myanmar. Available at: https://www.tni.org/files/publication-downloads/tni_p_shifting_cultivation_220518_online.pdf (accessed 5 August 2020).

Tran, B.N., Tanase, M.A., Bennett, L.T. and Aponte, C. (2020) High-severity wildfires in temperate Australian forests have increased in extent and aggregation in recent decades. *PLOS ONE* 15(11), e0242484. DOI: 10.1371/journal.pone.0242484

Vieira Nunhes, T., Viviani Garcia, E., Espuny, M., Homem de Mello Santos, V., Isaksson, R. *et al.* (2021) Where to go with corporate sustainability? Opening paths for sustainable businesses through the collaboration between universities, governments, and organizations. *Sustainability* 13(3), 1429. DOI: 10.3390/su13031429

Wester, P., Mishra, A., Mukherji, A. and Shrestha, A. (2019) *The Hindu Kush Himalaya Assessment: Mountains, Climate Change, Sustainability and People*. Springer Nature Switzerland AG, Cham, Switzerland.

Zeb, A. (2019) Spatial and temporal trends of forest cover as a response to policy interventions in the district Chitral, Pakistan. *Applied Geography* 102(2019), 39–46. DOI: 10.1016/j.apgeog.2018.12.002

Zhang-Turpeinen, H., Kivimäenpää, M., Aaltonen, H., Berninger, F., Koster, E. *et al.* (2020) Wildfire effects on BVOC emissions from boreal forest floor on permafrost soil in Siberia. *Science of the Total Environment* 711, 134851. DOI: 10.1016/j.scitotenv.2019.134851

Zheng, L., Gaire, N.P. and Shi, P. (2021) High-altitude tree growth responses to climate change across the Hindu Kush Himalaya. *Journal of Plant Ecology* 14(5), 829–842. DOI: 10.1093/jpe/rtab035

14 Conservation and Sustainable Utilization of Rangelands in the Hindu Kush Himalayan Region

Peipei Liu[1,*], Xiaoping Jing[1], Allan Degen[2] and Zhanhuan Shang[1]
[1]*College of Ecology, Lanzhou University, Lanzhou, China;* [2]*Institutes for Desert Research, Ben Gurion University of the Negev, Israel*

Abstract
Rangelands in the Hindu Kush Himalayan (HKH) region are distributed mainly in Afghanistan, Bhutan, China, India, Nepal and Pakistan and cover more than half the land area. They play vital roles in biodiversity hotspots, carbon sequestration, water regulation, local livelihoods and tourism development, and provide indispensable ecosystem services for 25–30 million people in the HKH countries. However, climate change, livestock over-grazing and increasing urbanization are causing widespread degradation and threaten the sustainable utilization of rangelands. To restore the degraded rangelands, strategies such as fencing, rotational grazing, artificial grassland and protected areas have been implemented. However, more efforts are needed to rescue the valuable rangelands in the HKH region.

Keywords: rangeland degradation and restoration, climate change, over-grazing, urbanization, fencing and rotational grazing, artificial grassland, protected area

14.1 Introduction

Rangelands represent the most extensive land cover type on Earth (Briske, 2017; Robinson *et al.*, 2019) and are defined as land that is covered by vegetation of herbs and/or shrubs and grazed by domestic and/or wild herbivores. They could include grassland, shrubland, desert and even forest used for grazing (Lund, 2007). In the Hindu Kush Himalayan (HKH) region, rangelands consist of mainly natural grassland, shrubland and scattered forest, and cover more than 50% of the land area (Sharma *et al.*, 2007; Joshi *et al.*, 2013a; Xu *et al.*, 2019; Wang *et al.*, 2019). The 2.29 million km^2 of rangelands are crucial for the livelihoods of 25–30 million pastoralists and agropastoralists in Afghanistan, Bhutan, China, India, Nepal and Pakistan (Ahmad *et al.*, 2021) by providing grazing land for livestock (Fig. 14.1). In addition, rangelands provide grazing land for wild herbivores (Fig. 14.2), and ecological services of biodiversity conservation, watershed management, carbon sequestration, climate regulation, ecotourism, and livelihoods for pastoralists (Gongbuzeren *et al.*, 2018; Wang *et al.*, 2022).

The traditional use of rangelands was sustainable; however, climate change, increased grazing pressure, rapid urbanization and improper

*Corresponding author: liupp@lzu.edu.cn

 Sustainable Ecological Restoration and Conservation in the Hindu Kush Himalayan Region (Z. Shang *et al.*)
DOI: 10.1079/9781800622579.0014

Fig. 14.1. Domestic yak grazing on the rangelands in Hongyuan, China. (Photos: Xiaoping Jing)

Fig. 14.2. Tibetan wild ass grazing on the rangelands in Gaize, China. (Photos: Xiaoping Jing)

management practices have caused widespread degradation in recent decades. For example, it was reported that 19–60% of alpine grassland has been degraded on the Tibetan Plateau since the 1980s (Wei *et al.*, 2022). Degraded rangelands are usually characterized by decreasing plant coverage and species richness, invasion of toxic weeds, changing soil texture, and declining soil nutrients (Bardgett *et al.*, 2021), which pose a threat to production and ecological functions. Governments have advocated several strategies to combat degradation and protect the rangelands, and these include fencing, rotational grazing, protected areas, forage-based grazing and settlement, and ecological restoration (Joshi *et al.*, 2013b; Wang *et al.*, 2022). These strategies were initiated mainly after 2000; however, more efforts are needed in rescuing the rangelands in the HKH region.

14.2 Distribution of Rangelands in the Hindu Kush Himalayan Countries

The HKH region covers all or part of eight countries, including Afghanistan, Bangladesh, Bhutan, China, India, Myanmar, Nepal and Pakistan. The rangelands are the most widely distributed ecosystem type in the HKH region, with the largest area in China and the smallest in Bhutan (Table 14.1). These countries, excluding Bangladesh and Myanmar, have vast high mountains ranging from 2500 m to 5500 m, which form variable and characteristic rangelands (Sharma *et al.*, 2007; Millar and Tenzing, 2021). The open woodlands and scrub, alpine dry scrub, temperate grassland and scrubland, alpine moist scrubland and meadows, and alpine marsh meadows are the most common rangelands (Table 14.1).

14.3 Functions and Services of Rangelands

The unique mountain rangelands ecosystem, the largest ecosystem type in the HKH region, is regarded as pivotal in terms of ecosystem functions and services (Dong *et al.*, 2020; Xu *et al.*, 2022). Biodiversity, carbon sequestration, water resources, local livelihoods and tourism in rangelands are discussed in the following section.

14.3.1 Biodiversity

Although the HKH is the youngest among global mountain biomes, the diverse rangelands habitats harbour many endemic plants and animals (Xu *et al.*, 2019). The rangelands cover all or part of four global biodiversity hotspots: Himalaya, Indo-Burma, mountains of south-west China and mountains of Central Asia (Chao *et al.*, 2023). These hotspots resulted from the geographic isolation and divergent selection under the complex evolution history of geological structure and changing climate conditions of the region (Wu *et al.*, 2022). Until 2011, the endemic species in the four biodiversity hotspots contributed to more than 50% of endemic plants and almost 43% of endemic birds, reptiles, amphibians and mammals in the world (Mittermeier *et al.*, 2011). Biodiversity provides myriad ecosystem services, but the HKH region is experiencing rapid biodiversity loss due to climate change and irrational human activities (Kotru *et al.*, 2020). This is especially

Table 14.1. The distribution and types of rangelands in the Hindu Kush Himalayan countries. (From Joshi *et al.*, 2013a)

Country	Area (km^2)	Main types of rangelands
Afghanistan	291,881	Open woodlands and scrubland, alpine dry scrubland
Bangladesh	–	–
Bhutan	17,420	Open woodlands and scrubland, alpine dry scrubland, temperate grassland and scrubland, alpine moist scrubland and meadows
China	1,545,542	Open woodlands and scrubland, alpine dry scrubland, temperate grassland and scrubland, alpine moist scrubland and meadows, alpine marsh meadows
India	169,381	Open woodlands and scrubland, alpine dry scrubland, temperate grassland and scrubland, alpine moist scrubland and meadows, alpine marsh meadows
Myanmar	–	
Nepal	77,827	Open woodlands and scrubland, alpine dry scrubland, temperate grassland and scrubland, alpine moist scrubland and meadows, alpine marsh meadows
Pakistan	188,118	Open woodlands and scrubland, alpine dry scrubland

true in the hotspots of Indo-Burma and mountains of south-west China, which are being identified as 'areas of imminent extinction' (Chaudhary *et al.*, 2022). Conserving biodiversity is urgent in the HKH region if the goal of the Convention on Biological Diversity is to be achieved.

14.3.2 Carbon sequestration

Rangelands are an important carbon sink, storing approximately 30% of the world's soil carbon. Furthermore, the HKH region contains the largest alpine permafrost in the world (Ding *et al.*, 2017), which is distributed mostly in the rangeland ecosystem at high elevation. Permafrost, described as soil with ground temperature at or below 0°C for two or more consecutive years, stores vast amounts of organic carbon mainly because of the slow carbon decomposition rate due to the low temperature and anaerobic environment (Strauss *et al.*, 2017). In addition, peat accumulation, cryoturbation, and accumulation of eolian, alluvial and colluvial soil contribute to the fast pedogenic processes for soil carbon sequestration (Ping *et al.*, 2015). The organic carbon pool of permafrost at the depth of 3 m is approximately 14.4 Pg C in 1.1 million km^2 (Mishra *et al.*, 2021). Accordingly, the soil carbon pool of the rangelands in the HKH region plays a vital role in achieving 'carbon neutrality' in the world. The rangelands are a continuous carbon sink due to the high plant carbon uptake and low soil carbon release, especially since the 21st century (Ding *et al.*, 2017; Wei *et al.*, 2021); however, spatial heterogeneity is evident. A recent study indicated that the rangelands ranging from 3000–5000 m should continue to be a carbon sink, but rangelands below 3000 m and above 5000 m might be carbon sources (Guo *et al.*, 2021).

14.3.3 Water resources

The HKH region is known as the water tower of Asia, being the source of ten main rivers, namely the Amu Darya, Brahmaputra, Ganges, Indus, Irrawaddy, Mekong, Salween, Tarim, Yellow and Yangtze (Table 14.2), which provide and regulate water cycling for much of the Asian continent (Rueff, 2014; Mukherji *et al.*, 2015). The water is vital for agriculture, and provides drinking water, food (e.g. fish), energy (hydropower generation) and environmental security for people along the river basins and the downstream areas (Joshi *et al.*, 2013a). The precipitation run-off, snowmelt and glacier melt are the main contributors to the river flow in the HKH region (Scott *et al.*, 2019). Rangelands degradation is widespread in the HKH region, which weakens the soil-water-holding capacity and ecosystem services (Dong *et al.*, 2012). The water resources and their proper management are vital for the sustainable development of rangelands.

14.3.4 Local livelihoods

The rangelands in the HKH region support millions of pastoralists who rely on livestock production for their livelihood (Sharma *et al.*, 2007; Wu *et al.*, 2014), as yaks and Tibetan sheep provide meat, milk (Fig. 14.3), leather, hides and dung (Fig. 14.4) (Ghazali *et al.*, 2023). The rangelands also provide household energy needs for cooking and heating, by providing fuelwood from trees, shrubs and bushes (Roomi and Khan, 2022). In addition, harvesting the yarsagumba or caterpillar fungus, *Cordyceps sinensis*, and the medicinal mushroom *Armillaria luteo-virens* in the rangelands have become very profitable for the residents (Fig. 14.5), as 100 g of *Cordyceps sinensis* provide yak herders with close to 50% of their annual income in Bhutan (Joshi *et al.*, 2013a). The rangelands are linked with the pastoralists' health because the raw materials of Tibetan medicines, such as *Saussurea involucrata*, *Sauromatum diversifolium* and *Rhodiola crenulata* are collected in rangelands. Meanwhile, mountain pastoralism is highly vulnerable (Hussain *et al.*, 2016), and global climate change and anthropogenic activities are depleting resources, putting pressure on pastoralists' livelihoods (Wu *et al.*, 2014).

14.3.5 Tourism

The mountain ranges with magnificent landscapes and ethnic groups with unique and diverse traditions have immense aesthetic value and attract many local and international tourists and pilgrims (Roomi and Khan, 2022). Wild birds and mammals, fishing, mountaineering, hiking, cycling and skiing attract tourists and increase the gross domestic product in the HKH countries (Gioli *et al.*, 2019).

Table 14.2. The basins of the ten main rivers in the Hindu Kush Himalayan region.

Rivers	Headstream	Location	Watershed	References
Amu Darya	Tigrovaya Balka Nature Reserve	Border between Afghanistan and Tajikistan	Afghanistan, Tajikistan, Kyrgyzstan, Turkmenistan, Uzbekistan	Salehie *et al.* (2022)
Brahmaputra	Kubi Kangri and Chemayundung glacier	Burang County of Tibet, China	China, India, Bangladesh	Pondari *et al.* (2020)
Ganges	Gangotri glacier	Devprayag Town, India	India, Bangladesh	Niloy *et al.* (2021)
Indus	Kailash Mountain	Western Tibet, China	China, India, Afghanistan, Pakistan	Jahan *et al.* (2023)
Irrawaddy	Nmai Hka and Mali River	South-east Tibet, China and northern Myanmar	China, India, Myanmar	Bhatpuria *et al.* (2022)
Mekong	Jifu Mountain	Zaduo town, China	Cambodia, China, Laos, Myanmar, Thailand, Vietnam	Zhong *et al.* (2022)
Salween	Tanggula Mountains	The border between Qinghai and Tibet, China	China, Myanmar, Thailand	Lamb *et al.* (2019)
Tarim	Karakoram Mountains	Southern Xinjiang, China	China	Kulaixi *et al.* (2023)
Yellow River	Bayan Har Mountains	Qinghai, China	China	Zhou *et al.* (2022)
Yangtze	Tanggula Range	Qinghai, China	China	Shen *et al.* (2009)

Fig. 14.3. A woman milking a yak in the morning on the Tibetan Plateau. (Photo: Mazhang Wei)

Fig. 14.4. Two women collecting yak dung on the Tibetan Plateau. (Photo: Peipei Liu)

Fig. 14.5. The trade in the yarsagumba or caterpillar fungus *Cordyceps sinensis* in Lhasa, China. (Photo: Peipei Liu)

There are many well-known tourist sites and activities, such as cycling the Sichuan-Tibetan trail in China and hiking the Annapurna trail in Nepal. These tourist activities provide employment opportunities and improve the living standard of pastoralists still relying on traditional livestock husbandry (Tuladhar *et al.*, 2023). However, there are many ecological disasters in the rangelands with the rapidly growing tourism (Gongbuzeren *et al.*, 2018), manifested mainly by environmental pollution and rangeland degradation. Much more attention must be paid to tourism management in the HKH countries in the future.

14.4 Threats to Rangelands Resources and Their Impacts

Mountain rangelands in the HKH region are storehouses of biodiversity and natural resources, but they are also one of the most fragile ecosystems in the world due to the harsh environmental conditions (Niu *et al.*, 2019). Both production and ecology functions of rangelands are under severe threat because of rapid climate change, over-grazing by livestock, widespread urbanization, and rapid population growth (Joshi *et al.*, 2013a; Wang *et al.*, 2022). To curb rangeland degradation, the process of rangeland degradation must be understood so that proper strategies of restoration can be initiated (Dong *et al.*, 2020). Here we review the key issues of climate change, over-grazing and urbanization as causes of rangelands degradation in the HKH region.

14.4.1 Climate change

The Intergovernmental Panel on Climate Change (IPCC) Special Report on the Cryosphere focused on the accelerating climate change and its impact in mountain regions (Mayewski *et al.*, 2020). Climate warming is greatly felt in the HKH region, especially on the Tibetan Plateau and the area south of Pakistan, high-elevation sites where the temperature increased at a rate of 0.2°C per decade from 1951 to 2014 (Ren and Shrestha, 2017). Although global warming slowed down or showed a hiatus during the past two decades, the annual mean air temperature has continued to increase in the HKH region (Sun *et al.*, 2017). Under the Paris Agreement, global air temperature is projected to increase by 1.5°C globally and by at least 1.8°C in the HKH region (Krishnan *et al.*, 2019). Annual total precipitation increased at a rate of 5.28% per decade during the last six decades, and more so in most parts of India and the northern Tibetan Plateau after the mid-1980s (Ren and Shrestha, 2017). Extreme climatic events (extreme drought and heavy rainfall, extreme warm and cold events) usually overlapped with chronic directional climate changes (e.g. rising temperature and elevating precipitation), and chronic directional climate changes increased the frequency and intensity of extreme climatic events (Ombadi *et al.*, 2023), especially on the Tibetan Plateau (Xu *et al.*, 2022). The climate has changed considerably in the past decades, and it is projected to change more in the near future in the HKH region (Krishnan *et al.*, 2019).

Climate change has been causing rangelands degradation and decreasing livestock productivity and pastoralists' livelihoods in rangelands (Hussain and Qamar, 2020; Wang *et al.*, 2022). Climate warming and increased precipitation in the HKH region have led to an increase in the normalized difference vegetation index (NDVI) since the 1980s (Panday and Ghimire, 2012). On the temporal scale, the NDVI increased from 1981 to 2000, decreased from 2000 to 2015 and increased again from 2015 to 2020, especially on the Tibetan Plateau (Wang *et al.*, 2022). On the spatial scale, the NDVI increased on the eastern Tibetan Plateau but decreased on the western Tibetan Plateau (Wang *et al.*, 2022). Overall, there was an expansion in grassland and shrubland in the HKH region because of the integrated hydrothermal environment change (Volder *et al.*, 2013), especially for areas above 5000 m (Anderson *et al.*, 2020). Of importance, the plant carbon uptake and soil carbon sequestration benefitted from rangelands greening (Guo *et al.*, 2021; Wei *et al.*, 2021). However, many rangelands have degraded in the past several decades due to climate change, with a change in plant community composition; climate change could contribute to more than 50% of the degradation (Wang *et al.*, 2019). Perennial grasses are replaced by xerophytic shrubs, semi-shrubs and some annual toxic plants during grassland degradation due to climate warming (Zhou *et al.*, 2023). Climate warming also reduced plant richness, sedge cover and soil nutrients, and increased forb cover in the mountain rangelands (Wang *et al.*, 2022). The precipitation, especially the extreme droughts, cause a lagging effect in carbon cycling, plant community succession, and ecosystem equilibrium (Guo *et al.*, 2021).

Rangelands are crucial for livestock production and pastoralists' livelihoods in the HKH region, and degradation of rangelands due to climate change results in reduced livestock production and income (Rasul *et al.*, 2019). Climate change also influences livestock grazing behaviour (Liu *et al.*, 2019), and climate warming can cause heat stress, which affects forage intake and metabolism (Jing *et al.*, 2022). Extreme climatic events could cause a severe and fast reduction in livestock productivity (Rasul *et al.*,

2019). For example, a multi-year drought from 1998 to 2002 in Balochistan province, Pakistan, led to serious shortages in water and fodder, resulting in the death of 1.76 million livestock (Hussain and Qamar, 2020). Heavy snowfall in rangelands caused high mortality in the Nyainqêntanglha range, Tanggula range, Bayankhar mountains and the region between the Kailas range and the neighbouring Himalayas (Ye *et al.*, 2019; Li *et al.*, 2022). Therefore, both the plant community and livestock productivity in rangelands have been seriously affected by climate change, threatening the ecological equilibrium and the sustainable development of livestock husbandry in the HKH region.

14.4.2 Over-grazing

Over-grazing – that is raising more livestock than the carrying capacity of the rangelands – is another main factor driving rangelands degradation (Feng *et al.*, 2023). Close to 52% of the herders and some ecological scholars believe that over-grazing is the primary cause of rangelands degradation in the HKH region (Wang *et al.*, 2015; Dorji, 2022). The rapid population growth, economic improvement, ownership and management of land, and the change of lifestyle can lead to over-grazing (Roomi and Khan, 2022). Degradation of rangelands due to over-grazing was reported in Gilgit-Baltistan in Pakistan; the three rivers headwaters, near Qinghai lake and Qilian mountains, northern Tibet; and Gannan region in China (Roomi and Khan, 2022). Large herds of herbivores can harm rangelands by trampling (Roomi and Khan, 2022), thus increasing soil bulk density (Li *et al.*, 2017) and reducing rangelands ecosystem services. Over-grazing by herbivores depletes the plants and litter in rangelands, which leaves soil patches exposed to the harsh environment and increases soil temperature while decreasing soil moisture (Liu *et al.*, 2021). Furthermore, over-grazing generally decreases soil carbon and nitrogen and thus affects carbon and nitrogen cycling (Wang *et al.*, 2022), which can also reduce the soil's capacity to maintain water and restrict the growth of plants (Roomi and Khan, 2022). As a consequence, over-grazing could lead to acute forage shortage particularly in the cold season (Dorji, 2022), which further decreases the livestock productivity and further impoverishes pastoral livelihoods relying on the rangeland system (Rasul *et al.*, 2019).

14.4.3 Urbanization

Small settlements are the typical urban centre, and they offer better economic opportunities, infrastructure, food security and education than rural areas (Han *et al.*, 2016). Under government policies, traditional herders are transforming from nomadic to sedentary or semi-sedentary herding practices to improve their living standards (Wang *et al.*, 2022). For example, between 1990 and 2020, the urbanization rate accelerated from 21% to 55%, and the area of urban lands increased from 202 km^2 to 797 km^2 on the Tibetan Plateau. Concomitantly, there was the construction of the Qinghai-Tibet railway, projects developing western China, and the west-to-east gas pipeline in China (Han *et al.*, 2016). As a result, rangelands were converted to buildings and highways to accommodate the rapidly growing population (Singh *et al.*, 2019). Although the local economy improved, urbanization is destroying the ecological environment, with rangelands degradation and the shrinkage of rangelands (Han *et al.*, 2016). Previous studies indicated that the area of degraded rangelands decreases with an increase in distance from roads. The degraded area within 10 km, 10–20 km and > 20 km from a road was approximately 74%, 18% and 8%, respectively (Liu and Lu, 2021). However, the area of rangelands degradation resulting from urbanization is still unclear in the HKH region, and future studies are warranted.

14.5 Management Strategies and Policies for Rangelands

Combating degradation is a central theme of the UN Decade on Ecosystem Restoration during 2021 to 2030. However, rangelands degradation in the HKH region has created environmental problems, biodiversity loss, imbalance of global biogeochemical cycles, and declining livelihoods and standard of living (Bardgett *et al.*, 2021). Ecologists attempted to combat rangelands degradation in the HKH region to achieve the Aichi Biodiversity Targets of the Convention on Biological Diversity and the United Nations Sustainable Development Goals. The key to combating degradation is promoting restoration projects (Feng *et al.*, 2023). In general, fencing and rotation grazing are used to restore lightly or moderately degraded grasslands, while artificial grassland is used to restore severely degraded rangelands, such as the 'black soil land' (or 'bare land') on the Tibetan Plateau (Dong *et al.*, 2020). Here we review fencing, rotational grazing, artificial grassland and a new technology for conservation of degraded rangelands in the HKH region.

14.5.1 Fencing

Currently, fencing is one of the most widely used management practices to protect degraded rangelands in the HKH region (Wu *et al.*, 2009; Yao *et al.*, 2019; Ingty, 2021) and to enable the natural recovery of vegetation, soil fertility, biodiversity, and water storage capacity (Cao *et al.*, 2019; Xu *et al.*, 2021). In several studies, fencing restored degraded grassland that had low coverage, biomass vegetation, soil nutrients and carbon stores (Yao *et al.*, 2019; Wang *et al.*, 2022). A grazing ban was enacted in degraded rangelands in the Indian state of Sikkim in 1998 and the project 'Returning Grazing Land to Grassland' was implemented in 2004 to protect degraded rangelands by fencing on the Tibetan Plateau in China. Although fencing improved the plant community and soil quality in the rangelands, there was a concomitant reduction in pastoralists' incomes and a perceived increase in human–wildlife conflict when the fencing duration was long-term (Ingty, 2021). Short-term fencing improved rangeland health, while long-term fencing favoured the growth of forage grasses and restrained the growth of noxious weeds (Wu *et al.*, 2009; Yao *et al.*, 2019). However, long-term fencing had some negative effects concerning plant community biodiversity, plant nutritional value (e.g. crude protein content), and soil carbon emission (Yao *et al.*, 2019; Luo *et al.*, 2020). A 13-year fencing study enabled alpine grassland to recover from over-grazing (Cao *et al.*, 2019); however, the effect of fencing period still requires further study.

14.5.2 Rotational grazing

Rotational grazing divides the rangelands into summer and winter areas to enable traditional transhumance in the high mountains (Fig. 14.6). This strategy allows the recovery of degraded rangelands for compensatory growth of forage during the resting period; enhances plant community biodiversity and biomass and soil carbon and nitrogen sequestration (Yao *et al.*, 2019; Dong *et al.*, 2020; Ingty, 2021); and lessens

Fig. 14.6. The rotational grazing management on the Tibetan Plateau. (Photo: Xiaoping Jing)

forage waste. A moderate stocking rate minimizes rangelands degradation, as it has a positive effect on plant community biodiversity, plant and soil carbon input, and carbon sequestration (Hafner *et al.*, 2012). Accordingly, seasonal rotational grazing with moderate grazing is generally practised by herders, as the rangelands remain sustainable and they can continuously benefit from livestock products and economic income from the rangelands.

14.5.3 Artificial grassland

It is very difficult and time-consuming to recover severely degraded rangelands. Based on a simulation quantitative model, the restoration of 'black soil land', a severely degraded grassland, requires longer than 50 years under non-grazing conditions and 115–500 years under light grazing (Li *et al.*, 2013). Active restoration with artificial vegetation is regarded as the most effective option to accelerate close-to-nature recovery (Shang *et al.*, 2018). In addition, artificial grassland provides greater above-ground biomass than natural grassland and could provide supplementary forage for livestock in the cold season. Apart from the production benefits, artificial grassland could provide ecological benefits, such as improved plant and soil carbon sequestration for removing atmospheric carbon and slowing climate warming (Dong *et al.*, 2022). Previous studies indicated that grass could be used as pioneer plants for restoring the 'black soil land' into sedge-dominated meadow, and, therefore, the revegetation of severely degraded grassland uses mainly cultivated grasses in current restoration practices (Dong *et al.*, 2020). Several native grass species, including *Stipa penicillata*, *Roegneria thoroldiana*, *Puccinellia distans*, *Elymus tangutorum*, *Stipa aliena* and *Deschampsia caespitosa* were domesticated and selected to restore degraded rangelands (Shang *et al.*, 2018). At present, many restoration specialists believe that mixed-species seeding is more stable than single-species seeding for the revegetation of extremely degraded rangelands (Dong *et al.*, 2020). However, artificial grassland usually suffers from arrested succession during the natural recovery process (Gao *et al.*, 2019). This raises the question of how to accelerate the succession process in artificial grassland.

We recommend three methods to accelerate the succession process. First, the artificial grassland should maintain a utilization rate of above-ground biomass between 40% and 60% in winter and should be fertilized, and weeds and rodents should be controlled to maintain stability (Shang *et al.*, 2018). Secondly, increased plant diversity improves vegetation production and soil carbon sequestration (Lange *et al.*, 2015), but the number of mixed species is usually below six in artificial grassland in the HKH region. Plant biodiversity is vital and should be increased to restore the degraded rangeland. Thirdly, biodiversity effects in the artificial grassland might become saturated during long-term restoration (Knops and Tilman, 2000), and thus new technology is required to overcome this problem. For example, biochar, a solid carbon-containing material, has a promising effect in biodiversity management (Lehmann *et al.*, 2021). Polycyclic and aromatic structures with high aryl C content and high C:N ratio could be sequestered directly in the soil carbon pool. The biochar addition could increase vegetation carbon capture and improve soil nutrients (Xia *et al.*, 2023) and carbon storage, with increased amount of soil aggregate and inhibition of soil organic mineralization (Lehmann *et al.*, 2021), thus improving ecological benefits after plant revegetation in degraded grassland.

14.5.4 Management policies for rangelands

Protected areas are created in designated areas via national policies for conservation objectives (Chaudhary *et al.*, 2022). They are important for biodiversity conservation and sustainable development in mountain rangelands as they reduce the negative impacts of human activities (Dong *et al.*, 2022). There were 575 protected areas in the HKH region in 2020, which covered approximately 40% of the land in the HKH countries and accounted for 8.5% of the global protected areas (Chaudhary *et al.*, 2022). Many of the important HKH rangeland areas, such as the Kekexili nature preserve and Sanjiangyuan National Park Reserve in China, are located within protected areas (Joshi *et al.*, 2013a). At present, protected areas have improved biodiversity, especially rare animals, and have curbed and even reversed rangelands degradation (Bardgett *et al.*, 2021). The plant community biomass and biodiversity in rangelands within protected areas are greater than the adjacent non-protected areas, as was observed in the protected areas in mid-Kunlun, Zoige and Selincuo in China (Wang *et al.*, 2022). In particular, fencing and artificial grassland in the protected areas benefitted rangelands conservation in the HKH region. Currently, 67% of eco-regions, 39% of hotspots, 69% of key biodiversity areas and 76% of important bird and biodiversity areas are still outside the protected areas in the HKH region (Chaudhary *et al.*, 2022), highlighting the need to reassess the boundaries of protected areas in the future.

The creation of protected areas has unintentional side effects such as the decrease of land for agriculture and livestock grazing, which leads to competition for land use between nature reserves and the growing population (Dong *et al.*, 2022). The loss of land by the residents could be mitigated by ecological compensation (Chaudhary *et al.*, 2022; Dong *et al.*, 2022). Attaining socio-economic development in harmony with the ecological environment is important to achieve a balance between livelihood and ecology (Feng *et al.*, 2023; Xue *et al.*, 2023), and, therefore, ecological compensation was developed in the HKH countries. Ecological compensation for rangelands is usually in the context of the grass-based livestock system, which determines the stocking rate based on the grassland carrying capacity and provides economic compensation for the pastoralist. Approximately 74% of the pastoralists are in favour of maintaining the ecological compensation system (Wang *et al.*, 2022). Payment for rangelands ecosystem services provides a feasible approach for supporting ecological compensation, in which downstream and global beneficiaries pay rangeland communities for supplying the ecosystem services in the HKH region (Joshi *et al.*, 2013a). For example, the payment transfer from central to local governments for land use is proving to be successful in China. Ecological compensation, which is essential to implement regional ecological protection, mobilizes all parties to participate in ecological management, and achieves regional green recovery and sustainable development in the HKH rangelands (Xue *et al.*, 2023).

14.6 Summary and Challenge

Although the causes and consequences of rangelands degradation and the developing restoration techniques are well documented, some barriers still exist in conserving rangelands and their sustainable utilization in the HKH region. The most important challenge is the vicious cycles of degradation following restoration. We argue that three actions should be deployed to curb degradation and promote restoration of rangelands in the HKH region. Solutions not only demand greater recognition of traditional restoration ecology but also more use of socio-ecology policies.

14.6.1 Technology innovation

Technology innovations in restoration should be centred mainly on improving methods of detecting rangelands degradation and increasing the success rate of rangelands restoration. We recommend that traditional ecological theories be used in rangelands ecosystems in this region. For example, the critical slowing-down theory has been used widely to monitor the resilience in detecting the degraded state in variable ecosystems, such as the forest and marine ecosystems (Rocha, 2022; Smith *et al.*, 2022). Theories of biodiversity and ecosystem functions to restore the degraded prairie are practised successfully in North America (Yang *et al.*, 2019) but have not been proved in alpine rangeland. In addition, developments in the technology of remote sensing offer options to evaluate the status of rangelands and aid in large-scale restoration efforts at a regional scale (Joshi *et al.*, 2013a).

14.6.2 Improving climate adaptive management

Over-grazing and climate change are the main threats to rangelands degradation; however, most restoration strategies are focused on over-grazing and neglect climate change in the HKH region. Improving the climate change adaptation strategy is imperative, because complex and variable climate change arrests succession in restoring rangelands in the long term (Shang *et al.*, 2018). More importantly, the chronic directional climate change such as rising temperature and increasing precipitation and discrete extreme climatic events are co-occurring in the HKH region (Xu *et al.*, 2022), but their combined effects on rangelands degradation and restoration are less studied, limiting our understanding on climate management in mountain rangelands.

14.6.3 Strengthening connectivity

Ecological values, such as carbon storage and economic value of rangelands have been underestimated globally (Bardgett *et al.*, 2021), so the international and national policies on rangelands protection are weak. More importantly, governance remains distributed unevenly across local, national and regional authorities in the HKH region, with limited connectivity across scales (Ojha *et al.*, 2019). For example, the HKH countries lack institutions to link upstream and downstream communities in river basins. Therefore, restoration practices are disjointed across countries, leading to a lack of communication on successful methods, which will undermine sustainable natural resource management in the mountain rangelands ecosystem.

References

Ahmad, Z., Postigo, J.C., Rahman, F. and Dittmann, A. (2021) Mountain pastoralism in the eastern Hindu Kush: the case of Lotkuh Valley, Pakistan. *Mountain Research and Development* 41(4), R16–R28. DOI: 10.1659/MRD-JOURNAL-D-21-00007.1

Anderson, K., Fawcett, D., Cugulliere, A., Benford, S., Jones, D. *et al.* (2020) Vegetation expansion in the subnival Hindu Kush Himalaya. *Global Change Biology* 26(3), 1608–1625. DOI: 10.1111/gcb.14919

Bardgett, R.D., Bullock, J.M., Lavorel, S., Manning, P., Schaffner, U. *et al.* (2021) Combatting global grassland degradation. *Nature Reviews Earth and Environment* 2(10), 720–735. DOI: 10.1038/s43017-021-00207-2

Bhatpuria, D., Matheswaran, K., Piman, T., Tha, T. and Towashiraporn, P. (2022) Assessment of large-scale seasonal river morphological changes in Ayeyarwady River using optical remote sensing data. *Remote Sensing* 14(14), 3393. DOI: 10.3390/rs14143393

Briske, D.D. (2017) Rangeland systems: foundation for a conceptual framework. In: Briske, D.D. (ed.) *Rangeland Systems: Processes, Management and Challenges.* Springer International Publishing, Cham, Switzerland, pp. 1–21.

Cao, J., Li, G., Adamowski, J.F., Holden, N.M., Deo, R.C. *et al.* (2019) Suitable exclosure duration for the restoration of degraded alpine grasslands on the Qinghai-Tibetan Plateau. *Land Use Policy* 86, 261–267. DOI: 10.1016/j.landusepol.2019.05.008

Chao, Z., Shang, Z., Fei, C., Zhuang, Z. and Zhou, M. (2023) Spatiotemporal analysis of urban expansion in the mountainous Hindu Kush Himalayas region. *Land* 12(3), 576. DOI: 10.3390/land12030576

Chaudhary, S., Uddin, K., Chettri, N., Thapa, R. and Sharma, E. (2022) Protected areas in the Hindu Kush Himalaya: a regional assessment of the status, distribution, and gaps. *Conservation Science and Practice* 4(10), e12793. DOI: 10.1111/csp2.12793

Ding, J., Chen, L., Ji, C., Hugelius, G., Li, Y. *et al.* (2017) Decadal soil carbon accumulation across Tibetan permafrost regions. *Nature Geoscience* 10(6), 420–424. DOI: 10.1038/NGEO2945

Dong, S., Wen, L., Li, Y., Wang, X., Zhu, L. *et al.* (2012) Soil-quality effects of grassland degradation and restoration on the Qinghai-Tibetan Plateau. *Soil Science Society of America Journal* 76(6), 2256–2264. DOI: 10.2136/sssaj2012.0092

Dong, S., Shang, Z., Gao, J., Boone, R.B. (2020) Enhancing sustainability of grassland ecosystems through ecological restoration and grazing management in an era of climate change on Qinghai-Tibetan Plateau. *Agriculture, Ecosystems and Environment* 287, 106684. DOI: 10.1016/j.agee.2019.106684

Dong, S., Shang, Z., Gao, J. and Boone, R. (2022) Enhancing the ecological services of the Qinghai-Tibetan Plateau's grasslands through sustainable restoration and management in era of global change. *Agriculture, Ecosystems and Environment* 326, 107756. DOI: 10.1016/j.agee.2021.107756

Dorji, K. (2022) Yak mortality causes in selected rangeland areas of Bhutan. *Bhutan Journal of Animal Science* 6(1), 56–62.

Feng, X., Zhao, Q., Li, J. and Qiu, H. (2023) The impact of political status and social capital on herders' overgrazing behaviors in the pastoral areas of China. *Journal of Environmental Management* 327, 116861. DOI: 10.1016/j.jenvman.2022.116861

Gao, X., Dong, S., Xu, Y., Wu, S., Wu, X. *et al.* (2019) Resilience of revegetated grassland for restoring severely degraded alpine meadows is driven by plant and soil quality along recovery time: a case study from the Three-river Headwater area of Qinghai-Tibetan Plateau. *Agriculture, Ecosystems and Environment* 279, 169–177. DOI: 10.1016/j.agee.2019.01.010

Ghazali, S., Zibaei, M. and Azadi, H. (2023) Impact of livelihood strategies and capitals on rangeland sustainability and nomads' poverty: a counterfactual analysis in southwest Iran. *Ecological Economics* 206, 107738. DOI: 10.1016/j.ecolecon.2023.107738

Gioli, G., Thapa, G., Khan, F., Dasgupta, P., Nathan, D. *et al.* (2019) Understanding and tackling poverty and vulnerability in mountain livelihoods in the Hindu Kush Himalaya. In: Wester, P., Mishra, A., Mukherji, A. and Shrestha, A. (eds) *The Hindu Kush Himalaya Assessment: Mountains, Climate Change, Sustainability and People*. Springer Nature, Cham, Switzerland, pp. 421–455.

Gongbuzeren, Huntsinger, L. and Li, W. (2018) Rebuilding pastoral social-ecological resilience on the Qinghai-Tibetan Plateau in response to changes in policy, economics, and climate. *Ecology and Society* 23(2), 21.

Guo, D., Song, X., Hu, R., Zhu, X., Jiang, Y. *et al.* (2021) Large-scale analysis of the spatiotemporal changes of net ecosystem production in Hindu Kush Himalayan Region. *Remote Sensing* 13(6), 1180. DOI: 10.3390/rs13061180

Hafner, S., Unteregelsbacher, S., Seeber, E., Lena, B., Xu, X. *et al.* (2012) Effect of grazing on carbon stocks and assimilate partitioning in a Tibetan montane pasture revealed by $13CO_2$ pulse labeling. *Global Change Biology* 18(2), 528–538. DOI: 10.1111/j.1365-2486.2011.02557.x

Han, Z., Song, W. and Deng, X. (2016) Responses of ecosystem service to land use change in Qinghai Province. *Energies* 9(4), 303. DOI: 10.3390/en9040303

Hussain, A. and Qamar, F.M. (2020) Dual challenge of climate change and agrobiodiversity loss in mountain food systems in the Hindu-Kush Himalaya. *One Earth* 3(5), 539–542. DOI: 10.1016/j.oneear.2020.10.016

Hussain, A., Rasul, G., Mahapatra, B. and Tuladhar, S. (2016) Household food security in the face of climate change in the Hindu-Kush Himalayan region. *Food Security* 8(5), 921–937. DOI: 10.1007/s12571-016-0607-5

Ingty, T. (2021) Pastoralism in the highest peaks: role of the traditional grazing systems in maintaining biodiversity and ecosystem function in the alpine Himalaya. *PLOS ONE* 16(1), e0245221. DOI: 10.1371/journal.pone.0245221

Jahan, A., Dar, T., Kumar, S. and Rai, N. (2023) Comprehensive isotopic characterization of the hydrological processes of the Indus river basin (IRB): a comparison between Upper Indus Basin (UIB) and Lower Indus Basin (LIB). *Journal of Mountain Science* 20(3), 705–723. DOI: 10.1007/s11629-022-7655-4

Jing, X., Ding, L., Zhou, J., Huang, X., Degen, A. *et al.* (2022) The adaptive strategies of yaks to live in the Asian highlands. *Animal Nutrition* 9, 249–258. DOI: 10.1016/j.aninu.2022.02.002

Joshi, L., Shrestha, R.M., Jasra, A.W., Joshi, S., Gilani, H. *et al.* (2013a) Rangeland ecosystem services in the Hindu Kush Himalayan region. In: Ning, W., Rawat, G.S., Joshi, S., Ismail, M., Sharma, E. *et al.* (eds) *High-altitude Rangelands and Their Interfaces in the Hindu Kush Himalayas*. ICIMOD, Kathmandu.

Joshi, S., Jasra, W.A., Ismail, M., Shrestha, R.M., Yi, S.L. *et al.* (2013b) Herders' perceptions of and responses to climate change in northern Pakistan. *Environmental Management* 52(3), 639–648. DOI: 10.1007/s00267-013-0062-4

Knops, J.M.H. and Tilman, D. (2000) Dynamics of soil nitrogen and carbon accumulation for 61 years after agricultural abandonment. *Ecology* 81(1), 88–98. DOI: 10.1890/0012-9658(2000)081[0088:DOSNAC]2.0.CO;2

Kotru, R.K., Shakya, B., Joshi, S., Gurung, J., Ali, G., Amatya, S. *et al.* (2020) Biodiversity conservation and management in the Hindu Kush Himalayan Region: Are transboundary landscapes a promising solution? *Mountain Research and Development* 40(2), A15. DOI: 10.1659/MRD-JOURNAL-D-19-00053.1

Krishnan, R., Shrestha, A.B., Ren, G., Rajbhandari, R., Saeed, S. *et al.* (2019) Unravelling climate change in the Hindu Kush Himalaya: rapid warming in the mountains and increasing extremes. In: Wester, P., Mishra, A., Mukherji, A. and Shrestha, A. (eds) *The Hindu Kush Himalaya Assessment: Mountains, Climate Change, Sustainability and People*. Springer Nature, Cham, Switzerland, pp. 57–97.

Kulaixi, Z., Chen, Y., Wang, C. and Xia, Q. (2023) Spatial differentiation of ecosystem service value in an arid region: a case study of the Tarim River Basin, Xinjiang. *Ecological Indicators* 151, 110249. DOI: 10.1016/j.ecolind.2023.110249

Lamb, V., Middleton, C. and Win, S. (2019) Introduction: resources politics and knowing the Salween River. In: Middleton, C. and Lamb, V. (eds) *Knowing the Salween River: Resource Politics of a Contested Transboundary River.* Springer International Publishing, Cham, Switzerland, pp. 1–15.

Lange, M., Eisenhauer, N., Sierra, C.A., Bessler, H., Engels, C. *et al.* (2015) Plant diversity increases soil microbial activity and soil carbon storage. *Nature Communications* 6(1), 6707. DOI: 10.1038/ncomms7707

Lehmann, J., Cowie, A., Masiello, C.A., Kammann, C., Woolf, D. *et al.* (2021). Biochar in climate change mitigation. *Nature Geoscience* 14(12), 883–892. DOI: 10.1038/s41561-021-00852-8

Li, J., Zou, Y., Zhang, Y., Sun, S. and Dong, X. (2022) Risk assessment of snow disasters for animal husbandry on the Qinghai-Tibetan plateau and influences of snow disasters on the well-being of farmers and pastoralists. *Remote Sensing* 14(14), 3358. DOI: 10.3390/rs14143358

Li, W., Cao, W., Wang, J., Li, X., Xu, C. *et al.* (2017) Effects of grazing regime on vegetation structure, productivity, soil quality, carbon and nitrogen storage of alpine meadow on the Qinghai-Tibetan Plateau. *Ecological Engineering* 98, 123–133. DOI: 10.1016/j.ecoleng.2016.10.026

Li, X.-L., Perry, G.L.W., Brierley, G., Gao, J., Zhang, J. *et al.* (2013) Restoration prospects for Heitutan degraded grassland in the Sanjiangyuan. *Journal of Mountain Science* 10(4), 687–698. DOI: 10.1007/s11629-013-2557-0

Liu, P., Ding, L., Zhou, Y., Jing, X. and Degen, A.A. (2019) Behavioural characteristics of yaks grazing summer and winter pastures on the Qinghai-Tibetan Plateau. *Applied Animal Behaviour Science* 218, 104826. DOI: 10.1016/j.applanim.2019.06.007

Liu, P., Lv, W., Sun, J., Luo, C., Zhang, Z. *et al.* (2021) Ambient climate determines the directional trend of community stability under warming and grazing. *Global Change Biology* 27(20), 5198–5210. DOI: 10.1111/gcb.15786

Liu, Y. and Lu, C. (2021) Quantifying grass coverage trends to identify the hot plots of grassland degradation in the Tibetan Plateau during 2000–2019. *International Journal of Environmental Research and Public Health* 18(2), 416. DOI: 10.3390/ijerph18020416

Lund, H.G. (2007) Accounting for the world's rangelands. *Rangelands* 29(1), 3–10.

Luo, C., Wang, S., Zhang, L., Wilkes, A., Zhao, L. *et al.* (2020) CO_2, CH_4 and N_2O fluxes in an alpine meadow on the Tibetan Plateau as affected by N-addition and grazing exclusion. *Nutrient Cycling in Agroecosystems* 117(1), 29–42. DOI: 10.1007/s10705-020-10062-0

Mayewski, P.A., Perry, L.B., Matthews, T. and Birkel, S.D. (2020) Climate change in the Hindu Kush Himalayas: basis and gaps. *One Earth* 3(5), 551–555. DOI: 10.1016/j.oneear.2020.10.007

Millar, J. and Tenzing, K. (2021) Transforming degraded rangelands and pastoralists' livelihoods in eastern Bhutan. *Mountain Research and Development* 41(4), D1–D7. DOI: 10.1659/MRD-JOURNAL-D-21-00025.1

Mishra, U., Hugelius, G., Shelef, E., Yang, Y., Strauss, J. *et al.* (2021) Spatial heterogeneity and environmental predictors of permafrost region soil organic carbon stocks. *Science Advances* 7(9), eaaz5236. DOI: 10.1126/sciadv.aaz5236

Mittermeier, R.A., Turner, W.R., Larsen, F.W., Brooks, T.M. and Gascon, C. (2011) Global biodiversity conservation: the critical role of hotspots. In: Zachos, F.E. and Habel, J. (eds) *Biodiversity Hotspots: Distribution and Protection of Conservation Priority Areas.* Springer, Berlin, Heidelberg, pp. 3–22.

Mukherji, A., Molden, D., Nepal, S., Rasul, G. and Wagnon, P. (2015) Himalayan waters at the crossroads: issues and challenges. *International Journal of Water Resources Development* 31(2), 151–160. DOI: 10.1080/07900627.2015.1040871

Niloy, N.M., Haque, M.M. and Tareq, S.M. (2021) Characteristics, sources, and seasonal variability of dissolved organic matter (DOM) in the Ganges River, Bangladesh. *Environmental Processes* 8(2), 593–613. DOI: 10.1007/s40710-021-00499-y

Niu, Y., Zhou, J., Yang, S., Chu, B., Ma, S. *et al.* (2019) The effects of topographical factors on the distribution of plant communities in a mountain meadow on the Tibetan Plateau as a foundation for target-oriented management. *Ecological Indicators* 106, 105532. DOI: 10.1016/j.ecolind.2019.105532

Ojha, H.R., Ghate, R., Dorji, L., Shrestha, A., Paudel, D. *et al.* (2019) Governance: key for environmental sustainability in the Hindu Kush Himalaya. In: Wester, P., Mishra, A., Mukherji, A. and Shrestha, A. (eds) *The Hindu Kush Himalaya Assessment: Mountains, Climate Change, Sustainability and People.* Springer Nature, Cham, Switzerland, pp. 545–578.

Ombadi, M., Risser, M.D., Rhoades, A.M. and Varadharajan, C. (2023) A warming-induced reduction in snow fraction amplifies rainfall extremes. *Nature* 619(7969), 305–310. DOI: 10.1038/s41586-023-06092-7

Panday, P.K. and Ghimire, B. (2012) Time-series analysis of NDVI from AVHRR data over the Hindu Kush-Himalayan region for the period 1982–2006. *International Journal of Remote Sensing* 33(21), 6710–6721. DOI: 10.1080/01431161.2012.692836

Ping, C.L., Jastrow, J.D., Jorgenson, M.T., Michaelson, G.J. and Shur, Y.L. (2015) Permafrost soils and carbon cycling. *Soil* 1(1), 147–171. DOI: 10.5194/soil-1-147-2015

Pondari, S., Dandabathula, G., Bera, A., Nagamani, P. and Amminedu, E. (2020) Characterization of drainage network of Brahmaputra river basin in Indian sub-continent using geospatial technologies. *Science, Technology and Environment Informatics* 8(01), 583–594.

Rasul, G., Saboor, A., Tiwari, P.C., Hussain, A., Ghosh, N. *et al.* (2019) Food and nutrition security in the Hindu Kush Himalaya: unique challenges and niche opportunities. In: Wester, P., Mishra, A., Mukherji, A. and Shrestha, A. (eds) *The Hindu Kush Himalaya Assessment: Mountains, Climate Change, Sustainability and People*. Springer Nature, Cham, Switzerland, pp. 301–338.

Ren, G.-Y. and Shrestha, A.B. (2017) Climate change in the Hindu Kush Himalaya. *Advances in Climate Change Research* 8(3), 137–140. DOI: 10.1016/j.accre.2017.09.001

Robinson, N.P., Allred, B.W., Naugle, D.E. and Jones, M.O. (2019) Patterns of rangeland productivity and land ownership: implications for conservation and management. *Ecological Applications* 29(3), e01862. DOI: 10.1002/eap.1862

Rocha, J.C. (2022) Ecosystems are showing symptoms of resilience loss. *Environmental Research Letters* 17(6), 065013. DOI: 10.1088/1748-9326/ac73a8

Roomi, I. and Khan, M.Z. (2022) *Mountain Studies: Understanding and Managing Mountains for People and Nature*. Karakoram International University, Gilgit, Pakistan.

Rueff, H. (2014) High-altitude rangelands and their interfaces in the Hindu Kush Himalayas. *Mountain Research and Development* 34(3), 305–306. DOI: 10.1659/mrd.mm141

Salehie, O., Ismail, T.B., Shahid, S., Hamed, M.M., Chinnasamy, P. *et al.* (2022) Assessment of water resources availability in Amu Darya River Basin using GRACE data. *Water* 14(4), 533. DOI: 10.3390/w14040533

Scott, C.A., Zhang, F., Mukherji, A., Immerzeel, W., Mustafa, D. *et al.* (2019) Water in the Hindu Kush Himalaya. In: Wester, P., Mishra, A., Mukherji, A. and Shrestha, A. (eds) *The Hindu Kush Himalaya Assessment: Mountains, Climate Change, Sustainability and People*. Springer Nature, Cham, Switzerland, pp. 257–299.

Shang, Z., Dong, Q., Shi, J., Zhou, H., Dong, S. *et al.* (2018) Research progress in recent ten years of ecological restoration for 'Black Soil Land' degraded grassland on Tibetan Plateau – concurrently discuss of ecological restoration in Sangjiangyuan Region. *Acta Agrectir Sinica* 26(1), 21. DOI: 10.11733/j.issn.1007-0435.2018.01.001 (in Chinese).

Sharma, E., Zhaoli, Y. and Sharma, B. (2007) ICIMOD's regional rangeland program for the Hindu Kush-Himalayas. *Mountain Research and Development* 27(2), 174–177. DOI: 10.1659/mrd.mp003

Sharma, E., Molden, D., Rahman, A., Khatiwada, Y.R., Zhang, L. *et al.* (2019) Introduction to the Hindu Kush Himalaya assessment. In: Wester, P., Mishra, A., Mukherji, A. and Shrestha, A. (eds) *The Hindu Kush Himalaya Assessment: Mountains, Climate Change, Sustainability and People*. Springer Nature, Cham, Switzerland, pp. 1–16.

Shen, Y.P., Wang, G., Wang, G., Pu, J. and Wang, X. (2009) Impacts of climate change on glacial water resources and hydrological cycles in the Yangtze River source region, the Qinghai-Tibetan Plateau, China: a progress report. *Sciences in Cold and Arid Regions* 1(6), 0475–0495.

Singh, S., Tanvir Hassan, S.M., Hassan, M. and Bharti, N. (2019) Urbanisation and water insecurity in the Hindu Kush Himalaya: insights from Bangladesh, India, Nepal and Pakistan. *Water Policy* 22(S1), 9–32. DOI: 10.2166/wp.2019.215

Smith, T., Traxl, D. and Boers, N. (2022) Empirical evidence for recent global shifts in vegetation resilience. *Nature Climate Change* 12(5), 477–484. DOI: 10.1038/s41558-022-01352-2

Strauss, J., Schirrmeister, L., Grosse, G., Fortier, D., Hugelius, G. *et al.* (2017) Deep Yedoma permafrost: a synthesis of depositional characteristics and carbon vulnerability. *Earth-Science Reviews* 172, 75–86. DOI: 10.1016/j.earscirev.2017.07.007

Sun, X., Ren, G., Xu, W., Li, Q. and Ren, Y. (2017) Global land-surface air temperature change based on the new CMA GLSAT data set. *Science Bulletin* 62(4), 236–238. DOI: 10.1016/j.scib.2017.01.017

Tuladhar, S., Hussain, A., Baig, S., Ali, A., Soheb, M. *et al.* (2023) Climate change, water and agriculture linkages in the upper Indus basin: a field study from Gilgit-Baltistan and Leh-Ladakh. *Frontiers in Sustainable Food Systems* 6, 1012363. DOI: 10.3389/fsufs.2022.1012363

Volder, A., Briske, D.D. and Tjoelker, M.G. (2013) Climate warming and precipitation redistribution modify tree–grass interactions and tree species establishment in a warm-temperate savanna. *Global Change Biology* 19(3), 843–857. DOI: 10.1111/gcb.12068

Wang, P., Lassoie, J.P., Morreale, S.J. and Dong, S. (2015) A critical review of socioeconomic and natural factors in ecological degradation on the Qinghai-Tibetan Plateau, China. *The Rangeland Journal* 37(1), 1–9. DOI: 10.1071/RJ14094

Wang, Y., Wu, N., Kunze, C., Long, R. and Perlik, M. (2019) Drivers of change to mountain sustainability in the Hindu Kush Himalaya. In: Wester, P., Mishra, A., Mukherji, A. and Shrestha, A. (eds) *The Hindu Kush Himalaya Assessment: Mountains, Climate Change, Sustainability and People*. Springer Nature, Cham, Switzerland, pp. 17–56.

Wang, Y., Lv, W., Xue, K., Wang, S., Zhang, L. *et al.* (2022) Grassland changes and adaptive management on the Qinghai-Tibetan Plateau. *Nature Reviews Earth and Environment* 3(10), 668–683. DOI: 10.1038/s43017-022-00330-8

Wei, D., Qi, Y., Ma, Y., Wang, X., Ma, W. *et al.* (2021) Plant uptake of CO_2 outpaces losses from permafrost and plant respiration on the Tibetan Plateau. *Proceedings of the National Academy of Sciences of the United States of America* 118(33), e2015283118. DOI: 10.1073/pnas.2015283118

Wei, Y., Lu, H., Wang, J., Wang, X. and Sun, J. (2022) Dual influence of climate change and anthropogenic activities on the spatiotemporal vegetation dynamics over the Qinghai-Tibetan Plateau from 1981 to 2015. *Earths Future* 10(5), e2021EF002566. DOI: 10.1029/2021EF002566

Wu, G.-L., Du, G.-Z., Liu, Z.-H. and Thirgood, S. (2009) Effect of fencing and grazing on a Kobresia-dominated meadow in the Qinghai-Tibetan Plateau. *Plant and Soil* 319(1), 115–126. DOI: 10.1007/s11104-008-9854-3

Wu, N., Ismail, M., Joshi, S., Yi, S.-L., Shrestha, R.M. *et al.* (2014) Livelihood diversification as an adaptation approach to change in the pastoral Hindu-Kush Himalayan region. *Journal of Mountain Science* 11(5), 1342–1355. DOI: 10.1007/s11629-014-3038-9

Wu, S., Wang, Y., Wang, Z., Shrestha, N. and Liu, J. (2022) Species divergence with gene flow and hybrid speciation on the Qinghai-Tibet Plateau. *New Phytologist* 234(2), 392–404. DOI: 10.1111/nph.17956

Xia, L., Cao, L., Yang, Y., Ti, C., Liu, Y. *et al.* (2023) Integrated biochar solutions can achieve carbon-neutral staple crop production. *Nature Food* 4(3), 236–246. DOI: 10.1038/s43016-023-00694-0

Xu, C., Liu, W., Zhao, D., Hao, Y., Xia, A., Yan, N. and Zeng, Y. (2022) Remote sensing-based spatiotemporal distribution of grassland aboveground biomass and its response to climate change in the Hindu Kush Himalayan Region. *Chinese Geographical Science* 32(5), 759–775. DOI: 10.1007/s11769-022-1299-8

Xu, J., Badola, R., Chettri, N., Chaudhary, R.P., Zomer, R. *et al.* (2019) Sustaining biodiversity and ecosystem services in the Hindu Kush Himalaya. In: Wester, P., Mishra, A., Mukherji, A. and Shrestha, A. (eds) *The Hindu Kush Himalaya Assessment: Mountains, Climate Change, Sustainability and People*. Springer Nature, Cham, Switzerland, pp. 127–165.

Xu, Y., Dong, S., Gao, X., Yang, M., Li, S. *et al.* (2021) Aboveground community composition and soil moisture play determining roles in restoring ecosystem multifunctionality of alpine steppe on Qinghai-Tibetan Plateau. *Agriculture, Ecosystems and Environment* 305, 107163. DOI: 10.1016/j.agee.2020.107163

Xue, J., Lang, J., Guan, Y. and Lu, S. (2023) Design and measurement of small-scale regional ecological compensation model. *Economic Analysis and Policy* 78, 1392–1405. DOI: 10.1016/j.eap.2023.05.016

Yang, Y., Tilman, D., Furey, G. and Lehman, C. (2019) Soil carbon sequestration accelerated by restoration of grassland biodiversity. *Nature Communications* 10(1), 718. DOI: 10.1038/s41467-019-08636-w

Yao, X., Wu, J., Gong, X., Lang, X., Wang, C. *et al.* (2019) Effects of long term fencing on biomass, coverage, density, biodiversity and nutritional values of vegetation community in an alpine meadow of the Qinghai-Tibet Plateau. *Ecological Engineering* 130, 80–93. DOI: 10.1016/j.ecoleng.2019.01.016

Ye, T., Liu, W., Wu, J., Li, Y., Shi, P. *et al.* (2019) Event-based probabilistic risk assessment of livestock snow disasters in the Qinghai-Tibetan Plateau. *Natural Hazards and Earth System Sciences* 19(3), 697–713. DOI: 10.5194/nhess-19-697-2019

Zhong, X., Li, J., Wang, J., Zhang, J., Liu, L. *et al.* (2022) Linear and nonlinear characteristics of long-term NDVI using trend analysis: a case study of Lancang-Mekong River Basin. *Remote Sensing* 14(24), 6271. DOI: 10.3390/rs14246271

Zhou, H., Yang, X., Zhou, C., Shao, X., Shi, Z. *et al.* (2023) Alpine grassland degradation and its restoration in the Qinghai-Tibet Plateau. *Grasses* 2(1), 31–46. DOI: 10.3390/grasses2010004

Zhou, Z., Sun, X., Zhang, X. and Wang, Y. (2022) Inter-regional ecological compensation in the Yellow River Basin based on the value of ecosystem services. *Journal of Environmental Management* 322, 116073. DOI: 10.1016/j.jenvman.2022.116073

15 Agriculture: Agrobiodiversity and Sustainable Approaches in the Hindu Kush Himalayan Region

Dongmei Li[1,*], Haonan Guo[1], Zhen Peng[1], Allan Degen[2], Madan Koirala[3], Devendra Gauchan[4] and Zhanhuan Shang[1]

[1]*College of Ecology, Lanzhou University, Lanzhou, China;* [2]*Institutes for Desert Research, Ben Gurion University of the Negev, Israel;* [3]*Central Department of Environmental Science, Tribhuvan University, Kathmandu, Nepal;* [4]*Alliance of Bioversity International and CIAT, Lalitpur, Nepal*

Abstract

Approximately 750 million people face severe food shortages on Earth. For these people, agriculture is the best option for national food security; however, drought, anthropogenic activities, urbanization, land fragmentation and industry have decreased arable land. This chapter describes: (i) the status of agricultural biodiversity in the Hindu Kush Himalayan (HKH) region; (ii) threats to agriculture and their impacts; (iii) local livelihoods and ecological restoration of agriculture; and (iv) shifts in management strategy, adaptations and pathways to improve agriculture in the HKH region.

Keywords: anthropogenic activities, agrobiodiversity conservation, agrobiodiversity restoration, food security, livelihood, subsistence agriculture

15.1 Introduction

Mountainous regions are important hubs of the world's biodiversity and have long provided local and international people access to abundant genetic resources for agriculture and animal husbandry (Perrigo *et al.*, 2020). However, mountains are also among the most ecologically fragile regions in the world, being increasingly impacted by human activities, political upheavals and unsustainable resource management (Tse-Ring *et al.*, 2010). As a result, it is urgent to adopt effective management techniques and policies to reduce or improve the region's ecological environment. In this chapter, we systematically review the threats and challenges facing agriculture in the Hindu Kush Himalayan (HKH) region, including climate change and anthropogenic activities, as well as management strategies and adaptations, including integrated crop and livestock production and efficient energy use. The aims of these strategies are to improve agricultural biodiversity and livelihoods of the local populations while developing a sustainable agro-ecosystem in the HKH region.

*Corresponding author: lidm21@lzu.edu.cn

DOI: 10.1079/9781800622579.0015

Due to its environmental heterogeneity, the HKH region is one of the world's major biodiversity hotspots (Paudel and Heinen, 2015), providing a variety of genetic resources for agriculture and animal husbandry (Singh *et al.*, 2011). For instance, from 1979 to 2013, approximately 44 nations throughout the world received rice germplasm from Bhutan despite Bhutan's landlocked, mountainous location (Ghimiray and Vernooy, 2017). This is of interest, as close to 50% of the people in the HKH countries are malnourished and a third of the population still experiences food insecurity (Rasul *et al.*, 2019). Surveys in Nepal revealed that agricultural biodiversity is much greater in self-sufficient communities (Kc *et al.*, 2016), demonstrating the importance of diversity for food security. The preservation of regional agricultural biodiversity is crucial for the sustainability of local food systems, as well as for the maintenance of food security.

The agrobiodiversity of the HKH mountain ranges is facing difficult obstacles. A vital resource for regional economic sustainability, agricultural biodiversity is a prerequisite for national food security, but is deteriorating in the HKH region. Traditional food crops, including buckwheat, barley, millet, sorghum, oats and beans, are progressively being phased out of the food chain, while monoculture farming is on the rise. For example, in Nepal, indigenous crops are increasingly being replaced by monocultures of potatoes (Thapa and Hussain, 2021), while most countries in the HKH region have experienced declines in traditional cereal cultivation in recent years. In India, the production share of traditional cereals declined from approximately 20% in 1961 to 2–5% in 2017, while in Pakistan they declined from 11% to 1% during this period (FAOSTAT, 2018; Rasul *et al.*, 2022). Moreover, the loss of agricultural biodiversity also threatens genetic diversity, which renders agriculture more vulnerable to biological and abiotic stressors, and affects crop resilience (Oliver *et al.*, 2015). The continual decline in soil fertility and the escalation of crop pests are just two examples of how the loss of agricultural biodiversity can impact ecosystems and ecosystem service functions. It can also alter carbon input and carbon decomposition, which can have an impact on how well ecosystems store carbon (Chen *et al.*, 2018). The preservation of biodiversity and the development of sustainable livelihoods in the region must contend with the effects of climate change and human activities.

15.2 Threats to Agriculture in the Hindu Kush Himalayan Region and Their Impacts

Climate change, including global warming, drought, erratic precipitation and snowfall, along with human activities like intensive agriculture, unsustainable resource management and infrastructure construction, have made the region's agriculture more vulnerable and are affecting livelihoods, social development, gender equality and regional culture.

15.2.1 Threat of climate change

The agro-ecosystem in the HKH region is faced with exceptionally difficult challenges since the region is substantially more sensitive to climate change than other regions in the world. Over the past three decades (1991–2020), climate change and the loss of agricultural biodiversity have posed serious challenges to the sustainability of food systems and overall food security in the region (Hussain and Qamar, 2020). Droughts and glacier shrinkage caused by warming, water shortages due to erratic snowfall and rainfall, and flooding brought on by glacier melt and lake outbursts have all increased the loss of agricultural biodiversity.

Climate warming

Increased crop susceptibility brought on by global warming poses a challenge to the development of agriculture in the area. During the past 50–60 years, the number of warm nights and days in the area has increased by 1.7 and 1.2 days, respectively, and most farmers are sensing this change, with 91% of respondents reporting that the daytime temperatures in summer are higher today than ten years ago. Further warming will be a severe threat to the environment in the area (Krishnan et al., 2019). Some plants may expand their ranges to higher elevations; however, species close to mountain peaks will have less capacity to do so, thus increasing crop susceptibility. Agricultural pest outbreaks brought on by rising temperature will lead to poor harvests and decreased yields. The rapid retreat of glaciers has led to landslides, glacial lake collapse, and has washed away downstream agricultural facilities and agricultural land, causing serious damage to downstream agriculture and communities (Khalid et al., 2021), while the retreat of glaciers has led to the abandonment of irrigation systems that rely on glacial snowmelt.

Unpredictable snowfall and rain

Floods and droughts have become more common as a result of inadequate water management and increasingly erratic rainfall. According to a survey of farmers in Nepal's three ecological regions – mountain, hill and tarai – 55.5% practise rain-fed farming and 44.5% use partial or complete irrigation (Paudel et al., 2020). Since the cropland in the Nepalese district of Burgaon depends entirely on rainfall, residents are forced to adjust to shifting precipitation patterns (Onta and Resurreccion, 2011). In India, where close to two-thirds of arable land still relies on monsoons, irrigation farming is largely dependent on groundwater and aqueducts (Dhawan, 2017), and farmers are already facing water shortages for irrigation due to reduced snowfall (Clouse et al., 2017). According to Yao et al. (2022), 23% of the people in western China depend on glacial meltwater, and most Sherpas use glacial meltwater to irrigate their crops and for drinking water (Gagné et al., 2014). Towns, agriculture, ecological security and sustainable growth rely on freshwater run-off, and consequently are severely impacted by erratic shifts in the amount of rain and snow. They must develop strategies to mitigate the uncertain supply of water.

15.2.2 Human activities

Intensive agricultural management

Several effects of agricultural intensification have impacted production negatively and have caused soil structure deterioration, soil organic matter loss, land degradation, salinization of irrigated areas, soil erosion and a decrease in agricultural biodiversity. For example, the uniformity of species and farming systems can compromise the resilience of agro-ecosystems to disturbances such as climate change, pests and diseases (Rasul et al., 2019). The heavy use of pesticides and drugs destroys non-target biodiversity (Ehrlich and Pringle, 2008), while chemical fertilizers cause the soil to harden and the nutrient circulation to be blocked. Frequent tilling of the soil not only reduces the carbon storage capacity but also leaves the soil more vulnerable to erosion and can cause degradation of the land. Insufficient fallow time depletes soil organic matter and degrades land, affecting nutrient cycling and the sustainability of agricultural systems.

Unsustainable natural resource management

Unsustainable management strategies for land, agriculture, food, water and other resources have threatened the sustainable development of agriculture in the HKH region. The over-use of natural resources can harm, and cause the loss of genetic diversity and the extinction of species. For instance, farmers in India depend on groundwater to cope with droughts, and irrigation has long been a key component of India's agricultural growth strategy. However, due to irresponsible water management, the groundwater table has decreased rapidly, and it is anticipated to continue to do so because of future climate change (Fishman, 2018). Some water-intensive crops had to be abandoned by farmers, which, undoubtedly, played a role in the decline of agricultural biodiversity. Illegal hunting and overfishing can destroy target species and affect the stability of entire ecosystems through the food chain. Excessive harvesting of fuelwood and commercial wood, excessive harvesting of non-wood forest products such as medicinal plants, and other destructive resource uses, challenge the sustainable development of agriculture in mountain areas.

Urbanization and infrastructure construction

Lands for urbanization and infrastructure development are generally converted from agricultural land, leading to the loss of agricultural biodiversity and jeopardizing the long-term sustainability of agricultural ecosystems. The Himalayan region has experienced rapid urbanization. Nagpur is one of the expanding cities in central India where unplanned development has led to the loss of agricultural land and groundwater resources (Dhyani *et al.*, 2018). In the Kailash Sacred Landscape of India, there is a high rate of human settlement expansion, especially in the valleys of the low-elevation range, which has increased by 6.8 times between 1976 and 2011, resulting in a substantial reduction in agricultural land (Singh *et al.*, 2018). Rapid urban development in Uttarakhand has encroached on productive agricultural land, reducing the land by 4.7% to 13.0% (Tiwari *et al.*, 2018). It is expected that by 2050 more than 50% of the population of HKH countries will live in cities (UN, 2014), which means that there will be more conversion of agricultural land to cities and infrastructure. It also means that a large number of farmers will abandon their original land, and agriculture on the periphery of cities will become more intensive, which will undoubtedly exacerbate the loss of agricultural biodiversity. In Uttarakhand, the road network increased from 44,332 km in 1991 to 46,836 km in 2001 and to 49,277 km in 2011 (Tiwari *et al.*, 2018), facilitating the emergence and development of many urban centres. However, it also occupies agricultural land.

15.2.3 Globalization

Globalization is a market-driven process that aims to promote global economic growth and prosperity. In the context of globalization, mountainous areas have become more 'market-orientated', in contrast to the traditional 'non-market'-oriented practices, which have restricted the development of mountainous agriculture (Jodha, 2004). To adapt to globalization, the intensive cultivation of individual cash crops such as mountain apples and buckwheat have replaced traditional crops, resulting in the breakdown of existing inter-related production and resource-use systems such as agriculture-forestry-livestock systems, resulting in a decline in agricultural biodiversity and an increase in the vulnerability of agro-ecosystems. Moreover, globalization has provided more opportunities for the introduction of invasive species. Coupled with frequent extreme climate events, the number of invasive plant species has been increasing (Seebens *et al.*, 2018), and invasive plants have reduced biodiversity and increased biological homogeneity (Ehrlich and Pringle 2008).

15.2.4 Other threats

In mountainous regions, particularly in smallholder areas, agricultural development is hampered by complex terrain, inadequate infrastructure and different regional cultures. In the Humla district of Nepal, more than 80% of the farmland is situated on high terrain, which, when combined with the lengthy growing season of the local crops, leads to low food production in the area (Kc *et al.*, 2016). Due to a lack of infrastructure, it is impractical to mechanize farming in isolated mountainous regions; instead, animals and humans transport products, and animals also plough fields (Figs 15.1 and 15.2). Complex regional cultures also pose a threat to the growth of agriculture. For instance, the extremely skewed distribution of arable land of rural Myanmar suggests that a quarter or possibly half the farmers are landless (Haggblade *et al.*, 2014). In Nepal, higher-caste people (Bahun, Chhetri, Thakuri and Lama) control much land, but the untouchable caste, Dalits, do not, and they must contribute half of their production to lease the land from upper-caste families. Alternatively, they could work as day labourers for the upper-caste families to receive payment (Onta and Resurreccion, 2011).

Fig. 15.1. Women bringing forage for livestock in the mid-hills of Nepal. (Photo: Allan Degen)

Fig. 15.2. Cattle ploughing rice paddies in the mid-hills of Nepal. (Photo: Allan Degen)

15.2.5 Impacts on agro-ecosystems

Land degradation has serious implications for people's livelihoods, regional economic development, social equity and cultural development (Table 15.1).

Factors that have serious negative impacts on people's livelihoods

Agriculture and livestock remain the main sources of livelihood in the HKH mountains, and crops are necessary to ensure a sustainable food supply (Massawe *et al.*, 2016). Extreme conditions, such as seasonal water shortages, droughts, floods, crop pests, livestock diseases, changes in agricultural production patterns and declines in agricultural biodiversity caused by climate change, have had serious negative impacts on people's livelihoods. For example, the Ganges basin region's key horticulture crop yields of peaches, apricots, plums and apples have decreased considerably because of a lack of prolonged low temperature during the dormant period (Dilshad

Table 15.1. Agro-ecological problems in the eight countries of the Hindu Kush Himalayan region.

	Agro-ecological problems
Afghanistan	Land degradation; soil erosion
Bangladesh	Land degradation
Bhutan	Limited arable land; soil erosion
China	Grassland degradation; the spread of poisonous weeds; soil erosion
India	Soil nutrient imbalance (decreased fertility); decrease of water table and water quality; soil erosion
Myanmar	Land salinization
Nepal	Soil erosion; reclamation of large areas of pasture
Pakistan	Soil salinization and acidification; water-quality deterioration; soil erosion

et al., 2019). The mortality of yaks and sheep due to climatic disasters, such as heavy snowfall, poses a serious threat to livestock productivity. According to villagers in the Humla district of Nepal, the yields of wheat, bare barley and buckwheat have decreased greatly due to climate change (Onta and Resurreccion, 2011).

Factors that have a serious negative impact on economic development

Mountainous locations are subjected to numerous disasters, which constrains regional economic development. For instance, in the Garhwal Himalayan region in 2013, heavy rains caused landslides and flash floods that caused the destruction of more than 200 structures and multiple fatalities (Bhambri *et al.*, 2016). Adequate funds were not available for restoration efforts. Additionally, the rise in food prices as a result of the reduction in agricultural production had a negative impact on social and economic growth. Degradation of grasslands and a decline in forage contributed to the decline in the number of yaks in India, Nepal and Bhutan (Ning *et al.*, 2016). Extreme events related to climate change are occurring frequently, with serious negative impacts on property security and economic development in the region.

Exacerbating the unequal gender and race relations

Crop productivity has decreased because of climate change and agricultural biodiversity loss, which has led to more unequal work division between sexes. For instance, to increase income, women must invest more time and effort in agricultural tasks, particularly when men leave the area in search of work elsewhere (Rasul *et al.*, 2014). Additionally, there are large gender disparities in resource ownership in the region (Goodrich *et al.*, 2017). When household food production declines, women are given less food, which affects their health. Another example is the lower-caste Dalit families in Nepal's Burgaon district, who were forced to increase their cultivation of crops (*fapar*) resistant to drought to adapt to arid conditions. However, this strategy increased the workload of women (Onta and Resurreccion, 2011), as women bear the main responsibility for agricultural production.

Exacerbating the loss of agricultural heritage

The intensification of agriculture and the decline of agricultural biodiversity can lead to the loss of cultural diversity and heritage. The Himalayan region has a wide range of traditional agricultural and pastoral systems and traditional agroforestry land uses. Mountain people have accumulated rich experience in the long-term adaptation to the region, such as rice–fish farming in south-west China, and have developed special human and geographical landscapes such as terraces, which are crucial for biodiversity conservation. They are also an integral aspect of the agricultural heritage system and contribute to sustaining socio-cultural diversity. The agricultural cultural heritage is not only a mode of production but also an ecologically friendly agricultural technology system. However, the decline in agricultural biodiversity caused by agricultural intensification can have a negative impact on local traditional agriculture and even values, and can lead to the loss of cultural diversity and heritage.

15.3 Agricultural Adaptation Management and Livelihood and Ecological Restoration Strategies

Climate change, anthropogenic activities and grassland degradation are having a negative impact on livelihoods and agro-ecosystems in the HKH region and will continue to do so (Mayewski *et al.*, 2020). It is vital to implement programmes to minimize or reverse the trend in agricultural biodiversity loss, while also supporting local economies and ecological recovery. To confront the crisis of climate change and agricultural biodiversity loss, mountain people and relevant authorities have implemented several strategies, including enhanced planting and farming systems, infrastructure adjustment, and climate-smart agriculture practices.

15.3.1 Agricultural adaptive management

Crop management

Agro-ecosystems are more stable and resilient when crops are well managed. For instance, crop diversification can increase the resiliency and stability of food and agricultural systems. Due to the frequent droughts caused by climate change and the overall decrease in agricultural water availability, some water-intensive crops are being abandoned. In addition to addressing climate change, these changes increase the diversity of agricultural systems and improve the diet composition. In addition, the time of planting of crops is being altered, and improved crop seeds are being sown to accommodate seasonal variations. For example, farmers in the Ganges basin who used to grow wheat in October or November ten years ago now opt to do so in December due to delayed rainfall (Dilshad *et al.*, 2019). Planting perennial betel nut trees in India has been recommended as a sustainable land-use strategy to improve the resilience of farms to climate change. Cotton farmers in Pakistan are switching from conventional to modern seed varieties that are more tolerant of moisture, aridity and extreme temperature (Ahmad and Afzal, 2020). Various actions are being taken by smallholder farming households in the HKH region to mitigate the hazards brought on by climate change (Table 15.2).

Grazing management

While preserving livelihoods, rational grazing management can support the long-term growth of crop production and animal husbandry. According to the change of seasonal climate conditions, and the requirement and availability of forage for livestock, the herders divide the grassland into winter and summer pastures, apply a rotational grazing system, and fence the area when the animals are not grazing. These grazing management techniques promote livestock productivity, health and income in marginal high-altitude populations and promote the virtuous cycle of grassland nutrients, the effective conversion of abiotic and biological resources, and, ultimately, maintain the versatility of grassland and the sustainability of the system.

Resource management

To ensure local livelihoods and ecological recovery, local communities, organizations and governments have implemented a variety of resource management strategies, including planning and adaptive management techniques, that combine land, agriculture, food, water and other resources. Returning organic fertilizer and straw to the field encourages the recycling of organic material, enhances soil fertility and improves crop yields (Fig. 15.3). In China's Qinghai-Tibetan Plateau, the implementation of the subsidy and reward

Table 15.2. Responses to climate change by smallholder farming households in the Hindu Kush Himalayan region. (Adapted from ICIMOD, 2014; Hussain *et al.*, 2016)

	Adopted strategies
Upper Indus (Pakistan)	Changed farming practices; introduced new crops; stopped planting certain crops; stopped rearing certain livestock; invested in disaster preparedness
Koshi (Nepal)	Stopped planting certain crops; introduced new crops; changed farming practices; stopped rearing certain livestock; changed grazing practices
Eastern Brahmaputra (India)	Initiated new off-farm activities; stopped planting certain crops; invested in disaster preparedness; changed farming practices; introduced new crops
Salween and Mekong (China)	Migrated for work; changed farming practices; introduced new crops; community jointly invested in irrigation; household invested in irrigation

Fig. 15.3. Women carrying livestock manure as fertilizer in mid-hills of Nepal. (Photo: Allan Degen)

policy for ecological protection has raised the herder's awareness of the need to use grassland sensibly while protecting the ecosystem. Natural resources and the environment (air, water, soil and biodiversity) are protected in Nagpur, India, by laws and regulations (Dhyani *et al.*, 2018). The water sector focuses on water needs at the local level by planning and improving recharge systems, such as solar irrigation. Resource management, in which people are actively involved in digging water cellars or collecting water in containers, is another potential way to alleviate inter-seasonal irrigation water shortages. These rational resource management strategies ensure the sustained growth of agriculture in the area.

15.3.2 Livelihood and ecological restoration strategies

Increase in agricultural biodiversity

Ecosystem services and people's overall nutritional security are impacted by the loss of agricultural biodiversity. Agricultural biodiversity adds to food diversity and reduced malnutrition in addition to lowering risks and boosting agricultural stability. According to a survey conducted in Gujarat and Haryana, India, crop diversity and adult dietary diversity were connected favourably (Singh *et al.*, 2020). Similar research in three ecological farming zones in Nepal concluded that better food diversity in farmer families is correlated with higher crop diversification (Kc *et al.*, 2016). In response to threats posed by climate change, farmers in the HKH region are choosing to grow more varieties of crops, rotate crops (Aase and Chapagain, 2017) and increase agricultural biodiversity. For example, by combining bee farming with farming systems, bees provide pollination services that increase crop productivity, while honey increases dietary diversity and income. High agricultural diversification can also help to decrease conflicts between livestock and wildlife reserves, as well as hasten the recovery of regional livelihoods and environment.

Maintaining traditional crop varieties

In general, traditional crops are more adaptable to climate stress and are more resilient to climate change than introduced crops, which is important for agricultural biodiversity and the resilience of agricultural systems (environmental sustainability). A wide variety of traditional herbs and forages provide services for the local population, including food, feed and medicine (Ali *et al.*, 2022). For example, during droughts, farmers abandon some water-intensive crops in favour of more drought-tolerant traditional crops; during droughts in India, locals made drinks using *Arenga obtusifolia*; and during the dry and cold seasons, some native plants, such as *Cousinia thomsonii* and *Cicer microphyllum*, can be good forage (Chandrasekhar *et al.*, 2007). Furthermore, farming management that preserves traditional crops increases the sustainability of agro-ecosystems.

Advancement of conservation agriculture

The major goal of conservation agriculture, which has advanced substantially in India (Bhan and Behera, 2014), is to preserve, enhance and effectively utilize natural resources to meet the rising need for food while reducing the impact of agriculture on the environment. Minimum soil disturbance, straw recycling and diverse crop rotation are the three crop-management tenets of conservation agriculture (FAO, 2014). Conservation agricultural methods have advanced significantly in India (Bhan and Behera, 2014). Mechanical tillage was reduced or not employed when conservation agriculture techniques were implemented in rice-wheat growing regions of the Indo-Gangetic plain. Organic mulch on the soil surface decreased the need of irrigation water by increasing water availability through rainfall interception. Costs of production have decreased, while soil quality has increased. In addition to rice–wheat farming systems, traditional agricultural crop management systems are gradually shifting from intensive to reduced/no-till farming systems (Bhan & Behera, 2014). Currently, the vigorous development of ecological agriculture in the HKH region is gradually replacing chemical agriculture, which tends to pollute the ecological environment. The implementation of conservation agriculture has successfully improved degraded land and supported the sustainability of agricultural production.

Rehabilitation of degraded land

In the HKH mountains, poverty is often linked to land degradation, which leads to poor harvests and further exacerbates the incidence of poverty. To improve local livelihoods, a number of land restoration measures have been implemented. For example, many foreign aid projects have focused on forage cultivation for livestock, thereby addressing local sources of livelihood and avoiding further over-use of already degraded land. The application of green fertilizer enhances soil health and fertility, increases crop yields and contributes to ecosystem sustainability. In India, green manure has been used extensively (Kumar *et al.*, 2014). Terraced farming can lessen run-off and soil nutrient losses, enabling vegetation to be regenerated in previously deforested places (Quincey *et al.*, 2007). Furthermore, adding legumes to the system can increase crop yield while preserving soil fertility and restoring degraded land. The appropriate departments must establish responsible agricultural management policies in addition to rehabilitating the currently degraded agricultural land to curb additional land degradation.

Bolstering the preservation of germplasm

Germplasm is a key resource for ensuring national food security and the supply of essential agricultural products. By combining the inherent benefits of agriculture and animal husbandry, forage has emerged as an important development path for alternative crops within the trend of adapting to climate change (Konvalina, 2016). Much research has been invested in the domestication and breeding of forage plants to address the lack of forage germplasm resources and to enhance production. Although forage germplasm requires extensive and complex assessment, a wide range of germplasm has been preserved for general agronomic traits, which has made possible impressive advances in genetic technology to produce high-yielding and disease-resistant crop and livestock breeds. For example, India has produced high-yield oats, and some research has been done on cultivating germplasm for leaf size, tillering, leaf protein and plant height (Roy and Choubey, 2005).

Improve fund guarantees

Supportive policies are crucial for economic development and ecological recovery in the HKH region. In India, a project was initiated in 2007 to enable farmers to access agro-meteorological data, which allows farmers to better manage their crops and safeguard them from adverse weather conditions (Aryal *et al.*, 2020). The Indian government offers minor incentives for cultivating new varieties of plants, while the Nepalese government encourages the private sector to participate in plant breeding and seed trade to promote new varieties (Joshi *et al.*, 2012). The Indian government also initiated other actions, such as providing free power and large subsidies for solar water pumps, ensuring irrigation water for farmers, and investing $4 billion in the public canal system between 1991 and 2007 (Dhawan, 2017). According to Hoda and Gulati (2013), Indian farmers received $13.7 billion in fertilizer subsidies in 2010–2011, $2.47 billion in loan subsidies, and a smaller sum of money from insurance. Various supporting policies have benefitted the sustainable development of livelihoods and ecosystem services in the HKH region.

Changes in livelihood strategies

Through unremitting efforts, the HKH region's livelihood strategy has shifted from traditional farming and pastoralism to an integrated labour livelihood strategy, and from total reliance on land to a combination of agricultural and non-agricultural livelihoods (Gioli *et al.*, 2019). For example, in Bhutan, China, India and Nepal, harvesting medicinal yarsagumba or caterpillar fungus (*Cordyceps sinensis*) has greatly increased the incomes of residents. People in Pakistan use non-agricultural income to restore their livelihoods by purchasing agricultural machinery and high-quality seeds (Hussain *et al.*, 2020). In Nepal, young people are becoming increasingly involved in the commercial raising of improved breeds of cattle, sheep and chickens (Pokharel, 2019). Diverse livelihoods increase the income of local people and therefore they are better equipped to withstand the threats of food insecurity. Furthermore, diverse livelihoods can reduce the area of land under cultivation and reduce the damage to the ecological environment because the expansion of agricultural land is still the main driver of deforestation. Nepal, for example, lost 9% of its forest cover and increased its arable land by 12% between 1990 and 2009, and is expected to add another 5% by 2023. In Myanmar, the expansion of agricultural land is the most fundamental driver of deforestation (Lim *et al.*, 2017). The majority of the population in the HKH region is still dependent on agriculture, which also means that the area of arable land will increase further at the expense of forest land (Uddin *et al.*, 2015). Thus, diversity in livelihood strategies replaces part of agricultural income and contributes to ecological restoration in mountain areas.

15.4 Changes in Management Strategies, Adaptations and Approaches to Improve Agriculture

To restore local livelihoods and agro-ecosystems, local communities, politicians and researchers

Table 15.3. Measures taken by relevant departments in the HKH region to restore ecosystem services and livelihoods.

	Adopted strategies
Agriculture sector	Integrated indigenous knowledge with advanced technology; developed new crop varieties; prevented diseases and pests; promoted biological nitrogen fixation; promoted technology and innovations in the areas of mechanization and intelligentization
Water conservancy department	Attached importance to local level planning on water use; constructed rainwater harvesting devices; improved supply system; built more resilient water systems; promoted technology and innovations in irrigation, such as solar energy for irrigation
Animal husbandry bureau	Increased research and development of drugs and vaccines for disease control; improved livestock production
Government	Provided free electricity for groundwater pumping; invested in agricultural infrastructure

Table 15.4. Sustainable agricultural practices in the Hindu Kush Himalayan region.

	Ecologically sustainable agricultural practices
Afghanistan	Integrated water resources management; diversification of crops and livelihoods; development of knowledge of irrigation systems; protect soil and save water
Bangladesh	Single-pass tillage (SPT) combined with herbicides
Bhutan	The Ministry of Agriculture and Forests attempted to increase domestic production; improved land management practices
China	Implemented the policy of reward and subsidy for grassland; converted farmland to forest and grassland; formed a core technical system for ecological grassland production and rational allocation of ecological functions
India	Adopted no-till practices; invested in sustainable crop management; formulated the task of protecting cultivated land; improved crops through molecular techniques and developed climate-resistant varieties; combated desertification; restored degraded lands and soils
Myanmar	Implemented climate-smart agriculture strategy
Pakistan	Grew a variety of crops; subsidized improved seed breeds
Nepal	Approved long-term Agricultural Development Strategy (2015–2035); formulated agricultural development strategy

implemented a variety of management strategies and initiatives. Future mountain agriculture development efforts ought to concentrate on the following directions.

15.4.1 Advancing ecological agricultural development

First, the residents in the HKH region adhere to mixed-crop and animal production, using animal dung as organic fertilizer and diverse grassland crops to feed livestock. They produce crops and forage near their homes and graze their livestock in summer grassland far from their homes. The herders manage livestock in a two-stage 'warm feeding' format, and process dairy and meat products to ensure that they can provide enough products to survive without external input. Secondly, they adhere to the combination of planting and breeding, following reasonable crop rotation, and appropriate fallow to achieve the sustainability of cultivated land. In addition, planting perennial crops minimizes soil erosion.

15.4.2 Increasing the resistance of agricultural systems

To achieve sustainability for the HKH mountain agricultural production system, it is necessary to improve the adaptability of crops by genetic selection to cope with climate (Hoban *et al.*, 2020). Under-utilized traditional crop breeds have much potential to increase food security since they are more capable of withstanding harsh environmental conditions, including climate change, than imported breeds (Mayes *et al.*, 2012; Cheng *et al.*, 2017). This can be achieved by properly balancing crop varieties, fully utilizing traditional crops, boosting agricultural genetic diversity, and paying attention to germplasm resource genetic enhancement. In addition, the improvement of agricultural infrastructure, such as the construction and improvement of irrigation systems, is also important for crops to cope with climate change.

15.4.3 Combining ecological and production functions

For agriculture and animal husbandry to thrive sustainably, it is crucial that there is a balance between production and ecology. According to Aryal *et al.* (2023), there is a negative correlation between crop production and soil and water conservation among ecosystem service functions in the HKH region, and, therefore, future development should not only focus on productivity and abundant material products but also on ecological benefits. Using the natural ecological environment as the foundation for development, it is important to allocate limited-space resources, combine conservation of biodiversity with intensification, and co-ordinate the relationship between development and ecological protection. Finally, it is important to integrate the overall effects of production development, ecological environment protection and economic benefits. The emphasis is placed, for instance, on the connections between productivity and biodiversity in agricultural environments and the connections between fodder availability and ecological function in grassland ecosystems.

15.4.4 Keeping it community-based

Local communities often possess valuable traditional knowledge and practices that can guide decisions on managing agriculture sustainably and building resilience to changing environmental conditions. Many farmers in the HKH region remain subsistence smallholders by employing traditional ecological knowledge and practices passed down from generation to generation. These include choosing the right planting season and crop varieties for sustainable development. Adaptation strategies should also consider the pressures, choices for livelihoods, and resources available to local communities. These choices should vary among communities and strategies should be established to safeguard smallholder economies, protect biodiversity and ensure food security. Conservation of agricultural biodiversity depends on the participation of people involved, so it is important to adhere to community-based efforts to achieve agricultural sustainability in the HKH region.

15.4.5 Enhancing departmental collaboration and communication

The co-ordination and co-operation of numerous departments are required for emergency rescue operations and the implementation of agricultural technology. Due to its complicated topography, the HKH region is subject to severe disaster risks, and climate change will make compounded disasters more frequent and intense. For instance, the rapid glacier melting brought on by global warming is anticipated to trigger calamities that cascade with the country and across borders. The most common example is how agriculture is being impacted in Nepal, Bhutan, India, Pakistan and other countries as a consequence of the rapidly melting glaciers. Fostering collaborative prevention, control and emergency rescue at the river basin level requires regional and inter-departmental co-operation. Co-ordination and co-operation among departments are also necessary for the extension and use of agricultural technology. Governments make decisions, invest in research, and develop and promote agricultural products; scientific research institutions are committed to research and development; the Department of Agriculture is responsible for promotion tasks and technical guidance; the Ministry of Water Resources ensures water for crops and livestock; while multi-departmental co-operation promotes agricultural development of the HKH region.

15.5 Conclusions

The HKH region is an important hub for biodiversity worldwide, but owing to anthropogenic activities, land use, and climate change, agricultural biodiversity is diminishing quickly in the area. The region's food security, sustainable economic development, ecological functions and services, social cohesion and agricultural cultural diversity are threatened by the loss of agricultural biodiversity. Local communities, organizations and governments have adopted a series of strategies, including integrated crop and livestock systems, energy planning and adaptive management practices, and implemented a series of livelihood and ecological restoration options to combat the negative effects of climate change and the loss of agricultural biodiversity. These strategies have aided the promotion of agro-ecological restoration and sustainable development in the HKH region. To ensure the long-term sustainability of agriculture in the region, it is necessary to continually advance, update and adapt current conservation and management strategies.

References

Aase, T. and Chapagain, P. (2017) Organic farmers on the ridge: the Hyolmo of Sermathang, Nepal. In: Aase, T. (ed.) *Climate Change and the Future of Himalayan Farming*. Oxford University Press, New Delhi, pp. 229–266.

Ahmad, D. and Afzal, M. (2020) Climate change adaptation impact on cash crop productivity and income in Punjab province of Pakistan. *Environmental Science and Pollution Research* 27, 30767–30777. DOI: 10.1007/s11356-020-09368-x

Ali, F., Khan, N. and Rahmonov, O. (2022) Ecosystem services and linkages of naturally managed *Monotheca buxifolia* (Falc.) A. DC. Forests with local communities across contiguous mountainous ranges in Pakistan. *Biology* 11(10), 1469. DOI: 10.3390/biology11101469

Aryal, J.P., Sapkota, T.B., Khurana, R., Khatri-Chhetri, A., Rahut, D.B. and Jat, M.L. (2020) Climate change and agriculture in South Asia: adaptation options in smallholder production systems. *Environment, Development and Sustainability* 22(6), 5045–5075. DOI: 10.1007/s10668-019-00414-4

Aryal, K., Maraseni, T. and Apan, A. (2023) Spatial dynamics of biophysical trade-offs and synergies among ecosystem services in the Himalayas. *Ecosystem Services* 59, 101503.

Bhambri, R., Mehta, M., Dobhal, D.P., Gupta, A.K., Pratap, B. *et al.* (2016) Devastation in the Kedarnath (Mandakini) Valley, Garhwal Himalaya, during 16–17 June 2013: a remote sensing and ground-based assessment. *Natural Hazards* 80, 1801–1822. DOI: 10.1007/s11069-015-2033-y

Bhan, S. and Behera, U.K. (2014) Conservation agriculture in India – problems, prospects and policy issues. *International Soil and Water Conservation Research* 2(4), 1–12. DOI: 10.1016/S2095-6339(15)30053-8

Chandrasekhar, K., Rao, K.S., Maikhuri, R.K. and Saxena, K.G. (2007) Ecological implications of traditional livestock husbandry and associated land use practices: a case study from the trans-Himalaya, India. *Journal of Arid Environments* 69, 299–314. DOI: 10.1016/j.jaridenv.2006.09.002

Chen, S., Wang, W., Xu, W., Wang, Y., Wan, H. *et al.* (2018) Plant diversity enhances productivity and soil carbon storage. *Proceedings of the National Academy of Sciences* 115(16), 4027–4032. DOI: 10.1073/pnas.1700298114

Cheng, A., Mayes, S., Dalle, G., Demissew, S. and Massawe, F. (2017) Diversifying crops for food and nutrition security – a case of teff. *Biological Reviews* 92(1), 188–198. DOI: 10.1111/brv.12225

Clouse, C., Anderson, N. and Shippling, T. (2017) Ladakh's artificial glaciers: climate-adaptive design for water scarcity. *Climate and Development* 9(5), 428–438. DOI: 10.1080/17565529.2016.1167664

Dhawan, V. (2017) Water and agriculture in India. Background paper for the South Asia expert panel during the Global Forum for Food and Agriculture (GFFA) (in German).

Dhimal, M., Bhandari, D., Dhimal, M.L., Kafle, N., Pyakurel, P. *et al.* (2021) Impact of climate change on health and well-being of people in Hindu Kush Himalayan region: a narrative review. *Frontiers in Physiology* 12, 651189. DOI: 10.3389/fphys.2021.651189

Dhyani, S., Lahoti, S., Khare, S., Pujari, P. and Verma, P. (2018) Ecosystem based disaster risk reduction approaches (EbDRR) as a prerequisite for inclusive urban transformation of Nagpur City, India. *International Journal of Disaster Risk Reduction* 32, 95–105. DOI: 10.1016/j.ijdrr.2018.01.018

Dilshad, T., Mallick, D., Udas, P.B., Goodrich, C.G., Prakash, A. *et al.* (2019) Growing social vulnerability in the river basins: evidence from the Hindu Kush Himalaya (HKH) region. *Environmental Development* 31, 19–33. DOI: 10.1016/j.envdev.2018.12.004

Ehrlich, P.R. and Pringle, R.M. (2008) Where does biodiversity go from here? A grim business-as-usual forecast and a hopeful portfolio of partial solutions. *Proceedings of the National Academy of Sciences* 105(S1), 11579–11586. DOI: 10.1073/pnas.0801911105

FAO (2014) *The 3 Principles of Conservation Agriculture.* Food and Agriculture Organization of the United Nations, Rome.

FAOSTAT (2018) *Data on Area under Cultivation of Cereals and Production of Cereals from 1961–2017.* Food and Agriculture Organizations of the United Nations, Rome.

Fishman, R. (2018) Groundwater depletion limits the scope for adaptation to increased rainfall variability in India. *Climatic Change* 147, 195–209. DOI: 10.1007/s10584-018-2146-x

Gagné, K., Rasmussen, M. and Orlove, B. (2014) Glaciers and society: attributions, perceptions, and valuations. *Wiley Interdisciplinary Reviews: Climate Change* 5(6), 793–808. DOI: 10.1002/wcc.315

Ghimiray, M. and Vernooy, R. (2017) The importance and challenges of crop germplasm interdependence: the case of Bhutan. *Food Security* 9(2), 301–310. DOI: 10.1007/s12571-017-0647-5

Gioli, G., Thapa, G., Khan, F., Dasgupta, P., Nathan, D. *et al.* (2019) Understanding and tackling poverty and vulnerability in mountain livelihoods in the Hindu Kush Himalaya. In: Wester, P., Mishra, A., Mukherji, A. and Shrestha, A. (eds) *The Hindu Kush Himalaya Assessment: Mountains, Climate Change, Sustainability and People*. Springer Nature, Cham, Switzerland, pp. 421–455.

Goodrich, C.G., Manjari, M. and Suman, B. (2017) Status of gender, vulnerabilities and adaptation to climate change in the Hindu Kush Himalaya: impacts and implications for livelihoods, and sustainable mountain development. ICIMOD Working Paper, Kathmandu.

Haggblade, S., Boughton, D., Cho, K.M., Denning, G., Kloeppinger-Todd, R. *et al.* (2014) Strategic choices shaping agricultural performance and food security in Myanmar. *Journal of International Affairs* 67(2), 55–71.

Hoban, S., Bruford, M., Jackson, J.D.U., Lopes-Fernandes, M., Heuertz, M. *et al.* (2020) Genetic diversity targets and indicators in the CBD post-2020 Global Biodiversity Framework must be improved. *Biological Conservation* 248, 108654. DOI: 10.1016/j.biocon.2020.108654

Hoda, A. and Gulati, A. (2013) India's agricultural trade policy and sustainable development. International Centre for Trade and Sustainable Development (ICTSD) issue paper 49. Geneva, Switzerland.

Hussain, A. and Qamar, F.M. (2020) Dual challenge of climate change and agrobiodiversity loss in mountain food systems in the Hindu-Kush Himalaya. *One Earth* 3(5), 539–542. DOI: 10.1016/j.oneear.2020.10.016

Hussain, A., Rasul, G., Mahapatra, B. and Tuladhar, S. (2016) Household food security in the face of climate change in the Hindu-Kush Himalayan region. *Food Security* 8, 921–937. DOI: 10.1007/s12571-016-0607-5

Hussain, A., Memon, J.A. and Hanif, S. (2020) Weather shocks, coping strategies and farmers' income: a case of rural areas of district Multan, Punjab. *Weather and Climate Extremes* 30, 100288. DOI: 10.1016/j.wace.2020.100288 (CC-BY 4.0).

ICIMOD (2014) *Results of Vulnerability and Adaptive Capacity Assessment* (VACA). International Centre for Integrated Mountain Development (ICIMOD), Kathmandu.

Jodha, N.S. (2004) Globalisation and mountain agriculture: a process of integration through disintegration. In: Adhikari, R. and Adhikari, K. (eds) *Evolving Sui Generis Options for the Hindu-Kush Himalayas.* South Asia Watch on Trade, Economics & Environment, pp.16–33.

Joshi, K.D., Conroy, C. and Witcombe, J.R. (2012) Agriculture, seed, and innovation in Nepal: industry and policy issues for the future. *Gates Open Research.* DOI: 10.21955/gatesopenres.1115308.1

Kc, K.B., Pant, L.P., Fraser, E.D., Shrestha, P.K., Shrestha, D. *et al.* (2016) Assessing links between crop diversity and food self-sufficiency in three agroecological regions of Nepal. *Regional Environmental Change* 16, 1239–1251. DOI: 10.1007/s10113-015-0851-9

Khalid, Z., Meng, X., Rana, I.A., Rehman, M.U. and Su, X. (2021) Holistic multidimensional vulnerability assessment: an empirical investigation on rural communities of the Hindu Kush Himalayan region, Northern Pakistan. *International Journal of Disaster Risk Reduction* 62, 102413. DOI: 10.1016/j.ijdrr.2021.102413

Konvalina, P. (2016) *Organic Farming: A Promising Way of Food Production.* InTechOpen.

Krishnan, R., Shrestha, A.B., Ren, G., Rajbhandari, R., Saeed, S. *et al.* (2019) Unravelling climate change in the Hindu Kush Himalaya: rapid warming in the mountains and increasing extremes. In: Wester, P., Mishra, A., Mukherji, A. and Shrestha, A. (eds) *The Hindu Kush Himalaya Assessment: Mountains, Climate Change, Sustainability and People*. Springer Nature, Cham, Switzerland, pp. 57–97.

Kumar, R., Mahajan, G., Srivastava, S. and Sinha, A. (2014) Green manuring: a boon for sustainable agriculture and pest management – a review. *Agricultural Reviews* 35(3), 196–206. DOI: 10.5958/0976-0741.2014.00906.4

Lim, C.L., Prescott, G.W., De Alban, J.D.T., Ziegler, A.D. and Webb, E. (2017) Untangling the proximate causes and underlying drivers of deforestation and forest degradation in Myanmar. *Conservation Biology* 31(6), 1362–1372. DOI: 10.1111/cobi.12984

Massawe, F., Mayes, S. and Cheng, A. (2016) Crop diversity: an unexploited treasure trove for food security. *Trends in Plant Science* 21(5), 365–368. DOI: 10.1016/j.tplants.2016.02.006

Mayes, S., Massawe, F.J., Alderson, P.G., Roberts, J.A., Azam-Ali, S.N. *et al.* (2012) The potential for underutilized crops to improve security of food production. *Journal of Experimental Botany* 63(3), 1075–1079. DOI: 10.1093/jxb/err396

Mayewski, P.A., Perry, L.B., Matthews, T. and Birkel, S.D. (2020) Climate change in the Hindu Kush Himalayas: basis and gaps. *One Earth* 3(5), 551–555. DOI: 10.1016/j.oneear.2020.10.007

Ning, W., Shaoliang, Y., Joshi, S. and Bisht, N. (2016) *Yak on the Move: Transboundary Challenges and Opportunities for Yak Raising in a Changing Hindu Kush Himalayan Region.* International Centre for Integrated Mountain Development (ICIMOD), Kathmandu.

Oliver, T.H., Isaac, N.J., August, T.A., Woodcock, B.A., Roy, D.B. *et al.* (2015) Declining resilience of ecosystem functions under biodiversity loss. *Nature Communications* 6(1), 1–8.

Onta, N. and Resurreccion, B.P. (2011) The role of gender and caste in climate adaptation strategies in Nepal emerging change and persistent inequalities in the far-western region. *Mountain Research and Development* 31(4), 351–356. DOI: 10.1659/MRD-JOURNAL-D-10-00085.1

Paudel, B., Zhang, Y., Yan, J., Rai, R., Li, L. *et al.* (2020) Farmers' understanding of climate change in Nepal Himalayas: important determinants and implications for developing adaptation strategies. *Climatic Change* 158, 485–502. DOI: 10.1007/s10584-019-02607-2

Paudel, P.K. and Heinen, J.T. (2015) Conservation planning in the Nepal Himalayas: effectively (re)designing reserves for heterogeneous landscapes. *Applied Geography* 56, 127–134. DOI: 10.1016/j.apgeog.2014.11.018

Perrigo, A., Hoorn, C. and Antonelli, A. (2020) Why mountains matter for biodiversity. *Journal of Biogeography* 47(2), 315–325. DOI: 10.1111/jbi.13731

Pokharel, C. (2019) Agricultural diversification in Nepal. In: Thapa, G., Kumar, A. and Joshi, P. (eds) *Agricultural Transformation in Nepal: Trends, Prospects, and Policy Options*. Springer, Singapore, pp. 323–384.

Quincey, D.J., Luckman, A., Hessel, R., Davies, R., Sankhayan, P.L. *et al.* (2007) Fine-resolution remote-sensing and modelling of Himalayan catchment sustainability. *Remote Sensing of Environment* 107(3), 430–439. DOI: 10.1016/j.rse.2006.09.021

Rasul, G., Hussain, A., Khan, M.A., Ahmad, F. and Jasra, A.W. (2014) *Towards a Framework for Achieving Food Security in the Mountains of Pakistan*. International Centre for Integrated Mountain Development (ICIMOD) Working Paper. Kathmandu.

Rasul, G., Saboor, A., Tiwari, P.C., Hussain, A., Ghosh, N. *et al.* (2019) Food and nutrition security in the Hindu Kush Himalaya: unique challenges and niche opportunities. In: Wester, P., Mishra, A., Mukherji, A. and Shrestha, A. (eds) *The Hindu Kush Himalaya Assessment: Mountains, Climate Change, Sustainability and People*. Springer Nature, Cham, Switzerland, pp. 301–338.

Rasul, G., Hussain, A., Adhikari, L. and Molden, D.J. (2022) Conserving agrobiodiversity for sustainable food systems in the Hindu Kush Himalaya. *International Journal of Agricultural Sustainability* 20(6), 1117–1135. DOI: 10.1080/14735903.2022.2057642

Roy, A.K. and Choubey, R.N. (2005) Utilization of gene pool in genetic improvement in forage oats achievements and future prospects. *Indian Journal of Plant Genetic Resources* 18(1), 108–109.

Seebens, H., Blackburn, T.M., Dyer, E.E., Genovesi, P., Hulme, P.E. *et al.* (2018) Global rise in emerging alien species results from increased accessibility of new source pools. *Proceedings of the National Academy of Sciences* 115(10), E2264–E2273. DOI: 10.1073/pnas.1719429115

Singh, G., Sarkar, M.S., Pandey, A., Lingwal, S., Rai, I.D. *et al.* (2018) Quantifying four decades of changes in land use and land cover in India's Kailash Sacred Landscape: suggested option for priority based patch level future forest conservation. *Journal of the Indian Society of Remote Sensing* 46, 1625–1635. DOI: 10.1007/s12524-018-0817-8

Singh, S., Bassignana-Khadka, I., Singh Karky, B. and Sharma, E. (2011) *Climate Change in the Hindu Kush-Himalayas: The State of Current Knowledge.* International Centre for Integrated Mountain Development (ICIMOD), Kathmandu.

Singh, S., Jones, A.D., DeFries, R.S. and Jain, M. (2020) The association between crop and income diversity and farmer intra-household dietary diversity in India. *Food Security* 12, 369–390. DOI: 10.1007/s12571-020-01012-3

Thapa, S. and Hussain, A. (2021) Climate change and high-altitude food security: a small-scale study from the Karnali region in Nepal. *Climate and Development* 13(8), 713–724. DOI: 10.1080/17565529.2020.1855099

Tiwari, P.C., Tiwari, A. and Joshi, B. (2018) Urban growth in Himalaya: understanding the process and options for sustainable development. *Journal of Urban and Regional Studies on Contemporary India* 4(2), 15–27.

Tse-Ring, K., Sharma, E., Chettri, N. and Shrestha, A. (2010) *Climate Change Vulnerability of Mountain Ecosystems in the Eastern Himalayas*. International Centre for Integrated Mountain Development (ICIMOD), Kathmandu.

Uddin, K., Chaudhary, S., Chettri, N., Kotru, R., Murthy, M. *et al.* (2015) The changing land cover and fragmenting forest on the Roof of the World: a case study in Nepal's Kailash Sacred Landscape. *Landscape and Urban Planning* 141, 1–10. DOI: 10.1016/j.landurbplan.2015.04.003

UN (2014) *World Urbanization Prospects: the 2011 Revision*. Department of Economic and Social Affairs, Population Division. United Nations Secretariat.

Yao, T., Bolch, T., Chen, D., Gao, J., Immerzeel, W. *et al.* (2022) The imbalance of the Asian water tower. *Nature Reviews Earth & Environment* 3(10), 618–632. DOI: 10.1038/s43017-022-00299-4

16 Function and Conservation of the Montane Desert Ecosystem in the Hindu Kush Himalayan Region

Rui Zhang[1,*], Allan Degen[2] and Jie Lian[1]

[1]*Naiman Desertification Research Station, Northwest Institute of Eco-Environment and Resources, Chinese Academy of Sciences (CAS), Lanzhou, China;*
[2]*Institutes for Desert Research, Ben Gurion University of the Negev, Israel*

Abstract

High-altitude montane deserts are prominent in the northern part of the Hindu Kush Himalayan (HKH) region; in particular in the Ladakh region of Jammu and Kashmir in northern India, and in the Kunlun mountains and the Tibetan Plateau in China. Many of these arid, very cold deserts have elevations exceeding 3000 m, an average annual precipitation of less than 40 mm, and a high rate of evapotranspiration. The montane desert ecosystem in the HKH is extremely fragile and is threatened by global warming and anthropogenic activities. This chapter describes: (i) threats to montane deserts and their impacts; (ii) strategies to conserve montane deserts; (iii) livelihoods of the local inhabitants; and (iv) ecological restoration of montane deserts in the HKH region.

Keywords: anthropogenic activities, arid and cold montane deserts, artificial grassland, desert conservation, global warming, livelihoods, water-saving irrigation

16.1 Introduction

Deserts are characterized by sparse vegetation cover and low and infrequent rainfall. With an average annual precipitation of less than 250 mm (Chen and Xiao, 1997), deserts often have a negative water balance because the high rate of evapotranspiration is greater than the rate of water input. These areas, which are collectively called 'drylands', can contain valuable mineral deposits that were formed in the arid environment or were exposed by erosion. Nearly all desert surfaces are plains where aeolian deflation, that is the removal of fine-grained material by the wind, has exposed loose gravel consisting predominantly of pebbles and stone fragments. Typical plant cover in deserts is sparse but diverse (Cheng *et al*., 2013; Lu *et al*., 2019), and although the plant communities have relatively low average productivity, they are widely distributed and exert a wide range of influence. These plants form a green barrier in desert areas, and are crucial for ecological services and local economic development (Liu *et al*., 2020).

Most desert plants, such as xerophytes, are drought- and salt-tolerant (Du and Sun, 1990;

*Corresponding author: zhangrui@lzb.ac.cn

 Sustainable Ecological Restoration and Conservation in the Hindu Kush Himalayan Region (Z. Shang *et al*.)
DOI: 10.1079/9781800622579.0016

Zhong *et al.*, 2014). Some desert plants store water in their leaves, roots and stems, while others possess long taproots that can penetrate to the water table. The stems and leaves of the plants lower the surface velocity of sand-carrying winds and protect the ground from erosion. Even small fungi and microscopic plant organisms on the soil surface, so-called cryptobiotic plants, can be important in reducing erosion and providing support for other living organisms. Cacti are often thought of as characteristic desert plants, but other plants, including the pea and sunflower families, have also adapted well to the arid environment. In general, grasses and shrubs are the dominant vegetation in cold deserts (Xu *et al.*, 2007). Oases, many of which are artificial, are often the only areas in deserts that can support crops and permanent habitation.

Deserts account for approximately 18.5% of the total land area (Houghton *et al.*, 2009). High-altitude montane deserts, many exceeding 3000 m, are cold and arid and are prominent in the northern part of the HKH region; in particular in the Ladakh region of Jammu and Kashmir in northern India, and in the Kunlun mountains and the Tibetan Plateau in China (Fig. 16.1). Agriculture is the main livelihood in the cold deserts of Jammu and Kashmir, even though the average annual precipitation is less than 40 mm. Humidity is low, the rate of evapotranspiration is high, and the soils are alkaline and poor in nutrients. In addition, soil erosion is widespread, and glacial erosion is typical in the cold, arid region of Ladakh (Gupta and Arora, 2016).

The Qaidam basin, located on the north-east slope of the Kunlun mountains in the transition zone between monsoon and westerly regions, has a continental climate and is affected by the arid climate and complex terrain. The basin is very sensitive to global climate change and is an important ecological barrier and living gene bank of the Qinghai-Tibetan Plateau (Xin *et al.*, 2021). The Gobi desert, a large, cold desert and grassland region in northern China and southern Mongolia, and the sixth largest desert in the world, is located partially in the basin and is characterized by the widespread distribution of desert steppe, dwarf and semi-shrub ecosystems (Fig. 16.2). The alpine deserts on the Qinghai-Tibetan Plateau are prominent in north/north-west Tibet, the Naqu Valley, the Yarlung Zangbo river area and the valley of the Jinsha river, and have extremely important strategic and ecological functions (Liu *et al.*, 2020).

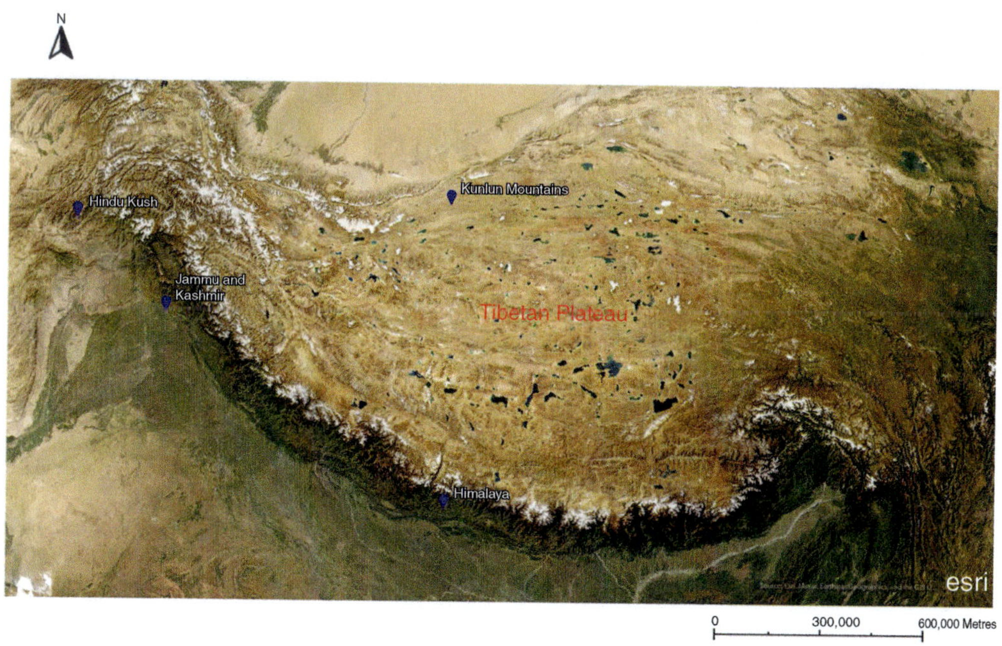

Fig. 16.1. The northern part of the Hindu Kush Himalayan region.

Fig. 16.2. Gobi desert (A); desert steppe (B); and dwarf and semi-shrub ecosystems (C).

Approximately 10×10^{13} kg of total carbon is sequestered in the desert areas, and therefore knowledge on the spatial distribution of the vegetation biomass and soil properties is important in understanding the carbon storage of desert systems. In addition, information on the biodiversity and ecosystem stability at the regional scale can provide a basis for maintaining biodiversity, restoring degraded lands and developing sustainable regional economies in desert regions (Zhong *et al.*, 2014). At present, the fragile desert ecosystem is being challenged mainly by anthropogenic activities and climate change.

16.2 Threats to Montane Deserts and Their Impacts

The desert ecosystem possesses a number of functions, such as soil conservation, windbreak, sand consolidation, carbon storage and biodiversity maintenance, which play important roles in maintaining sustainable economic and social development, improving the livelihoods of farmers and herdsmen and reducing poverty (Chen and Xiao, 1997; Cheng *et al.*, 2013). Desert ecosystems are influenced by climate change and anthropogenic activities in different ways. For example, the biomass of annual plants and perennial shrubs in desert ecosystems are affected by changes in the intensity and pattern in rainfall, air temperature, wind speed and solar radiation, which have an impact on ecosystem services. The effects of anthropogenic activities, however, are more complex. Under the influence of macroeconomies, policies and micro-markets, desert ecosystem services are affected by land-use changes; in particular, by urbanization (Lu *et al.*, 2019). The stability and sustainable development of the montane desert ecosystem, with its special geographical location and climatic conditions, are threatened continually by harsh environmental conditions, global climate change and increasing anthropogenic activities.

16.2.1 Threats to montane deserts from natural factors

Wind, water and glacial erosions are common occurrences in montane deserts (Gupta and Arora, 2016). Because of the low rainfall and water resources, and harsh climatic conditions, the yield of natural plant biomass is low and unstable. The lack of water and forage restricts the sustainable development of local animal husbandry and the regional economy. Traditional flooding is used widely for irrigating artificial grassland on the Qinghai-Tibetan Plateau, and increases the forage available to the grazing livestock (Wang *et al.*, 2021).

The Ngari alpine desert on the Tibetan Plateau, a typical montane desert, is characterized by strong solar radiation, a large range in diurnal air temperature and annual fluctuations in precipitation, strong winds, and a lack of native trees and grasses, which restrict the social, economic and ecological developments of the region. Desertification in Ngari is still in the developing stages due to multiple factors, including the fragile ecological environment, the hydrothermal conditions and anthropogenic activities (Liu *et al.*, 2022). The Qiangtang Plateau in north-west Tibet has a plateau desert ecosystem and provides habitats for rare and unique fauna such as the Tibetan antelope, Tibetan wild donkey, black-necked crane and wild yak. However, desertification is expanding, and disasters caused by diseases, invasion of harmful insects and cave collapse have increased due to climate warming and anthropogenic activities. Hence, the biodiversity in this region is being seriously threatened (Ba *et al.*, 2012; Gong and Ba, 2013).

Although the high-altitude montane desert grasslands in Qaidam have sparse vegetation, they are an important site for animal husbandry and they act as a green barrier to prevent the expansion of the desert towards the Qinghai lake area, the headwater of Three Rivers and the Hexi Corridor of Gansu province. Strong winds occur frequently with the change in temperature, leading to soil erosion, while rocks undergo thermal expansion and contraction, and physical weathering. Because of low precipitation and air temperature, and poor soil nutrients, plant biomass production is low. In recent years, the extensive mode of production and management, and the non-controlled exploitation of resources have caused serious soil erosion and grasslands degradation, which directly affects the local economy and the life of local residents (Zheng *et al.*, 2005; Liang and Hou, 2007).

16.2.2 Threats to montane deserts from anthropogenic activities

Anthropogenic activities, including fire, livestock grazing, land reclamation, mining, land-use change and urbanization, expand desertification and affect desert ecosystem functions and services. Desertification threatens the livelihood of inhabitants in the region, as well as the wild plants and animals. Over-grazing, agricultural reclamation and deforestation are the main anthropogenic factors affecting montane deserts in the Qaidam area. Since the late 1950s, the grasslands have been over-grazed, as the stocking rate is more than 20% above the number of livestock that the grasslands can support. Between the 1950s and 1990s, 180 km^2 of grasslands were reclaimed as farmland and approximately 67 km^2 of cultivated farmland, that was converted from grasslands, were abandoned due to water shortage and high soil salinity. Psammophyte plant species account for a large proportion of the grasslands' vegetation. They have few, dry leaves that are easy to ignite, and are used by local people and migrant workers for fuel. As a result, the reclamation and abandonment of farmland, and the deforestation of montane desert grasslands in the Qaidam area, are very common. With the removal and destruction of the vegetation, the soil has been eroded by wind and water, and soil nutrients have been reduced (Zheng *et al.*, 2005).

Roads are an important infrastructure for sustainable development of the economy and society, especially in desert areas where the distance between sites can be substantial. However, the construction of roads and the expansion of the road network have had a great impact on the environment and wildlife, as they cause environmental pollution, habitat destruction of wild animals and land fragmentation. Bridges, channels and culverts are constructed in areas along the highway in nature reserves to allow migration, movement and foraging by wild and domestic animals. Wildlife protection has to be taken into account in future development, especially in the areas of road construction (Li *et al.*, 2018).

16.3 Conservation of Montane Deserts and Strategies in the Hindu Kush Himalayan Region

The montane desert ecosystem is being threatened and the grasslands have been seriously degraded by climate change and human activities. Major issues that require attention in this region include the loss of vegetative cover due to over-grazing in the watersheds and rangelands, and the high use of wood and shrubs for fuel by the local and nomadic populations (IUCN Pakistan, 2000). The loss of soil and soil nutrients due to wind, water and glacial erosions, the limited water availability and the degraded grassland affect the livestock production and agricultural economy of the area (Harpinder, 2017). Consequently, projects must be implemented to curb desertification and restore degraded land, and to promote stability and livelihood improvement in this region (Liang and Hou, 2007). A series of protection measures have been initiated to improve desert ecosystem functions and services, and these include ecological management projects, application of water-saving irrigation methods in artificial grasslands, policy incentives and the establishment of protected areas.

16.3.1 Implementing ecological engineering in montane deserts

To improve the ecological environment in arid areas and curb the expansion of desertification, a series of key ecological projects have been initiated by the Chinese government, including the construction of shelterbelts in the three northern regions, the protection of natural forests, the conversion of farmland to forest, the control of sandstorms at the sources in Beijing and Tianjin, and the provision of subsidies for the protection of montane desert ecosystems. Due to the effects of these projects and to urbanization, the land-use and desert ecosystem services are improving (Wang, 2018). The grasslands area has increased, the unused land area has decreased, and the urban centres and cultivated land area have been altered dramatically between 2000 and 2015. In addition, the restoration of degraded grasslands and the conversion of unused land into grasslands by air seeding, have effectively increased and improved the grazing land available to livestock. The implementation of the 'ecological migration policy' has transformed unused land into cultivated land and has developed water channels and urban centres. These changes are restoring and enhancing the ecosystem services gradually, which are vital for regional development (Lu *et al.*, 2019).

In the 1950s and 1960s, desertification was very widespread in the Qaidam basin due to the rapidly increasing population, removal of a large number of sand-fixing plants that were used for fuel, conversion of grasslands to cropland (Fig. 16.3) and over-grazing by livestock. Desertification of grasslands reached one third of the total grasslands in this region, which poses a threat to the livelihood of the local population and to sustainable development. In 2001, the Forestry Department of China introduced projects of converting farmland into forests or grasslands on barren hills and slopes. However, establishing forests and grasslands in arid deserts is difficult due to the lack of precipitation and water for irrigation and high evapotranspiration rates. Therefore, special management measures

Fig. 16.3. The cropland in the Hindu Kush Himalayan region. (www.icimod.org)

should be implemented for this to be successful. For example, hole seeding, that is the sowing of seeds in small holes in degraded grasslands, is especially important in areas with more than 150 mm of precipitation and little wind-blown sand. The proportion of seedling emergence of five kinds of herbage using this method displayed the following order: *Elymus dahuricus* > *Artemisia desertorum* > *Onobrychis viciifolia* > *Caragana korshinskii* > *Nitraria tangutorum*. The seedling growth rate was fastest in *Artemisia desertorum* followed by *Nitraria tangutorum*, *Caragana korshinskii*, *Elymus dahuricus* and *Onobrychis viciifolia*. When preparing the holes, the native vegetation should be maintained and not be destroyed. Grass species that are tolerant of drought and of poor soil nutrition, with good overwintering ability and high yield, should be selected. *Artemisia desertorum*, in particular, fits these criteria well, but other plants such as *Nitraria tangutorum*, *Caragana korshinskii* and *Elymus dahuricus* are also suitable. The number of seeds sown should be high to ensure the desired rate of seedling emergence. Reseeding should be carried out at the end of May to the middle of June before the rainy season, and no later than the middle of July; otherwise it is difficult for the plants to overwinter. The artificial seeding of herbage in holes reduces soil turnover and maintains the existing native vegetation and the original soil structure while minimizing further desertification caused by wind erosion. Consequently, the artificial hole-seeding method in desert areas can improve the ecological environment, increase the productivity of desert grasslands and enhance the social and economic benefits of the inhabitants (Xu *et al.*, 2007). By adopting the grass grid sand fixation method for the protection of native vegetation, the survival rate of trees in degraded grasslands increased by 70%, the emergence rate of pasture increased by 50% and the vegetation coverage increased by greater than 20%. Grass can overwinter safely and soil erosion is controlled, thereby reversing the passive situation of degradation and desertification of grasslands (Zheng *et al.*, 2005).

There are a number of plants that are well adapted to the harsh conditions of montane deserts and that can contribute beneficial functions. *Hedysarum scoparium* and *Hedysarum mongolicum* are leguminous shrubs resistant to drought, poor soil nutrients and extreme air temperatures, and they prevent soil, wind and water erosion by stabilizing the sand dunes. These fast-growing shrubs can be harvested for three years after sowing. The leaves and branches provide nutritious feed for livestock and can be dried and stored and offered to livestock when pasture is scarce (Liu *et al.*, 2022). *Helianthus tuberosus* is a perennial plant of the chrysanthemum family with an irregular underground tuber. This plant is well-adapted to harsh conditions, being resistant to cold, drought and saline and alkaline soils. In addition, it has low soil-nutrient requirements, strong reproductive ability, fast growth rate, and stabilizes the sand which acts as a windbreak and prevents erosion (Fan and Ma, 2008). It is rich in inulin, glucose, alcohol and starch, which can be used in food and pickling, and can also be processed for inulin and other health products. The foliage provides forage for livestock, especially in times of sparse vegetation.

16.3.2 Water-saving irrigation for artificial grasslands in montane deserts

With the rise in population and improvement in the economy, there is an increased demand for meat and meat products. This situation has increased the pressure on the grazing land, and in combination with global warming and anthropogenic activities has caused the deterioration of the alpine desert ecosystem, including widespread degradation of the grasslands (Liu *et al.*, 2020). The large imbalance between livestock feed requirements and herbage availability can be mitigated by establishing artificial grasslands. Besides providing forage for livestock, artificial grasslands also improve soil nutrients, promote the development of grasslands animal husbandry, enhance the livelihood and income of local inhabitants, and provide ecological service functions (Gu *et al.*, 2018; Zhang *et al.*, 2019). The mixed sowing of gramineous and leguminous herbage, which have become the main plants in artificial grasslands in alpine deserts, not only improves the yield and quality of herbage, but also improves soil nutrients and quality (Liu *et al.*, 2017). *Avena sativa* and *Vicia sativa* are resistant to drought, cold and alkaline soils, and have been sown widely in artificial grasslands on the Qinghai-Tibetan Plateau.

Water has been the main limiting factor in establishing artificial grasslands in alpine deserts. Traditional flood irrigation is generally adopted for artificial grasslands in alpine desert grasslands areas; however, more efficient irrigation systems must be implemented for artificial grasslands to be successful (Yu *et al.*, 2020). Micro-spray irrigation, in which a belt sprays water evenly in the field, is a relatively new method of water-saving irrigation. It has the advantages of being simple, relatively cheap, and can be installed and disassembled easily. Regulated deficit irrigation saves water and increases production by applying some water stress at certain growth stages and then relies on the adaptability of the crops to overcome the stress (Kou *et al.*, 2014). Micro-sprinkler irrigation combined with regulated deficit irrigation can reduce soil-water loss and improve water-use efficiency. Studies have demonstrated that moderate water deficit at the early stage of crop growth and rehydration at a later stage can achieve compensatory growth and improve crop yield and water-use efficiency (Ali *et al.*, 2019; Pardo *et al.*, 2020). With full irrigation at the joint stage and moderate water deficiency at the flowering stage, the mixed sowing of *Vicia avena* and *Vicia sativa* saved water and improved the yield and quality of the crop (Wang *et al.*, 2022).

16.3.3 Establishment of protected areas in montane deserts

The alpine desert region of the Tibetan Plateau is the main habitat for some endangered wild ungulates, such as Tibetan antelopes, Tibetan wild donkeys and wild yaks. The fragile alpine desert ecosystem and biological chain are easily disturbed and unbalanced by climate change and human activities (Schaller, 1998). In recent years, increased livestock grazing, urban expansion, road construction and mining has degraded the grasslands and encroached upon and fragmented the habitats of the endangered animal species (Wu and Zhang, 2004). Human activities such as expansion of roads and settlements, which are widespread in the Qiangtang National and Hoh Xil reserves, resulted in a loss of 16% of the wildlife habitat. The loss of habitat is actually greater as grazing and mining were not considered in the assessment. Human activities can be mitigated by better planning of the road system, managing the grasslands more efficiently and providing corridors for the animals to enable them to migrate without danger. The establishment of key potential habitats and protected areas could improve the survival of endangered species and ensure the long-term continuation of the population (Zhuge *et al.*, 2014). The current protected areas are not conducive to effective overall protection of the endangered ungulates, in particular during their migrations. In the future, a network system of protected areas with connecting corridors should be established across the alpine deserts, which would enable the safe movement of wild animals between administrative boundaries and, therefore, expand their habitats. A unified mechanism of information and resource sharing among protected areas should be initiated, which would co-ordinate conservation measures for the protection and management of the entire Himalayan ecosystem and endangered species (Zhuge *et al.*, 2015).

16.4 Local Livelihoods and Ecological Restoration of Montane Deserts

Agriculture in the montane deserts is challenging because of the low air temperature, scant rainfall, low relative humidity, strong winds and radiation, high evapotranspiration rate, soil erosion and short growing season. Mixed farming, consisting of the production of food crops, vegetables and fruit, and the rearing of livestock, is the main livelihood in the region (Gupta and Arora, 2016). The application of organic manures, compost and biofertilizer is common, and these additives not only improve the nutrient content of the soil, but they also recycle wastes. Agroforestry, medicinal plants, fruits and vegetables can be integrated in the management of wastelands and the restoration of degraded agricultural lands. Soil and water conservation, including rainwater harvesting, and agronomic options, such as crop rotation, mulching and nutrient management, especially green manuring, and mixed use of organic manures and fertilizers, can be incorporated in the management of soil and water resources for sustainable agriculture and the improvement of the economic status of the residents in the region.

Implementing the grasslands ecological reward policy was designed for ecological restoration and livelihood enhancement of herdsmen in montane desert areas (Ma *et al.*, 2021). The education level of herdsmen in desert areas is generally low. Compulsory primary education should be implemented, vocational education and skill training should be strengthened, and the sustainable livelihood of herdsmen should be improved. The resettlement policies should be designed so that the herdsmen have a clear understanding that grasslands ecological protection is meant to achieve sustainable development of the pastoral areas, which is related directly to their livelihood and living environment in the future. The housing communities for herders should include community culture centres with grasslands traditions to improve the quality of life of the inhabitants. Based on the current industrial chain of 'grassland animal husbandry – processing industry of animal products', grasslands tourism can be extended to include horse, camel and yak riding (Fig. 16.4). An increase in income from tourism would reduce the pressure on grasslands and would broaden the livelihood options. A three-way win-win-win situation could be realized consisting of ecological protection, transformation of animal husbandry, and an increase in income of herdsmen in desert areas (Yang *et al.*, 2020).

Fig. 16.4. Grasslands tourism: horse, camel and yak riding. (Photos: Dongmei Li and Di Lu)

16.5 Shifts in Management Strategy, Adaptations and Pathways to Conserve Deserts in the Hindu Kush Himalaya Region

Infrastructure is necessary for the development of desert areas, but the construction of roads, railways and airports, and industry pose a threat to the stability of desert ecosystems. Therefore, the proper balance between development and ecological protection is crucial for the sustainable development of the desert ecosystem (Zheng *et al.*, 2005). For example, potential sources of heavy metal pollution in copper mining areas, such as tailings ponds, which contain waste material left over from ore extraction processes, and waste rock landfill, should be planned carefully. With the passage of time, the rocks and wastes from tailings ponds can be broken down by such processes as weathering and leaching, which accelerate the release of heavy metals. The main factors affecting the diffusion of heavy metals in soil are wind and water flow. Therefore, in the process of copper mine development, waste material should not be removed from the pit. The tailings ponds should be treated with anti-seepage material, and the dry waste material should be near dust-proof to minimize dust pollution (Cao *et al.*, 2019).

Biological soil crust, an important component of desert landscape, is composed of green algae, lichens, mosses and other cryptic plants, soil microorganisms, mycelium and a complex of root secretions and surface soil particles. As the 'engineer' of a desert ecosystem, biological soil crust plays an important role in maintaining soil stability, promoting soil development, enhancing material circulation and energy flow in sandy land, regulating soil temperature and humidity, and improving and maintaining biological diversity (Liu *et al.*, 2022). Consequently, biological soil crust can be used for the ecological restoration of abandoned mines, which can be important for desert landscape management and sustainable development of the desert ecosystem. High-quality sustainable development of the desert area can be achieved by implementing innovative desert protection measures, such as developing ecological agriculture and ecological tourism and the establishment of national parks, to improve the ecological environment and enhance the livelihood and income of herders.

16.5.1 Development of ecological pastoral areas in montane deserts

The global mean sea and land surface temperatures increased linearly by 0.85°C from 1880 to 2012. The average annual air temperature between 2003 and 2012 was 0.78°C greater than between 1850 and 1900. The HKH region has experienced the most rapid rate of climate warming worldwide (Zheng *et al.*, 2021). This trend is particularly true for the Kunlun mountains, resulting in frequent meteorological disasters and extreme climate events in recent years, which has affected the local economic and social development. Therefore, knowledge of the spatio-temporal patterns of the climate in ecologically fragile areas would improve our ability to predict climate trends so that decision makers would have the tools to avert or reduce climatic disasters (Gong and Ba, 2013; Du, 2018).

Developing ecological agriculture should make full use of light and heat resources and improve the utilization efficiency of water resources by combining land cultivation, agroforestry and animal husbandry. Production bases for commodity grain, livestock products and non-staple foods should be established, and the local ecological environment should be improved. There is a need to focus on restoring abandoned land, increasing the yield of crops, and improving ecological agriculture (Du, 1989). Specifically, low-yield fields should be transformed, water conservation should be improved, land should be cleared for cultivation, and shelter around croplands should be strengthened. Facility agriculture, which supplies vegetables and fruits out of season, should be developed in montane desert as this could be quite profitable for the farmers. For example, the use of the well-known vertical production and cultivation technology that is common in the lowlands to produce muskmelons could be increased in montane deserts, as it can achieve the off-season planting of two crops in one shed within one year (Mu, 2015). The alpine desert ecosystem is characterized by simple structure, low productivity, poor stability, weak restoration ability, and fragility; that is, it is easily disturbed by external factors. Artificial vegetation restoration is an effective way to prevent ecosystem degradation and achieve community stability. Measures such as sealing, direct afforestation and seedling planting can be implemented. Shrub species that can resist drought and sand can be planted on the windward side of dunes, and mixed forest with trees and shrubs can be planted on the inter-hill land (flat depression between sand dunes), which has greater soil-water content. In the management of artificial forests, proper management is necessary to promote the positive succession of the community and to improve ecological efficiency (Yu *et al.*, 2022).

Ecotourism has been suggested as a strategy for conserving and managing the desert's natural and cultural heritages. As one of the fastest-growing sectors of the global tourism industry, ecotourism has the potential to serve as a viable option for promoting sustainable development of the desert biome. Many factors can provide opportunities for the development of ecotourism in deserts, but they should be considered carefully for both nature conservation and local socio-economic well-being, as they may deplete the desert's natural and cultural features. Ecotourism should be developed leading to sustainable governance because, if managed properly, it can lessen some of the threats to the desert biome. Ultimately, desert ecotourism can act as an agent to promote sustainable development goals and can contribute to the well-being of desert inhabitants and to threatened desert heritage (Santarém *et al.*, 2019). The development of ecological facility agriculture and ecotourism can be achieved by sustainable use of deserts for human activities by implementing strategies to harvest and use water more efficiently, to produce sustainable crops and to graze livestock in an integrated management system. Once these actions are functioning, the ultimate goal of building an ecologically viable pastoral system could be realized in the HKH region.

16.5.2 Building montane desert national parks across regions or borders in the Hindu Kush Himalayan region

Montane deserts are disturbed relatively less by anthropogenic activities than are other areas. At present, the main strategy to protect the sustainable development of montane deserts is to prohibit all human activities, except for nature patrols and scientific research. In the core areas of the reserve, close-to-nature restoration should be implemented. In ecological conservation, appropriate artificial measures should be taken to restore traditional utilization areas. The existing conservation measures cannot guarantee the sustainable development of the area. National parks across regions or borders have been recognized internationally as a management option that co-ordinates ecological environmental protection and resource development and utilization. Therefore, montane desert national parks across regions or borders, based on ecological integrity and corridor connectivity, should be established to remove the administrative boundary fragmentation and management system segmentation of the existing reserves. This would improve the protection and management of the entire montane desert ecosystem and the overall efficiency of conservation of montane desert areas in the HKH region (Zhuge *et al.*, 2015).

16.6 Conclusions

The montane desert ecosystem in the HKH is extremely fragile and is threatened by global warming and anthropogenic activities. The loss of soil and soil nutrients due to wind, water and glacial erosions, limited water availability and the degraded grasslands affect ecosystem functions and services, livestock production and the agricultural economy, and livelihoods of the inhabitants. Ecological management projects, application of water-saving irrigation methods in artificial grasslands, policy incentives, and the establishment of protected areas have been initiated to conserve desert ecosystem functions and services. In future, innovative desert protection programmes, development of ecological agriculture and ecotourism, and the establishment of national parks should be implemented to improve the environment and enhance the livelihood and income of herders. Once accomplished, high-quality sustainable development of the montane deserts in the HKH region could be achieved.

References

Ali, S., Xu, Y., Ma, X., Ahmad, I., Jia, Q. *et al.* (2019) Deficit irrigation strategies to improve winter wheat productivity and regulating root growth under different planting patterns. *Agricultural Water Management* 219(6), 1–11. DOI 10.1016/j.agwat.2019.03.038

Ba, S.T.R., Da, Q., Chi, S.J., La, Z. and Qimei, Y.Z. (2012) Temporal and spatial distribution of precipitation in Changtang Steppe Desert ecological functional area in northern Tibet in recent 40 years. *Xizang Science and Technology* 8, 59–62 (in Chinese).

Cao, Y.H., Zhang, Y.K., Gao, Z.G., Liu, C.Y., Liu, L. *et al.* (2019) Analysis of soil heavy metal pollution in a copper mining area in Alpine desert. *Protection and Utilization of Mineral Resources* 39(2), 95–100 (in Chinese).

Chen, J.W. and Xiao, J. (1997) Desert, desertified land resources and nature protection. *Forestry Resource Management* 2, 31–35 (in Chinese).

Cheng, L.L., Guo, H. and Lu, Q. (2013) Review on the evaluation of desert ecosystem services. *Journal of Desert Research* 33(1), 281–287 (in Chinese).

Du, Q. (1989) Preliminary study on ecological agriculture in desert oasis of Qaidam. *Agricultural Research in the Arid Areas* 4, 30–35 (in Chinese).

Du, Q. and Sun, S.Z. (1990) *Plant Ecology. Vegetation and Its Utilization in Qaidam Area*. Science Press, Beijing.

Du, Y.E. (2018) Temporal and spatial patterns of vegetation and lakes in Qaidam Basin and their responses to climate change. Lanzhou University, Lanzhou, China.

Fan, G.H. and Ma, Y.L. (2008) Cultivation experiment of Jerusalem artichoke in desert land of Qaidam Basin. *Qinghai Agriculture and Forestry Science and Technology* 3,16–17 (in Chinese).

Gong, J.Q.P. and Ba, Z. (2013) Spatio-temporal distribution of gales in Changtang Steppe Desert ecological functional area in Northwest Tibet in the past 40 years. *Anhui Agricultural Sciences* 41(36), 14171–14172 (in Chinese).

Gu, Y.J., Han, C.L., Kong, M., Shi, X.Y., Zdruli, P. *et al.* (2018) Plastic film mulch promotes high alfalfa production with phosphorus-saving and low risk of soil nitrogen loss. *Field Crops Research* 229, 44–54. DOI: 10.1016/j.fcr.2018.09.011

Gupta, R.D. and Arora, S. (2016) Ecology, soil and crop management for livelihoods in Ladakh region: an overview. *Journal of Soil and Water Conservation* 15(2), 178–185.

Harpinder, S. (2017) *Ecosystem Functions and Management: Theory and Practice*. Routledge, London.

Houghton, R.A., Hall, F. and Goetz, S.J. (2009) Importance of biomass in the global carbon cycle. *Journal of Geophysical Research* 114(3), 101–113. DOI: 10.1029/2009JG000935

IUCN Pakistan (2000) Balochistan partnerships for sustainable development Balochistan Programme Office Marker Cottage. Zarghoon Road, Quetta, Pakistan.

Kou, D., Su, D.R., Wu, D. and Li, Y. (2014) Effects of subsoil deficit drip irrigation on water consumption, yield and quality of alfalfa. *Transactions of the Chinese Society of Agricultural Engineering* 30(2), 116–123 (in Chinese).

Li, Y., Zhu, Y.F. and Dong, B.C. (2018) Evaluation on utilization effect of highway wildlife passage in desert area. *Journal of Highway and Transportation Science and Technology* (Applied Technology Edition) 14(9), 316–318 (in Chinese).

Liang, Z.S. and Hou, X.M. (2007) Ecological planning and construction model design of desert steppe in Qaidam Basin: a case study of desert degraded steppe in Dulan County. *Pratacultural and Animal Husbandry* 12, 31–34 (in Chinese).

Liu, J.Y., Wu, P.X., Li, J.B., Ma, Y.D. and Zhao, G.P. (2022) Feasibility analysis on the introduction and cultivation of flower stick and poplar wood in alpine desert and semi-desert area. *Journal of Anhui Agricultural Sciences* 50(18), 104–108 (in Chinese).

Liu, S.J., Wei, X.H., Zheng, Q.Q, Lin, Q.X., Luo, X.L. *et al.* (2020) Soil organic carbon content in Ngari Alpine desert of Tibet. *Journal of Desert Research* 40(4), 234–240 (in Chinese).

Liu, W.H., Zhang, Y.J., Shi, S.L., Sun, J., Wei, X.-X. *et al.* (2017) Effect of variety, fertilization and mixture sowing on oat biomass allocation in the alpine cultivated pasture. *Journal of Plant Nutrition and Fertilizers* 23(2), 398–407 (in Chinese).

Liu, Y.M., Yang, H.Y., Liu, F.L., Wang, J.J., Wang, L.H. *et al.* (2022) Effects of bacteriophagous nematodes on soil enzyme activity under soil nodules in desert area. *Journal of Desert Research* 42(2), 77–84 (in Chinese).

Lu, Z.Y.F., Xu, D.Y., Zhang, X.J., Zhao, Z.R. and Zhang, X.Y. (2019) Effects of land use change on ecosystem services in Alxa arid desert region. *Research of Soil and Water Conservation* 26(6), 296–302 (in Chinese).

Ma, H.L., Lin, H.L., Xiong, X.Y. and Han, X.P. (2021) Effects of grassland ecological compensation policies on grassland vegetation status in the Qinghai-Tibet Plateau. *Acta Agrariae Sinica* 29(3), 545–554 (in Chinese).

Mu, R. (2015) Study on green and efficient vertical cultivation technology of melon in off-season facilities in Alpine desert area. *Friends of Farmers Getting Rich* 10, 60–61 (in Chinese).

Pardo, J.J., Martínez-Romero, A., Léllis, B.C., Tarjuelo, J.M. and Domínguez, A. (2020) Effect of the optimized regulated deficit irrigation methodology on water use in barley under semiarid conditions. *Agricultural Water Management* 228(2), 105925. DOI: 10.1016/j.agwat.2019.105925

Santarém, F., Saarinen, J. and Brito, J.C. (2020) Desert conservation and management: ecotourism. *Encyclopedia of the World's Biomes* 259–273. DOI: 10.1016/B978-0-12-409548-9.11827-5

Schaller, G.B. (1998) *Wildlife of the Tibetan Steppe.* Wildlife Conservation Society and Tibetan Plateau Project, University of Chicago, Chicago, Illinois.

Wang, C.P. (2018) Dynamic change and driving force analysis of desertified land in the middle part of Kubuqi Desert. *Forest Resources Management* 1, 63–71, 154 (in Chinese).

Wang, J.H., Li, G., Yu, X.X., Kang, Y.X., Yin, M.H. *et al.* (2021) Effects of regulated deficit micro-sprinkler irrigation on yield, quality and water utilization of artificial grassland in Alpine desert. *Journal of Soil and Water Conservation* 35(4), 200–207, 216. DOI: 10.13870/j.cnki.stbcxb.2021.04.028 (in Chinese).

Wang, J.H., Li, G., Yin, M.H., Qi, G.P., Kang, Y.X. *et al.* (2022) Effects of regulated deficit irrigation on soil environment and grass growth of artificial mixed grassland in alpine desert region. *Journal of Pratacultural Science* 31(1), 95–106 (in Chinese).

Wu, X.M. and Zhang, H.F. (2004) Resources regarding populations of Tibetan antelope (*Pantholops hodgsoni*) and the status of its protection. *Chinese Journal of Nature* 3(3), 143–154 (in Chinese).

Xin, Y., Shu, K., Dai, L.C., Qian, D.W., Guo, X.W. *et al.* (2021) Spatial characteristics of soil organic carbon in the alpine desert of Qaidam Basin, Qinghai-Tibet Plateau. *Chinese Journal of Grassland* 43(11), 113–118 (in Chinese).

Xu, G.F., Cang, S.H., Na, M.J., Ren, Q. and Wang, H.X. (2007) Study on hole-like supplementary seeding of desert grassland in Qaidam Basin. *Grassland and Turf* 4, 77–79 (in Chinese).

Yang, Q., Nan, Z.B., Chen, Q.Q. and Tang, Z. Study on herdsmen's satisfaction degree and influencing factors of grassland ecological subsidy and reward policy based on the empirical study of Qinghai-Tibet Plateau and western desert in Gansu Province. *Acta Ecologica Sinica* 40(4), 1436–1444 (in Chinese).

Yu, T.L., Xu, Y., Zhou, H.Y., Wang, L., Ou, E.H. *et al.* (2022) Succession law of typical mixed artificial communities in alpine desert region. *Journal of Qinghai University* 40(6), 1–8 (in Chinese).

Yu, X.X., Qi, G.P., Kang, Y.X. and Yin, M.H. (2020) Effects of irrigation patterns on yield and water consumption characteristics of mixed planting artificial grassland in Alpine desert. *Water Conservancy Planning and Design* 3, 130–134, 178 (in Chinese).

Zhang, X.M., Ma, Q.H., Zhang, Z.L., Wang, Z., Yang, H. and Shen, Y. (2019) Effects of fertilization rate on forage yield and water use efficiency of artificial grassland in an alpine arid area. *Scientia Agricultura Sinica* 52(8), 1368–1379 (in Chinese).

Zheng, J.Z., Liu, L.M., Qin, H.S., Ma, D.S., Guo, J.F. *et al.* (2005) Planning and model design of ecological construction in Qaidam Desert Zone: a case study of desert degradation grassland construction in Delingha City. *Soil and Water Conservation in China* 11, 16–18, 60 (in Chinese).

Zheng, L., Shi, P., Song, M., Zhou, T., Zong, N. *et al.* (2021) Climate sensitivity of high altitude tree growth across the Hindu Kush Himalaya. *Forest Ecology and Management* 486(9), 118963. DOI: 10.1016/j.foreco.2021.118963

Zhong, Z.B., Zhou, G.Y., Yang, L.S., Liu, H.C. and Song, W.Z. (2014) Biomass distribution pattern of several desert shrub vegetation in Qaidam Basin. *Journal of Desert Research* 34(4), 1042–1048 (in Chinese).

Zhuge, H.J., Li, X.W., Zhang, X., Gao, F. and Xu, D.H. (2014) Appropriate habitat identification and conservation assessment of Tibetan antelope in the alpine desert of the Tibetan Plateau. *Chinese Journal of Applied Ecology* 25(12), 3483–3490 (in Chinese).

Zhuge, H.J., Lin, D.Q. and Li, X.W. (2015) Ecological corridor identification and conservation assessment of Tibetan antelope in the alpine desert area of Qinghai-Tibet Plateau. *Chinese Journal of Applied Ecology* 26(8), 2504–2510 (in Chinese).

17 Wetlands in the Hindu Kush Himalayan Region: Eco-economic Function and Conservation Strategies

Ritika K.C.[1], Usha Rai[2], Monika Ghimire[2] and Udhab Raj Khadka[2,*]
[1]*Central Department of Environmental Science, Tribhuvan University, Kathmandu, Nepal;* [2]*Department of Environmental Science, Amrit Campus, Tribhuvan University, Kathmandu, Nepal*

Abstract
Wetland is the transition between aquatic and terrestrial ecosystems, and is known as one of the most diverse and productive ecosystems. The Hindu Kush Himalaya (HKH) region encompasses several wetlands of ecological and economic relevance; however, they are threatened by natural and human factors. For the sustained eco-economic functioning of these wetlands, the HKH member countries (except Afghanistan) have committed to the conservation and sustainable use of wetlands. To achieve this, the HKH member countries should work together through a transboundary landscape approach, with China and India, which are emerging as global leaders in scientific knowledge and technological advancement, playing vital roles. For the better management of wetlands, it is crucial to assess their eco-economic functions and the conservation efforts. This chapter provides an insight on wetlands, their importance, and conservation strategies in the HKH region.

Key words: HKH, livelihood, mountain, water management

17.1 Introduction

Mountains are very important landforms that support diverse ecosystems. They occupy 22% of the global land surface and are home to 13% of the world's population (FAO, 2015; Wester *et al.*, 2019). In terms of livelihood, about 50% of the people depend directly on mountain resources, primarily water. The mountains support 25% of the global terrestrial biodiversity and possess close to 50% of the biodiversity hotspots. Out of the plant species that supply 80% of the world's food, six of them, namely apple, barley, maize, potatoes, sorghum and tomatoes, originated in mountains (Fleury, 1999; Wester *et al.*, 2019). Mountains provide goods and services, including water, hydroelectricity, timber, biodiversity, niche products, mineral resources, recreation and flood management (Schild and Sharma, 2011; Molden and Sharma, 2013; Wester *et al.*, 2019). In humid parts of the world, mountains provide 30–60% of the freshwater downstream and in semi-arid and arid regions, they provide 70–95% (Kapos *et al.*, 2000). Besides natural resources, mountains are rich in ethnicity and cultural diversity. Despite being rich in resources and culture, poverty is higher in mountains and people are at higher risks than in other regions.

*Corresponding author: ukhadka@cdes.edu.np

DOI: 10.1079/9781800622579.0017

According to FAO (2015), 39% of mountain people (including both urban and rural) in developing countries are vulnerable to food insecurity (Wester *et al.*, 2019).

Among the mountain ranges, the Hindu Kush Himalayan (HKH) region covers over 4 million km^2, approximately 2.9% of the global land area and 18% of the mountain area. The region extends 3500 km encompassing the whole or part of eight countries, including Afghanistan, Bangladesh, Bhutan, China, India, Myanmar, Nepal and Pakistan. As the HKH region is dominated by high mountains, vast glaciers and large rivers, it has earned many names. For instance, because the region contains 14 of the world's highest mountains over 8000 m and most of the peaks over 7000 m, it has been named the 'roof of the world', and, as it is the world's most glaciated area outside the polar regions, it is also known as the 'Third Pole'. In addition, as the HKH region stores and supplies much water, it is known as the 'freshwater tower' of South Asia (Immerzeel *et al.*, 2010). Water originating from the HKH region as snow and/or glacial-melt, and rainfall, feed ten large river systems, namely the Amu Darya, Indus, Ganges, Brahmaputra (Yarlungtsanpo), Irrawaddy, Salween (Nu), Mekong (Lancang), Yangtse (Jinsha), Yellow (Huanghe) and Tarim (Dayan) in Asia. These river systems provide water, ecosystem services and the basis of livelihoods to more than 240 million people upstream in the mountains and 1.65 billion people, a fifth of the world's population, downstream (Karki, 2012; Shrestha *et al.*, 2015, ICIMOD, 2023). With freezing and thawing, the glaciers generate and recharge the high-altitude wetlands that provide various eco-economic functions (Khan and Baig, 2017). The high-altitude wetlands serve as a habitat for species with trans-Himalayan ranges (Hawkes *et al.*, 2011; Ragettli *et al.*, 2016). Furthermore, these wetlands are crucial to religious traditions and local economies (Acharya and Dokham, 1998; Sharma *et al.*, 2015). Additionally, wetlands play an important role in purifying water and decontaminating wastewater and, thus, are considered as the 'kidneys of Earth' (Ye *et al.*, 2022). The wetlands also play an important role in mitigating climate change by acting as carbon sinks. The peatlands in China are one of the most important stores of carbon in the mountain regions storing 1500–4000 tonnes per ha, or up to 8–20 times more than mountain forests and 50–100 times more than mountain grasslands. Despite this importance, the high-altitude wetlands are considered to be largely unmanaged (ICIMOD, 2010). Until a few decades ago, wetlands were among the low-priority resources in terms of protection and conservation. Over the past half-century, wetlands have undergone serious degradation due to climate change and anthropogenic activities that have threatened regional ecological security and sustainable socio-economic development. In the last century, more than 50% of the global wetlands have been reported to be degraded (Millennium Ecosystem Assessment, 2005). This chapter discusses the status of wetlands in the HKH region, their conservation, and management strategies.

17.2 Water Resources in the HKH

Snow and ice are dominant features in the HKH region. There are over 54,000 glaciers in the region covering a total area of more than 60,000 km^2. Altogether, the glaciers comprise over 6000 km^3 of ice reserves, acting as freshwater reservoirs for the greater region. In the drier part of Asia, more than 10% of local river flow originates from ice and snow melt (Sivakumar and Stefanski, 2011; Shrestha *et al.*, 2015). The HKH region, being a source of water for the major rivers in Asia, provides water and ecosystem services for both upstream and downstream communities. The mountains in the HKH region have been reported to be covered with 112,767 km^2 of glaciated area (Dyurgerov and Meier, 2005).

17.3 Status of Wetlands in HKH Region

Wetlands are an important, productive ecosystem, with their value equalling 47% of the total value of the global ecosystems (Costanza *et al.*, 1997; Mitsch *et al.*, 2013). Wetlands provide many ecological services and functions and play major roles in regulating global climate and sustaining species (Hu *et al.*, 2017). However, wetlands are under constant threat from human and natural activities. Temperatures in the Himalayan region are rising faster than the global average causing the receding of glaciers and decline in reserves of snow, ice and water in the high mountains, which ultimately affects wetlands and their services (Chaudhry, 2010). Shrestha *et al.* (1999) reported that warming increases with an increase in altitude. Based on the temperature data from 1971 to 2014, according to the Department of Hydrology and Meteorology, the annual maximum temperature in Nepal increased at 0.056°C per annum (DHM, 2017).

Global climate change, unsustainable human activities, extraction and diversion of water for agricultural use, waste disposal, urbanization and land-use change, pollution, prolonged drought, increased tourism, and fisheries are threats that are common to all countries in the HKH region (Murray, 2009; Chaudhry, 2010; Ye *et al.*, 2022). The HKH region possesses diverse types of wetlands including peatlands, glaciated lakes, springs, marshes, floodplains and man-made reservoirs, which provide ecosystem and economic services. However, wetlands are degrading due to climatic and anthropogenic factors. In this connection, the status of wetlands is discussed by countries in the HKH region.

17.3.1 Afghanistan

Climatically, most of Afghanistan is dry; thus, wetlands in Afghanistan are of great significance. In Afghanistan, there are three types of wetlands – river courses, lakes and marshes, and artificial lakes (Malikyar, 2017).

Main threats to wetlands in Afghanistan include:

i. conversion into agricultural lands, which resulted in the loss of thousands of ha;
ii. encroachment and conversion to residential areas;
iii. water used for irrigation and other purposes;
iv. drought and decrease in water table;
v. grazing and fuel collection; and
vi. lack of policies and regulations.

The impacts of wetlands degradation in Afghanistan include:

i. loss of groundwater recharge;
ii. decrease in natural flood control system;
iii. rapid water contamination, pollution increase and lack of natural control;
iv. loss of food sources; and
v. decrease in environmental stability.

There are many other impacts besides the above.

17.3.2 Bangladesh

Bangladesh lies in the world's largest delta, the Bengal basin, formed by the Ganges, Brahmaputra and Meghana rivers. The basin is a vast lowland, which was once thought to be the largest wetlands, most of which has been converted to rice growing (Chowdhury, 1998). In Bangladesh, the wide range of wetlands includes more than 700 rivers and streams, thousands of shallow freshwater lakes, marshes, floodplains, inshore coastal areas, extensive estuarine systems, water storage reservoirs, fishponds and flooded cultivated lands. The total wetland area is estimated to be 7–8 million ha, about 50% of the total land area of the country (Khan, 1993). This includes 5.4 million ha of open and closed lakes on floodplains that are inundated every year (Ali, 1990). Approximately 90% of the wetlands are dependent on flow of water from three major rivers, and most flow from India; they are now threatened by diversion of waters, e.g. Ganges-Padma river (Gopal and Wetzel, 1995). Wetlands in Bangladesh are the major natural resource on which the fishing industry and agriculture depend. Thus, water resources or wetlands in Bangladesh are considered to be the central pillar of the natural resource base. However, these wetlands are also threatened by several climatic and anthropogenic drivers.

The major threats to wetlands in Bangladesh include:

i. population pressure;
ii. ill-planned national, local and rural infrastructures such as roads and narrow culverts;
iii. extensive agriculture and urbanization;
iv. land reclamation and expansion of agriculture;
v. conversion of wetlands into rice fields;
vi. flood control drainage and irrigation (FCDI) projects for enhancement of agricultural productivity and flood control;
vii. over-fishing and introduction of alien invasive species and extensive shrimp cultivation;
viii. harvesting of wetlands vegetation without replacement;
ix. siltation due to degradation of the associated watersheds;

x. increased use of fertilizer and pesticide for rice monoculture;
xi. indiscriminate regulation of flows of the river systems in the upper riparian areas;
xii. pollution of water due to industrial, urban, agrochemical and other pollutants, including pollution from transboundary sources;
xiii. change of hydrological conditions in the watershed and climate change; and
xiv. gas and oil exploration in the wetlands.

Degradation of wetlands is alarming in urban areas such as Dhaka with its impact visible mainly in:

i. natural environment;
ii. ecosystem; and
iii. drainage congestion.

17.3.3 Bhutan

Wetlands in Bhutan include high-altitude wetlands, marshes, bogs, lakes, swamps, paddy fields, springs, rivers, ponds and streams. There are 3027 high-altitude wetlands, which cover about 0.26% of the total land cover (Tariq *et al.*, 2021). In Bhutanese culture, lakes are considered sacred and holy sites, and thus conservation is prioritized (Tariq *et al.*, 2021). Despite this, there are several water/wetland-related issues/threats in Bhutan.

The major threats include:

i. water pollution;
ii. land-use change;
iii. climate change;
iv. drought;
v. over-grazing;
vi. human encroachment;
vii. over-fishing/hunting;
viii. stone/sand quarrying; and
ix. lack of conservation initiative.

The impacts of wetland degradation include:

i. loss in biodiversity;
ii. change in hydrological cycles; and
iii. loss of livelihoods.

17.3.4 China

China possesses many rivers, numerous lakes and a long coastline, creating a variety of wetlands. The second Nationwide Wetland Survey (2009–2013) recorded 53.5 million ha of wetlands (excluding paddy fields, and wetlands of Hongkong, Macao, and Taiwan) accounting for 5.6% of the total land area of China (Ye *et al.*, 2022). Out of the total area, 46.7 million ha (87.4%) are natural wetlands. There are 21.7 million ha (40.7%) of inland marshes, 10.6 million ha (19.8%) of riverside wetlands, 8.59 million ha (16.1%) of lakeside wetlands, and 5.80 million ha (10.9%) of coastal wetlands (Ye *et al.*, 2022). Between 1950 and 2010, 57% of coastal wetlands in China were degraded (Ye *et al.*, 2022), and from 1975 to 2017 they shrunk by 54% (Ye *et al.*, 2022).

According to Ye *et al.* (2022), the following are the major threats to wetlands in China:

i. climate change;
ii. urbanization and change in land use;
iii. environmental pollution; and
iv. human encroachment.

Causes of wetland degradation include (Song *et al.* 2014):

i. enhanced soil carbon emission;
ii. changed hydrological cycling; and
iii. regional temperature increase.

17.3.5 India

In the HKH region, India is one of the vast countries with varied topography and climatic regimes supporting diverse and unique wetland habitats (Prasad *et al.*, 2002). According to Space Administration Center, SAC, India has about 757,060 wetlands with a total wetland (open water, aquatic vegetation and surrounding hydric soils) area of 15.3 million ha accounting for nearly 4.6% of the total area. Inland wetlands account for 69%, coastal wetlands 27% and other wetlands (smaller than 2.25 ha) 4% (SAC, 2011) of the wetlands. The water-spread area of wetlands in India varies with the seasons. The inland wetlands cover 7.4 million ha post-monsoon and 4.8 million ha pre-monsoon; and coastal wetlands cover 1.2 million ha post-monsoon and 1 million ha pre-monsoon (SAC, 2011). These wetlands are facing climate-induced and anthropogenic stresses. For example, the high-altitude wetlands of HKH regions (Sikkim Eastern Himalaya) are threatened by tourism, waste and non-native species (Joshi and Dhyani, 2009). In addition, sedimentation due to vehicle traffic, road construction and military installations fragment and degrade wetlands along the Indian trans-boundary border (Islam and Rahmani, 2008).

Overall, the major threats to wetlands in India include:

i. urbanization and land-use change;
ii. agriculture residue – fertilizer in India has increased from 2.8 million tons in 1973–1974 to 28.3 million tons in 2010–2011;
iii. municipal and industrial pollution;
iv. climate change – it has been estimated that India will lose about 84% of the coastal wetlands and 13% of saline wetlands with climate change-induced seawater rise of 1 m;
v. immersion of idols and religious ritual wastes;
vi. introduction of exotic species;
vii. encroachment and unregulated aquaculture; and
viii. dredging

Causes of wetland degradation in India include:

i. changes in hydrological cycle;
ii. biodiversity loss;
iii. disruption in the livelihoods of the people in coastal areas; and
iv. change in regional temperature.

17.3.6 Myanmar

Myanmar, with an area of 667,553 km^2, is endowed with vast areas of forests and wetlands. The principal wetlands consist of mangroves, swamp forests, lakes and marshes. Wetlands are an important natural resource of Myanmar and play a crucial role in the country's economy.

Threats associated with wetland conservation in Myanmar include:

i. supply of freshwater to agriculture, industries and housing can be a problem for long-term existence of wetlands;
ii. development of dams and unsustainable agricultural practices cause negative impact on the regular flow of water into the wetlands;
iii. pollution, particularly the release of industrial waste into rivers, streams and lakes, is a serious threat to sustainable fisheries;
iv. over-fishing and use of unauthorized fishing methods;
v. increased deposition of silt in wetlands due to inappropriate land-use methods in upland areas;
vi. hunting and netting of water birds;
vii. weak awareness of the importance of wetlands and their conservation;
viii. inadequate qualified staff to work in wetland conservation and management; and
ix. insufficient financial resources and equipment for wetland conservation and management.

17.3.7 Nepal

Wetlands in Nepal occupy roughly 5% of the total land. Rivers are the most common wetlands in Nepal with more than 6000 rivers and rivulets. Lakes in Nepal cover 3% of the total available water of the country. Most geothermal springs are located in the mid-hills and mountains. Marshes are near streams and rivers in the Terai flood plains. The extent of shrub-dominated wetlands, shrub swamps, shrub-dominated freshwater marshes, shrub carr and alder thicket on inorganic soils in Nepal is unknown.

According to Conservation and Sustainable Use of Wetlands in Nepal (CSUWN, 2010), the major threats to wetlands include:

i. conversion of wetlands to other land use;
ii. over-exploitation;
iii. pollution;
iv. biodiversity loss;
v. invasion of alien species;
vi. encroachment of alien species;
vii. sedimentation; and
viii. disasters.

Causes of wetland degradation include:

i. change in regional temperature;
ii. alteration in water volumes;
iii. loss of biodiversity; and
iv. river siltation.

17.3.8 Pakistan

Despite having an arid climate, Pakistan supports over 780,000 ha of wetlands covering 9.7% of the total land area, with 225 nationally significant wetlands, of which 19 have been listed in Ramsar sites for their global significance.

The major causes of wetlands degradation in Pakistan include (Khan and Ashrad, 2014):

i. conversion of wetlands for agriculture and other land uses;
ii. changes in water-flow regime;
iii. over-harvesting of wetland resources;
iv. organic and inorganic pollutants; and
v. deficient policies and inadequate management.

The impacts of wetland degradation in Pakistan include:

i. erosion in wetlands;
ii. siltation in reservoirs;
iii. affecting the livelihood of people living in the inland and coastal areas;
iv. diminishing biodiversity; and
v. decrease in fish stocks.

17.4 Conservation of Wetlands in the HKH Region

Wetlands are degrading with serious implications for the environment. Conservation of wetlands is not only a national issue but also a global concern. The Convention on Wetlands of International Importance Especially as Waterfall Habitat (Ramsar Convention) held in February 1971 in Ramsar, Iran, was an inter-governmental initiative for promoting the conservation and wise use of wetlands through local and national actions, and international co-operation. The convention entered into force in December 1975. At present, the number of parties participating in the convention has reached 172, with 2493 declared sites and a total area of 256,759,538 ha. The HKH region has 197 sites covering 11,270,044 ha (Table 17.1).

The wetlands across the HKH region are threatened by climatic and non-climatic factors. The countries in the HKH region declared their wetlands of international importance, and formulated policy/legal frames for the wetlands conservation. The country-wide initiatives for the conservation of wetlands are discussed here.

17.4.1 Afghanistan

In Afghanistan, wetlands are important natural heritages. However, there appear to be no officially declared wetlands of international importance (Ramsar sites). As the country is not a party to the Ramsar Convention, only a part of the Hamom wetland (a Ramsar site), which is shared by Afghanistan and Iran, is considered as a Ramsar site (Rosenblum and Schmeier, 2022). Recognizing the importance of wetlands, the government of Afghanistan has declared wetlands (Kol-e-Hasmat Khan) as a protected area in 2017 (https://www.unep.org/news-&-stories/story/kabul-wetl&-declared-new-protected-area-migrating-birds [accessed 21 February 2024]). The National Environmental Protection Agency of Afghanistan is the primary authority responsible for the development and management of protected areas in Afghanistan (National Environmental Protection Agency of the Islamic Republic of Afghanistan, 2017). Afghanistan has formulated policies and legal instruments

Table 17.1. Country-wise Ramsar sites in the HKH region. (From Mi and Xie, 2002)

S.N.	Name of the country	No. of wetlands	Area (ha)	Date of entry
1	Afghanistan	n.a	n.a.	n.a.
2	Bangladesh	02	611,200	21 September 1992
3	Bhutan	03	1225	7 September 2012
4	China	82	7,647,895	31 July 1992
5	India	75	1,326,677	01 February 1982
6	Myanmar	06	278,679	17 March 2005
7	Nepal	10	60,561	17 April 1988
8	Pakistan	19	1,343,807	23 November 1976
	Total	**197**	**11,270,044**	

n.a., not available (From: https://www.ramsar.org/country-profiles [accessed 5 June 2023]) for the conservation of the natural heritages including wetlands within the country.

The major policies include:

i. Afghanistan National Development Strategy (2009); and
ii. National Protected Area System Plan (NPSAP).

The legal framework of the country includes:

i. Environmental Law (2007) for the sustainable use and management of natural resources, and the conservation and rehabilitation of the environment towards achieving specified social, economic, reconstruction and ecological goals; and
ii. Water Law in Afghanistan (2009) for the conservation, equitable distribution, and efficient and sustainable use of water resources, strengthening the national economy and securing the rights of the water users, in accordance with the principles of Islamic jurisprudence and the praiseworthy customs and traditions of the people.

17.4.2 Bangladesh

Bangladesh signed the Ramsar Convention in September 1992 and has two Ramsar sites, including Sundarbans, the world's largest mangrove forest, and Tanguar Haor, covering 611,200 ha. There is no specific policy on wetlands and water, but the protection and management come under another resource law.

Policy frameworks of Bangladesh include:

i. Environmental Policy (1992);
ii. Environment Action Plan (1992);
iii. National Fisheries Policy (1998); and
iv. National Water Policy (1999) – controls wetlands management and ensures progress towards fulfilling national goals of economic development, poverty alleviation, food security, public health and safety, a decent standard of living, and protection of the natural environment (GOB, 1999);
v. National Water Management Plan (2004);
vi. Coastal Zone Policy (2005);
vii. Fisheries Strategy and Inland Capture Fisheries Strategy (2006);
viii. Jalmohal Management Policy (2009);
ix. Master Plan of Haor Areas (2012); and
x. National Policies on Water, Environment and Fisheries.

The legal provisions include:

i. The Canal Act (1864);
ii. Local Government (Pourashava) Act (2009), which gave Pourashava wide responsibilities in town planning and development, public health and sanitation, water supply and sewage disposal, maintenance of public infrastructure and amenities;
iii. Local Government (City Corporation) Act (2009);
iv. Protection and Conservation of Fish Act 1950 (amended 1982);
v. Tanks Improvement Act (1939);
vi. Bangladesh Environmental Conservation Act (1995 revised 2010), which provides for conservation of the environment, improvement of environmental standards, and control and mitigation of environmental pollution;
vii. Water Reservoir Conservation Act (2000), which prohibits water bodies being developed into high land;
viii. Environment Court Act (2010) for the amendment and consolidation of the existing law in establishing courts and related matters to expedite the trial of environmental crimes;
ix. National River Protection Commission Act (2013) to establish a Commission for preventing illegal occupation of rivers, pollution of water and environment, pollution of rivers caused by industrial factories, illegal constructions and various irregularities, and ensuring multidimensional use of rivers for socio-economic development including restoration of the normal flow of rivers, proper maintenance thereof and making them navigable; and
x. Bangladesh Water Act (2013), which provides for integrated development, management, abstraction, distribution, use, protection and conservation of water resources.

17.4.3 Bhutan

Bhutan became party to the Ramsar Convention on 10 January 2012, enlisting the country as the 161st contracting party. In Bhutan, Ramsar sites cover almost 1225 ha of the country. It has three wetlands of international importance:

Bumdeling and Khotoka (first Ramsar sites), and Gangtey-Phobji, the largest wetlands of the Ramsar sites. Bhutan established policies and acts for the conservation and management of wetlands. Some of these policies and acts include Water Policy (2003), Water Act of Bhutan (2011), Water Regulation (2012) (Gawel and Ahsan, 2014), and National Environment Protection Act (NEPA) (2007). Until 2003, when Bhutan's first water policy was developed, there was no water policy, but water issues were addressed by Land Act 1979 (now Land Act of Bhutan, 2007) and Forest and Nature Conservation Act (1995) (Tariq *et al.*, 2021). Since then, many policies and laws relating to water have been developed. The following are the policies and legal instruments relevant to the protection and conservation of water resources in Bhutan (Tariq *et al.*, 2021):

i. Bhutan Water Policy (2003) recognizes water as a valuable natural resource and asset essential to all social, financial and natural prosperity. The policy provides a framework for water rights, water conservation, water user interest, water priorities, water-related issues, principles for water resource development and management, issues and resolution related to transboundary water, and delineates institutional roles and strategies to improve institutional management (Tariq *et al.*, 2021);
ii. National Irrigation Policy (2012), developed for achieving self-sufficiency and food security through development of irrigation. It focuses on new sustainable approaches to irrigation infrastructure development and maintenance (Tariq *et al.*, 2021);
iii. National Integrated Water Resource Management (IWRM) Plan (2006), focuses on the strong Bhutanese custom of water sharing and co-ordinated management. It serves as the guideline in the planning and implementation of climate change adaptation to achieve sustainable water security and provides a holistic approach in water management at the abasi level, incorporating both water and land resources. Moreover, it optimizes water supply by assessing groundwater and surface water, incorporating reuse of wastewater, and also assessing environmental impacts (Tariq *et al.*, 2021);
iv. Forest and Nature Conservation Rules (2006);
v. Biodiversity Action Plan (2009);
vi. Second National Communication to the UNFCCC (2011); and
vii. National Strategies and Action Plan for Low Carbon Development (2012).

Legal provisions of Bhutan include:

i. Forest and Nature Conservation Act (1995), which provides for the protection and sustainable use of forests, wildlife and related natural resources of Bhutan for the benefit of present and future generations;
ii. Mines and Minerals Management Act (1995), which regulates the exploitation of minerals in Bhutan and provides a legal framework for regulating the mining industry;
iii. Pesticides Act of Bhutan (2000), which regulates the safe use and handling of chemicals to prevent public health and environmental hazards;
iv. Environmental Assessment Act (2000), establishes procedures for the assessment of potential effects of strategic plans, policies, programmes and projects on the environment, and for the determination of policies and measures to reduce potential adverse effects and promote environmental benefits;
v. Water Act (2011), which aims to ensure that the water resources of Bhutan are protected, conserved and managed in an economically efficient and environmentally sustainable manner and to establish suitable institutions for this task;
vi. Electricity Act (2001), which enables the restructuring of the power supply industry and the possible participation of the private sector, by providing mechanisms for licensing and regulating the operations of power companies;
vii. Regulation on Environmental Clearance of Projects (2002); and
viii. Biodiversity Act (2003).

17.4.4 China

China entered into the Ramsar Convention in 1992. China has declared 82 wetlands as Ramsar sites covering 7,647,895 ha, making it the largest in the region and the fourth largest in the world (Ye *et al.*, 2022). After becoming party to the convention, China has committed to the conservation and management of the wetlands by implementing policies and legal provisions (Sun *et al.*, 2014).

Since its accession to the Ramsar Convention, China has made continuous efforts to strengthen the protection and management of wetlands. The State Forestry Administration was established for the implementation of wetland protection. China established 316 new wetland/nature reserves at the national, provincial and county levels. After about 30 years, a wetland protection system was established with nature reserves as the mainstay and complementary wetlands parks, protected areas, forest parks and water source reserves.

The Chinese government has given great importance to wetland protection and has launched several relevant policies including:

i. Action Plan for China's Wetland Protection (2000);
ii. National Catalogue of Wetlands (2003);
iii. Circular on Strengthening the Management of Wetland Protection (2004);
iv. Circular on Developing Wetland Parks (2005);
v. Guidelines for Promoting Ecological Civilization (2013);
vi. Nation Wetland Conservation Action Plan (2013);
vii. National Forestry and Grassland Administration (2017), which implemented the National Wetland Protection and Restoration Scheme and revised the regulations on wetland protection and management;
viii. Wetland Protection Law of the People's Republic of China (2021), the first special law that provided a legal basis and paved the way for China's wetland protection; and
ix. National Wetland Protection Project Plan (2002–2030). In 2003, the approval by the State Council of the National Wetlands Project Plan (2002–2030) marked the beginning of a new historical stage of China's wetland protection. Since then, wetland protection was included in five-year plans (12th Five-Year Plan, 2012; 13th Five-Year Plan, 2016; 14th Five-Year Plan, 2021) and national economic and social development (Ye *et al.*, 2022).

Legal provisions of People's Republic of China include:

i. Agricultural Law of the People's Republic of China (1993) enacts principles for the development, management and preservation of agriculture;
ii. Nature Reserve Regulation of the People's Republic of China (1994);
iii. Marine Environmental Protection Law of the People's Republic of China (2000) to protect marine environment and resources, prevent pollution, maintain ecological balance, safeguard human health and promote the development of marine programmes;
iv. Provisional Stipulations of the Mandatory Content in Urban Planning of the People's Republic of China (2002);
v. Breeding and Protection Regulation of Aquatic Culture Resources of the People's Republic of China (1979);
vi. Water Pollution Prevention Law of the People's Republic of China (1984) for protecting and improving the environment, preventing and controlling water pollution, protecting water ecology, guaranteeing the safety of drinking water, protecting the health of the public, promoting the construction of ecological civilization, and promoting the sustainable economic and social development;
vii. Land Management Law of the People's Republic of China (1986);
viii. Water Law of the People's Republic of China (1988) for rationally developing, utilizing and protecting water resources, preventing and controlling water disasters, fully exploiting the derivation of comprehensive benefits from water resources so as to satisfy the needs in national economic development and in the livelihoods of the people;
ix. Implementation Rules of the Soil and Water Conservation Law of the People's Republic of China (1993);
x. Flood Prevention Law of the People's Republic of China (1997);
xi. Management Methods of the Inshore Marine Environmental Function Zone;
xii. Regulation on Wetland Protection of the People's Republic of China (2005); and
xiii. Wetland Conservation Law of the People's Republic of China (2020) to strengthen wetland protection, maintain wetland ecological functions and biodiversity, ensure ecological security, promote the construction of ecological civilization, and achieve harmonious coexistence between man and nature.

17.4.5 India

India committed to the Ramsar Convention in 1982 enlisting Chilka lake and Keoladeo Ghana

National Park as the first wetlands of India as Ramsar sites in October 1981. At present, India has 75 Ramsar sites covering 1,326,677 ha. The Ministry of Environment, Forest and Climate Change (MoEFCC) is the central authority for wetland resource management. However, MoEFCC, via the Wetlands Conservation and Management Rules 2017, has abdicated its enforcement authority to the states for delineating wetlands and ensuring activities that meet Ramsar's principle of 'wise use'. Recognizing the cultural importance of high-altitude wetlands, Sikkim (state) government executed its own legal provision for the conservation of wetlands in its jurisdiction. For example, the Places of Worship Act 1992 of Sikkim (India) prohibits the conversion of any place of worship and provides for the maintenance of the religious character of the sacred wetlands. There are several laws or legal provisions in India which directly or indirectly help protect and conserve high-altitude wetlands.

The national level policies of relevance include:

i. Coastal Zone Regulation Notification (1991);
ii. National Conservation Strategy and Policy Statement on Environment and Development (1992);
iii. National Lake Conservation Plan (1993);
iv. National River Conservation Plan (NRCP) (1995);
v. National Policy and Macro-level Action Strategy on Biodiversity (1999);
vi. National Water Policy (2002);
vii. National Environment Policy (NEP) (2006) generates a legally enforceable regulatory mechanism for identified wetlands to prevent their degradation, enhance their conservation and wise use by all the stakeholders;
viii. National Action Plan on Climate Change (2008) recognizes the important role of wetlands in the context of climate change; and
ix. Wetland Conservation and Management Rules (2017).

Laws formulated for the conservation of wetlands include:

i. Indian Forest Act (1927), which consolidates the law relating to forests, the transit of forest produce and the duty leviable on timber and other forest produce;
ii. Wildlife (Protection) Act (1972), provides a legal framework for the protection of wild animals and plants, management of their habitats, regulation, and control of trade in wild animals, plants and products made from them;
iii. Water (Prevention & Control of Pollution) Act (1974) provides for the prevention and control of water pollution and for the maintaining or restoring of wholesomeness of water;
iv. Forest (Conservation) Act (1980) to prevent large-scale deforestation;
v. Environmental (Protection) Act (1986) authorizes the central government to protect and improve environmental quality, control and reduce pollution from all sources, and prohibit or restrict the setting and/or operation of any industrial facility on environmental grounds;
vi. Wildlife (Protection) Amendment Act (2022);
vii. Biodiversity Act (2002) for conservation of biological diversity, sustainable use of its components and fair and equitable sharing of the benefits arising out of the use of biological resources, knowledge and for matters connected therewith or incidental thereto;
viii. Scheduled Tribes and Other Traditional Forest Dwellers (recognition of Forest Rights) Act (2006).

17.4.6 Myanmar

Myanmar entered the Ramsar Convention on wetlands in 2005 with Moeyungyi Wetland Wildlife Sanctuary as the first Ramsar site. Currently, it has six Ramsar sites covering 278,679 ha. Since its entry to the convention, Myanmar has been continuously working on the protection and management of wetlands. Recently, Myanmar has endorsed National Wetland Policy 2019.

In Myanmar, wetlands are important natural resources, having a vital role in both ecology and economy. The major wetlands of Myanmar include mangroves, swamp forests, lakes and marshes. The coastal vegetations (forests) provide services such as fishing, forest products and medicinal plants. In addition, they act as a natural barrier against the seepage of seawater into agricultural areas and protect the coastal land from erosion. Mangroves are the main wetlands in the coastal area. They are extremely productive and are important for a large number of aquatic fauna, being the home to many native and migratory waterfowl. Genera such as *Rhizophora*, *Bruguiera*, *Sonneratia* and *Avicennia* are unique to the region and are provided with effective protection. A network of freshwater wetlands, rivers and adjacent wetlands is important for water supply, transport and habitats for freshwater fish. The natural lakes in Myanmar are mainly associated with river systems and provide breeding sites for water birds. These wetlands are important for the fishery industry and form the basis for national and local fisheries. Freshwater fish from the inland wetlands are the major protein sources in Myanmar. Despite these functions, the natural integrity and value of many inland wetlands (lakes) are degraded due to human activities, reducing their natural value. This suggests that the lakes in Myanmar need to be preserved to maintain the wetland integrity, biological diversity and sustainability of eco-economic services. Yet, in many areas, the conservation values of the wetlands have not been adequately recognized in land-use planning, resulting in the loss and degradation of a large proportion of wetlands and mangroves. The Forest Department of Myanmar is collaborating with relevant departments for the conservation of wetlands and coastal areas to provide habitats for migratory birds.

The country established the National Wetland Committee in 2016 and has the following relevant policies for the conservation and sustainable utilization of wetlands:

i. National Water Policy (2015), focuses on two aspects: water resource management and water resource use. Access to water of equitable quantity and quality for all social, environmental and economic needs of the present and future generations is the priority of this policy;
ii. National Wetland Policy (2019), includes six wetland policies and 17 objectives. It focuses on the conservation, management and sustainable use of wetlands; and
iii. Environmental Conservation Law (2012), addresses environmental protection, including water resources. It emphasizes the sustainable management and conservation of natural resources, which include water bodies.

17.4.7 Nepal

Nepal entered the Ramsar Convention in 1988 declaring the Koshi Tappu as the first Ramsar site and expressing its commitment to the conservation and sustainable utilization of wetlands. Currently, it has ten wetlands of international importance covering 60,561 ha. The wetlands are distributed across the ecological regions of the country, i.e. high-altitude, mid-hills and Terai (Shrestha *et al.*, 2020). In Nepal, wetland conservation and management laws include National Conservation Strategy (1988) and Wetland Conservation and Management Policy (2016). The government of Nepal implemented National Wetland Policy (2003) with the objective of involving the local people for the management of wetlands (Poudel, 2009). A memorandum of understanding was signed between the National Lake Conservation Development Committee and World Wildlife Fund (WWF) Nepal to share expertise, knowledge and resources for the conservation and sustainable management of lakes and wetlands of Nepal. Nepal has formulated and executed several relevant policies/legal provisions for the protection and conservation of wetlands, which include:
i. National Park and Wildlife Conservation Regulation (1974);
ii. Chitwan National Park Regulations (1974);
iii. Himalayan National Parks Regulations (1979);
iv. National Conservation Strategy (1988 or 1989) for sustainable use of wetlands;
v. The Master Plan for the Forestry Sector (1989) for people's participation in natural resource management;
vi. Nepal Environmental Policy and Action Plan (NEPAP) (1993, further elaborated in 1998) prioritized the need to identify and protect biologically significant wetlands and water bodies;
vii. Water Resources Regulation (1993);
viii. Forest Regulation (1995);
ix. Bardia National Park Regulations (1995);
x. Buffer Zone Regulations (1996);
xi. Buffer Zone Management Guidelines (1999);
xii. Environmental Protection Rules (1997 updated in 2019);
xiii. Buffer Zone Management Guidelines (1999);
xiv. Irrigation Rules (2000);
xv. Hydropower Development Policy (2001);
xvi. Nepal's Biodiversity Strategy (NBS) (2002) identified wetlands as one of the key ecosystems and stressed the need for institutional arrangements for wise use and conservation;
xvii. Environmental Impact Assessment Review Guidelines (2002);

xviii. Water Resources Strategy Nepal (2002);
xix. National Wetland Policy (2003) outlines the need for a co-ordinated approach to conserve, manage and promote wise use of the national wetlands;
xx. Irrigation Policy (2003);
xxi. Initial Environmental Examination Manual (2004);
xxii. National Water Plan (2005);
xxiii. Koshi River Basin Management Strategic Plan (2011–2021);
xxiv. National Climate Change Policy (2019); and
xxv. National Water Resources Policy (2020).

The legal frameworks for wetland conservation include:

i. The National Parks and Wildlife Conservation Act (1973) outlined the establishment of the protected areas and wildlife protection;
ii. Plant Protection Act (1977);
iii. Tourism Act (1977);
iv. Soil Conservation and Watershed Management Act (1982) and its Regulation;
v. National Trust for Nature Conservation Act (1982) for the establishment of the National Trust for Nature Conservation to conserve and manage nature and natural resources;
vi. Pesticide Act (1991) regulates the import, export, production, purchase, sale and use of pesticides used to destroy pests of seeds, plants, trees, animals and birds;
vii. The Forest Act (1993) provided the basis for collaborative wetland management inside the national forest areas, particularly under the community forest;
viii. Water Resources Act (1992) forms the base of the legal framework for equitable use, protection, development and management of surface or groundwater resources; for identifying beneficial uses of water resources and to control pollution and adverse impacts that may arise by such uses, and for keeping water resources free of pollution.
ix. The National Environment Protection Act (1996 updated in 2019) provisioned for an environmental impact assessment (EIA), which is important for wetlands conservation.

17.4.8 Pakistan

Pakistan entered into the Ramsar Convention in 1976. Currently, it has 19 Ramsar sites covering 1,343,627 ha. For the wise use of wetlands, Pakistan implemented policies like National Wetland Policy (2019) and National Climate Change Policy (2012). Water management falls under the Water and Power Development Authority (WAPDA).

The policy frameworks of Pakistan include:

i. Pakistan National Wetlands Policy (2009);
ii. National Climate Change Policy (2012);
iii. Biodiversity Action Plan (2015);
iv. National Forest Policy (2015);
v. National Food Security Policy (2017); and
vi. National Water Policy of Pakistan (2018).

The legal frameworks of Pakistan include:

i. Water and Power Development Authority Act (1958) to provide for the unified and co-ordinated development of the water and power resources of Pakistan;
ii. Indus River System Authority (IRSA) Act (1992), established for regulating and monitoring the distribution of water resources of the Indus River system among the provinces, in accordance with provisions of the water accord; and
iii. Environmental Protection Act (1997), which principally makes provisions for the administration of matters affecting the environment, environmental impact assessment, and the handling of hazardous matters.

17.4.9 Regional policies, institutions and agreements

To achieve the goal of wetlands conservation in the region, the HKH countries have to consider the cross-border mechanisms and networks related to social, ecological and economic issues linked to biodiversity use and conservation (Kotru *et al.*, 2020). To institutionalize the concept of long-term transboundary co-operation to enhance ecological integrity and socio-cultural resilience in the region, as well as to help countries in the HKH link their national commitment to global conservation goals, the ICIMOD partnered with the eight member counties to discuss the concept of trans-Himalayan transect and embedded landscapes (Kotru *et al.*, 2020) and provide forums for knowledge exchange, agreements and management decisions (Kotru *et al.*, 2020).

According to the IPCC fourth Assessment Report, HKH is defined as a data-deficit region and, hence, ICIMOD is actively working to generate interest for transboundary collaboration for research, monitoring and implementation of action-research projects (Kotru *et al.*, 2020). Long-term environmental and socio-ecological monitoring (LTESM) is a concept developed and simultaneously used in transboundary landscape programmes for planning and collecting data from a range of ecosystems linked at the landscape scale (agriculture, forest, rangeland and wetland). The HKH Regional Cooperation Framework (RCF) highlights biodiversity and environmental and cultural conservation through scientific and technical co-operation (Kotru *et al.*, 2020). Transboundary co-operation is a concept for the future – one that bridges nations for peace and stability based on balancing conservation and development with common understanding and acceptance.

17.5 Livelihoods and Ecological Restoration of Wetlands

The Himalayan range alone has 35,110 km^2 of snow and ice cover containing 3,735 km^3 of eternal snow and ice. The HKH region is the origin of ten major river basins along with being home to four global biodiversity hotspots and 330 important bird areas. Hills and mountains, particularly the Hindu Kush Himalaya mountain system, have always constituted places where adaptation, mitigation and resilience are hallmarks of the people and the landscape they inhabit. The people of the Himalaya have always maintained a rich cultural identity and have maintained food security and biogenetic diversity within the parameters of their own traditions. Diverse cultures, languages, religions and traditional knowledge systems thrive in the HKH region (Wester *et al.*, 2019).

The south-Asian countries, including Afghanistan, Bangladesh, Bhutan, India, Maldives, Nepal, Pakistan and Sri Lanka, are home to more than 40% of the world's poor. The region includes only about 3% of the world's land area but houses a quarter of the world's population (1.6 billion people). Rice and wheat are the staple foods in this region and they require huge amounts of water and energy. Overall, approximately 3 billion people benefit from food and energy produced in these river basins. Despite having the world's largest water resources, many inhabitants are the poorest of the poor, and many are marginalized subsistence farmers of diverse ethnic groups and minorities not benefitting from these vast resources. Water resources in the HKH could be used for economic prosperity in the region, maximizing the benefits of this renewable resource. Therefore, better water management should be the major goal in poverty reduction and for improving food security. This section discusses water resource use for poverty reduction, and pertinent institutional arrangements and policy provision amidst the sustainable development agenda, underlining the need for international and regional co-operation.

Livelihood and ecological restoration of wetlands are important because wetlands provide valuable ecosystem services, such as flood control, water purification and habitats for wildlife. They also support the livelihoods of millions of people who depend on wetlands for fishing, agriculture and tourism. Restoring degraded wetlands can mitigate the impacts of climate change, protect biodiversity and enhance resilience of the communities. Aquatic resource restoration projects are complex and require a broad range of principles and practices to ensure their success. In Nepal, despite having policies and international treaties related to biodiversity conservation, wetlands are not defined as a separate category of ecologically important areas in the National Parks and Wildlife Conservation Act (1973). However, wetlands play a crucial role in maintaining water quality and supporting wildlife habitats. Although the policy documents lack specific provisions related to wetlands, they are regarded as additional values for wetland conservation, including the protection of aquatic animals, catchment areas, national forests and the environment. These Acts prohibit activities such as introducing harmful substances, blocking or diverting water sources, clearing forests, quarrying, establishing industries and generating pollution (Poudel, 2009).

In Nepal, the regional policies, institutions and agreements for the livelihoods and ecological restoration of wetlands include the National Wetland Policy (2012), the Wetland Conservation and Management Regulation (2017), National

Ramsar Strategy and Action Plan, Nepal (2018–2024), the Department of National Parks and Wildlife Conservation, the Department of Forests and Soil Conservation, and the Ramsar Convention on Wetlands of International Importance.

The National Ramsar Strategy and Action Plan for Nepal for 2018–2024 ensures that Nepal's wetlands are conserved effectively, used wisely and their benefits are recognized and valued nationally and globally. There are five objectives, including effectively conserving and managing the Ramsar site network; managing wetlands, including the Ramsar sites according to the principles of wise use; engaging federal, state and local stakeholders; enhancing Ramsar implementation through national and international co-operation; and monitoring and evaluating the implementation of the National Ramsar Strategy and Action Plan. There are also several key actions identified to achieve each objective, such as an inventory of wetlands, updating the RIS (Ramsar Information Sheet) of Ramsar sites, developing guidelines for wise use of wetlands, and securing financial resources for implementing the plan. The plan also includes strategies for demonstrating wise use of wetlands, recognizing wetland functions, services and benefits, and mitigating threats to wetland sustainability.

17.6 Management Strategy, Adaptations and Pathways in Wetland Conservation in the HKH Region

The HKH faces several critical regional issues related to climate change, degradation of natural resources, flood-related transboundary humanitarian impacts, social changes, out-migration and globalization, all of which provide an opportunity for the HKH countries to co-operate to address the challenges (Kotru *et al.*, 2020). Governments in the region are being strongly advocated to combine and accelerate efforts to advance sustainable mountain development, with a focus on global conservation and development, such as the UN Sustainable Development Goals 2030.

China and India are emerging as world leaders in scientific and technological research and can co-ordinate future collaborations among HKH countries to counter the growing impacts of climate change and other environmental challenges (Sharma, 2017). Regional institutions such as ICIMOD, working to generate scientific evidence and future scenarios, have a crucial role for the betterment of the HKH region. All the HKH countries are committed to global targets such as the UN Sustainable Development Goals, the Paris Agreement within the UNFCCC, the Sendai Framework for Disaster Risk Reduction (SFDRR), and the Aichi Biodiversity Targets of the Convention on Biological Diversity (CBD). A shared framework and understanding on natural resources governance at the regional scale, supported by appropriate institutions, can be fruitful in achieving the broader goal of regional co-operation among the HKH countries (Kotru *et al.*, 2020). Long-term piloting and monitoring of innovative approaches, and meeting regional capacity building, should be the aim of transboundary co-operation. The Transboundary Landscapes Programme designed by ICIMOD rests on the following fundamental pillars:

i. promoting integrated and participatory approaches to ecosystem management in mountain development throughout Regional Member Countries (RMCs), starting from the premise that ecosystem services flow beyond administrative boundaries and thus require regional and multilateral co-operation in transboundary landscapes;
ii. adopting approaches that integrate multi-stakeholder priorities with scientific data in such a way that trade-offs between conservation and development can be assessed and addressed;
iii. promoting regional acceptance of common frameworks, long-term research and monitoring, common methodologies, information sharing, and collaborative management for the purpose of converting science into conservation and development policies and practices;
iv. contributing scientific input to global policies such as the Ramsar Convention, the CBD, the Intergovernmental Science-Policy Platform on Biodiversity and Ecosystem Services (IPBES) and the UNFCCC-Global Landscapes Forum; and
v. demonstrating improved transboundary co-operation among member countries through regional policies and strategic partnerships, leading to sustainable mountain ecosystem services and equitable livelihood benefits at regional landscape levels.

17.7 Strategies at the National Level on Ecological Compensation

The HKH region is rich in natural and cultural heritage. Its diverse ecosystems provide multiple goods and services benefitting billions of people in the HKH mountains and downstream. However, global environmental changes and growing economies, like rapid urbanization and industrial development, are detrimental to ecosystem health and services, and, ultimately affect the livelihoods and the standard of living of the local mountain communities. The HKH region is experiencing rapid biodiversity loss and habitat degradation. The mega-investments, such as the Belt and Road Initiative, hydropower development, and regional waterways for economic development and food security present new challenges for conserving the biodiversity resources in the region (Kotru *et al.*, 2020). Thus, in the HKH region, sustaining ecosystem services in the future is a key challenge.

China has also established and implemented a wetland ecological compensation system, which has greatly enhanced the protection and management of wetlands. Excessive utilization of wetland resources leads to the loss of wetland services while minimum utilization hinders economic development. Therefore, before the development and utilization of wetland resources, a scientific assessment of each wetland resource is essential to determine a reasonable development and utilization scale (Ye *et al.*, 2022).

17.8 Conclusions

Wetlands in the HKH region serve a dual purpose: ecological benefit and economic benefit. The largest rivers in south-east Asia, which provide water, ecosystem services and livelihoods, are fed by the HKH region. Despite having high eco-economic value, the high-altitude wetlands are degraded due to climatic and anthropogenic factors, which are threatening regional ecological security and sustainable socio-economic development. Considering the eco-economic importance of the wetlands, the HKH member countries are committed to wetlands conservation and management. Except for Afghanistan, all the member countries have become parties to the Ramsar Convention, and thus far have declared 197 Ramsar sites in the HKH region. To protect and conserve wetlands in HKH region, all member countries have formulated policies and legal frameworks. Conservation of wetlands by individual HKH member countries would be inadequate. Thus, to achieve the goal of conservation of wetlands, the HKH member countries should establish a cross-border mechanism and network related to eco-economic issues linked to natural resources and their conservation. For the proper management and conservation of wetlands in the KHK region, a transboundary landscape approach should be prioritized and countries in HKH region should work together. China and India, which are emerging as leaders in scientific and technological advancement, could take the lead in countering the impacts of climate change and environmental challenges. By having a common understanding, shared responsibilities, and transboundary co-operation on natural resource governance, the broader goal of regional co-operation can be achieved. For effective wetland management in the HKH region, the following key points should be considered:

i. Strengthen existing policies and institutional frameworks for wetland management and conservation, enhancing co-ordination between government agencies and stakeholders, and relevant institutions.
ii. Wetland management should be integrated and developed with other sectors such as agriculture, forestry and water management.
iii. Implement community-based conservation approaches.
iv. Support sustainable livelihoods such as ecotourism, sustainable fisheries and other income-generating activities, which can reduce the pressure on wetland resources.

References

Acharya, C. and Dokham, A.S.G. (1998) Sikkim: the hidden holy land and its sacred lakes. *Bull Tibetol* 3, 10e15.

Ali, M.Y. (1990) Open water fisheries and environmental changes. In: Rahman, A.A., Huq, S. and Conway, G.R. (eds) *Environmental Aspects of Surface Water Systems of Bangladesh*. University Press Ltd, Dhaka.

Chaudhry, A.A. (2010) *Wetlands in Pakistan: What Is Happening to Them*. World Environment Day, 5.
Chowdhury, S. (1998) Saving our oceans: some thoughts on fisheries conservation. *The Daily Star* 2(100), Dhaka.
Costanza, R., d'Arge, R., De Groot, R., Farber, S., Grasso, M. *et al.* (1997) The value of the world's ecosystem services and natural capital. *Nature* 387(6630), 253–260. DOI: 10.1016/S0921-8009(98)00020-2
CSUWN (2010) *Simsar Varnamala: Conservation and Sustainable Use of Wetland in Nepal*. Government of Nepal, GEF, UNDP.
DHM (2017) Observed climate trend analysis in the districts and physiographic regions of Nepal (1971–2014). Department of Hydrology and Meteorology, Government of Nepal, Kathmandu.
Dyurgerov, M.D. and Meier, M.F. (2005) *Glaciers and Changing Earth System: A 2004 Snapshot*. Institute of Arctic and Alpine Research, University of Colorado, Boulder, Colorado.
FAO (2015) *Mapping the Vulnerability of Mountain Peoples to Food Insecurity*. Food and Agriculture Organization of the United Nations, Rome.
Fleury, J.M. (1999) Mountain biodiversity at risk. *IDRC Briefing* 2, 1–6.
Gawel, A. and Ahsan, I. (2014) *Review and Compendium of Environmental Policies and Laws in Bhutan: Input to the Asian Judges Network on Environment.* Asian Development Bank.
GOB (1999) National Water Policy. Government of Bhutan Ministry of Water Resources, Dhaka.
Gopal, B. and Wetzel, R.G. (1995) Limnology in developing countries, V.I. New Delhi. *International Association for Limnology* 1–230.
Hawkes, L.A., Balachandran, S., Batbayar, N., Butler, P.J. and Frappell, P.B. (2011) The trans-Himalayan flights of bar-headed geese (*Anser indicus*). *Proceedings of the National Academy of Sciences* 108(23), 9516–9519. DOI: 10.1073/pnas.1017295108
Hu, S., Niu, Z., Chen, Y., Li, L. and Zhang, H. (2017) Global wetlands: potential distribution, wetland loss, and status. *Science of the Total Environment* 586, 319–327. DOI: 10.1016/j.scitotenv.2017.02.001
ICIMOD (International Centre for Integrated Mountain Development) (2010) *Functions and Services of Wetlands in the Eastern Himalayas: Impacts of Climate Change.* Hill Side Press, Kathmandu.
ICIMOD (2023) Ecosystems in the Hindu Kush Himalaya. Available at: https://hkh.icimod.org/hi-wise/ecosystems/ (accessed 20 September 2023).
Immerzeel, W.W., Van Beek, L.P. and Bierkens, M.F.P. (2010) Climate change will affect the Asian water towers. *Science* 328 (5984), 1382–1385. DOI: 10.1126/science.1183188
Islam, Z.A. and Rahmani, A.R. (2008) *Potential and Existing Ramsar Sites in India.* Indian Bird Conservation Network: Bombay Natural History Society, Birdlife International and Royal Society for the Protection of Birds. Oxford University Press.
Joshi, R. and Dhyani, P.P. (2009) Environmental sustainability and tourism – implications of trend synergies of tourism in Sikkim Himalaya. *Current Science* 97(1), 33–41.
Kapos, V., Rhind, J., Edwards, M., Price, M.F. and Ravilious, C. (2000) Developing a map of the world's mountain forests. In: Price, M.F. and Butt, N. (eds) *Forests in Sustainable Mountain Development: A State-of-Knowledge Report for 2000.* CAB International, Wallingford, UK, pp. 4–9.
Karki, G. (2015) *Climate Change Vulnerability Assessment and Potential Adaptation Measures for Local Communities of the Chure Project Area.* WWF Nepal/Sustainable Land Management Project.
Khan, A.A. (1993) Freshwater wetlands in Bangladesh: opportunities and options. In: Nishat, A., Hussain, Z., Roy, M.K. and Karim, A. (eds) *Freshwater Wetlands in Bangladesh – Issues and Approaches for Management.* IUCN, The World Conservation Union, Dhaka
Khan, A.A. and Ashrad, S. (2014) Wetlands of Pakistan: distribution, degradation and management. *Pakistan Geographical Review* 69(1), 28–45.
Khan, H. and Baig, S. (2017) High altitude wetlands of the HKH Region of Northern Pakistan – status of current knowledge, challenges and research opportunities. *Wetlands* 37, 371–380. DOI: 10.1007/s13157-016-0868-yt
Kotru, R., Pradhan, N., Shakya, B. and Amatya, S. (2020) *Beyond Boundaries: Contouring Transboundary Landscapes in the Hindu Kush Himalaya.* ICIMOD, Kathmandu.
Malikyar, G.M. (2017) *The Environment of Afghanistan (2010–2017): Pressures, Progress, Challenges/Gaps.* National Environment Protection Agency of Afghanistan, Kabul.
Mi, D. and Xie, Z. (2002) *Glacier Inventory of China*. Xi'an Cartographic Publishing House.
Millennium Ecosystem Assessment (2005) *Ecosystems and Human Wellbeing: Wetlands and Water Synthesis.* World Resources Institute, Washington, DC.
Mitsch, W.J., Bernal, B., Nahlik, A.M., Mander, Ü., Zhang, L. *et al.* (2013) Wetlands, carbon, and climate change. *Landscape Ecology* 28(4), 583–597. DOI: 10.1007/s10980-012-9758-8

Molden, D. and Sharma, E. (2013) ICIMOD's strategy for delivering high-quality research and achieving impact for sustainable mountain development. *Mountain Research and Development* 33(2), 179–183. DOI: 10.1659/MRD-JOURNAL-D-13-00018.1

Murray, A.B. (2009) *A Manual for an Inventory of Greater Himalayan Wetlands*. International Centre for Integrated Mountain Development, Kathmandu.

National Environmental Protection Agency of the Islamic Republic of Afghanistan (2017) National Biodiversity Strategy & Action Plan. Available at: https://www.cbd.int/doc/world/af/af-nbsap-01-en.pdf (accessed 20 September 2023).

Poudel, B.S. (2009) Wetland conservation in Nepal: policies, practices, problems and possibilities. *Banko Janakari* 19(3), 5–9. DOI: 10.3126/banko.v19i3.2205

Prasad, S.N., Ramachandra, T.V., Ahalya, N., Sengupta, T., Kumar, A. *et al.* (2002) Conservation of wetlands of India: a review. *Tropical Ecology* 43(1), 173–186.

Ragettli, S., Immerzeel, W.W. and Pellicciotti, F. (2016) Contrasting climate impact on river flows from high altitude catchments in the Himalayan and Andes Mountains. *The Proceedings of the National Academy of Sciences* 113(33), 9222–9227. DOI: 10.1073/pnas.1606526113

Rosenblum, Z.H. and Schmeier, S. (2022) Global wetland governance: introducing the transboundary wetlands database. *Water* 14(19), 3077. DOI: 10.3390/w14193077

SAC (Space Applications Centre) (2011) *National Wetland Atlas*. Indian Space Research Organisation, Ahmedabad.

Schild, A. and Sharma, E. (2011) Sustainable mountain development revisited. *Mountain Research and Development* 31(3), 237–241. DOI: 10.1659/MRD-JOURNAL-D-11-00069.1

Sharma, B., Rasul, G. and Chettri, N. (2015) The economic value of wetland ecosystem services: evidence from the Koshi Tappu Wildlife Reserve, Nepal. *Ecosystem Services* 12, 84–93. DOI: 10.1016/j.ecoser.2015.02.007

Sharma, E. (2017) China–India collaborative opportunities on science and technology for sustainable development of the Hindu Kush Himalaya. *Science Bulletin* 62(10), 673–674. DOI: 10.1016/j.scib.2017.03.007

Shrestha, A.B., Wake, C.P., Mayewski, P.A. and Dibb, J.E. (1999) Maximum temperature trends in the Himalaya and its vicinity: an analysis based on temperature records from Nepal for the period 1971–94. *Journal of Climate* 12(9), 2775–2786. DOI: 10.1175/1520-0442(1999)012<2775:MTTITH>2.0.CO;2

Shrestha, A.B., Agrawal, N.K., Alfthan, B., Bajracharya, S.R., Maréchal, J. *et al.* (2015) *The Himalayan Climate and Water Atlas: Impact of Climate Change on Water Resources in Five of Asia's Major River Basins.* ICIMOD, GRID-Arendal and CICERO, Kathmandu.

Shrestha, B., Shrestha, S., Shrestha, A. and Khadka, U.R. (2020) Ramsar sites in Nepal: conservation, present scenario, biodiversity value and threats. *Journal of Wetlands Ecology* 1–15. DOI: 10.3126/jowe.v2020i0.24782

Sivakumar, M.V.K. and Stefanski, R. (2011) Climate change in South Asia. In: Lal, R. (ed.) *Climate Change and Food Security in South Asia.* Springer, Dordrecht, The Netherlands, pp. 13–30.

Song, K., Wang, Z., Du, J., Liu, L., Zeng, L. *et al.* (2014) Wetland degradation: its driving forces and environmental impacts in the Sanjiang Plain, China. *Environmental Management* 54(2), 255–271. DOI: 10.1007/s00267-014-0278-y

Sun, Z., Zhang, L., Sun, W., Jiang, H., Mou, X. *et al.* (2014) China's wetlands conservation: achievements in the eleventh 5-year plan (2006–2010) and challenges in the twelfth 5-year plan (2011–2015). *Environmental Engineering & Management Journal* 13(2), 379–394. DOI: 10.30638/eemj.2014.043

Tariq, M.A.U.R., Wangchuk, K. and Muttil, N.A. (2021) A critical review of water resources and their management in Bhutan. *Hydrology* 8(1), 31. DOI: 10.3390/hydrology8010031

Wester, P., Mishra, A., Mukherji, A. and Shrestha, A.B. (2019) *The Hindu Kush Himalaya Assessment – Mountains, Climate, Sustainability and People*. Springer Nature, Cham, Switzerland. ISBN 978-3-319-92288-1 (eBook).

Ye, S., Pei, L., He, L., Xie, L., Zhao, G. *et al.* (2022) Wetlands in China: evolution, carbon sequestrations and services, threats, and preservation/restoration. *Water* 14(7), 1152. DOI: 10.3390/w14071152

18 Glaciers and Permafrost: Function and Conservation

Tianyun Qi[1,*], Allan Degen[2] and Zhanhuan Shang[1]
[1]*College of Ecology, Lanzhou University, Lanzhou, China;* [2]*Institutes for Desert Research, Ben Gurion University of the Negev, Israel*

Abstract
This chapter presents an overview of glaciers and permafrost in the Hindu Kush Himalayan (HKH) region. The importance of glaciers and permafrost is described, along with threats to these natural resources. There is an urgent need for action to preserve glaciers and permafrost, and case studies on ecological restoration and local livelihoods related to glacial permafrost are presented. Furthermore, strategies for protecting and managing glaciers and permafrost are proposed. Finally, there is a discussion on the adaptation and transformation of methods adopted to mitigate the impact of climate change and how to protect these essential natural resources for future generations. Overall, the emphasis is on the urgent need to study and protect glaciers and permafrost in the HKH area for the well-being of the millions of residents and the sustainability of the environment.

Keywords: Hindu Kush Himalayan (HKH) region, ecosystem, climate change, community livelihoods, protection and management strategy

18.1 Introduction

Glaciers and permafrost serve as vital components in the ecosystem of the Hindu Kush Himalaya (HKH) region, providing fresh water to millions of people and supporting various ecosystems. Yet, climate change continues to cause substantial damage to these natural resources, leading to their rapid melting and posing serious threats to the livelihoods of local communities. To better understand the threats to this ecosystem, we reviewed the research on glaciers and permafrost in the HKH region from different perspectives, including functions, challenges and potential consequences. Furthermore, we systematically examined protective and managerial strategies, adaptation measures, and case studies for restoring and preserving glacier and permafrost ecosystems in the HKH region, and thus improving the quality of life for local residents. This chapter provides valuable information pertinent to policy-making decisions in promoting the preservation of glaciers and permafrost in the HKH region for future generations.

18.2 Glaciers and Permafrost

In recent years, there has been a growing awareness of the crucial role that glaciers and permafrost

*Corresponding author: qity21@lzu.edu.cn

DOI: 10.1079/9781800622579.0018

play in climate regulation and change (Mackelprang *et al.*, 2016; Margesin and Collins, 2019; Karjalainen *et al.*, 2020), and their contribution to the livelihoods of highland communities (Chen *et al.*, 2019; Li *et al.*, 2023). Furthermore, glaciers and permafrost serve as critical indicators of climate change (Kääb *et al.*, 2007).

18.2.1 Glaciers

The term 'glacier' originates from the Latin word *glaciēs* which means ice. Glacier formation, growth and movement are referred to as glacial processes, while the processes and characteristics related to glaciers are collectively termed 'glaciation' (Ding *et al.*, 2021).

Glaciers cover approximately 3% of the Earth's land surface. Approximately 99% of the glaciers exist within enormous ice sheets called continental glaciers, located in polar regions, while the remaining 1% is scattered throughout mountain ranges on various continents (Benn and Evans, 2014). Notably, Pakistan has the greatest number of glaciers (Giese *et al.*, 2022).

Glaciers are among the most sensitive indicators of climate change, being affected by long-term changes in precipitation, air temperature and cloud cover. Additionally, they play a critical role in sea-level fluctuations. Glaciers can be categorized based on their morphology, thermal properties and behaviour (Giese *et al.*, 2022). High-mountain glaciers are located on mountain peaks and slopes, valley glaciers fill up mountain valleys, while massive glaciers that extend across mountain ranges, ridges or volcanoes are called ice sheets or ice caps (Giese *et al.*, 2022). The ablation zone refers to the area where glacier mass is lost, whereas the accumulation zone represents the upper portion of the glacier where accumulation exceeds ablation (Giese *et al.*, 2022). Healthy glaciers usually showcase larger accumulation zones and robust termini. The mass and size of the terminus are used to assess the health of glaciers (Fig.18.1; Giese *et al.*, 2022). Over time, and in response to environmental changes, glacier retreat and the loss of mass has become increasingly prevalent. For instance, following the end of the Little Ice Age, around 1850, glaciers across the planet receded substantially (Heckmann and Morche, 2019). A slight cooling trend between 1950 and 1985 advanced many high-mountain glaciers, but, since 1985, the rates of glacier retreat and mass loss have escalated rapidly (O'Connor and Costa, 1993).

18.2.2 Permafrost

Permafrost refers to soil, rocks and sediment that have been continuously frozen for more than two years and is characterized by temperatures below 0°C (Cochand *et al.*, 2019). In areas without ice cover, permafrost lies beneath the soil, rock and sediment layers and forms an 'active layer' (Cochand *et al.*, 2019). Permafrost can occur at depths ranging from several centimetres to several hundred metres and is not necessarily located on the top layer of the ground (Siewert *et al.*, 2021). The distribution of permafrost is divided into four types: continuous, discontinuous, sporadic and isolated patches. Approximately 15% of the northern hemisphere, or 11% of the Earth's surface, is covered with permafrost, covering a total area of approximately 180,000 km^2 (Heginbottom *et al.*, 1993). Melting permafrost causes strong geomorphic, hydrological and ecological changes and is an important factor in landscape evolution (Bouchard *et al.*, 2016). As climate warming causes permafrost to melt, vast amounts of stored organic carbon are released into the atmosphere (Schuur *et al.*, 2015).

18.2.3 Glaciers and permafrost in the Hindu Kush Himalayan region

Glaciers

The HKH region, with its high concentration of snow and glaciers, is often referred to as the 'Third Pole' (Bajracharya and Shrestha, 2011). These glaciers are important freshwater sources for millions of people in the region, and their meltwater performs a critical role in sustaining downstream arid river basins (Bajracharya *et al.*, 2015; Prakash, 2020), while also regulating global climate (Bolch *et al.*, 2019). Glaciers in different regions of the HKH respond differently due to the diverse topography and climates (Schmid *et al.*, 2015). As a result, the climate changes across the region, ranging from cold and dry in the west to warm and humid in the east. The glaciers in the region have undergone unprecedented shrinkage, affecting hundreds of millions of people who depend on them (Zheng

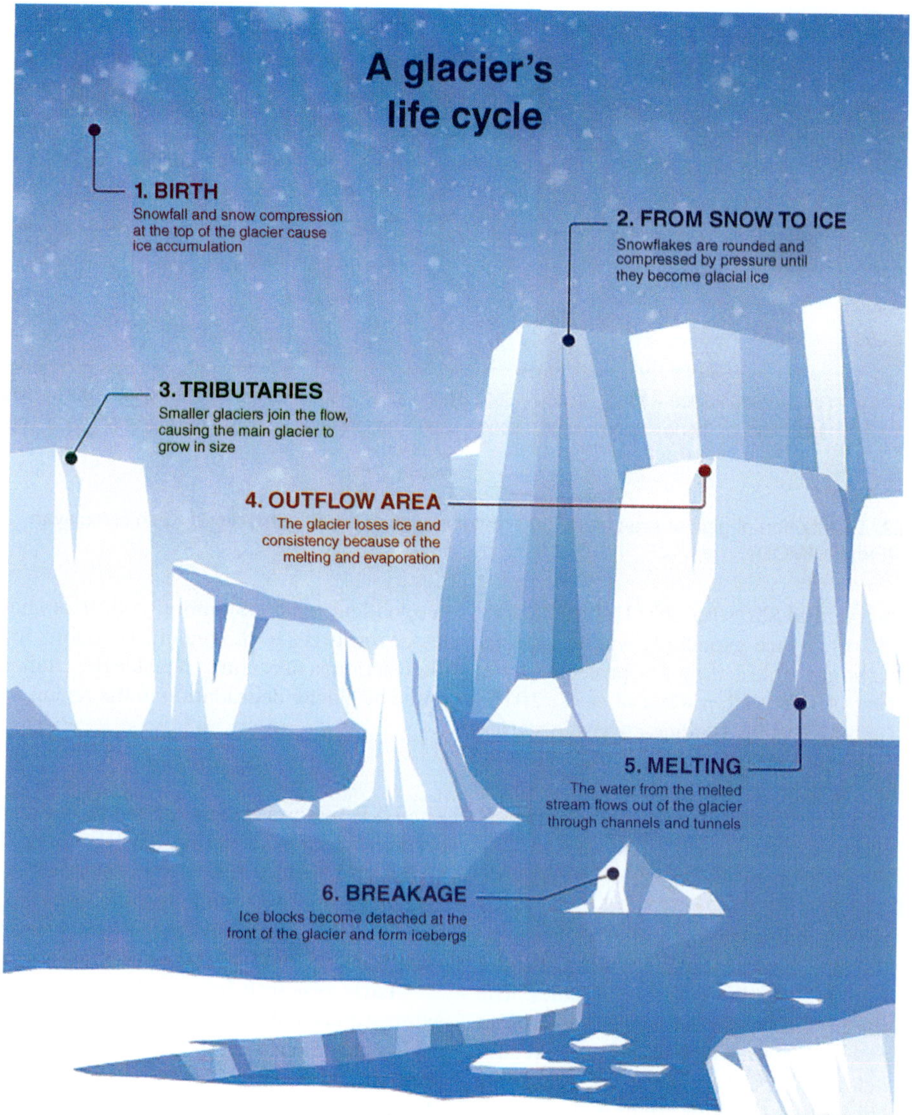

Fig. 18.1. The life cycle of glaciers. (From World Glacier Monitoring Service)

et al., 2021). Furthermore, the formation and expansion of glacier lakes increase the likelihood of glacial lake outburst floods, posing a threat to downstream communities (Zheng *et al.*, 2021).

Up to a decade ago, there were only a few peer-reviewed studies on the impact of climate change on the cryosphere components within the HKH region (Mukherji *et al.*, 2019). Since then, however, numerous research papers have documented the impacts of climate change on glaciers, river flow, and even, to some extent, snow and permafrost across different parts of the HKH region (Gruber *et al.*, 2017; You and Ren, 2017; Bolch *et al.*, 2019; Dimri *et al.*, 2021; Satti *et al.*, 2023).

In the HKH region, there are 54,252 glaciers covering a total area of 60,054 km^2 and containing an estimated ice reserve of 6127 km^3 (Bajracharya and Shrestha, 2011). Only 1.4% of the region is covered with glaciers, but their combined ice reserves are approximately triple the annual precipitation of the area. Over 60% of the total glacier area is located between 5000 m and 6000 m above sea level (Fig. 18.2;

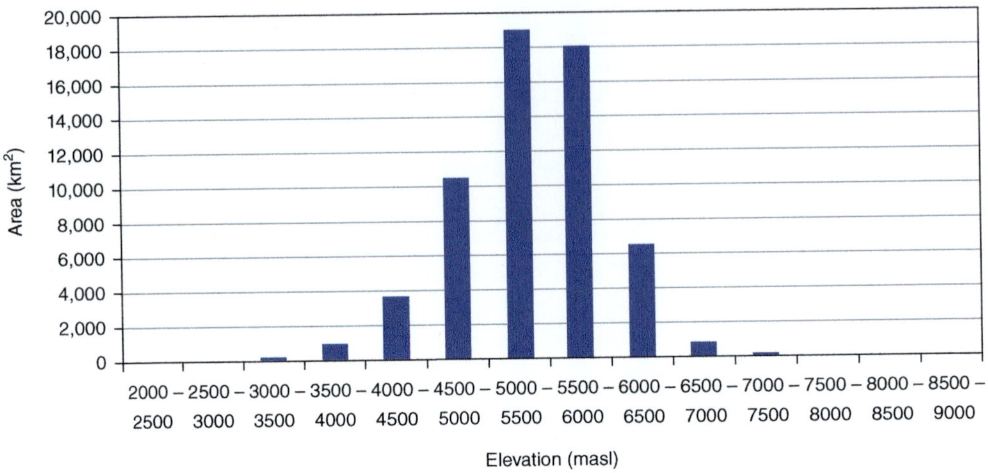

Fig. 18.2. Distribution of glacier area in the 500-metre elevation band in the Hindu Kush Himalayan region. (From ICIMOD)

Bajracharya and Shrestha, 2011), but glaciers below 5700 m are especially vulnerable to climate change unless they are protected by a thick layer of debris (Bajracharya *et al.*, 2015; Rowan *et al.*, 2015). Glacier coverage varies considerably among river basins, with the Indus, Yarlung Tsangpo/Brahmaputra and Ganges basins having the largest glacier areas (Bajracharya and Shrestha, 2011), but 79%, 60% and 77% of the areas, respectively, fall below the altitude of 5700 m (Li *et al.*, 2022c).

Due to global warming and other environmental factors, glaciers in the HKH region have been shrinking rapidly in recent decades. The International Centre for Integrated Mountain Development (ICIMOD) reported that the region has lost close to 15% of its glacier area since the 1970s, projecting that this trend will continue (Bajracharya and Shrestha, 2011). The report also warns that if current trends persist, two-thirds of the remaining glaciers could vanish by the end of this century. The meltwater from these glaciers has already led to numerous natural disasters such as floods and landslides, declining hydroelectricity production, reduced agricultural water use, and changes in river flow for downstream communities. In addition, besides the direct effects, the long-term repercussions of glacier loss in the HKH region could be devastating. Glacier melt could lead to insecure access to water and food supplies, and even displacement of entire communities. It could exacerbate geopolitical tensions among countries sharing water resources in the region. Therefore, immediate action is necessary to tackle the challenges posed by glacier degradation in the region.

Permafrost

The HKH region is characterized by widespread permafrost areas, covering approximately 15% of its total surface (Fig. 18.3; Anderson *et al.*, 2020). Permafrost regulates water and nutrient cycles, vital functions in the ecosystem, and provides habitats for animal species.

Early research indicated the presence of extensive permafrost in the HKH region, based on insights from nearby mountain ranges and limited local evidence (Gruber *et al.*, 2017; Bolch *et al.*, 2019). Today, however, we have a more refined understanding of where permafrost is likely to occur in the mountains. With an area of about 1 million km^2, permafrost covers 14 times as much ground as glaciers in the HKH region (Table 18.1). Two-thirds of the Tibetan Plateau is overlain by permafrost, with China claiming the largest share. All countries, except Bhutan, have considerably more estimated permafrost than glacier area (Table 18.1). Given variations in temperature and terrain across the region, the range of elevations where perennial ice can form is wide (Fig. 18.2), with permafrost anticipated at altitudes as low as 3500 masl in some western parts of the HKH region. Meanwhile,

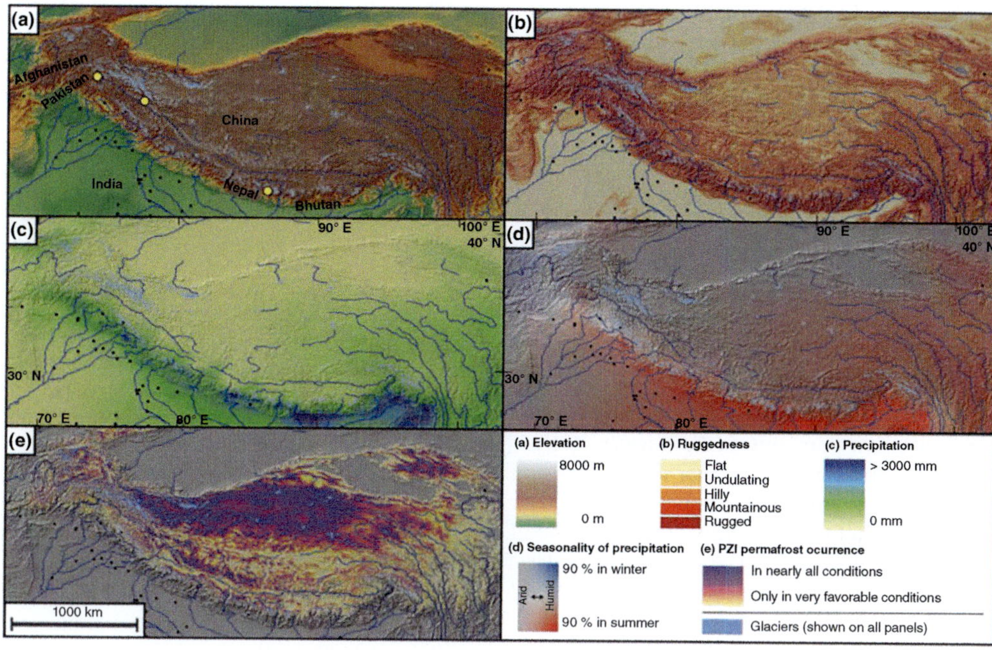

Fig. 18.3. The Hindu Kush Himalayan region: (a) topographic map; (b) relief map; (c) map of the average annual precipitation distribution; (d) percentage of seasonal precipitation; and (e) permafrost zone. (From Gruber *et al.*, 2017)

Table 18.1. Permafrost and glacier areas in the Hindu Kush Himalayan Region. (From ICIMOD)

Country	Permafrost [10^3 km^2]	Glacier [10^3 km^2]
Afghanistan	17.5 (5.9–31.6)	2.6–3.1
Bhutan	1.2 (0.3–3.5)	1.0–1.5
China	906 (486–1356)	29.2–33.6
India	40.1 (10.5–77.4)	11.8–19.6
Myanmar	0.1 (0.03–0.5)	0.04–0.06
Nepal	11.1 (6.4–16.1)	4.1–4.8
Pakistan	26.6 (7.2–53.3)	11.3–17.0
Total	**1003 (516–1538)**	**60.1–79.8**

the central and eastern HKH region have rocky slopes without permafrost at southern exposures well above 6000 m.

Like glaciers, permafrost in the HKH region is also threatened by global warming (Mayewski *et al.*, 2020). As temperature rises, permafrost melts at an alarming rate, which could have severe consequences for the environment and local communities. Melting permafrost can lead to various environmental issues, such as intensified landslides and soil erosion, changes to hydrological conditions, alterations in vegetation patterns and loss of wildlife habitats (Kohler and Maselli, 2009; Prăvălie, 2021). Additionally, large amounts of greenhouse gases, such as carbon dioxide and methane, could be released (Karl *et al.*, 2009; Chen *et al.*, 2013).

The melting of permafrost also has an important impact on local communities: many people in the HKH region depend on permafrost for their livelihoods, including grazing livestock, agriculture and tourism (Sharma *et al.*, 2010; Mukherji *et al.*, 2019). Melting permafrost could make it difficult for these communities to maintain their traditional way of life, leading to economic and social disruptions. Understanding the glaciers and permafrost in the HKH region is essential because of their vast area (Fig. 18.2) and the dense population, as well as the potential thaw-related impacts. Immediate action is warranted to address the challenges posed by melting glaciers and permafrost in the HKH region.

18.3 Functions of Glaciers and Permafrost

18.3.1 Storage and regulation

Glaciers and permafrost serve as natural reservoirs, storing vast amounts of fresh water. As the temperature rises during warmer seasons, water seeps into the ground to form underground aquifers, which gradually release water to regulate and maintain downstream river flow during dry seasons. The HKH region relies heavily on these water resources, and, consequently, with the disappearance of glaciers and permafrost, water availability is reduced and risk of drought is increased (Bolch *et al.*, 2019).

The HKH region has many natural water storage systems, including the largest glaciers outside of the polar regions, snow and permafrost, over 3000 glacial lakes with an area totalling close to 802 km^2 (Shugar *et al.*, 2020), Himalayan wetlands that support various ecological processes, such as river flow, sediment capture, nutrient recycling and groundwater recharge, and vast underground aquifers, including unconsolidated Bhabar aquifers in the south of the Siwalik range and Terai plains (Pathak, 2016; Shrestha *et al.*, 2018). Groundwater in mountainous areas serves as the source of many springs. However, there is mounting evidence that springs in the foothills of the Himalayas are undergoing a reduction in water volume (Tambe *et al.*, 2020). These natural water storage systems are predicted to play a crucial yet often overlooked role in supplying water in the future.

18.3.2 Ecosystem support

Glaciers and permafrost in the HKH region play a crucial role in supporting ecosystems by maintaining biodiversity, soil fertility and important ecological processes. Glaciers sustain various ecosystems, including alpine meadows, wetlands, forests and farmland. The HKH region is highly regarded for its biodiversity, as it serves as part of four global biodiversity hotspots, with diverse species, high endemism and unique ecosystems. It is also recognized as one of the 200 ecoregions in the world that offer valuable goods and services such as food, water, habitat, pollination and climate regulation to support the livelihoods of millions of people both in the region and downstream.

Meltwater from glaciers feeds streams and rivers downstream, provides vital water sources for plants and animals, and supports fisheries and aquatic life. Glacier melting also increases the soil-moisture levels during drought periods, which is crucial for plant growth and nutrient cycling. Permafrost in high-altitude regions also supports numerous ecosystems. When permafrost thaws, it releases nutrients into the soil, promotes vegetation growth and supports wildlife. Additionally, permafrost regulates water flow by controlling the rate at which water seeps into the soil and groundwater system, affecting vegetation distribution and essential ecological processes. Overall, the ecosystems support provided by glaciers and permafrost in the HKH region is fundamental to maintain biodiversity, enable agriculture and other economic activities, and provide various ecological services.

18.3.3 Carbon storage and climate regulation

The HKH region is characterized by climate diversity, resulting in substantial variations in glacier mass balance and river discharge across the region (Bolch *et al.*, 2012). The weakening of monsoons from east to west restricts the penetration of the Karakoram range and leads to reduced summer precipitation in the west. Annual rainfall gradient ranges from over 3000 mm in the east to less than 300 mm in the arid west. Over 70% of the eastern and central watersheds receive their annual rainfall during the summer monsoon season. Consequently, in the west, glaciers undergo winter accumulation and summer melting, while in the east there is synchronous summer accumulation and melting.

Glaciers, particularly snow, affect the climate by reflecting most of the incoming solar radiation back into the atmosphere (Gurung *et al.*, 2017) and by providing local cooling services (Mukherji *et al.*, 2019). Permafrost acts as a carbon sink or carbon source, influencing greenhouse gas emissions. Large, deep carbon reservoirs located in permafrost areas play an essential role in the global land carbon cycle. Projected permafrost melting will lead to increased release of deep permafrost and soil-surface emissions (such as organic carbon released by microbial decomposition) into the atmosphere (Voigt *et al.*, 2017; Lacroix *et al.*, 2022). In summary, glaciers and permafrost store enormous amounts of carbon and greenhouse gases, and release methane and carbon dioxide when permafrost melts, causing global warming and further accelerating permafrost degradation.

18.3.4 Disaster reduction

Glaciers play a vital role in disaster reduction by regulating river flow. Acting as natural reservoirs, they store water in winter and gradually release it during peak demand in summer, thus reducing floods in the monsoon season and ensuring a constant supply of water for agriculture and other needs all year round (Nie *et al.*, 2021). However, climate warming has led to increased glacier melting and intensified the frequency and damage caused by glacial lake outburst floods (GLOFs), resulting in serious damage to infrastructure and communities in the HKH region (Khalid *et al.*, 2021; Chen *et al.*, 2023). When ice dams rupture due to melting water from glacial lakes, a large amount of water flows downstream, resulting in flooding, which has caused considerable damage in recent years.

Permafrost is also critical in disaster reduction, playing a role in stabilizing mountain slopes and preventing landslides (Patton *et al.*, 2019; Hjort *et al.*, 2022). However, with rising surface temperatures, permafrost is melting rapidly, which increases slope instability and the frequency and severity of landslides (Savi *et al.*, 2021). For example, the Khumbu region of Nepal experienced a devastating landslide in 2014 caused by permafrost melting, which destabilized the slope and caused it to collapse; 18 people died and many homes were destroyed.

18.3.5 Maintaining soil stability

Soil stability in the HKH region is linked to the roles played by glaciers and permafrost. The melting of glaciers and permafrost can compromise soil stability, leading to elevated risks of natural disasters such as landslides and erosion (Crozier, 2010; Huggel *et al.*, 2012). The glaciers form a protective layer of snow and ice under slopes, which aids in soil stabilization (Singh *et al.*, 2011). However, as glacier ice melts with warming temperatures, this safeguarding layer diminishes, thereby increasing the frequency and severity of damage caused by erosions and landslides (Singh *et al.*, 2011). Additionally, sediment from eroded slopes damages the water quality and quantity in rivers, affecting communities downstream (Nepal *et al.*, 2018). In certain areas, structures and infrastructures are built atop permafrost and rely heavily on stable permafrost layers beneath them for support (Li *et al.*, 2022a).

Permafrost also plays a crucial stabilizing role in the soils of the HKH region. As permafrost thaws, it destabilizes the soil, increasing the risk of landslides and erosion (Li *et al.*, 2022a; Svennevig *et al.*, 2022). Furthermore, permafrost functions like a natural sponge that absorbs water in the rainy season and releases it during droughts (Douglas *et al.*, 2020). However, the melting of permafrost could lead to the destruction of this natural storage system, ultimately impacting the water supply for agricultural and other purposes.

18.3.6 Landscape shaping

Glaciers and permafrost are crucial in shaping the landscape of the HKH region. Glaciers, through slow movement and erosional transport, sculpt the mountains in distinctive landforms like U-shaped valleys, cirques and moraines on the Earth's surface. The erosion and transportation facilitated by glaciers have contributed to the formation of many valleys, ridges and peaks in the region. Permafrost also plays a role in landscape shaping by acting as an adhesive, binding the landscape together, enhancing slope stability and reducing erosion. The ice-cored moraines, which occur due to sediment deposition from melting glaciers, serve as a prime example of landscape shaping in the HKH region. Over time, these moraines can form complex patterns and structures, creating unique landscapes.

18.3.7 Cultural formation and dissemination

The HKH region is the cradle of various cultures that are linked to the natural environment. Glaciers and permafrost in this area play a pivotal role in shaping and sustaining these cultures, providing vital resources such as water, food and other essentials for human survival (Mukherji *et al.*, 2019; Talukder *et al.*, 2021). The Sherpa people, who inhabit the high-altitude areas of the HKH region, are a notable example of a community with a unique culture closely linked with the mountainous environment (Carey *et al.*, 2014; Tuladhar *et al.*, 2021). For centuries, the Sherpa have relied on glacier meltwater for irrigation and drinking water (Gagné *et al.*, 2014) and glacier ice and snow to construct houses and temples (Allison, 2015). Another example is the Ladakhi people, who reside in the northernmost region of India's Jammu and Kashmir state. They practice 'glacier farming', a form of agriculture that involves using glacier meltwater to irrigate crops (Saxena *et al.*, 2021). This technique has been in use for thousands of years and is vital for ensuring food security in the area.

Apart from supporting human cultures, glaciers and permafrost also support spiritual traditions in the HKH region. Frozen features in the cryosphere, such as snow, glaciers and glacial lakes, are often deemed sacred and symbolic, influencing how local communities perceive and interact with their surroundings (Gagné *et al.*, 2014). Many local communities view these features as sacred and deserving of respect and protection. For instance, the Khumbu Sherpa people of Nepal believe that the Khumbu glacier is the home of the goddess Nupzangma and must be appeased with offerings and prayers to ensure a good harvest (McDowell *et al.*, 2012). Buddhists in Bhutan and Nepal also recognize various natural features such as snow peaks, rivers and wildlife as sacred and draw inspiration from them (Allison, 2015).

18.4 Threats to Glaciers and Permafrost

The HKH region boasts the largest snow and ice reserves outside the polar regions, providing water resources for approximately 100 million people. The ecosystems not only support the survival of mountain communities directly but also affect downstream areas indirectly. However, the HKH region's socio-ecological system is threatened by frequent extreme weather events, including erratic rainfall, droughts, avalanches and landslides, as well as the increasing frequency of glacier expansion and retreat. Despite their importance, the glaciers and permafrost in the HKH region face multiple threats from climate change and human activities.

18.4.1 Climate change

The HKH region is highly sensitive to climate change and faces glacier retreat, shrinkage and degradation (Miller *et al.*, 2012; Dhimal *et al.*, 2021). Global climate warming increases the already present risks of natural disasters such as floods, landslides and glacial lake outburst floods caused by the accelerating melting of ice and snow (Nie *et al.*, 2021). The long-term consequences include gradual weakening of water supply as ice and snow cover decrease, leading to potential consequences for communities, agriculture and hydropower stations. The glaciers and glacial lakes in the region are the major sources of rivers such as the Ganges, Yarlung Zangbo and Indus in south Asia, which provide 70% of the water from glacier melting (Yao *et al.*, 2022), and many of these regions face the risks associated with reduced freshwater run-off. Rising temperatures can cause earlier and faster snow melting in spring, changing the time and distribution of run-off in the region, and increasing glacier melt substantially (Nie *et al.*, 2021). Thus, climate warming poses far-reaching consequences to the environment, society and economy of the HKH region, including special administrative regions and nearby territories. Water scarcity, conflicts, natural disasters, loss of life and property, and maintenance challenges should be anticipated and addressed.

18.4.2 Mining and construction activities

Mining and construction in the HKH region have greatly impacted the glaciers and permafrost. Construction of roads and infrastructure can lead to increased erosion rates, landslides and accelerated melting of glaciers and permafrost (Ding *et al.*, 2021; Li *et al.*, 2022b). Mining operations typically require heavy machinery, which can cause landscape disruption and increase pollution. This has direct and indirect impacts on the environment and water sources. For example, the construction of the Karakoram Highway has caused erosion and melting of glaciers and permafrost (Nepal *et al.*, 2014).

18.4.3 Over-grazing and deforestation

Over-grazing and deforestation are two major factors contributing to environmental degradation in the HKH region that have potential impacts on glaciers and permafrost. Over-grazing results in soil erosion, reduced vegetation cover, and increased run-off, which ultimately accelerates glacier and permafrost melting (Liu *et al.*, 2014; Ehlers *et al.*, 2022). Animal trampling can destroy permafrost layers, while loss of vegetation cover reduces soil moisture (Beer *et al.*, 2020; Yang *et al.*, 2022). In the Indus river basin in Pakistan, over-grazing by livestock led to soil erosion and loss of vegetation cover, thereby increasing river run-off and sedimentation rates, but reducing downstream water resources and posing threats to local communities' livelihoods (Anjum *et al.*, 2010; Abbas *et al.*, 2022).

Deforestation, like over-grazing, causes soil erosion and loss of vegetation cover, and increases run-off and erosion rates (Kavian *et al.*, 2014). Deforestation also weakens trees' capacity to absorb carbon dioxide, contributing to greenhouse gas emissions and amplifying climate change (Tinker *et al.*, 1996; Yoro and Daramola, 2020). Forest loss is a grave issue in the HKH region, notably in areas like Chitwan National Park in Nepal where it has triggered soil erosion, biodiversity loss, landslides and floods.

18.4.4 Infrastructure development

Infrastructure development in the HKH region has the potential to impact glaciers and permafrost. Construction of roads, dams, hydropower plants and urbanization can alter the natural water cycle, leading to changes in river flow and affecting the stability of the ecosystem (Kohler *et al.*, 2014; Rasul and Molden, 2019). With glaciers in the HKH region already experiencing retreat due to climate warming, infrastructure development can increase solar radiation absorption and accelerate glacier melting (Kohler *et al.*, 2014). This has been observed in the construction of roads and infrastructure in Khardung La pass, India, where increased sunlight reflection on nearby glaciers increased the melting rate. Large dam and hydropower projects can also cause changes in river flow and sediment transport, impacting downstream communities which rely on glacial meltwater (Li *et al.*, 2022c). Additionally, the construction of dams on Himalayan rivers raises critical questions about modern engineering safety in building dams in young and fragile mountainous areas, which are susceptible to erosion (Wester *et al.*, 2019).

Permafrost has high compressive strength, good load capacity and low hydraulic permeability below 0°C, but as it thaws, ground ice turns into water, resulting in dramatic changes (Krautblatter *et al.*, 2013; Wu *et al.*, 2020). As permafrost continues to thaw, maintenance and repair of structures such as dams, roads, railways and buildings increase. For example, the construction of the China-Pakistan Economic Corridor resulted in landslides and rockfall in the Gilgit-Baltistan region of Pakistan (Suman, 2022).

18.4.5 Tourism development

The HKH region boasts a unique combination of geographical features, cultural diversity, comfortable climate and captivating natural beauty, making it a booming tourist destination with popular spots like Rangamati in Bangladesh, Leh in India, Kathmandu in Nepal, and Murree in Pakistan. The region's tourism industry is growing rapidly (Singh *et al.*, 2020), which could increase glacier melting, as new infrastructure such as roads and hotels leads to increased carbon emissions and deforestation, and ultimately results in temperature rise and further glacier melting (Stefanica, 2017; Gulzar *et al.*, 2021).

Tourism development poses a considerable risk to permafrost, a crucial foundation for building, roads and other infrastructure in high-altitude areas. Infrastructure construction in this region could interfere with the permafrost layers, causing rapid degradation (Street, 2020). For instance, the famous Annapurna Circuit Trek in Nepal crosses high-altitude villages and the high-altitude Thorong La pass, standing at 5416 m (Skach, 2016). In recent years, tourism on this trail has grown immensely, with thousands of trekkers visiting annually, leading to an upsurge in demand for accommodations, food and other services, and driving new infrastructure construction. These activities, in turn, are causing deforestation, soil erosion and increased carbon emissions, all contributing to nearby glacier melting and permafrost degradation. Therefore, while tourism brings vital economic benefits to communities in the HKH region, it is crucial to consider its potential environmental impact on the glaciers and permafrost.

18.4.6 Natural disasters

Natural disasters, such as earthquakes, floods and avalanches, can have severe impacts on glaciers and permafrost in the HKH region. These events can trigger landslides and rockfalls that can damage glaciers and permafrost. In 2015, Nepal was hit by a massive earthquake measuring 7.8 on the Richter scale that caused 19 deaths and a series of avalanches and landslides (Kargel *et al.*, 2016) which damaged glaciers and permafrost. The seasonal monsoon rains in Pakistan in 2010 caused nationwide flooding which resulted in several landslides and rockfalls and also damaged glaciers and permafrost areas (Shah *et al.*, 2017). These natural disasters intensified erosion and instability in high-altitude regions, posing new dangers.

Other natural disasters could also affect glaciers and permafrost in the HKH region. Wildfires, for instance, could reduce soil-water-holding capacity, increase run-off and downstream floods (Shakesby and Doerr, 2006) and expose soil to sunlight and warm air temperatures, leading to permafrost melting (Gibson *et al.*, 2019). Droughts could lower soil-moisture content, destabilize permafrost regions and intensify erosion (Nilsson *et al.*, 2015).

18.4.7 Atmospheric pollution

Air pollution has become a significant environmental issue in the HKH region, with its impact on glaciers and permafrost being one of the most prominent concerns. Air pollution accelerates the melting of glaciers and permafrost, with black carbon being identified as a major contributor (Kang *et al.*, 2019). Black carbon absorbs sunlight efficiently and, when deposited on snow or ice, reduces the surface reflectivity, causing more heat to be absorbed and the acceleration of melting (Hansen and Nazarenko, 2004; Chen *et al.*, 2022b). This phenomenon is known as the 'albedo effect'. The black carbon emitted from South Asia's urbanization and agricultural activities is a crucial factor contributing to glacier melting, as reported for the Gangotri glacier in India (Negi *et al.*, 2019). Air pollution also has a negative impact on microbial communities residing in glaciers and permafrost, which play a vital role in maintaining healthy and balanced ecosystems (Suyal *et al.*, 2021).

18.5 Consequences of Retreating Glaciers and Permafrost

Glaciers and permafrost play a crucial role in preserving ecological and environmental harmony. Nevertheless, these natural resources are increasingly vulnerable to deterioration, owing to climate change and human activities, which could have severe environmental consequences. Some of the detrimental effects of glacier and permafrost degradation in the HKH region are discussed below.

18.5.1 Mountain landslides

Glacial and permafrost melting can increase the risk of landslides (Gruber *et al.*, 2017). As glaciers melt, surface and subsurface water increase soil and rock erosion (Crozier, 2010; Huggel *et al.*, 2012). Similarly, permafrost thawing can disrupt soil stability, further raising the likelihood of landslides (Singh *et al.*, 2011; Li *et al.*, 2022a). For example, a massive landslide in India's Kedarnath region in 2018, in which over 200 lives were lost and the infrastructure was damaged greatly, was caused by melting glaciers and permafrost (Pradhan *et al.*, 2018). In 2021, a large-scale landslide occurred in India's Chamoli region due to glacial melting (Li *et al.*, 2022c), which caused flash floods that caused at least 80 deaths and widespread damage to the area's infrastructure.

18.5.2 Glacier retreat

Glacier retreat and permafrost melting are among the most important occurrences in the HKH region. Since the 1930s, India's Gangotri glacier has retreated more than 1500 m, while in the past 40 years, the Dokriani glacier lost 26% of its area (Negi *et al.*, 2012). These changes are affecting the flow of rivers and streams, which are crucial sources of water for agriculture, hydropower and other activities. As temperatures continue to rise, permafrost is thawing, making soil unstable and increasing the risk of natural disasters such as landslides and floods (Singh *et al.*, 2011; Li *et al.*, 2022c).

18.5.3 Release of greenhouse gases

The melting of glaciers and permafrost in the HKH region has a substantial impact on greenhouse gas emissions, either directly by release of carbon and methane from the melting of organic matter or indirectly by changes in hydrological cycles. Permafrost contains a vast amount of frozen organic matter. As the temperature rises and permafrost melts, this organic matter decomposes, releasing carbon dioxide and methane into the atmosphere (Davidson and Janssens, 2006; Schuur *et al.*, 2015). A report by the Intergovernmental Panel on Climate Change (IPCC) in 2019 warned that if warming rates continue at the current pace, permafrost may release up to 280 billion tons of carbon into the atmosphere by 2100 (Engelbrecht and Monteiro, 2021).

Changes in hydrological cycles also affect greenhouse gas emissions. Glaciers act as a natural carbon sink through photosynthesis by algae and other organisms, absorbing carbon dioxide from the atmosphere. However, as glaciers melt, they release this carbon back into the atmosphere (Stibal *et al.*, 2012; Wang *et al.*, 2021). Changes in water flow rate and volume resulting from glacier retreat impact carbon dioxide exchange between the atmosphere and water bodies (Kundzewicz, 2008; Milner *et al.*, 2017). In the HKH region, wetlands and lakes on the Qinghai-Tibetan Plateau release methane, while permafrost melting in Nepal's Himalayan region releases carbon dioxide and methane, highlighting the impact of glacier and permafrost melting on greenhouse gas emissions (Chen *et al.*, 2022a; Saikia *et al.*, 2023). These impacts do not only contribute to global climate warming but they also have local effects on the region's ecosystems and communities.

18.5.4 Flood outbreaks

With the melting of glaciers and permafrost in the HKH region, there is an increasing risk of new water channels forming, which may lead to flooding downstream. Destabilization of lakes is also increasing the threat of glacial lake outburst floods (Veh *et al.*, 2020), which occurs when a glacial or moraine dam bursts (Harrison *et al.*, 2018). In recent years, the area has experienced several such incidents (Table 18.2). For example, the 2013 flood in north India's Uttarakhand state was a result of glacier melting-induced flooding in the HKH region (Lorenz and Dittmer, 2021; Molden *et al.*, 2022). The flood caused widespread damage to infrastructure, livelihoods and ecosystems and resulted in over 5700 deaths or disappearances of people. In 2018, rains triggered landslides and flash floods in the Garhwal Himalayas, causing the destruction of buildings, evacuation of thousands of people, and several fatalities (Sati, 2018). In 2021, a flood in the Chamoli district, north India, left 100 people dead and many injured (Khanduri, 2022; Li *et al.*, 2022b). Initially thought to be caused by a glacial lake outburst, remote sensing determined that it was caused by the rapid flow of ice due to the kinetic energy of rockfall combined with downstream flow (Prakash *et al.*, 2021).

The melting of permafrost can trigger flooding and other natural disasters by destroying slope stability, leading to landslides, floods and mudflows (Singh *et al.*, 2011; Li *et al.*, 2022c). For example, in 2018, in Kedarnath in India, permafrost melting contributed to flooding and

Table 18.2. Glacier flood events across the Hindu Kush Himalayan region. (Data from Ashraf *et al.*, 2021)

Date	Description
2013	Glacial lake outburst floods (GLOFs) in Reshun Valley damaged infrastructure and property, and severely affected livelihoods.
2015, 2019, 2020	GLOFs in Golen Valley near Chitral city swept away four suspension bridges, a small hydropower station, roads, irrigation systems and standing crops.
1994, 2002, 2005, 2012, 2013	GLOFs and accompanying debris flows in Sosot resulted in damage to 65 households, infrastructure and agricultural land, including numerous fruit trees, and the loss of several human lives.
1978, 1991, 2010	GLOF and flash-flood events in Darkut damaged cultivated land, forest area and over 300 houses. The land was buried under thick debris cover during one of the flood events.
1978, 1992, 2009, 2010	GLOFs on Hinarchi glacier damaged natural forest and agricultural lands of Khama, Bulchi and Chira villages.
1974, 1992, 1997, 2010	Dubani glacier/lakes outburst damaged natural forest and Taisote and Massingot villages. Loss of several human lives and 60 goats were reported.
1997, 2000	GLOFs and debris flow occurred at Kande from a tributary of the Hushe river, which partially destroyed the village in Shyok basin.
2007, 2008	GLOFs due to Passu glacier in Upper Hunza Valley.
2008–2013, 2015, 2017	Series of GLOF events due to fill and breach of the supraglacial lakes over Ghulkin glacier in Upper Hunza Valley.

landslides that caused substantial damage and loss of life (Li *et al.*, 2022c). Flood-prone areas like west Champaran, Bihar, India, and other communities near rivers or within embankments cannot carry out agricultural activities for three to four months each year as the fields are eroded by rivers or completely submerged due to floods (Shekhar *et al.*, 2010; Kishore *et al.*, 2019).

18.5.5 Changes in ecosystems

The melting of glaciers and permafrost in the HKH region can have important ecological impacts, resulting in changes to wildlife habitats, biodiversity, as well as affecting the availability of critical ecosystem services such as water and carbon storage. These impacts could lead to alterations in ecosystem functioning and a decline in biodiversity. Glacier melting in the HKH region has the potential to affect freshwater ecosystems such as rivers and lakes (Rasul, 2014; Nepal *et al.*, 2018). As glaciers melt, the quantity and rate of water flow are affected, which may have implications for freshwater supplies and ecosystem functions (Jeelani *et al.*, 2012). Additionally, changes in water temperature and nutrient levels could impact the distribution and abundance of aquatic organisms, including fish, invertebrates and algae (Neale *et al.*, 2023). The melting of permafrost can impact terrestrial ecosystems such as forests and grasslands. As permafrost melts, the soil becomes more waterlogged, reducing support for some animal and plant species (Chen *et al.*, 2021). Furthermore, permafrost melting changes soil nutrients and microbial activity, reducing the growth and distribution of plant species (Pedersen *et al.*, 2020; Ramm *et al.*, 2022). Large vegetation changes, such as meadows turning into shrubland, have already occurred in the eastern Himalayas (Manish *et al.*, 2016). These changes could lower the ecosystem's capacity to support livestock-based pastoralism. The eastern fringe of the Tibetan Plateau is undergoing changes due freeze-thaw processes, permafrost degradation, extreme precipitation events, and over-grazing by livestock (Wang *et al.*, 2018; Niu *et al.*, 2021; Chen *et al.*, 2022a). These feedbacks reduce vegetation cover, increase soil erosion, and decrease key animal species such as the Tibetan Plateau pika (Liu *et al.*, 2012; Fayiah *et al.*, 2020).

18.5.6 Infrastructure damage

The melting of glaciers and permafrost can cause floods and landslides that could damage infrastructure, including roads, bridges, buildings and other essential facilities (Martha *et al.*, 2021).

Permafrost melting can affect buildings and other infrastructure as they depend on stable foundations. However, with permafrost melting, buildings and other infrastructure can develop cracks or even collapse, resulting in higher maintenance and repair costs (Li *et al.*, 2022a). The 2015 Nepal earthquake is an example where the HKH infrastructure sustained severe damage, with glacier melting exacerbating the destruction and increasing losses and casualties (Molden *et al.*, 2016). Overall, the infrastructure damage that could result from glacier and permafrost melting in the HKH region may have substantial economic and social impacts, such as disruptions to transportation networks, higher infrastructure repair costs, and threats to public safety.

18.5.7 Transmission of diseases

The melting of glaciers and permafrost can impact human and animal health indirectly through disease transmission (Table 18.3). Snowmelt and glacier run-off affect the occurrence and prevalence of water-related diseases directly and indirectly (Ebi *et al.*, 2007). A shift towards more rainy and dry seasons is predicted in the future, with increased rainfall during the rainy season causing flash floods, and drier conditions during the dry season leading to water scarcity and greater health risks. During the rainy season, flooding may cause faecal contamination of water sources, while water storage could heighten the risk of water-borne illnesses like cholera and typhoid fever (Ercumen *et al.*, 2015).

Melting glaciers and permafrost affect water quality in rivers and other freshwater sources due to the release of debris, sediments and pollutants, which could increase the incidence of water-borne diseases particularly in downstream communities that depend on these water sources (Bhateria and Jain, 2016; Talukder *et al.*, 2021). Permafrost melting increases zoonotic disease risks between humans and animals, and changes in environmental conditions such as temperature, precipitation or vegetation could have a profound impact on the dynamics of pathogen life cycles, vectors and hosts (Harvell *et al.*, 2002; Elad and Pertot, 2014). Permafrost also has a long-term influence on preserving microorganisms and viruses that have been inactive for thousands of years (Miner *et al.*, 2021; Wu *et al.*, 2022). As permafrost melts, these microorganisms and viruses may become active, posing a risk to both animals and humans (Yarzábal *et al.*, 2021). The anthrax outbreak in Siberia in 2016 was an example of the relationship between permafrost melting and the release of anthrax spores that had been frozen in the soil for over 70 years (Ezhova *et al.*, 2021).

Table 18.3. Current climate-related health determinants and outcomes in the HKH region; synthesis of national reports. (From Ebi *et al.*, 2007)

Country	Afghanistan	Bangladesh	Bhutan	China	Nepal	India
Heatwaves	M–P	P	—	P	P	P
Flood deaths/morbidity	—	—	—	—	—	—
Glacial lake floods	M–P	—	M–P	M–P	M–P	M–P
Flash	M–P	P	M–P	M–P	M–P	M–P
Riverine (plain)	P	P	—	P	P	P
Vector-borne diseases	P	P	P	P	P	P
Malaria	P	P	P	P	M–P	P
Japanese encephalitis	—	P	—	P	P	P
Kala-azar	P	—	—	—	P	P
Dengue	—	P	P	P	—	P
Water-borne diseases	M–P	P	M–P	M–P	M–P	M–P
Water scarcity, quality	M–P	P	P	M–P	M–P	M–P
Drought-related food insecurity	M–P	P	—	M–P	—	M–P

M–P, health determinants or outcomes that occur in mountainous and non-mountainous (i.e. plain) areas; P, health determinants or outcomes that occur only in non-mountainous (i.e. plain) areas; –, health determinants or outcomes that do not exist in the country.

18.5.8 Culture

The glaciers and permafrost are changing, which can have substantial impacts on the cultural traditions and practices of local communities and their relationship with the natural environment. One key example of cultural change resulting from glacier melt is the impact on traditional water management systems. Many communities in the HKH region have developed sustainable, complex water management systems that rely on the timing and availability of glacier and snow melt (Bolch *et al.*, 2019). However, as glaciers melt, the quantity and timing of water flow changes, which can render these traditional water management systems ineffective, and result in a loss of cultural heritage and knowledge (Motschmann *et al.*, 2020).

Another example is the effect of permafrost thawing on traditional land-use practices. Permafrost has been a critical part of the indigenous and local community culture and economy in the HKH region, supporting activities like grazing, hunting and reindeer herding (Dong *et al.*, 2016). As permafrost thaws, these practices may become unsustainable or difficult, causing changes in traditional livelihoods and cultural customs (Wang *et al.*, 2019).

Yak production, a traditional livelihood for many communities in this area, provides an example of cultural change resulting from the melting of glaciers and permafrost in the HKH region (Wu *et al.*, 2014). Permafrost thawing and unstable pastures may make it difficult to feed yaks, leading to economic and cultural impacts (Bhadwal *et al.*, 2019). Overall, melting glaciers and permafrost in the HKH region can result in cultural changes affecting the identity, traditions and values of local communities, and ultimately in a loss of cultural diversity and heritage.

18.6 Protection of Glaciers and Permafrost

18.6.1 Reducing greenhouse gas emissions

The HKH region's glacier and permafrost ecosystems are threatened by climate warming. One of the most effective methods to protect these ecosystems is to reduce greenhouse gas emissions, which are a major cause of global warming. There are various ways to achieve this, including transitioning from fossil fuels to renewable energy sources such as solar, wind or hydropower (Nielsen *et al.*, 2021; Pimm *et al.*, 2021; Sri Shalini *et al.*, 2021), promoting energy efficiency and conservation measures in buildings and transportation systems (Lv and Shang, 2022; Tsai and Tsai, 2022), encouraging sustainable land-use practices such as reforestation, and prohibiting deforestation (Harper *et al.*, 2018; Picoli and Machado, 2021), and adopting circular-economy principles to decrease waste and improve resource efficiency (Sawhney, 2021).

18.6.2 Tree planting and afforestation

Afforestation can be crucial in protecting glaciers and permafrost. Trees mediate temperature and precipitation (Grießinger *et al.*, 2021), curb global warming and reduce greenhouse gas emissions by absorbing carbon dioxide from the atmosphere. To optimize the effectiveness of afforestation for glacier and permafrost protection, tree species, planting density and management practices should be considered. Locally adapted tree species offer high survival rates and provide valuable ecological benefits (Klisz *et al.*, 2023). Reasonable planting densities should avoid inter-tree competition while promoting optimal ecosystem services such as water regulation and biodiversity enhancement (Di Sacco *et al.*, 2021). Noteworthy examples of afforestation projects in the HKH region are Pakistan's Ten Billion Tree Tsunami programme, which aims to plant 10 billion trees nationwide, including in mountainous areas, to combat climate warming and restore ecosystems (Baig *et al.*, 2021), India's National Afforestation Programme, which plants trees on barren and uncultivable land to prevent soil erosion, enhance ecological resilience and support community livelihoods (Bhatt *et al.*, 2020), and Bhutan's advocacy for reforestation projects in the eastern Himalayan river basins to contribute to global climate goals while promoting sustainable development and poverty reduction (Dhyani *et al.*, 2020).

18.6.3 Sustainable land use

To ensure the lasting health and stability of ecosystems in the HKH region, sustainable land-use practices such as terracing and agroforestry (Rasul, 2014; Wang *et al.*, 2019; Wester *et al.*, 2019) provide ecological, social and economic benefits. Land, agricultural and water resource managements are crucial when protecting glaciers and permafrost, and can also promote soil conservation, improve local ecological functions, reduce soil erosion, and protect local biodiversity. In areas susceptible to drought, sustainable irrigation practices maintain healthy vegetation cover, while decreasing soil salinization and nutrient loss (Shrivastava and Kumar, 2015). By adopting sustainable land-use practices, the HKH region not only safeguards glaciers and permafrost but also enhances ecosystem health, supports local communities and economies, and ensures long-term environmental sustainability.

18.6.4 Protection of vegetation cover

Vegetation cover is essential in stabilizing permafrost and preventing erosion, making it effective in protecting glaciers and permafrost. Plants help regulate temperature and precipitation, reduce soil erosion, and promote watershed health. Some methods to preserve vegetation cover that have been implemented in the region are as follows:

1. Curbing deforestation. Deforestation releases carbon stored in trees, reduces biodiversity, and alters weather patterns leading to climate change. Sustainable land-use practices such as agroforestry, community forest management and timber plantations should be promoted to counter deforestation (Rasul, 2014; Wang *et al.*, 2019; Wester *et al.*, 2019). Bhutan, for instance, has pledged to maintain at least 60% forest cover nationwide by preventing illegal logging and implementing sustainable forestry practices (Brunet *et al.*, 2001).

2. Promoting reforestation. Reforestation helps restore degraded ecosystems, recover critical habitats and improve landscape connectivity. Crucial factors for successful reforestation include species selection and planting density (Kunstler *et al.*, 2016; Klisz *et al.*, 2023). India's 'Green India Mission' aims to increase the country's forest cover by 5 million ha through afforestation and reforestation by 2030 (Davis and Robbins, 2018).

3. Encouraging natural regeneration. Natural regeneration occurs when native tree and shrub species grow again on areas where vegetation has been removed or degraded. Encouraging this process incurs minimal investment (Chazdon and Guariguata, 2016). The 'Nepal Forest Action Plan' includes provisions for protecting and strengthening natural forests to promote natural regeneration (Di Sacco *et al.*, 2021).

4. Limiting grazing. Grazing damages young plants, removes ground cover and leads to soil compaction. To limit the negative impact of grazing, it is crucial to establish protected areas, grazing quotas and seasonal grazing bans (Adhikari *et al.*, 2021; Poudel *et al.*, 2022). The Indian Nature Conservation Foundation collaborates with local communities to implement comprehensive land management plans, including restriction of grazing in sensitive areas and promotion of sustainable livestock management practices (Smith *et al.*, 2016). Adopting these measures to protect vegetation cover maintains healthy ecosystems and promotes long-term stability of glaciers and permafrost.

18.6.5 Installation of insulation layer

Installing insulation is a highly effective strategy to protect glaciers and permafrost. The insulation layer reduces heat transfer between the atmosphere and the ground, shielding the ice and permafrost beneath. These barriers, made of reflective surfaces, vegetation cover or synthetic materials, maintain a cooler ground and decrease the melting of permafrost. The insulation not only protects glaciers and permafrost but also improves the health of the ecosystem and rural livelihoods, and promotes long-term environmental sustainability. There are a number of methods for installing insulation, and the following have been implemented successfully in the region:

1. Artificial snow. By adding a layer of artificial snow, glaciers can reflect more sunlight and prevent rapid melting. In Nepal's Everest region, thick artificial ice has been deployed experimentally to cover small sections of glaciers and thereby reduce melting and conserve ice (Rounce *et al.*, 2018).

2. Surface insulation. Insulating the ground with materials such as straw or leaves reduces heat absorption. Bhutan's 'Green Roads' project insulates soil on embankments and adjacent areas using retaining walls and roadside vegetation cover, which lowers surface temperature and reduces snow-melting on mountains (Pender, 2010).

3. Groundwater management. Appropriate groundwater management can reduce the amount of heat transferred from the ground into the atmosphere, thereby functioning as an insulation layer. Underground water storage provides a steady supply of water for the surrounding ecosystems.

18.6.6 Monitoring and research

Monitoring and research are crucial for protecting glaciers and permafrost. By collecting data on ecosystem changes over time, we can gain a better understanding of how to safeguard them. Here are several methods of monitoring and research along with examples of measures undertaken in the area:

1. Glacier mass balance monitoring. ICIMOD established a network to monitor glacier mass balance, and offers training to technical personnel and researchers in the Himalayan region (Taylor *et al.*, 2021). This information can guide decisions related to water resource management, infrastructure planning and natural-disaster mitigation.

2. Measuring soil temperature profiles. Permafrost is particularly sensitive to climate change, such as temperature fluctuations and increases in precipitation (Teufel and Sushama, 2019; Wang and Liu, 2022). Monitoring soil temperature profiles provides information on permafrost changes over time, which could identify the areas susceptible to degradation (Wu *et al.*, 2023). The Asian High Mountain Remote Sensing and Multi-scale Modeling Program was implemented by partners from the University of Alaska, Fairbanks, and Nepal, China, Bhutan and India to measure permafrost temperature using satellite imagery (Bhattacharya *et al.*, 2021).

3. Studying glacier ecosystems. Studying glacier ecosystems provides valuable insights into how they change over time. A project led by Kathmandu University, Nepal, and ICIMOD uses traditional ecological knowledge and modern scientific techniques to understand the biodiversity of glaciers and permafrost in the region (Carey *et al.*, 2014).

4. Developing early-warning systems. Early-warning systems can mitigate the impact of natural disasters such as flash floods and landslides caused by glacier or permafrost melt (Sharma *et al.*, 2022). These systems typically rely on real-time monitoring of meteorological conditions and river flow. In Pakistan, the Glacier Monitoring and Research Programme regularly monitors major glaciers in the Karakoram, Hindu Kush and Himalayan regions to alert communities to risks related to flash floods (Ashraf *et al.*, 2012).

18.6.7 Infrastructure design

Infrastructure design is an effective method of protecting glaciers and permafrost. By combining natural cooling features, energy-efficient solutions, and minimizing soil disturbance, these fragile ecosystems can be safeguarded. Here are several ways to incorporate glacier and permafrost protection by infrastructure design, including some measures implemented in the area:

1. Passive cooling. Proper ventilation, shading, insulation and soil-moisture management can limit heat absorption by buildings, keeping them cool during hot summers. India has developed a new building code promoting passive cooling for both newly constructed and existing buildings (Gokarakonda *et al.*, 2019; Graham and Rawal, 2019).

2. Seismic-resistant design. Earthquakes can destroy an infrastructure, especially in mountainous regions where buildings are more prone to collapse. Incorporating seismic-resistant designs can reduce catastrophes and the need for constant reconstruction. The Nepali government emphasizes implementing earthquake-resistant technologies and developing shelters in urban centres to address climate change and environmental degradation (Klein *et al.*, 2019).

3. Green roofs. Green roofs provide an insulating layer that helps absorb rainfall, thereby reducing water volume that could erode soil and harm nearby ecosystems. They also enhance energy efficiency by decreasing air-conditioning demand. The Bhutanese government promotes using green roofs in urban infrastructure construction (Chimi *et al.*, 2023).
4. Bridge design. Bridges can act as barriers to water flow, thereby increasing floods and soil erosion. Designing bridges with minimal environmental impact and considering weather and hydrological patterns mitigates these effects. In Afghanistan's Bamyan province, two bridges with ramps were constructed to avoid seasonal water flow. The project was supported by the World Bank rural access improvement programme to withstand floods (Moffett *et al.*, 2003).

18.6.8 Community participation

Community participation is crucial in protecting glaciers and permafrost. By involving local communities in conservation efforts, long-term sustainability can be achieved while ensuring that the initiatives meet the needs of the region's residents. Following are several ways to engage communities in conservation efforts, as well as examples of measures already implemented in the area:

1. Community-led conservation. These programmes prioritize the role of locals in decision making and provide them with opportunities to collaborate on implementing conservation measures. An example of successful community-led conservation can be found in Nepal's Langtang Valley, where a restoration initiative was undertaken to improve degraded permafrost areas (Tuladhar *et al.*, 2021). Measures included constructing check dams and other structures to control water flow, stabilize slopes and reduce erosion.
2. Sustainable livelihoods. Developing sustainable livelihoods reduces dependence on income-generating activities harmful to glaciers and permafrost. By providing training and support for sustainable agricultural practices, ecotourism and handicraft production, communities can develop alternative sources of income that align with environmental protection (Anand *et al.*, 2012).
3. Education and awareness-raising. Education and raising awareness through school-based education programmes, community workshops and outreach activities can emphasize the importance of protecting glaciers and permafrost for future generations (Shrestha *et al.*, 2021).
4. Participatory research. Involving local communities in scientific research produces better data, increases ownership and strengthens collaborative conservation projects (Biswas, 2020).

18.7 Restoration of Glaciers and Permafrost and Local Livelihoods

The ecological restoration of glaciers and permafrost aims to mitigate the impacts of climate change. Here are some examples and cases of ecological restoration measures for glaciers and permafrost.

18.7.1 Protection measures for glaciers and permafrost

Monitoring glacier dynamics

Monitoring glacier dynamics is important for protecting glaciers and permafrost. Researchers and policy-makers can better understand how these ecosystems change over time by collecting data on glacier movement, mass balance and other variables, and then develop appropriate response measures to mitigate human impacts. The following are some methods for the successful monitoring of glacier dynamics in the region:

1. Ground-based monitoring. This includes installing sensor networks inside and around glaciers to measure characteristics such as temperature, precipitation and ice thickness. The Department of Hydrology and Meteorology (DHM) in Nepal installed an automatic weather station network at selected sites across the country, including the Himalayan region, to provide real-time data on weather patterns and glacier movement (Talchabhadel *et al.*, 2022).
2. Remote sensing. Remote sensing uses satellite imagery or aerial photography to track changes in glacier movement and mass loss over time. ICIMOD collaborates with regional partners to

develop and use free remote-sensing tools such as hydrological forecasting, glacier modelling, and climate-data analysis to enable efficient decision making (Johnston and Smakhtin, 2014; Bajracharya *et al.*, 2021).

3. Modelling. This involves using mathematical models to simulate glacier behaviour under different scenarios, which can help predict future dynamics of glaciers and quantify the effects of different intervention measures. In Pakistan, the United Nations Development Programme (UNDP) supports the Glacier Monitoring and Research Programme to expand research in mountainous areas, allow scientists to monitor water levels in the country's most critical glaciers, and communicate information to relevant government agencies (Wasi, 2022; Melis *et al.*, 2023).

4. Citizen science. Citizen science initiatives train volunteers to collect glacier dynamics data using low-cost sensors, digital cameras, and even smartphones, to promote public participation. The Mountain Ultra Runners Organization in Kathmandu provides a platform for citizen scientists around the world to collect mountain data.

Control of land use

Controlling land use is an effective method to protect glaciers and permafrost. Through zoning regulations, development plans and other measures, land-use practices can be managed to reduce soil erosion, prevent vegetation degradation and improve overall ecosystem health. Successful examples include:

1. Forest conservation. Deforestation causes negative impacts, such as soil erosion and loss of wildlife habitat. Implementing forest conservation measures, like afforestation or restoration of closed-canopy forests, can increase vegetation cover, improve soil stability, and reduce erosion. For example, in India's Kedarnath Wildlife Sanctuary, the REDD+ Himalayas project manages degraded forests through community management to enhance the long-term carbon storage protection potential (Singh *et al.*, 2017; Shukla *et al.*, 2022).

2. Soil conservation. Measures such as terracing, check dams and infiltration corridors can prevent soil erosion and sedimentation that accelerate glacier melting and harm downstream agricultural ecosystems. Bhutan is experimenting with terraced organic farming to reduce soil erosion in fragile landscapes and develop sustainable livelihoods for farmers (Neuhoff *et al.*, 2014; Sati, 2014).

3. Livestock grazing management. Uncontrolled grazing can damage vegetation, leading to soil erosion and degradation. By controlling the stocking rate, implementing restorative grazing practices, and choosing suitable fodder crops, agricultural biodiversity can be enhanced, soil moisture protected, and sustainable livestock production maintained. In Nepal, the Green Livelihoods project is encouraging grassroots communities to participate in pilot activities showcasing good livestock production practices and modern feed-production technologies (Allen, 2019; Maharjan *et al.*, 2021).

4. Land-use zoning. Land-use zoning provides guidance on how to use different areas of land and minimize the impact on permafrost and glacier areas. Bhutan is protecting wetland areas, covering 10% of the country's land, to conserve natural ecosystems and ecological niches for bird, animal and fish populations (Banerjee and Bandopadhyay, 2016).

Restricting infrastructure development

Infrastructure development can have negative impacts on the stability of the permafrost. Limiting infrastructure development is a key strategy to protect glaciers and permafrost, reduce human impact on these fragile ecosystems, and prevent environmental degradation. Here are some successful methods implemented in the area to restrict infrastructure development:

1. Conducting an environmental impact assessment (EIA). Before approving an infrastructure, third-party evaluators assess the potential environmental impact, which allows public participation and ensures that sustainability issues are considered in decision making.

2. Establishing zoning restrictions. Development in sensitive areas such as steep slopes, wetlands, earthquake-prone areas, and areas with permafrost intrusion should be prohibited. For example, infrastructure development and commercial activities are prohibited to maintain biodiversity in the Manaslu Conservation Area in Nepal (WWW Nepal, 2018; Bhandari, 2022).

3. Developing renewable energy alternatives. Replacing traditional and harmful infrastructure with renewable energy sources like solar power can meet the growing demand for electricity in high-altitude areas, reduce negative impacts on glaciers and permafrost, and support urban-rural integration.
4. Ensuring compliance monitoring and enforcement. The practice of protecting sensitive ecosystems by adhering to robust institutional arrangements, effective monitoring and enforcement mechanisms while abiding by democratic values and upholding restricted measures should be followed. China introduced the National Park System to protect ecologically delicate areas, aiming to enhance ecological adaptability and safeguard species diversity (Gu *et al.*, 2023).

Adaptive management strategy

Adopting adaptive management strategies is a critical approach in safeguarding glaciers and permafrost. Adaptive management incorporates stakeholder feedback and newly acquired information to modify management measures and adapt to changing circumstances. Several successful adaptive management strategies in the region include:

1. Participatory decision making. Stakeholders developed sustainable resource management strategies, making decisions related to natural resources utilization and conservation. The community-led governance approach in Nepal's Kanchenjunga Conservation Area strengthened co-operation between indigenous communities and government authorities, effectively protecting biodiversity while promoting ecotourism through landscape restoration (Nyaupane *et al.*, 2020).
2. Flexible policies. Policies could be adjusted by implementing new policies or modifying existing ones to accommodate environmental changes. Bhutan has made significant progress in providing a comprehensive framework for environmental protection under its Gross National Happiness development ideology, empowering local-level collaboration, corporate social responsibility programmes, and integrating community needs and preferences into planning processes (Devkota *et al.*, 2023).
3. Early-warning systems. It is important to monitor environmental conditions, alert stakeholders, and provide preventive interventions to protect the ecosystems. China's Frozen Soil Inventory and Monitoring System provides data on frozen soil characteristics and establishes early-warning mechanisms for ice dams and glacier debris flows, thereby reducing damage risks for new hydroelectric power plants (Li *et al.*, 2022b).
4. Climate-adaptive livelihood strategies. Communities should enhance resilience against climate-change impacts while promoting the health of ecosystems, ecological services and unique identities. ICIMOD's Integrated Livelihood Support Project in India supports the residents of Pangi Valley, a high-altitude area in Himachal Pradesh, potentially affected by unstable glaciers and permafrost, by developing alternative livelihood options such as beekeeping, terrace agriculture and animal husbandry (Nautiyal *et al.*, 2022; Ali *et al.*, 2023).

18.7.2 Ecological restoration measures for glaciers and permafrost

Habitat restoration

Habitat restoration is an effective method for protecting glaciers and permafrost. By restoring degraded habitats, ecological resilience can be enhanced, biodiversity protected, and the livelihoods of natural resource-dependent communities supported. The following are successful habitat restoration methods in the region:

1. Reforestation. Trees should be planted on degraded land to prevent soil erosion, enhance carbon sequestration, increase plant species diversity, and restore wildlife habitats. Nepal's community-led tree planting movement complements agriculture, forestry and forest rangeland systems, promotes integrated watershed management, protects fragile high-mountain ecosystems, and supports local livelihoods (Romeo *et al.*, 2021; Aryal *et al.*, 2022).
2. Wetland restoration. Water storage capacity should be improved by reducing surface run-off and sedimentation, and regulating groundwater recharge. The Three Rivers Source Basin Restoration Project aims to restore degraded, dry upstream or sloping moist grasslands on the Qinghai-Tibetan Plateau by removing invasive

species, introducing native grasses, and establishing non-woody wetlands as potential carbon sinks and biodiversity hotspots (Han *et al.*, 2018; Shao *et al.*, (2017).

3. Grassland management. Sustainable practices should be implemented to protect and restore degraded grasslands, including rotational grazing, improved livestock feed quality, protecting easement rights, and voluntary pasture treatment. Bhutan's community green yak enterprise project, initiated by Gross National Happiness ideology, manages pastures through standardized grazing cycles and allocates areas based on animal needs while minimizing environmental damage (Nkonya *et al.*, 2014; 2016).

4. Alpine ecosystem restoration. Changes in climate conditions, including temperature, humidity and geological disasters have led to the deterioration of alpine ecosystems, and therefore they require urgent protection measures. ICIMOD's Kailash Sacred Landscape Protection and Development Initiative aims to establish long-term ecosystem vitality through participatory methods, research and innovative protection measures, and active community participation (Singh, 2018).

Wetland restoration

Wetland restoration is crucial for the protection of glaciers and permafrost as it can enhance water storage capacity, improve ecosystem services, support biodiversity, and strengthen carbon sequestration. There are several methods that have been successfully implemented to restore wetlands in the region:

1. Rehumidification. This method reconstructs the original hydrological conditions by modifying the water management system, and thereby raising the groundwater level to aid the restoration of damaged peatlands. In China, downward sloping perforated subsoil pipelines have been installed to improve yak grazing quality, avoid upstream water accumulation, and enhance carbon capture (Regmi and Huettmann, 2020).

2. Reforestation and gully blockage. Gully blockage involves damming or filling with organic material to prevent soil degradation and retain more water in the landscape. In Bhutan, the Royal Society for Protection of Nature (RSPN) implemented a channel blockage plan to restore the degraded watershed of the Thimphu Valley by constructing a dam and planting vegetation to reduce soil erosion and maintain soil moisture (Inskipp *et al.*, 2020).

3. Artificial wetlands. Constructing artificial wetlands replicates natural processes such as sedimentation, biofiltration, denitrification and carbon storage and can prevent harmful metals from infiltrating rivers, provide protected habitats for waterfowl migration, and improve water quality through better irrigation facilities (Lytton *et al.*, 2021).

4. Sustainable land use practices. Emphasizing sustainable land-use practices, like halting wetland grazing and improving agricultural irrigation, can aid wetland restoration. Bhutan implemented sustainable pasture management practices to restore degraded high-altitude pastures, promoted rotational grazing, isolated sensitive areas, and constructed ponds to avoid over-grazing and soil degradation (Millar and Tenzing, 2023).

Wildlife protection

Permafrost provides crucial habitats for wildlife in the HKH area. Protecting these habitats through conservation measures is key to maintaining ecological resilience, promoting biodiversity and preserving the livelihoods of communities reliant on natural resources (Beer *et al.*, 2020; Hjort *et al.*, 2022). Following are some methods for protecting wildlife in the region:

1. Protected areas. Establishing protected areas safeguards the habitats of threatened species. In Bhutan, a network of 11 protected areas has contributed to protecting rare and endangered wildlife such as Bengal tigers and snow leopards (Tshewang *et al.*, 2021).

2. Community protection. Empowering local communities to manage natural resources is essential. In Nepal, the Annapurna Reserve Project succeeded in implementing community-led conservation measures to protect the snow leopard population, while supporting ecotourism through habitat restoration and waste-management practices (Hanson *et al.*, 2022).

3. Specific species protection. Implementing specialized conservation plans focused on specific endangered species is an effective approach.

The Snow Leopard Project in India involved specialized conservation actions aimed at conserving snow leopards and their key ecosystems in northern India (Singh *et al.*, 2020a).

4. Habitat restoration. Habitats can be restored by reconnecting fragmented habitats for migratory species, restoring key foraging and breeding sites, and improving water management systems. China's Sanjiangyuan National Park conservation plan relocates pastoralist settlements from sensitive wetlands, promotes native grassland restoration, maintains ecologically sensitive grazing practices and supports unique biodiversity within its vast territory (Cao *et al.*, 2020; Wu *et al.*, 2020; Ma *et al.*, 2022).

Indigenous knowledge and practices

The use of indigenous knowledge and practices is essential for protecting glaciers and permafrost. Recognizing and adopting traditional knowledge and practices passed down through generations can enhance resilience to environmental changes and promote sustainable development. Below are some effective methods in protecting glaciers and permafrost through local traditional knowledge and practices:

1. Traditional agriculture. The use of traditional agriculture helps maintain soil-moisture content and prevent erosion, and thereby preserve the water holding capacity of the landscape. In Bhutan, terraced agriculture maintains ecosystem services that support crop yield while preventing landslides, increasing plant biodiversity and promoting carbon sequestration (Aryal *et al.*, 2020).

2. Livestock management. Proper livestock management improves grassland and increases soil-nutrient cycling, while reducing soil compaction. In Kyrgyzstan, community-managed pastures rotate during the annual sheep migration period, which protects mountain meadows and wetlands (Schoch *et al.*, 2010; Kasymov and Thiel, 2019).

3. Traditional water harvesting. Local communities in mountain areas use traditional water harvesting, such as diversion channels and stone spouts, to slowly recharge groundwater systems and distribute water to highland areas for household use and downstream irrigation. Dhikur Pokhari village in Nepal showcases how collective governance principles lead to efficient management of water sources, forest restoration and indigenous grazing practices (Thapa *et al.*, 2020).

Education and awareness

Education and awareness are crucial in protecting glaciers and permafrost. By raising public awareness of climate change, environmental protection and sustainable development, actions can be taken to mitigate their impacts on fragile mountain ecosystems. Here are some effective methods of protecting glaciers and permafrost through education and publicity in the region:

1. Environmental education programme. This programme raises awareness of ecosystem services and provides practical training for sustainable waste management, eco-friendly food, hiking practices, and community projects that promote local livelihoods. For example, Adventure Consultants organizes an Environmental Management and Mountaineering Training Programme in Nepal (Kc, 2013).

2. Public outreach initiatives. The non-profit organization, GlacierWorks, uses high-resolution gigabit pixel images of the Himalaya mountains to educate people on the impact of climate change through exhibitions, educational projects and outreach activities.

3. Ecotourism. Responsible tourism and reduction of ecological footprints should be encouraged by setting limits on the maximum number of tourists and promoting low-density travel in areas with exemplary mountains, which is practised in Bhutan's ecotourism industry (Emmett and Nye, 2017).

4. Capacity building and training. Community awareness, skills and confidence to carry out projects related to protection through capacity building and training programmes, such as the IHR2012 project in India, should be encouraged (Negi and Maikhuri, 2012).

18.8 Glacier and Permafrost Protection Management

Strong collaboration between researchers and local communities and governments to develop new methods for managing resources, strategies, adaptations, and methods for the protection and management of glaciers and permafrost is needed. Below are some effective strategies and innovative methods for protecting and managing glaciers and permafrost in the region.

18.8.1 Integrated watershed management

To improve water management practices, an integrated approach should be adopted that considers ecological, social and economic factors. Such integration helps to protect vital ecosystems while addressing the various pressures related to water demand and usage. The Watershed Management Training and Research Institute at Kathmandu University, Nepal, has implemented an Integrated Watershed Management Plan for the Rapti basin, which provides capacity-building support to stakeholders and promotes businesses, governmental entities and community groups (Achet and Fleming, 2006).

18.8.2 Renewable energy

Local renewable energy alternatives, such as micro-hydropower projects and solar and wind energy, represent solutions for boosting the existing energy infrastructure, while minimizing carbon emissions and reducing reliance on natural resources such as wood and fuel. Bhutan has implemented a national policy of relying solely on hydropower as its main energy source, which has helped to reduce greenhouse gas emissions while preserving the country's rich biodiversity (Kamei *et al.*, 2021).

18.8.3 Reducing disaster risk

Reduction in disasters is crucial, as catastrophic events such as floods and landslides can cause tremendous damage to mountain infrastructure, local communities and the natural environment. The Kochi basin programme, supported by ICIMOD, has assisted stakeholders in developing satellite- and sensor-based technological inputs for effective disaster risk reduction policies within its watershed (Sharma *et al.*, 2019).

18.8.4 Community involvement

Community participation is important in managing glaciers and permafrost. Local communities possess valuable traditional knowledge and practices that provide solutions for management decisions and help build resilience to changing environmental conditions. The local communities in the HKH region have a close relationship with socio-cultural norms and beliefs, political factors and lifestyle choices. Bottom-up participation has facilitated the establishment of a business priority growth model and stimulated activities aimed at achieving common protection goals. For example, Bhutan has exhibited an exemplary leadership role in collective tree planting across ten villages and improved spring-water management through the implementation of community-led protection guidelines (Gurung *et al.*, 2006). Through community involvement in decision-making processes, coupled with access to resources and tools necessary for sustainable management, it is possible to enhance the long-term viability of these ecosystems (Molden *et al.*, 2016; Zimmer *et al.*, 2022).

18.8.5 Research and monitoring

Scientifically led research and monitoring are essential for tracking the impacts on glaciers and permafrost to establish early warning systems for upcoming threats. China has launched a special project called 'Third Pole Environment', emphasizing scientific co-operation among international researchers, forming interdisciplinary teams, and monitoring and analyzing the impact of climate change and the dynamics of cryosphere changes in some key watersheds in the HKH region (Marsden, 2017). It is also necessary to monitor the physical characteristics of glaciers and permafrost, such as thickness, range and distribution. This information can be utilized to identify areas at risk of degradation or collapse, and enable intervention measures. Advancements in remote-sensing technology have made it easier and more cost-effective to monitor glaciers and permafrost (Niu *et al.*, 2018; Pandey *et al.*, 2023). As awareness of the importance of glaciers and permafrost increases, so does the need for improving management strategies.

In recent years, comprehensive methods have been developed that consider both the physical and ecological aspects of these natural resources.

18.8.6 Ecological sustainability

Communities, policy-makers and researchers aim to safeguard glaciers and permafrost in the HKH area while promoting sustainable development. For this to occur efficiently, an understanding of the processes that support glacier and permafrost ecosystems is crucial. The ecosystems are highly complex and interdependent, with soil microorganisms playing an essential role in nutrient cycling and ecosystem functions. The inclusion of traditional knowledge and practices should be encouraged, as they provide valuable insights into these ecosystems (Pedersen *et al.*, 2020; Ramm *et al.*, 2022).

However, implementing these measures throughout the region presents challenges such as inadequate funding from governments and international institutions, insufficient institutional capacity, unequal sharing of knowledge, poor utilization of best practices, and using measurement standards that emphasize economic benefits over social equality. Therefore, to protect glaciers and permafrost effectively, conservation and management strategies should be reviewed and updated regularly, based on reliable science and input from local communities and stakeholders.

References

Abbas, S., Dastgeer, G., Yaseen, M. and Latif, Y. (2022) Land-use change impacts on soil and vegetation attributes in the Kanshi River basin, Potohar Plateau, Pakistan. *Land Degradation and Development* 33(15), 2649–2662. DOI: 10.1002/ldr.4252

Achet, S.H. and Fleming, B. (2006) A watershed management framework for mountain areas: lessons from 25 years of watershed conservation in Nepal. *Journal of Environmental Planning and Management* 49(5), 675–694. DOI: 10.1080/09640560600849962

Adhikari, L., Khan, B., Joshi, S., Ruijun, L., Ali, G. *et al.* (2021) Community-based trophy hunting programs secure biodiversity and livelihoods: learnings from Asia's high mountain communities and landscapes. *Environmental Challenges* 4: 100175. DOI: 10.1016/j.envc.2021.100175

Ali, S.N., Pandey, P., Singh, P., Mishra, S., Shekhar, M. *et al.* (2023) Intimidating evidences of climate change from the higher Himalaya: a case study from Lahaul, Himachal Pradesh, India. *Journal of the Indian Society of Remote Sensing* 1–14. DOI: 10.1007/s12524-023-01686-0

Allen, R. (2019) Strategies to address land degradation issues in the Hindu Kush – Himalayas. In: Bridges, E.M. (ed.) *Response to Land Degradation*. CRC Press, Boca Raton, Florida, pp. 304–313.

Allison, E.A. (2015) The spiritual significance of glaciers in an age of climate change. *Wiley Interdisciplinary Reviews: Climate Change* 6(5), 493–508. DOI: 10.1002/wcc.354

Anand, A., Chandan, P. and Singh, R.B. (2012) Homestays at Korzok: supplementing rural livelihoods and supporting green tourism in the Indian Himalayas. *Mountain Research and Development* 32(2), 126–136. DOI: 10.1659/MRD-JOURNAL-D-11-00109.1

Anderson, K., Fawcett, D., Cugulliere, A., Benford, S., Jones, D. *et al.* (2020) Vegetation expansion in the subnival Hindu Kush Himalaya. *Global Change Biology* 26(3), 1608–1625. DOI: 10.1111/gcb.14919

Anjum, S.A., Wang, L., Xue, L.L., Saleem, M.F., Wang, G.X. *et al.* (2010) Desertification in Pakistan: causes, impacts and management. *Journal of Food Agriculture & Environment* 8, 1203–1208.

Aryal, J.P., Sapkota, T.B., Khurana, R., Khatri-Chhetri, A., Rahut, D.B. *et al.* (2020) Climate change and agriculture in South Asia: adaptation options in smallholder production systems. *Environment, Development and Sustainability* 22(6), 5045–5075. DOI: 10.1007/s10668-019-00414-4

Aryal, S., Maraseni, T.N. and Cockfield, G. (2022) Changes in transhumance systems in Nepal: analysing socio-ecological impacts using driver-pressure-state-impact-response framework. In: Timsina, J., Maraseni, T.N., Gauchan, D., Adhikari, J. and Ojha, H. (eds) *Agriculture, Natural Resources and Food Security: Lessons from Nepal.* Springer International Publishing, Cham, Switzerland, pp. 297–314.

Ashraf, A., Naz, R. and Roohi, R. (2012) Glacial lake outburst flood hazards in Hindukush, Karakoram and Himalayan Ranges of Pakistan: implications and risk analysis. *Geomatics, Natural Hazards and Risk* 3(2), 113–132. DOI: 10.1080/19475705.2011.615344

Ashraf, A., Iqbal, M.B., Mustafa, N., Naz, R. and Ahmad, B. (2021) Prevalent risk of glacial lake outburst flood hazard in the Hindu Kush-Karakoram-Himalaya region of Pakistan. *Environmental Earth Sciences* 80(12), 451. DOI: 10.1007/s12665-021-09740-1

Baig, M.B., Burgess, P.J. and Fike, J.H. (2021) Agroforestry for healthy ecosystems: constraints, improvement strategies and extension in Pakistan. *Agroforestry Systems* 95, 995–1013. DOI: 10.1007/s10457-019-00467-4

Bajracharya, B., Thapa, R.B. and Matin, M.A. (2021) *Earth Observation Science and Applications for Risk Reduction and Enhanced Resilience in Hindu Kush Himalaya Region: A Decade of Experience from SERVIR*. Springer Nature, Cham, Switzerland.

Bajracharya, S.R. and Shrestha, B.R. (2011) *The Status of Glaciers in the Hindu Kush Himalayan Region*. International Centre for Integrated Mountain Development (ICIMOD), Kathmandu.

Bajracharya, S.R., Maharjan, S.B., Shrestha, F., Guo, W., Liu, S. *et al.* (2015) The glaciers of the Hindu Kush Himalayas: current status and observed changes from the 1980s to 2010. *International Journal of Water Resources Development* 31(2), 161–173. DOI: 10.1080/07900627.2015.1005731

Banerjee, A. and Bandopadhyay, R. (2016) Biodiversity hotspot of Bhutan and its sustainability. *Current Science* 521–527.

Beer, C., Zimov, N., Olofsson, J., Porada, P. and Zimov, S. (2020) Protection of permafrost soils from thawing by increasing herbivore density. *Scientific Reports* 10(1), 1–10. DOI: 10.1038/s41598-020-60938-y

Benn, D. and Evans, D.J.A. (2014) *Glaciers and Glaciation*. Routledge, London.

Bhadwal, S., Sharma, G., Gorti, G. and Sen, S.M. (2019) Livelihoods, gender and climate change in the Eastern Himalayas. *Environmental Development* 31, 68–77. DOI: 10.1016/j.envdev.2019.04.008

Bhandari, M.P. (2022) *Green Web-II: Standards and Perspectives from the IUCN Program/Policy Development in Environment Conservation Domain-with reference to India, Pakistan, Nepal, and Bangladesh.* CRC Press, Boca Raton, Florida.

Bhateria, R. and Jain, D. (2016) Water quality assessment of lake water: a review. *Sustainable Water Resources Management* 2, 161–173. DOI: 10.1007/s40899-015-0014-7

Bhatt, I.D., Negi. V.S. and Rawal, R.S. (2020) Promoting nature-based solution (NbS) through restoration of degraded landscapes in the Indian Himalayan Region. *Nature-based Solutions for Resilient Ecosystems and Societies*, 197–211. DOI: 10.1007/978-981-15-4712-6_12

Bhattacharya, A., Bolch, T., Mukherjee, K., King, O., Menounos, B. *et al.* (2021) High Mountain Asian glacier response to climate revealed by multi-temporal satellite observations since the 1960s. *Nature Communications* 12(1), 4133. DOI: 10.1038/s41467-021-24180-y

Biswas, M.K. (2020) Human population growth: impacts on the environment, human health welfare. Available at: https://surendranathcollege.ac.in/new/upload/INDRANIL_KARUNIT%2072020-05-17ENVS%20UNIT%207%20Human%20Communities%20and%20the%20Environment.pdf (accessed 28 September 2023).

Bolch, T., Kulkarni, A., Kääb, A., Huggel, C., Paul, F. *et al.* (2012) The state and fate of Himalayan glaciers. *Science* 336(6079), 310–314. DOI: 10.1126/science.1215828

Bolch, T., Shea, J.M., Liu, S., Azam, F.M., Gao, Y. *et al.* (2019) Status and change of the cryosphere in the extended Hindu Kush Himalaya region. In: Wester, P., Mishra, A., Mukherji, A. Shrestha, A. (eds) *The Hindu Kush Himalaya Assessment: Mountains, Climate Change, Sustainability and People.* Springer Nature, Cham, Switzerland, pp. 209–255.

Bouchard, F., MacDonald, L.A., Turner, K.W., Thienpont, J.R., Medeiros, A.S. *et al.* (2016) Paleolimnology of thermokarst lakes: a window into permafrost landscape evolution. *Arctic Science* 3(2), 91–117. DOI: 10.1139/as-2016-0022

Brunet, S., Bauer, J., De Lacy, T. and Tshering, K. (2001) Tourism development in Bhutan: tensions between tradition and modernity. *Journal of Sustainable Tourism* 9(3), 243–263. DOI: 10.1080/09669580108667401

Cao, W., Wu, D., Huang, L. and Liu, L. (2020) Spatial and temporal variations and significance identification of ecosystem services in the Sanjiangyuan National Park, China. *Scientific Reports* 10(1), 6151. DOI: 10.1038/s41598-020-63137-x

Carey, M., Baraer, M., Mark, B.G., French, A., Bury, J. *et al.* (2014) Toward hydro-social modeling: merging human variables and the social sciences with climate-glacier run-off models (Santa River, Peru). *Journal of Hydrology* 518, 60–70. DOI: 10.1016/j.jhydrol.2013.11.006

Chazdon, R.L. and Guariguata, M.R. (2016) Natural regeneration as a tool for large-scale forest restoration in the tropics: prospects and challenges. *Biotropica* 48(6), 716–730. DOI: 10.1111/btp.12381

Chen, H., Zhu, Q., Peng, C., Wu, N., Wang, Y. *et al.* (2013) The impacts of climate change and human activities on biogeochemical cycles on the Qinghai-Tibetan Plateau. *Global Change Biology* 19(10), 2940–2955. DOI: 10.1111/gcb.12277

Chen, H., Chen, Y., Li, W. and Li, Z. (2019) Quantifying the contributions of snow/glacier meltwater to river runoff in the Tianshan Mountains, Central Asia. *Global and Planetary Change* 174, 47–57. DOI: 10.1016/j.gloplacha.2019.01.002

Chen, H., Ju, P., Zhu, Q., Xu, X., Wu, N. *et al.* (2022a) Carbon and nitrogen cycling on the Qinghai-Tibetan plateau. *Nature Reviews Earth and Environment* 3(10), 701–716. DOI: 10.1038/s43017-022-00344-2

Chen, N., Liu, M., Allen, S., Deng, M., Khanal, N.R. *et al.* (2023) Small outbursts into big disasters: earthquakes exacerbate climate-driven cascade processes of the glacial lakes failure in the Himalayas. *Geomorphology* 422, 108539. DOI: 10.1016/j.geomorph.2022.108539

Chen, S., Zhang, R., Mao, R., Zhang, Y., Chen, Y. *et al.* (2022b) Sources, characteristics and climate impact of light-absorbing aerosols over the Tibetan Plateau. *Earth-Science Reviews* 232, 104111. DOI: 10.1016/j.earscirev.2022.104111

Chen, Z., Ge, S., Zhang, Z., Du, Y., Yao, B. *et al.* (2021) Soil moisture but not warming dominates nitrous oxide emissions during freeze–thaw cycles in a Qinghai-Tibetan plateau alpine meadow with discontinuous permafrost. *Frontiers in Ecology and Evolution* 9, 676027. DOI: 10.3389/fevo.2021.676027

Chimi, K.M.A.A., Thinley, J., Cheki, T., Chettri, A. and Chettri, N. (2023) *Architectural Insights and City Profile of Phuentsholing Bhutan*. DOI: 10.13189/cea.2023.110220

Cochand, M., Molson, J. and Lemieux, J.M. (2019) Groundwater hydrogeochemistry in permafrost regions. *Permafrost and Periglacial Processes* 30(2), 90–103. DOI: 10.1002/ppp.1998

Crozier, M.J. (2010) Deciphering the effect of climate change on landslide activity: a review. *Geomorphology* 124(3–4), 260–267. DOI: 10.1016/j.geomorph.2010.04.009

Davidson, E.A. and Janssens, I.A. (2006) Temperature sensitivity of soil carbon decomposition and feedbacks to climate change. *Nature* 440(7081), 165–173. DOI: 10.1038/nature04514

Davis, D.K. and Robbins, P. (2018) Ecologies of the colonial present: pathological forestry from the taux de boisement to civilized plantations. *Environment and Planning E: Nature and Space* 1(4), 447–469. DOI: 10.1177/2514848618812029

Devkota, D., Miller, D.C., Wang, S.W. and Brooks, J.S. (2023) Biodiversity conservation funding in Bhutan: thematic, temporal, and spatial trends over four decades. *Conservation Science and Practice* e12757. DOI: 10.1111/csp2.12757

Dhimal, M., Bhandari, D., Dhimal, M.L., Kafle, N., Pyakurel, P. *et al.* (2021) Impact of climate change on health and well-being of people in Hindu Kush Himalayan region: a narrative review. *Frontiers in Physiology* 12, 651189. DOI: 10.3389/fphys.2021.651189

Dhyani, S., Gupta, A.K. and Karki, M. (2020) *Nature-based Solutions for Resilient Ecosystems and Societies*. Springer Singapore.

Di Sacco, A., Hardwick, K.A., Blakesley, D., Brancalion, P.H., Breman, E. *et al.* (2021) Ten golden rules for reforestation to optimize carbon sequestration, biodiversity recovery and livelihood benefits. *Global Change Biology* 27(7), 1328–1348. DOI: 10.1111/gcb.15498

Dimri, A.P., Allen, S., Huggel, C., Mal, S., Ballesteros-Canovas, J.A. *et al.* (2021) Climate change, cryosphere and impacts in the Indian Himalayan Region. *Current Science.* DOI: 10.18520/cs/v120/i5/774-790

Ding, Y., Mu, C., Wu, T., Hu, G., Zou, D. *et al.* (2021) Increasing cryospheric hazards in a warming climate. *Earth-Science Reviews* 213, 103500. DOI: 10.1016/j.earscirev.2020.103500

Dong, S., Liu, S. and Wen, L. (2016) Vulnerability and resilience of human-natural systems of pastoralism worldwide. *Building Resilience of Human-Natural Systems of Pastoralism in the Developing World: Interdisciplinary Perspectives* 39–92. DOI: 10.1007/978-3-030-20591-1_6

Douglas, T.A., Turetsky, M.R. and Koven, C.D. (2020) Increased rainfall stimulates permafrost thaw across a variety of Interior Alaskan boreal ecosystems. *NPJ Climate and Atmospheric Science* 3(1), 28. DOI: 10.1038/s41612-020-0130-4

Ebi, K.L., Woodruff, R., von Hildebrand, A. and Corvalan, C. (2007) Climate change-related health impacts in the Hindu Kush-Himalayas. *EcoHealth* 4(3), 264–270. DOI: 10.1007/s10393-007-0119-z

Ehlers, T.A., Chen, D., Appel, E., Bolch, T., Chen, F. *et al.* (2022) Past, present, and future geo-biosphere interactions on the Tibetan Plateau and implications for permafrost. *Earth-Science Reviews* 104197. DOI: 10.1016/j.earscirev.2022.104197

Elad, Y. and Pertot, I. (2014) Climate change impacts on plant pathogens and plant diseases. *Journal of Crop Improvement* 28(1), 99–139. DOI: 10.1080/15427528.2014.865412

Emmett, R.S. and Nye, D.E. (2017) *The Environmental Humanities: A Critical Introduction.* MIT Press, Massachusetts.

Engelbrecht, F. and Monteiro, P. (2021) *Climate Change: The IPCC's Latest Assessment Report.* Available at: https://hdl.handle.net/10520/ejc-quest-v17-n3-a12 (accessed 28 September 2023).

Ercumen, A., Arnold, B.F., Kumpel, E., Burt, Z., Ray, I. *et al.* (2015) Upgrading a piped water supply from intermittent to continuous delivery and association with waterborne illness: a matched cohort study in urban India. *PLOS Medicine* 12(10), e1001892. DOI: 10.1371/journal.pmed.1001892

Ezhova, E., Orlov, D., Suhonen, E., Kaverin, D., Mahura, A. *et al.* (2021) Climatic factors influencing the anthrax outbreak of 2016 in Siberia, Russia. *EcoHealth* 18(2), 217–228. DOI: 10.1007/s10393-021-01549-5

Fayiah, M., Dong, S.K., Khomera, S.W., Ur Rehman, S.A., Yang, M. *et al.* (2020) Status and challenges of Qinghai-Tibet Plateau's grasslands: an analysis of causes, mitigation measures, and way forward. *Sustainability* 12(3), 1099. DOI: 10.3390/su12031099

Gagné, K., Rasmussen, M.B. and Orlove, B. (2014) Glaciers and society: attributions, perceptions, and valuations. *Wiley Interdisciplinary Reviews: Climate Change* 5(6), 793–808. DOI: 10.1002/wcc.315

Gibson, C.M., Estop-Aragonés, C., Flannigan, M., Thompson, D.K. and Olefeldt, D. (2019) Increased deep soil respiration detected despite reduced overall respiration in permafrost peat plateaus following wildfire. *Environmental Research Letters* 14(12), 125001. DOI: 10.1088/1748-9326/ab4f8d

Giese, A., Rupper, S., Johnson, E. and Forster, R. (2022) Indus River Basin glacier melt at the subbasin scale. *Frontiers in Earth Science* 10, 352. DOI: 10.3389/feart.2022.767411

Gokarakonda, S., Shrestha, S., Caleb, P.R., Rathi, V., Jain, R. *et al.* (2019) Decoupling in India's building construction sector: trends, technologies and policies. *Building Research and Information* 47(1), 91–107. DOI: 10.1080/09613218.2018.1490054

Graham, P. and Rawal, R. (2019) Achieving the 2 C goal: the potential of India's building sector. *Building Research and Information* 47(1), 108–122. DOI: 10.1080/09613218.2018.1495803

Grießinger, J., Meier, W.J.H., Bast, A., Debel, A., Gärtner-Roer, I. *et al.* (2021) Permafrost biases climate signals in $\delta^{18}O_{\text{tree-ring}}$ series from a sub-alpine tree stand in Val Bever, Switzerland. *Atmosphere* 12(7), 836. DOI: 10.3390/atmos12070836

Gruber, S., Fleiner, R., Guegan, E., Panday, P., Schmid, M.O. *et al.* (2017) Inferring permafrost and permafrost thaw in the mountains of the Hindu Kush Himalaya region. *The Cryosphere* 11(1), 81–99. DOI: 10.5194/tc-11-81-2017

Gu, Z., Gong, J. and Wang, Y. (2023) Construction and evaluation of ecological networks among natural protected areas based on 'quality-structure-function': a case study of the Qinghai-Tibet area. *Ecological Indicators* 151, 110228. DOI: 10.1016/j.ecolind.2023.110228

Gulzar, A., Islam, T., Gulzar, R. and Hassan, T. (2021) Climate change and impacts of extreme events on human health: an overview. *Indonesian Journal of Social and Environmental Issues (IJSEI)* 2(1), 68–77. DOI: 10.47540/ijsei.v2i1.180

Gurung, D.R., Maharjan, S.B., Shrestha, A.B., Shrestha, M.S., Bajracharya, S.R. *et al.* (2017) Climate and topographic controls on snow cover dynamics in the Hindu Kush Himalaya. *International Journal of Climatology* 37(10), 3873–3882. DOI: 10.1002/joc.4961

Gurung, T.R., Bousquet, F. and Trébuil, G. (2006) Companion modeling, conflict resolution, and institution building: sharing irrigation water in the Lingmuteychu Watershed, Bhutan. *Ecology and Society* 11(2).

Han, Z., Song, W., Deng, X. and Xu, X. (2018) Grassland ecosystem responses to climate change and human activities within the Three-River Headwaters region of China. *Scientific Reports* 8(1), 1–13. DOI: 10.1038/s41598-018-27150-5

Hansen, J. and Nazarenko, L. (2004) Soot climate forcing via snow and ice albedos. *Proceedings of the National Academy of Sciences* 101(2), 423–428. DOI: 10.1073/pnas.2237157100

Hanson, J.H., Schutgens, M., Baral, N. and Leader-Williams, N. (2022) Assessing the potential of snow leopard tourism-related products and services in the Annapurna Conservation Area, Nepal. *Tourism Planning and Development* 1–21.

Harper, A.B., Powell, T., Cox, P.M., House, J., Huntingford, C. *et al.* (2018) Land-use emissions play a critical role in land-based mitigation for Paris climate targets. *Nature Communications* 9(1), 2938. DOI: 10.1038/s41467-018-05340-z

Harrison, S., Kargel, J.S., Huggel, C., Reynolds, J., Shugar, D.H. *et al.* (2018) Climate change and the global pattern of moraine-dammed glacial lake outburst floods. *The Cryosphere* 12(4), 1195–1209. DOI: 10.5194/tc-12-1195-2018

Harvell, C.D., Mitchell, C.E., Ward, J.R., Altizer, S., Dobson, A.P. *et al.* (2002) Climate warming and disease risks for terrestrial and marine biota. *Science* 296(5576), 2158–2162. DOI: 10.1126/science.1063699

Heckmann, T. and Morche, D. (2019) Glacier changes since the Little Ice Age. In: Heckmann, T. and Morche, D. (eds) *Geomorphology of Proglacial Systems: Landform and Sediment Dynamics in Recently Deglaciated Alpine Landscapes*. Springer, Cham, Switzerland, pp. 23–42.

Heginbottom, J.A., Brown, J., Melnikov, E.S., Ferrians Jr, O.J. (1993) Circumarctic map of permafrost and ground ice conditions. *Proceedings of the Sixth International Conference on Permafrost* 2, 1132–1136. South China University of Technology Press, Wushan Guangzhou, China.

Hjort, J., Streletskiy, D., Doré, G., Wu, Q., Bjella, K. *et al.* (2022) Impacts of permafrost degradation on infrastructure. *Nature Reviews Earth and Environment* 3(1), 24–38. DOI: 10.1038/s43017-021-00247-8

Huggel, C., Clague, J.J. and Korup, O. (2012) Is climate change responsible for changing landslide activity in high mountains? *Earth Surface Processes and Landforms*, 37(1), 77–91. DOI: 10.1002/esp.2223

Inskipp, C., Grimmett, R. and Inskipp, T. (2020) *Birds of Bhutan and the Eastern Himalayas*. Bloomsbury Publishing, London.

Jeelani, G., Feddema, J.J., van der Veen, C.J. and Stearns, L. (2012) Role of snow and glacier melt in controlling river hydrology in Liddar watershed (western Himalaya) under current and future climate. *Water Resources Research* 48(12). DOI: 10.1029/2011WR011590

Johnston, R. and Smakhtin, V. (2014) Hydrological modeling of large river basins: How much is enough? *Water Resources Management* 28, 2695–2730. DOI: 10.1007/s11269-014-0637-8

Kääb, A., Frauenfelder, R. and Roer, I. (2007) On the response of rockglacier creep to surface temperature increase. *Global and Planetary Change*, 56(1–2), 172–187. DOI: 10.1016/j.gloplacha.2006.07.005

Kamei, M., Wangmo, T., Leibowicz, B.D. and Nishioka, S. (2021) Urbanization, carbon neutrality, and Gross National Happiness: sustainable development pathways for Bhutan. *Cities* 111, 102972. DOI: 10.1016/j.cities.2020.102972

Kang, S., Zhang, Q., Qian, Y., Ji, Z., Li, C. *et al.* (2019) Linking atmospheric pollution to cryospheric change in the Third Pole region: current progress and future prospects. *National Science Review* 6(4), 796–809. DOI: 10.1093/nsr/nwz031

Kargel, J.S., Leonard, G.J., Shugar, D.H., Haritashya, U.K., Bevington, A. *et al.* (2016) Geomorphic and geologic controls of geohazards induced by Nepal's 2015 Gorkha earthquake. *Science* 351(6269), aac8353. DOI: 10.1126/science.aac8353

Karjalainen, O., Luoto, M., Aalto, J., Etzelmüller, B., Grosse, G. *et al.* (2020) High potential for loss of permafrost landforms in a changing climate. *Environmental Research Letters* 15(10), 104065. DOI: 10.1088/1748-9326/abafd5

Karl, T.R., Melillo, J.M. and Peterson, T.C. (2009) *Global Climate Change Impacts in the United States: A State of Knowledge Report from the US Global Change Research Program*. Cambridge University Press, UK.

Kasymov, U. and Thiel, A. (2019) Understanding the role of power in changes to pastoral institutions in Kyrgyzstan. *International Journal of the Commons* 13(2), 931–948.

Kavian, A., Azmoodeh, A. and Solaimani, K. (2014) Deforestation effects on soil properties, runoff and erosion in northern Iran. *Arabian Journal of Geosciences* 7, 1941–1950. DOI: 10.1007/s12517-013-0853-1

Kc, G. (2013) Environmental management and sustainable tourism development in the Annapurna region, Nepal. Doctoral thesis, Buckinghamshire New University, UK.

Khalid, Z., Meng, X., Rana, I.A., ur Rehman, M. and Su, X. (2021) Holistic multidimensional vulnerability assessment: an empirical investigation on rural communities of the Hindu Kush Himalayan region, northern Pakistan. *International Journal of Disaster Risk Reduction* 62, 102413. DOI: 10.1016/j.ijdrr.2021.102413

Khanduri, S. (2022) Disastrous events of 2021 in Uttarakhand Province of India: causes, consequences and suggestions for disaster risk reduction (DRR). *International Journal of Earth Sciences Knowledge and Applications* 4(2), 178–188.

Kishore, A., Joshi, P.K. and Pandey, D. (2019). Droughts, distress, and policies for drought-proofing agricul ture in Bihar, India. In: Pal, B., Kishore, A., Joshi, P. and Tyagi, N. (eds) *Climate Smart Agriculture in South Asia.* Springer, Singapore.

Klein, J.A., Tucker, C.M., Steger, C.E., Nolin, A., Reid, R. *et al.* (2019) An integrated community and ecosystem-based approach to disaster risk reduction in mountain systems. *Environmental Science and Policy* 94, 143–152. DOI: 10.1016/j.envsci.2018.12.034

Klisz, M., Chakraborty, D., Cvjetković, B., Grabner, M., Lintunen, A. *et al.* (2023) Functional traits of boreal species and adaptation to local conditions. In: Girona, M.M., Morin, H., Gauthier, S. and Bergeron, Y. (eds) *Boreal Forests in the Face of Climate Change: Sustainable Management.* Springer, Cham, Switzerland, pp. 323–355.

Kohler, T. and Maselli, D. (2009) *Mountains and Climate Change: From Understanding to Action.* Geographica Bernensia, Bern.

Kohler, T., Wehrli, A. and Jurek, M. (2014) *Mountains and Climate Change: A Global Concern.* Sustainable Mountain Development Series. Centre for Development and Environment (CDE), Swiss Agency for Development and Cooperation (SDC) and Geographica Bernensia, Bern.

Krautblatter, M., Funk, D. and Günzel, F.K. (2013) Why permafrost rocks become unstable: a rock-ice-mechanical model in time and space. *Earth Surface Processes and Landforms* 38(8), 876–887. DOI: 10.1002/esp.3374

Kundzewicz, Z.W. (2008) Climate change impacts on the hydrological cycle. *Ecohydrology and Hydrobiology* 8(2–4), 195–203. DOI: org/10.2478/v10104-009-0015-y

Kunstler, G., Falster, D., Coomes, D.A., Hui, F., Kooyman, R.M. *et al.* (2016) Plant functional traits have globally consistent effects on competition. *Nature* 529(7585), 204–207. DOI: 10.1038/nature16476

Lacroix, F., Zaehle, S., Caldararu, S., Schaller, J., Stimmler, P. *et al.* (2022). Mismatch of N release from the permafrost and vegetative uptake opens pathways of increasing nitrous oxide emissions in the high Arctic. *Global Change Biology* 28(20), 5973–5990. DOI: 10.1111/gcb.16345

Li, D.F., Lu, X.X., Walling, D.E., Zhang, T., Steiner, J.F. *et al.* (2022a) High Mountain Asia hydropower systems threatened by climate-driven landscape instability. *Nature Geoscience* 15(7), 520–530. DOI:10.1038/s41561-022-00953-y

Li, K.L., Chen, R.S., Liu, G.H. and Yong, Y. (2022b) Cryosphere water resources supply function and service in China. *Advances in Climate Change Research* 13(3), 408–420. DOI: 10.1016/j.accre.2022.03.004

Li, W.P., Wang, W., Gao, X., Wang, X.C., Wang, R.H. *et al.* (2022c) Inventory and spatiotemporal patterns of glacial lakes in the HKH-TMHA region from 1990 to 2020. *Remote Sensing* 14(6), 1351. DOI: 10.3390/rs14061351

Li, X.P., Wang, L., Hu, B.Y., Chen, D.L., Liu, R.S. *et al.* (2023) Contribution of vanishing mountain glaciers to global and regional terrestrial water storage changes. *Frontiers in Earth Science* 11, 269. DOI: 10.3389/feart.2023.1134910

Liu, Y.S., Fan, J.W., Harris, W., Shao, Q.Q., Zhou, Y.C. *et al.* (2012) Effects of plateau pika (*Ochotona curzoniae*) on net ecosystem carbon exchange of grassland in the Three Rivers Headwaters region, Qinghai-Tibet, China. *Plant and Soil* 366, 491–504. DOI: 10.1007/s11104-012-1442-x

Liu, Z.F., Yao, Z., Huang, H.Q., Wu, S., Liu, G. *et al.* (2014) Land use and climate changes and their impacts on runoff in the Yarlung Zangbo river basin, China. *Land Degradation and Development* 25(3), 203–215. DOI: 10.1002/ldr.1159

Lorenz, D.F. and Dittmer, C. (2021) Disasters in the 'abode of gods': vulnerabilities and tourism in the Indian Himalaya. *International Journal of Disaster Risk Reduction* 55, 102054. DOI: 10.1016/j.ijdrr.2021.102054

Lv, Z.H. and Shang, W.L. (2022) Impacts of intelligent transportation systems on energy conservation and emission reduction of transport systems: a comprehensive review. *Green Technologies and Sustainability* 1(6), 100002. DOI: 10.1016/j.grets.2022.100002

Lytton, L., Ali, A., Garthwaite, B., Punthakey, J.F. and Saeed, B. (2021) *Groundwater in Pakistan's Indus Basin: Present and Future Prospects.* World Bank Group, Washington, DC.

Ma, T., Swallow, B., Zhong, L.S., Xu, K., Sang, W.G. *et al.* (2022) Local perspectives on social-ecological transformation: China's Sanjiangyuan National Park. *Environment Development and Sustainability* 1–21. DOI: 10.1007/s10668-022-02786-6

Mackelprang, R., Saleska, S., Jacobsen, C.S., Jansson, J.K. and Tas, N. (2016) Permafrost meta-omics and climate change. *Annual Review of Earth and Planetary Sciences* 44, 439–462. DOI: 10.1146/annurev-earth-060614-105126

Maharjan, S.B., Steiner, J.F., Shrestha, A.B., Maharjan, A., Nepal, S. *et al.* (2021) The Melamchi flood disaster: cascading hazard and the need for multihazard risk management. *International Centre for Integrated Mountain Development,* Kathmandu.

Manish, K., Telwala, Y., Nautiyal, D.C. and Pandit, M.K. (2016) Modelling the impacts of future climate change on plant communities in the Himalaya: a case study from Eastern Himalaya, India. *Modeling Earth Systems and Environment* 2, 1–12. DOI: 10.1007/s40808-016-0163-1

Margesin, R. and Collins, T. (2019) Microbial ecology of the cryosphere (glacial and permafrost habitats): current knowledge. *Applied Microbiology and Biotechnology* 103, 2537–2549. DOI: 10.1007/s00253-019-09631-3

Marsden, S. (2017) *Environmental Regimes in Asian Subregions: China and the Third Pole*. Edward Elgar Publishing, Cheltenham, UK.

Martha, T.R., Roy, P., Jain, N., Kumar, K.V., Reddy, S. *et al.* (2021) Rock avalanche induced flash flood on 07 February 2021 in Uttarakhand, India: a photogeological reconstruction of the event. *Landslides* 18(8), 2881–2893. DOI: 10.1007/s10346-021-01691-9

Mayewski, P.A., Perry, L.B., Matthews, T. and Birkel, S.D. (2020) Climate change in the Hindu Kush Himalayas: basis and gaps. *One Earth* 3(5), 551–555. DOI: 10.1016/j.oneear.2020.10.007

McDowell, G., Ford, J.D., Lehner, B., Berrang-Ford, L. and Sherpa, A. (2012) Climate-related hydrological change and human vulnerability in remote mountain regions: a case study from Khumbu, Nepal. *Regional Environmental Change* 13(2), 299–310. DOI: 10.1007/s10113-012-0333-2

Melis, M.T., Fugazza, D., Dessì, F.G., Gallo, M., Naitza, L. *et al.* (2023) Glaciers & students: the new glaciers inventory of Pakistan. *EGU General Assembly Conference Abstracts* (pp. EGU–13561). DOI: 10.5194/egusphere-egu23-13561

Millar, J. and Tenzing, K. (2023) High altitude rangelands and pastoralism in Bhutan: using sustainable development goals to address land degradation and poverty. Available at: https://uknowledge.uky.edu/igc/24/7-2/9/ (accessed 28 September 2023).

Miller, J.D., Immerzeel, W.W. and Rees, G. (2012) Climate change impacts on glacier hydrology and river discharge in the Hindu Kush-Himalayas. *Mountain Research and Development* 32(4), 461–467. DOI: 10.1659/MRD-JOURNAL-D-12-00027.1

Milner, A.M., Khamis, K., Battin, T.J., Brittain, J.J., Barrand, N.E. *et al.* (2027) Glacier shrinkage driving global changes in downstream systems. *Proceedings of the National Academy of Sciences* 114(37), 9770–9778. DOI: 10.1073/pnas.1619807114

Miner, K.R., D'Andrilli, J., Mackelprang, R., Edwards, A., Malaska, M.J. *et al.* (2021) Emergent biogeochemical risks from Arctic permafrost degradation. *Nature Climate Change* 11(10), 809–819. DOI: 10.1038/s41558-021-01162-y

Moffett, M., Fazio, M.W. and Wodehouse, L. (2003) *A World History of Architecture*. Laurence King Publishing, London.

Molden, D., Sharma, E. and Acharya, G. (2016) Lessons from Nepal's Gorkha earthquake 2015. In: Singh, S.P., Khanal, S.C., Joshi, M. (eds) *Lessons from Nepal's Earthquake for the Indian Himalayas and the Gangetic Plains*. Central Himalayan Environment Association (CHEA), Nainital, pp. 1–14.

Molden, D.J., Shrestha, A.B., Immerzeel, W.W., Maharjan, A., Rasul, G. *et al.* (2022) The great glacier and snow-dependent rivers of Asia and climate change: heading for troubled waters. *Water Security Under Climate Change* 223–250. DOI: 10.1007/978-981-16-5493-0_12

Motschmann, A., Huggel, C., Carey, M., Moulton, H., Walker-Crawford, N. *et al.* (2020) Losses and damages connected to glacier retreat in the Cordillera Blanca, Peru. *Climatic Change* 162, 837–858. DOI: 10.1007/s10584-020-02770-x

Mukherji, A., Sinisalo, A., Nüsser, M., Garrad, R. and Eriksson, M. (2019) Contributions of the cryosphere to mountain communities in the Hindu Kush Himalaya: a review. *Regional Environmental Change* 19(2), 1311–1326. DOI: 10.1007/s10113-019-01484-w

Nautiyal, S., Goswami, M., Prakash, S., Rao, K.S., Maikhuri, R.K. *et al.* (2022) Spatio-temporal variations of geo-climatic environment in a high-altitude landscape of Central Himalaya: an assessment from the perspective of vulnerability of glacial lakes. *Natural Hazards Research* 2(17), DOI: 10.1016/j.nhres.2022.07.003

Neale, P.J., Williamson, C.E., Banaszak, A., Häder, D., Hylander, S. *et al.* (2023) The response of aquatic ecosystems to the interactive effects of stratospheric ozone depletion, UV radiation, and climate change. *Photochemical and Photobiological Sciences* 22(3), 1–35. DOI: 10.1007/s43630-023-00370-z

Negi, G.C.S., Samal, P.K., Kuniyal, J.C., Sharma, R.K. and Dhyani, P.P. (2012) Impact of climate change on the western Himalayan mountain ecosystems: an overview. *Tropical Ecology,* 53(3), 345–356.

Negi, P.S., Pandey, C.P. and Singh, N. (2019) Black carbon aerosols in the ambient air of Gangotri Glacier valley of north-western Himalaya in India. *Atmospheric Environment* 214, 116879. DOI: 10.1016/j.atmosenv.2019.116879

Negi, V.S. and Maikhuri, R.K. (2012) Socio-ecological and religious perspective of agrobiodiversity conservation: issues, concern and priority for sustainable agriculture, Central Himalaya. *Journal of Agricultural and Environmental Ethics,* 26(2), 491–512. DOI: 10.1007/s10806-012-9386-y

Nepal, S., Flügel, W.A. and Shrestha, A.B. (2014) Upstream-downstream linkages of hydrological processes in the Himalayan region. *Ecological Processes* 3(19), 1–16. DOI: 10.1186/s13717-014-0019-4

Nepal, S., Pandey, A., Shrestha, A.B. and Mukherji, A. (2018) Revisiting key questions regarding upstream-downstream linkages of land and water management in the Hindu Kush Himalaya (HKH) Region. Himalayan Adaptation, Water and Resilience Research, Kathmandu.

Neuhoff, D., Tashi, S., Rahmann, G. and Denich, M. (2014) Organic agriculture in Bhutan: potential and challenges. *Organic Agriculture* 4, 209–221. DOI: 10.1007/s13165-014-0075-1

Nie, Y., Pritchard, H., Liu, Q., Hennig, T., Wang, W.L. *et al.* (2021) Glacial change and hydrological implications in the Himalaya and Karakoram. *Nature Reviews Earth and Environment* 2(2), 91–106. DOI: 10.1038/s43017-020-00124-w

Nielsen, K.S., Nicholas, K.A., Creutzig, F., Dietz, T. and Stern, P. (2021) The role of high-socioeconomic-status people in locking in or rapidly reducing energy-driven greenhouse gas emissions. *Nature Energy* 6(11), 1011–1016. DOI: 10.1038/s41560-021-00900-y

Nilsson, C., Polvi, L. and Lind, L. (2015) Extreme events in streams and rivers in arctic and subarctic regions in an uncertain future. *Freshwater Biology* 60(12), 2535–2546. DOI: 10.1111/fwb.12477

Niu, F.J., Yin, G.A., Luo, J., Lin, Z.J. and Liu, M.H. (2018) Permafrost distribution along the Qinghai-Tibet Engineering Corridor, China using high-resolution statistical mapping and modeling integrated with remote sensing and GIS. *Remote Sensing* 10(2), 215. DOI: 10.3390/rs10020215

Niu, Y.L., Li, S.Y., Liu, Y., Shi, J.J., Wang, Y.L. *et al.* (2021) Regulation of alpine meadow patch coverage on runoff and sediment under natural rainfall on the eastern Qinghai-Tibetan Plateau. *Journal of Hydrology* 603, 127101. DOI: 10.1016/j.jhydrol.2021.127101

Nkonya, E., Srinivasan, R., Anderson, W. and Edward, K. (2014) *Assessing the Economic Benefits of Sustainable Land Management Practices in Bhutan*. International Food Policy Research Institute (IFPRI), Washington, DC.

Nkonya, E., Mirzabaev, A. and Braun, J.V. (2016) *Economics of Land Degradation and Improvement: A Global Assessment for Sustainable Development*. Springer International Publishing, Cham, Switzerland.

Nyaupane, G.P., Poudel, S. and York, A. (2020) Governance of protected areas: an institutional analysis of conservation, community, and tourism outcomes. *Journal of Sustainable Tourism* 30(2), 2686–2705. DOI: 10.1080/09669582.2020.1858089

O'Connor, J.E. and Costa, J.E. (1993) Geologic and hydrologic hazards in glacierized basins in North America resulting from 19th and 20th century global warming. *Natural Hazards* 8(2), 121–140. DOI: 10.1007/BF00605437

Pandey, M., Pandey, P.C., Ray, Y., Arora, A., Jawak, S. *et al.* (2023) *Advances in Remote Sensing Technology and the Three Poles*. John Wiley and Sons, Hoboken, New Jersey.

Pathak, D. (2016) Water availability and hydrogeological condition in the Siwalik foothill of east Nepal. *Nepal Journal of Science and Technology* 17(1), 31–38. DOI: 10.3126/njst.v17i1.25061

Patton, A., Rathburn, S.L. and Capps, D.M. (2019) Landslide response to climate change in permafrost regions. *Geomorphology* 340, 116–128. DOI: 10.1016/j.geomorph.2019.04.029

Pedersen, E.P., Elberling, B. and Michelsen, A. (2020) Foraging deeply: depth-specific plant nitrogen uptake in response to climate-induced N-release and permafrost thaw in the High Arctic. *Global Change Biology* 26(11), 6523–6536. DOI: 10.1111/gcb.15306

Pender, J.S. (2010) *Climate Change, Its Impacts and Possible Community Based Responses in Bangladesh*. Church of Bangladesh, Dhaka.

Picoli, M.C.A. and Machado, P.G. (2021) Land use change: the barrier for sugarcane sustainability. *Biofuels, Bioproducts and Biorefining* 15(6), 1591–1603. DOI: 10.1002/bbb.2270

Pimm, A.J., Palczewski, J., Barbour, E.R. and Cockerill, T.T. (2021) Using electricity storage to reduce greenhouse gas emissions. *Applied Energy* 282, 116199. DOI: https://doi.org/10.1016/j.apenergy.2020.116199

Poudel, T.R., Aryal, P.C., Thapa-Parajuli, R., Thapa, A., Yadav, S.K. *et al.* (2022) Depredation loss drives human–wildlife conflict perception in the Trans-Himalayas. *Journal of Environmental Management* 311, 114763. DOI: 10.1016/j.jenvman.2022.114763

Pradhan, S.P., Vishal, V. and Singh, T.N. (2018) Finite element modelling of landslide prone slopes around Rudraprayag and Agastyamuni in Uttarakhand Himalayan terrain. *Natural Hazards* 94, 181–200. DOI: 10.1007/s11069-018-3381-1

Prakash, A. (2020) Retreating glaciers and water flows in the Himalayas: implications for governance. Observer Research Foundation, New Delhi.

Prakash, B., Kumar, K., Joshi, V. and Goyal, D. (2021) Causes of the triggering of Chamoli glacier burst of 7th February 2021 in Uttarakhand, India. *Disaster Advances* 14(7), 60–67. DOI: 10.25303/147da6021

Prăvălie, R. (2021) Exploring the multiple land degradation pathways across the planet. *Earth-Science Reviews* 220, 103689. DOI: 10.1016/j.earscirev.2021.103689

Ramm, E., Liu, C.Y., Ambus, P., Butterbach-Bahl, K., Hu, B. *et al.* (2022) A review of the importance of mineral nitrogen cycling in the plant-soil-microbe system of permafrost-affected soils: changing the paradigm. *Environmental Research Letters* 17(1), 013004. DOI: 10.1088/1748-9326/ac417e

Rasul, G. (2014) Food, water, and energy security in South Asia: a nexus perspective from the Hindu Kush Himalayan region. *Environmental Science and Policy* 39, 35–48. DOI: 10.1016/j.envsci.2014.01.010

Rasul, G. and Molden, D.J. (2019) The global social and economic consequences of mountain cryospheric change. *Frontiers in Environmental Science* 7, 91. DOI: 10.3389/fenvs.2019.00091

Regmi, G.R. and Huettmann, F. (2020) *Hindu Kush Himalaya Watersheds Downhill: Landscape Ecology and Conservation Perspectives.* Springer International Publishing, Cham, Switzerland.

Romeo, R., Manuelli, S., Geringer, M. and Barchiesi, V. (2021) *Mountain Farming Systems: Seeds for the Future: Sustainable Agricultural Practices for Resilient Mountain Livelihoods*. Food and Agriculture Organization of the United Nations, Rome.

Rounce, D.R., King, O., McCarthy, M., Shean, D.E. and Salerno, F. (2018) Quantifying debris thickness of debris-covered glaciers in the Everest region of Nepal through inversion of a subdebris melt model. *Journal of Geophysical Research: Earth Surface* 123(5), 1094–1115. DOI: 10.1029/2017JF004395

Rowan, A.V., Egholm, D.L., Quincey, D.J. and Glasser, N.F. (2015) Modelling the feedbacks between mass balance, ice flow and debris transport to predict the response to climate change of debris-covered glaciers in the Himalaya. *Earth and Planetary Science Letters* 430, 427–438. DOI: 10.1016/j.epsl.2015.09.004

Saikia, A., Pathak, B., Singh, P.K. and Pradip, B. (2023) Decadal changes in atmospheric methane emissions in the eastern Himalayan region: source apportionment and impact assessment. *International Journal of Environmental Research* 17(1), 10. DOI: 10.1007/s41742-022-00501-x

Sati, V.P. (2014) *Towards Sustainable Livelihoods and Ecosystems in Mountain Regions*. Springer International Publishing, Cham, Switzerland.

Sati, V.P. (2018) Cloudburst-triggered natural hazards in Uttarakhand Himalaya: mechanism, prevention, and mitigation. *International Journal of Geological and Environmental Engineering*,12(1), 45–48.

Satti, Z., Naveed, M., Shafeeque, M., Ali, S., Abdullaev, F. *et al.* (2023) Effects of climate change on vegetation and snow cover area in Gilgit Baltistan using MODIS data. *Environmental Science and Pollution Research* 30(7), 19149–19166. DOI: 10.1007/s11356-022-23445-3

Savi, S., Comiti, F. and Strecker, M.R. (2021) Pronounced increase in slope instability linked to global warming: a case study from the eastern European Alps. *Earth Surface Processes and Landforms* 46(7), 1328–1347. DOI: 10.1002/esp.5100

Sawhney, A. (2021) Striving towards a circular economy: climate policy and renewable energy in India. *Clean Technologies and Environmental Policy* 23, 491–499. DOI:10.1007/s10098-020-01935-7

Saxena, A., Raghuvanshi, M.S. and Suna, T. (2021) Traditional water management of subsistence agriculture system in cold arid Ladakh: a review. *International Journal of Scientific and Engineering Research* 12(4).

Schmid, M.O., Baral, P., Gruber, S., Shahi, S., Shrestha, T. *et al.* (2015) Assessment of permafrost distribution maps in the Hindu Kush Himalayan region using rock glaciers mapped in Google Earth. *The Cryosphere* 9(6), 2089–2099. DOI: 10.5194/tc-9-2089-2015

Schoch, N., Steimann, B. and Thieme, S. (2010) Migration and animal husbandry: competing or complementary livelihood strategies. *Natural Resources Forum* 34(3), 211–221. DOI: 10.1111/j.1477-8947.2010.01306.x

Schuur, E.A.G., McGuire, A.D., Schädel, C., Grosse, G., Harden, J.W. *et al.* (2015) Climate change and the permafrost carbon feedback. *Nature* 520(7546), 171–179. DOI: 10.1038/nature14338

Shah, A.A., Ye, J., Abid, M. and Ullah, R. (2017) Determinants of flood risk mitigation strategies at household level: a case of Khyber Pakhtunkhwa (KP) province, Pakistan. *Natural Hazards* 88, 415–430. DOI: 10.1007/s11069-017-2872-9

Shakesby, R.A. and Doerr, S.H. (2006) Wildfire as a hydrological and geomorphological agent. *Earth-Science Reviews* 74(3–4), 269–307. DOI: 10.1016/j.earscirev.2005.10.006

Shao, Q., Cao, W., Fan, J., Huang, L. and Xu, X. (2017) Effects of an ecological conservation and restoration project in the Three-River Source Region, China. *Journal of Geographical Sciences* 27, 183–204. DOI: 10.1007/s11442-017-1371-y

Sharma, E., Chettri, N. and Oli, K.P. (2010) Mountain biodiversity conservation and management: a paradigm shift in policies and practices in the Hindu Kush Himalayas. *Ecological Research* 25, 909–923. DOI: 10.1007/s11284-010-0747-6

Sharma, S., Talchabhadel, R., Nepal, S., Ghimire, G.R., Rakhal, B. *et al.* (2022) Increasing risk of cascading hazards in the central Himalayas. *Natural Hazards* 1–10. DOI: 10.1007/s11069-022-05462-0

Sharma, T.P.P., Zhang, J., Koju, U.A., Zhang, Z.,Yun, B. *et al.* (2019) Review of flood disaster studies in Nepal: a remote sensing perspective. *International Journal of Disaster Risk Reduction* 34, 18–27. DOI: 10.1016/j.ijdrr.2018.11.022

Shekhar, A., Dwivedi, S. and Bhagwat, I. (2010) Ensuring safe water and sanitation during floods in rural communities of Bihar State, India. *Waterlines* 204–208. DOI: 10.3362/1756-3488.2010.021

Shrestha, A.B., Shukla, D., Pradhan, N.S., Sharmila, D., Fayezurahman, A. *et al.* (2021) Developing a science-based policy network over the Upper Indus Basin. *Science of the Total Environment* 784, 147067. DOI: 10.1016/j.scitotenv.2021.147067

Shrestha, S.R., Tripathi, G.N. and Laudari, D. (2018) Groundwater resources of Nepal: an overview. *Groundwater of South Asia* 169–193. DOI: 10.1007/978-981-10-3889-1_11

Shrivastava, P. and Kumar, R. (2015) Soil salinity: a serious environmental issue and plant growth promoting bacteria as one of the tools for its alleviation. *Saudi Journal of Biological Sciences* 22(2), 123–131. DOI: 10.1016/j.sjbs.2014.12.001

Shugar, D.H., Burr, A., Haritashya, U.K., Kargel, J.S., Waston, C.S. *et al.* (2020) Rapid worldwide growth of glacial lakes since 1990. *Nature Climate Change* 10(10), 939–945. DOI: 10.1038/s41558-020-0855-4

Shukla, J., Dhyani, S., Pujari, P., Mishra, A.K. and Verma, P. (2022) Impact of agriculture intensification on forest degradation and tree carbon stock: promoting multi-criteria optimization for restoration in Central India. *Land Degradation and Development* 33(16), 3103–3117. DOI: 10.1002/ldr.4375

Siewert, M.B, Lantuit, H., Richter, A. and Hugelius, G. (2021) Permafrost causes unique fine-scale spatial variability across tundra soils. *Global Biogeochemical Cycles* 35(3), e2020GB006659. DOI: 10.1029/2020GB006659

Singh, N. (2018) The Kailash Sacred Landscape Conservation Initiative: assessment and potential of cooperation in the Himalayas. Available at: https://www.semanticscholar.org/paper/The-Kailash-Scared-Landscape-Conservation-and-of-in-Singh/ed8f1d622f6c8307c4ba189f8d5d91119a02a8e2 (accessed 28 September 2023).

Singh, R., Krausman, P.R., Pandey, P., Maheshwari, A., Rawal, R.S. *et al.* (2020) Predicting habitat suitability of snow leopards in the western Himalayan mountains, India. *Biology Bulletin* 47, 655–664. DOI: 10.1134/S106235902101012X

Singh, S., Youssouf, M., Malik, Z.A. and Bussmann, R.W. (2017) Sacred groves: myths, beliefs, and biodiversity conservation: a case study from Western Himalaya, India. *International Journal of Ecology* (9), 1–12. DOI: 10.1155/2017/3828609

Singh, S., Tanvir Hassan, S.M, Hassan, M. and Bharti, N. (2020) Urbanisation and water insecurity in the Hindu Kush Himalaya: insights from Bangladesh, India, Nepal and Pakistan. *Water Policy* 22(S1), 9–32. DOI: 10.2166/wp.2019.215

Singh, S.P., Bassignana-Khadka, I., Singh, K.B. and Sharma, E. (2011) *Climate Change in the Hindu Kush Himalayas: The State of Current Knowledge*. International Centre for Integrated Mountain Development (ICIMOD), Kathmandu.

Skach, B. (2016) The Annapurna Road: development and tourism on the Annapurna circuit. Independent Study Project (ISP) Collection. 2402. Available at: https://digitalcollections.sit.edu/isp_collection/2402 (accessed 28 September 2023).

Smith, P., House, J.I., Bustamante, M., Sobocká, J. and Harper, R. (2016) Global change pressures on soils from land use and management. *Global Change Biology* 22(3), 1008–1028. DOI: 10.1111/gcb.13068

Sri Shalini, S., Palanivelu, K., Ramachandran, A. and Vijaya, G.S.R. (2021) Biochar from biomass waste as a renewable carbon material for climate change mitigation in reducing greenhouse gas emissions: a review. *Biomass Conversion and Biorefinery* 11, 2247–2267. DOI: 10.1007/s13399-020-00604-5

Stefanica, M. (2017) Environmental impact of transportation in the tourism industry: dimensions and actions. *Studies and Scientific Researches. Economics Edition* (25). DOI: 10.29358/sceco.v0i25.377

Stibal, M., Šabacká, M. and Žárský, J. (2012) Biological processes on glacier and ice sheet surfaces. *Nature Geoscience* 5(11), 771–774. DOI: 10.1038/ngeo1611

Street, E. (2020) When 'Perma' is no longer 'perma': investigating permafrost degradation in Churchill, Manitoba. Available at: https://yorkspace.library.yorku.ca/items/ab60ae88-8760-45f3-b637-a6fadcdd3d7e (accessed 28 September 2023).

Suman, S. (2022) Water security and regime building: a middle power perspective. Available at: https://www.sadf.eu/wp-content/uploads/2022/09/Working-Paper.29.pdf (accessed 28 September 2023).

Suyal, D.C., Joshi, D., Kumar, S., Bhatt, P., Narayan, A. *et al.* (2021) Himalayan microbiomes for agro-environmental sustainability: current perspectives and future challenges. *Microbial Ecology* 83(3), 1–33. DOI: 10.1007/s00248-021-01849-x

Svennevig, K., Hermanns, R.L., Keiding, M., Binder, D., Citterio, M. *et al.* (2022) A large frozen debris avalanche entraining warming permafrost ground: the June 2021 Assapaat landslide, West Greenland. *Landslides* 19(11), 2549–2567. DOI: 10.1007/s10346-022-01922-7

Talchabhadel, R., Ghimire, G.R., Sharma, S., Dahal, P., Panthi, J. *et al.* (2022) Weather radar in Nepal: opportunities and challenges in a mountainous region. *Weather* 77(5), 160–164. DOI: 10.1002/wea.3994

Talukder, B., Matthew, R., Bunch, M.J., Hipel, K.W. *et al.* (2021) Melting of Himalayan glaciers and planetary health. *Current Opinion in Environmental Sustainability* 50, 98–108. DOI: 10.1016/j.cosust.2021.02.002

Tambe, S., Dhakal, S., Dhakal, D., Sharma, G., Sherpa, P.N. *et al.* (2020) Scaling up spring revival in the Himalaya: graduating from spring-centric to aquifer-centric nature-based solutions. *Nature-based Solutions for Resilient Ecosystems and Societies*, 29–50. DOI: 10.1007/978-981-15-4712-6_2

Taylor, L.S., Quincey, D.J., Smith, M.W., Baumhoer, C.A., McMillan, M. *et al.* (2021) Remote sensing of the mountain cryosphere: current capabilities and future opportunities for research. *Progress in Physical Geography: Earth and Environment* 45(6), 931–964. DOI: 10.1177/03091333211023690

Teufel, B. and Sushama, L. (2019) Abrupt changes across the Arctic permafrost region endanger northern development. *Nature Climate Change* 9(11), 858–862. DOI: 10.1038/s41558-019-0614-6

Thapa, S., Shrestha, S., Bhattarai, S., Miya, M.S. and Tutunga, D.G. (2020) Assessment of residents' understanding of potential ecosystem services of Phewa Watershed. *Journal of Forest and Natural Resource Management* 2(1), 61–78. DOI: 10.3126/jfnrm.v2i1.40223

Tinker, P.B., Ingram, J.S.I. and Struwe, S. (1996) Effects of slash-and-burn agriculture and deforestation on climate change. *Agriculture, Ecosystems and Environment* 58(1), 13–22. DOI: 10.1016/0167-8809(95)00651-6

Tsai, W.T. and Tsai, C.H. (2022) Interactive analysis of green building materials promotion with relevance to energy consumption and greenhouse gas emissions from Taiwan's building sector. *Energy and Buildings* 261, 111959. DOI: 10.1016/j.enbuild.2022.111959

Tshewang, U., Tobias, M.C. and Morrison, J.G. (2021) State of environment in Bhutan. In: Tshewang, U., Tobias, M.C. and Morrison, J.G. (eds) *Bhutan: Conservation and Environmental Protection in the Himalayas*. Springer, pp. 1–24.

Tuladhar, S., Pasakhala, B., Maharjan, A. and Mishra, A. (2021) Unravelling the linkages of cryosphere and mountain livelihood systems: a case study of Langtang, Nepal. *Advances in Climate Change Research* 12(1), 119–131. DOI: 10.1016/j.accre.2020.12.004

Veh, G., Korup, O. and Walz, A. (2020) Hazard from Himalayan glacier lake outburst floods. *Proceedings of the National Academy of Sciences* 117(2), 907–912. DOI: 10.1073/pnas.1914898117

Voigt, C., Marushchak, M.E., Lamprecht, R.E., Jackowicz-Korczynski, M., Lindgren, A. *et al.* (2017) Increased nitrous oxide emissions from Arctic peatlands after permafrost thaw. *Proceedings of the National Academy of Sciences* 114(24), 6238–6243. DOI: 10.1073/pnas.1702902114/-/DCSupplemental

Wang, J. and Liu, D. (2022) Vegetation green-up date is more sensitive to permafrost degradation than climate change in spring across the northern permafrost region. *Global Change Biology* 28(4), 1569–1582. DOI: 10.1111/gcb.16011

Wang, L.K., Wang, M.H.S., Wang, N.Y. and Wong, W. (2021) Effect of global warming and climate change on glaciers and salmons. *Integrated Natural Resources Management* 1–36. DOI: 10.1007/978-3-030-55172-8_1

Wang, Y., Spencer, R.G.M., Podgorski, D.C., Kellerman, A.M., Rashid, H. *et al.* (2018) Spatiotemporal transformation of dissolved organic matter along an alpine stream flow path on the Qinghai-Tibet Plateau: importance of source and permafrost degradation. *Biogeosciences* 15(21), 6637–6648. DOI: 10.5194/bg-15-6637-2018

Wang, Y., Wu, N., Kunze, C., Long, R., Perlik, M. (2019) Drivers of change to mountain sustainability in the Hindu Kush Himalaya. In: Wester, P., Mishra, A., Mukherji, A. and Shrestha, A. (eds) *The Hindu Kush Himalaya Assessment: Mountains, Climate Change, Sustainability and People.* Springer Nature, Cham, Switzerland, pp. 17–56.

Wasi, N. (2022) The place and role of Pakistan in United Nations Organizations. In: Hippler, A. and Ahmed, V. (eds) *Global Pakistan: Pakistan's Role in the International System*. Friedrich Ebert Stiftung, Pakistan.

Wester, P., Mishra, A., Mukherji, A., Shrestha, A.B. (2019) *The Hindu Kush Himalaya Assessment: Mountains, Climate Change, Sustainability and People*. Springer Nature, Cham, Switzerland.

Wu, G., Li, G., Cao, Y., Chen, D., Qi, S. *et al.* (2023) Distribution and degradation processes of isolated permafrost near buried oil pipelines by means of electrical resistivity tomography and ground temperature monitoring: a case study of Da Xing'anling Mountains, northeast China. *Remote Sensing* 15(3), 707. DOI: 10.3390/rs15030707

Wu, J., Wu, G., Zheng, T., Zhang, X. and Zhou, K. (2020) Value capture mechanisms, transaction costs, and heritage conservation: a case study of Sanjiangyuan National Park, China. *Land Use Policy* 90, 104246. DOI: 10.1016/j.landusepol.2019.104246

Wu, N., Ismail, M., Joshi, S., Liang, Y., Shrestha, R. *et al.* (2014) Livelihood diversification as an adaptation approach to change in the pastoral Hindu-Kush Himalayan region. *Journal of Mountain Science* 11, 1342–1355.

Wu, R., Trubl, G., Taş, N. and Jansson, J.K. (2022) Permafrost as a potential pathogen reservoir. *One Earth* 5(4), 351–360. DOI: 10.1016/j.oneear.2022.03.010

WWF Nepal (2018) Infrastructure assessment in snow leopard habitat of Nepal. WWF Nepal, Kathmandu.

Yang, J., Wu, G., Jiao, J., Dyck, M. and He, H. (2022) Freeze-thaw induced landslides on grasslands in cold regions. *Catena* 219, 106650. DOI: 10.1016/j.catena.2022.106650

Yao, T., Bolch, T., Chen, D., Gao J., Immerzeel, W.W. *et al.* (2022) The imbalance of the Asian water tower. *Nature Reviews Earth and Environment* 3(10), 618–632. DOI: 10.1038/s43017-022-00299-4

Yarzábal, L.A., Salazar, L.M.B. and Batista-García, R.A. (2021) Climate change, melting cryosphere and frozen pathogens: Should we worry…? *Environmental Sustainability* 4(3), 489–501. DOI: 10.1007/s42398-021-00184-8

Yoro, K.O. and Daramola, M.O. (2020) *CO_2 Emission Sources, Greenhouse Gases, and the Global Warming Effect*. Woodhead Publishing 3–28. DOI: 10.1016/B978-0-12-819657-1.00001-3

You, Q.L. and Ren, G.Y. (2017) An overview of studies of observed climate change in the Hindu Kush Himalayan (HKH) region. *Advances in Climate Change Research* 8(3), 141–147. DOI: 10.1016/j.accre.2017.04.001

Zheng, G., Allen, S.K., Bao, A., Ballesteros-Canovas, J.A., Huss, M. *et al.* (2021) Increasing risk of glacial lake outburst floods from future Third Pole deglaciation. *Nature Climate Change* 11(5), 411–417. DOI: 10.1038/s41558-021-01028-3

Zimmer, A., Beach, T., Klein, J.A. and Jorge, R.B. (2022) The need for stewardship of lands exposed by deglaciation from climate change. *Wiley Interdisciplinary Reviews: Climate Change* 13(2), e75. DOI: 10.1002/wcc.753

19 Gender Roles and Practices in Ecological Conservation and Restoration

Youyan Liu[1,*], Allan Degen[2] and Zhanhuan Shang[1]
[1]*College of Ecology, Lanzhou University, Lanzhou, China;* [2]*Institutes for Desert Research, Ben Gurion University of the Negev, Israel*

Abstract

Gender issues are deeply linked with ecological conservation and restoration. The international community has recognized the importance of women in promoting harmonious coexistence between people and nature; however, women's participation in ecological conservation in the Hindu Kush Himalaya (HKH) region, where patriarchy is prevalent, has not yet reached the level that affects decision- and policy-making. The contribution of women to ecological conservation and restoration has gone unnoticed, and, consequently, steps should be taken to improve their status, provide them with a platform to express their ideas and offer them more important positions. This chapter describes: (i) gender inequalities in the HKH and suggestions on how to mitigate them; and (ii) the importance of women in ecological conservation, livelihoods and carbon management.

Keywords: education, Hindu Kush Himalayas (HKH) region, inequalities, resource ecological conservation and restoration, status of women

19.1 Introduction

Protecting the ecological environment and achieving sustainable development are common challenges faced by the international community. However, it is important to include gender in the decisions, as men and women bear different roles and social responsibilities in the conservation and restoration of ecological resources (Broeckhoven and Cliquet, 2015; Liu *et al.*, 2017). Gender differences in behaviours, roles, rights and services have different impacts on ecosystems.

After entering the 21st century, the international community is increasingly aware of the importance and urgency of gender issues (Sen and Nayar, 2015; Castañeda *et al.*, 2020). International ecological conferences and reports have stressed the importance of women's contributions in ecological environmental conservation and restoration, and the need to include women in decision-making. The Rio Declaration on Environment and Development in 1992 stated that women play a key role in environmental management and development, and that their full participation is essential for achieving sustainable development (Ma, 2022). The 2005 Human Development Report of the United Nations Development Programme (UNDP) stressed that the elimination of gender discrimination is an important step to achieve the Millennium Development Goals (MDGs), and linked gender equality with poverty reduction (UNDP, 2005). In 2015,

*Corresponding author: liuyouy@lzu.edu.cn

DOI: 10.1079/9781800622579.0019

the 2030 Agenda for Sustainable Development by the UN called for the empowerment of all women and girls (UN, 2015). Institutions such as the World Bank, the Asian Development Bank (ADB) and the International Union for Conservation of Nature (IUCN) actively promote the implementation of gender policies. Unfortunately, most ecological restoration programmes today still ignore gender issues, which suggests that they do not consider gender to be an important factor in restoration practices (Broeckhoven & Cliquet, 2015).

Women have always been pioneers, innovators and participants in the cause of ecological environmental conservation. The famous Green Belt movement in Africa and the Chipko movement in India are the results of early women who participated in ecological conservation movements due to the destruction of their living environment (Ai and Liu, 2015). In 1974, the French feminist Francoise d'Eaubonne proposed the concept of 'ecofeminism', which dealt with ecological problems from a gender perspective and emphasized that solving ecological problems must include a woman's perspective (Kathryn, 2023). She believed that there is a relationship between the oppression of women and the degradation of nature (Zeng *et al.*, 1999), that a special natural connection exists between nature and women (Bryant and Pini, 2006; Wang *et al.*, 2011), and she criticized the domination of men over women. Men have always been regarded as conquerors of nature, whereas women have a deeper sense of identity with nature (Griffin, 1978; Hayes, 2001; Hunter *et al.*, 2004).

Gender is an important component of ecosystem health research, as gender inequality exacerbates the deterioration of the ecological environment and leads to the decline of ecosystem services. Concomitantly, the decline of ecosystem services, in turn, increases the uneven distribution of resources, resulting in more serious gender inequality, forming a vicious circle (Ai and Liu, 2015). In areas of the HKH region, such as China's Qinghai-Tibetan Plateau and Nepal, that are influenced by traditional culture and religious beliefs, women are responsible for the household and raising children as well as heavy labour in agriculture and livestock production. Their activities are related to ecological conservation and reducing carbon emissions, which are a crucial part of the carbon cycle, more so than are men (Dong *et al.*, 2003; Shang *et al.*, 2016). However, women are often outside the decision-making circle, as they lack equal voice in issues related to the sustainable use of natural resources.

The relationship between gender and ecological conservation and restoration should be understood better, and the role of gender in the relationship should be examined. Women should be allowed to express their ideas and should be provided with more opportunities to participate in decision-making, as their perspectives could be beneficial in ecological conservation and restoration.

19.2 The Role and Contribution of Women in Ecological Conservation

Since ancient times, women have played important roles in ecological conservation and family life. Women's natural 'maternal instincts' make them more sensitive to the surrounding environment. They experience the greatest impacts of climate change and ecological environment deterioration (UN, 2022), as 80 % of people who left their homes due to climate change are women, while their role as caregivers and food and fuel providers leaves them more vulnerable to floods and droughts (Guangming Daily, 2018). Men and women differ when using, allocating and consuming ecological resources. Compared with men, women are more likely to adopt 'clean technologies and clean energy' such as solar and wind energy, and their activities are low-carbon and environmentally friendly (Liu, 2007; Ma, 2022). Women have stronger requirements for the sustainable use of ecological resources and stronger awareness of ecological environment maintenance than men (Yan and Wang, 2000). Furthermore, women are better spokespersons and are more proactive in protecting the ecological environment than men (Xiao and McCright, 2014; Long and Liu, 2019; Shen and Tang, 2023); however, their contribution to ecological resource conservation, family livelihoods and carbon management is often unnoticed.

19.2.1 Women's role in forest ecological conservation

Forests are vital resources for people living in mountainous areas, are related to their livelihoods and daily lives, and are important safety nets against natural disasters. The Chipko movement in India, a non-violent protest dominated by women, was one of the most important ecological conservation actions in India's forest conservation. In 1973, due to large-scale deforestation, the livelihoods of the residents were at risk. The local women used their bodies to prevent deforestation and, thus, initiated a global conservation movement. The government promised to ban logging within 15 years and the Chipko movement protected 5000 km^2 of forest in the Himalayan region (https://chinadialogue.org.cn [accessed 25 February 2024]).

19.2.2 Women's role in water resource conservation

Women are indispensable in water conservation, water environment management and water conservation as they undertake more than 70% of water management and water collection (Fig. 19.1) worldwide (https://china.un.org [accessed 25 February 2024]). In the HKH region, this is especially true in wetland conservation, shoreline greening, and river and lake maintenance (Manase *et al.*, 2003; Shen and Tang, 2023). For example, Tibetan women play a vital role in combating the ecological degradation of wetlands in the upper reaches of the Yellow river in China (Li and Wang, 2004), while the waterfall landscape cleaning groups in the Munigou Scenic Area of Songpan County, Aba Tibetan Autonomous Prefecture, and Sichuan Province, China, are all women from the surrounding villages. These women spend a third of their time in maintaining the sanitation of the water landscape (Shen and Tang, 2023). A study of local councils in India reported that the number of drinking water programmes was 62% higher in women- than in men-led council areas (UN, 2023).

19.2.3 Women's role in biodiversity conservation

Women promote plant and animal diversity. In a survey by Malika *et al.* (2008) on agricultural activities and agricultural biodiversity conservation in western Asia, including Syria, Lebanon and Jordan, women's knowledge of local conditions and of medicinal and spice plants played a key role in the conservation and sustainable use of agricultural biodiversity (Liu *et al.*, 2017). In Mexico, women increased local plant diversity through the use of food crops and medicinal plants. A study in India reported that women often use plants related to family livelihoods, such as medicines, health products, cleaning products, fibres, food and tools (Liu *et al.*, 2017).

Fig. 19.1. Tibetan women bringing water to the home. (Photo: Allan Degen)

19.2.4 Women's role in grasslands conservation

Grasslands are important natural resources (Wang *et al.*, 2011), and there is a high degree of compatibility between grassland ecological conservation and women's activities. In alpine grassland areas, many men leave the area to work for their livelihoods and, consequently, much of the work rests with women. Women shepherd the livestock, collect livestock dung, and are directly related to the management of grassland. Research in the Yellow river headwater region on the Qinghai-Tibetan Plateau concluded that women are willing to participate in grassland co-operative management and conservation, but due to their low status in the family and society, they cannot make decisions in grassland management (Gou *et al.*, 2015; Yang *et al.*, 2020).

19.2.5 Women's role in ecological restoration

The World Resources Institute recognizes women as active promoters of restoration and conservation (Elwell, 2016), and that women have made outstanding contributions in ecological restoration programmes. This is particularly true in forest, dryland, grassland and oasis restorations. India's forest management organization and the wasteland environmental conservation programmes require women members and their participation (Broeckhoven and Cliquet, 2015), as was also reported in some organizations in Nepal. The Dumritumka Adarsha Women's Community Forest User Group (DAWCFUG) has been successful in restoring degraded forests in Rauta Village, Udayapur, Nepal, and was awarded the BirdlifeNature Hero Award. Neighbouring forest communities have successfully replicated this award-winning model (Sellmyer *et al.*, 2019). To restore the Azraq Oasis Ramsar site in Jordan, the Monitoring CEDAW programme, which ensures women's participation and planning and supports women's decision-making, was implemented successfully (IUCN, 2008; Broeckhoven and Cliquet, 2015).

19.2.6 Environmental litigation in gender issues

Decision-making by women has proved to be more environmentally friendly than decision-making by men. Decisions by women leaders reduce the probability of environmental immoral behaviour and the level of corporate environmental misconduct, compared to men. Liu (2018) and Adams *et al.* (2011) concluded that women chief executive officers (CEOs) are less prosecuted for environmental violations and are more concerned about the welfare of stakeholders than men, which indicates that women are more likely to take action to pre-empt environmental risks that may harm the community. Women's leadership offers different perspectives and provides broader environmental solutions, which improves decision-making on environmental issues. Furthermore, women executives are less confident and more likely to seek advice than are male executives (Adams *et al.*, 2011; Huang and Kisgen, 2013; Cumming *et al.*, 2015; Liu, 2018). It can be concluded that women protect ecological resources in diverse ways, and play a key role in generating harmony between people and nature.

19.3 Contributions of Women to Livelihoods and Household Economy

Under the influence of traditional 'patriarchal ideology' and culture, housework is the responsibility of women. This is especially true in many remote mountainous regions, where females, from young girls (Fig. 19.2) to old women (Fig. 19.3), handle household chores and much of the hard labour. Many Tibetan traditions exacerbate the intensity of Tibetan women's housework. For example, it is taboo for Tibetan men to carry water and wash laundry. The global society has made efforts to promote gender equality and improve women's social status; however, women's housework time is still much greater than that of men in alpine regions (Shang *et al.*, 2016).

Grazing in pastoral regions and farming in agricultural regions are basic livelihoods, and women are the main workers (Dong *et al.*, 2003). The British botanist Wilson described the Jiarong Tibetan women in the Badi-Bawang regions of Danba as follows: '...they farmed, herded sheep, transported agricultural products to the market, felled wood and transported water'. In Tagong grassland of Kangding in Tibet, women are responsible for cutting winter grass, storing winter feed and milking animals,

Fig. 19.2. Young girls carrying wood in lower Mustang, Nepal. (Photo: Allan Degen)

Fig. 19.3. An old woman grinding corn in the mid-hills of Nepal. (Photograph: Allan Degen)

activities that usually require much labour (Shen and Tang, 2023).

In the Gartran region of Nepal, Deshar and Koirala (2020) concluded that most of the work in agropastoral systems, such as preparing food for the household and for livestock (Fig. 19.4), collecting dung, milking livestock and processing milk (Fig. 19.5), grinding cereals and legumes, fetching water and fuel, and planting (Fig. 19.6) and weeding, were done by women alone. There were 14 activities, such as grazing livestock, sheep shearing and harvesting, that were handled jointly by men and women, while only ploughing was done by men alone. A survey on the Qinghai-Tibetan Plateau in China arrived at similar conclusions, while a study on the Qilian mountains region also concluded that women were the main workforce for livestock production, undertaking nearly 80% of production management and 74% of daily housework (Gou *et al.*, 2016).

Women in the alpine region also contribute to the economic income of the household. They earn additional income by producing handicrafts, specialty jewellery and fabrics, and by using wool to produce clothes, shoes and blankets.

Fig. 19.4. A Tibetan woman preparing feed for yaks. (Photo: Allan Degen)

Fig. 19.5. A woman churning buffalo milk in the mid-hills of Nepal. (Photo: Allan Degen)

Fig. 19.6. Women planting rice in the mid-hills of Nepal. (Photo: Allan Degen)

In India, women's income from forests and public land accounts for 33–45% of the total household income, while men account for only 13% (Liu *et al.*, 2017). In China's Qinghai-Tibet Plateau, the main source of income is from livestock and dairy products, which are handled mainly by women. Nearly 30% of the family income in the region is from the work of women. It was estimated that a family with 50 yaks can earn 3000 yuan a month from May to December (Shang *et al.*, 2016). In addition, women can earn money from collecting the very sought-after medicinal fungus (*Cordyceps sinensis*).

According to Norgaard and York (2005), when women's income increases, more environmental treaties are passed. Thus, the following steps should be taken: (i) promote women's economic development and empower them economically, as this enhances their status in ecological conservation and restoration programmes; (ii) expand women's control over property, land and resources, and offer more incentives and compensation to woman-headed households; (iii) increase the quota of women for all job positions, and ensure fair labour rights and pay; (iv) support women from poor families financially; (v) encourage the establishment of women's industrial parks; and (vi) provide more help and financial support to women's enterprises.

19.4 Role of Women in Carbon Management

Carbon dioxide (CO_2) is one of the main components and contributors of greenhouse gas (GHG) emissions. Concentration of CO_2 reached 413.2 parts per million in 2020, which was 149% of the pre-industrial level (WMO, 2021). Livestock are important contributors of CO_2 emissions in the carbon cycle (IPCC, 2007; WMO, 2010; Shang *et al.*, 2016), while livestock products are the key link of carbon output. Through activities such as shepherding livestock, milking animals and producing milk products, women and men differ greatly in carbon input and output (Ma, 2022). Livestock production inputs carbon through products such as meat, milk and wool (Deshar and Koirala, 2020). The output of carbon products can eliminate carbon redundancy through the accumulation of excess animal carbon products in the alpine grassland carbon cycle, which is an important part of the negative feedback mechanism. Moderate animal carbon output is necessary to maintain the normal carbon cycle of grassland ecosystems.

In addition, preparing meals by women and food preferences influence carbon emissions. It was reported that countries with women of high political status emit less CO_2 than countries with women of low political status (Ergas and York, 2012). Women's activities are the key to family carbon management, as they are highly correlated with the ecosystem carbon cycle in terms of fuel use and management (Shang *et al.*, 2016). In some under-developed and remote rural regions of China, women are also energy users in housework and, thus, they have an impact on climate change. Livestock dung (Fig. 19.7) is an essential energy source for cooking and heating (Liu, 2007), and, in the plateau, collecting, drying and stacking dung, and cooking are daily activities for women. In China's alpine regions, yak dung accounts for 91.4% of total fuel use (Sharma, 2010). Using dung as fuel reduces the damage to shrubs and trees and the use of fossil fuels. This greatly reduces CO_2 emissions and accelerates the decomposition of livestock carbon into the atmosphere or ash into the soil, which is beneficial for maintaining atmospheric carbon balance. The use of more environmentally friendly energy sources, like solar energy, can reduce deforestation. In this way, carbon emissions can be effectively controlled and GHG can be further reduced.

Fig. 19.7. A Tibetan woman using yak dung to cook and heat the house. (Photo: Allan Degen)

19.5 Current Situation of Women in the Hindu Kush Himalayan Region

The Hindu Kush Himalayan region is home to people from different races, cultures, religions and social structures (https://www.icimod.org/ [accessed 25 February 2024]). Due to its special geographical location, the region faces a double challenge: extreme harsh environmental conditions and poverty. As a result, women living in the HKH region have more responsibilities than women in other regions. With socio-economic development, men often migrate in search of employment, but long-established religions, cultures, social norms and gender ideologies limit women's mobility within the area of the household. Most women in the HKH region have lesser access to land, enterprises, education, skills, information and decision-making rights (Gurung *et al.*, 2011). The women also have to cope with the fragile and harsh environment in the HKH region, as well as health problems and violence linked to poverty.

While women in the HKH region have little education, they are intuitive in maximizing natural resources through long-standing indigenous knowledge. Existing policies and strategies in the HKH region consider women as a vulnerable or targeted group, and their contributions and ability to act as agents of change are not considered (Goodrich *et al.*, 2021). Evidence has demonstrated that gender inequality is a serious impediment to sustainable mountain development. International communities and organizations, and regional governments and institutions, such as ICIMOD, have recognized the importance of gender in addressing climate change, and ecological conservation and restoration. Policies to improve the status of women and protect their rights have been proposed to reduce inequality between genders in the HKH region.

19.5.1 Women and education

Traditionally, women do not attend school in the HKH region, and women's obligations to the household leave them with very little time to study. The high outmigration of men increases the workload of women, resulting in low enrolment rates and high drop-out rates for girls. In the Qinghai-Tibet Plateau, the illiteracy rate of women is two or three times that of men, while in Nepal, most women have little or no formal education. In the report *The State of Gender Equality and Climate Change in Nepal*, the gender equality index for primary and secondary school enrolment has improved in recent years, but girls' enrolment in technical and vocational schools remained low at 0.53 in 2015, well below the target of 0.66. Low level of education impedes women's ability to understand information. For example, in 2016, a survey in Nepal found that 60.9% of women did not know about or had not heard of climate change (CBS, 2016; Goodrich *et al.*, 2021). In Pakistan, growing economic vulnerability has forced parents to arrange early marriages for daughters. Women's inequality in education is seriously affecting the co-ordinated and sustainable development of the region's society, economy and culture.

To improve the education level of women and to narrow the gap between men and women in the HKH region, ICIMOD trained women from Nepal, Pakistan, Afghanistan, Bhutan and Bangladesh in global observation and geospatial information technology. In regions such as Gansu and Sichuan in China, the Rural Women Empowerment Programme was implemented to empower women and help them start businesses. Women in Khyber, Pakistan, received technical training in the operation and maintenance of hydraulic piston pumps, and now manage irrigated land, make market decisions, and have greater financial decision-making rights. In India, a policy of free bus travel for all women laid the foundation for better education and job opportunities for them.

More measures are needed to improve women's education and programmes should be provided for women with ecological conservation training. Education should be compulsory for women and they should be offered financial help to attend school, as this would reduce illiteracy rates and early marriages. In addition, women should attend professional and technical training programmes, as this would help them acquire skills to achieve economic, living and career improvements.

19.5.2 Health and medical care

In the HKH region, due to economic constraints, most families live in tents and stone houses with no chimneys or exhaust vents; however, because of low temperatures, households are forced to heat by burning dung all year round. As a result, the region has one of the highest indoor emissions of black carbon and fine particulate matter in the world. Women are particularly affected by indoor air pollution as they do all the cooking, and they are at much greater risk of illnesses such as cancer, and cardiovascular and respiratory diseases than men (Pope and Dockery, 2006; Holthaus, 2015). Approximately 8700 people, mainly women and children, die prematurely each year from indoor air pollution in Nepal (Fig. 19.8; WHO, 2009; ICIMOD, 2021). In Pakistan, certain social practices increase the possibility that women would contact diseases such as COVID-19. Due to the lack of toilets, and the cultural stigma of women using toilets in the presence of a man's family, many women are forced to defecate in the fields. The faeces remain in the field and the women are unable to wash their hands thoroughly (ICIMOD, 2020).

Heavy domestic and agricultural work also endanger women's health in the HKH region. Women have higher mortality and morbidity; however, women's healthcare has been largely neglected, as few policies address women's health. Some positive steps have been implemented for women's health, but many more improvements are needed. ICIMOD and Megh Pyne Abhiyan (MPA) set up Eco-San toilets, which are built on a raised platform to provide healthy and hygienic options for women affected by floods. Through Eco-San, women were mobilized to raise awareness of the need to install toilets, spread health knowledge and improve their health.

19.5.3 Status of resource acquisition and allocation

Women from poor and marginalized groups are highly dependent on natural resources such as forests, land and water, and participate in the management of these resources. However, they are not treated equally in the use and allocation of natural resources. In many parts of the HKH region, such as the Koshi river basin and Afghanistan, women are faced with the dilemma of the use and management of water resources. They are not entitled to fair water rights and there is no effective guarantee in terms of safe drinking water. Some measures are being taken to improve the conditions. ICIMOD's 'Koshi Watershed Programme', in partnership with HELVETAS Swiss-Nepal International Cooperation Partner, launched an initiative in three different eco-regions in the Koshi Watershed to help promote effective, efficient and equitable water management at the local level. The initiative aims to improve access to water and ensure equal access to water for all people, including women.

19.5.4 Status of decision-making rights

With the continuous development of social economy, the importance of women is becoming

Fig. 19.8. A woman cooks at home on a fire in the mid-hills of Nepal. (Photo: Allan Degen)

more noticeable in political, social and family life. However, in the HKH region, women are marginalized and excluded from key decision-making positions, opportunities and resources. In the agricultural sector, the representation of women in federal agencies is only 11%, and in official positions is only 9%. Only 6% of the community forest user groups in Nepal's forestry sector are chaired by women (Goodrich *et al.*, 2021; ICIMOD, UNEP & UN Women, 2021).

Women also have limited decision-making rights in the family due to their lack of land and property. In the high-altitude pastoral region of Nepal, women are free to decide on inexpensive purchases for everyday life. But for expensive and major purchases related to family livelihoods, such as land and yaks, women need their husbands' consent (Deshar and Koirala, 2020). In the Tibetan Plateau region, most family decisions and household income are controlled mainly by men (Gou *et al.*, 2015). However, Tibetan families with women controlling income have more knowledge in productive activities and make better decisions for the family and on social resources (Yang, 2019).

In 2019, gender equality and social inclusion (GESI) projects were initiated in two rural municipalities in Nijhang (Saipal in Bajhang district and Byas in Darchula district) to involve women in local resource management plans. In Maipokari district of eastern Nepal, the management committee only had women members, and they effectively improved the management and sustainability of local resources and increased women's right to participate in decision-making. In Mithinkot, in Kabuhepalanchok, Nepal, women increased their influence on farm and family decision-making by participating in hybrid seed production. The co-operatives are led and managed by woman farmers, which enhances women's leadership.

19.5.5 Status and prejudice

In recent years, more policies are considering gender issues, improving the status of women and striving for equal rights at all levels. However, prejudice and discrimination against women persist in countries around the world, resulting in their disadvantaged position in the family and society. In the HKH region, women are stereotyped in their social institutions, religions and traditions (Acharya and Bennett, 1981). Women themselves are often not aware of the injustice against them. In the HKH region, 'patriarchal regimes' are strongly biased in favour of men, regardless of caste, class, race and religion (ICIMOD, UNEP and UN Women, 2021). Under the long-term influence of religious beliefs, women are considered 'unclean' and they are not allowed to participate in some religious activities. The feudal traditional concept of men superiority restricts women's daily life and the right to make their own choices. In terms of the labour force, women are often perceived as inferior to men and, therefore, receive lower wages.

The Hindu Kush Himalayan Renewable Energy and Energy Efficiency Capacity (REEECH) programme established the Agricultural Renewable Energy Solutions (RESA) Incubation programme to lessen the many biases against women in entrepreneurship, innovation and technological capabilities by placing women from diverse backgrounds in unconventional roles in all businesses. ICIMOD's Himalica initiative is committed to empowering local communities, especially women, and diversifying their sources of income through the establishment of Kangchenjunga Himalica Agricultural Industry (KHAI). Ten Nepalese women members received key positions in the production, processing and marketing of vegetables and cardamom products.

19.6 Enhancing the Status of Women in Ecological Conservation and Restoration

Many programmes and policies were initiated to alleviate women's inequality in the HKH region, but, due to regional cultural and institutional differences, they were difficult to promote and implement, especially in remote mountainous regions. The 'deep-rooted' gender differences are difficult to change. Empowering women in social, economic and ecological aspects, and improving their status in ecological conservation and ecological restoration programmes are directions that need to be taken for the sustainable development of the Hindu Kush Himalayan region in the future.

Women's contributions should be impressed upon men, and women's decision-making rights should be ensured. The government should encourage women to participate in decision-making, development and implementation of policies in ecological environment and restoration. Organizations should be established to monitor and penalize gender inequalities in ecological resource protection and restoration programmes. A women's ecological research institute, which would enable women to participate in ecological conservation and restoration and receive equal salary, should be established. This would allow women to express their ideas from their perspectives, and their contributions to ecological conservation and restoration could be recognized.

19.7 Conclusions

Women play important roles in the conservation of forests, water, grasslands and biodiversity, in family livelihoods and in carbon management. However, in the HKH region, many women are not educated, receive poor healthcare, and are limited in land ownership and decision-making because of the social pressures and long-standing traditions in the area. There are many cases demonstrating that the elimination of gender inequality would lessen environmental problems and promote ecological restoration. Strong programmes should be implemented in the HKH region to improve the rights, education, healthcare and status of women, and empower them with key positions in conservation and restoration of the ecological environment.

References

Acharya, M. and Bennett, L. (1981) *Status of Women in Nepal*. Centre for Economic Development and Administration, Kathmandu, Nepal.

Adams, R.B., Licht, A.N. and Sagiv, L. (2011) Shareholders and stakeholders: How do directors decide? *Strategic Management Journal* 32(12), 1331–1355. DOI: 10.1002/smj.940

Ai, J. and Liu, S. (2015) Ecological civilization new idea on gender: male participation from eco-feminist. *Ecological Economy* 31(8), 183–186.

Broeckhoven, N. and Cliquet, A. (2015) Gender and ecological restoration: time to connect the dots. *Restoration Ecology* 23(6), 729–736. DOI: 10.1111/rec.12270

Bryant, L. and Pini, B. (2006) Towards an understanding of gender and capital in constituting biotechnologies in agriculture. *Sociologia Ruralis* 46(4), 261–279. DOI: 10.1111/j.1467-9523.2006.00417.x

Castañeda, C.I., Sabater, L., Owren, C., Boyer, A.E. and Wen, J. (2020) *Gender-based Violence and Environment Linkages: The Violence of Inequality*. IUCN, Gland, Switzerland.

CBS (Central Bureau of Statistics) National Planning Commission (2016) *National Climate Change Impact Survey 2016: A Statistical Report*. Government of Nepal, Kathmandu.

Cumming, D.J., Leung, T.Y. and Rui, O.M. (2015) Gender diversity and securities fraud. *Social Science Electronic Publishing* 58(5), 1572–1593. DOI: 10.5465/amj.2013.0750

Dehui, Z., Fengqi, J., Zhiping, F. and Xiaojun, D. (1999) Ecosystem health and sustainable development for human. *Chinese Journal of Applied Ecology* 10(6), 751–756.

Deshar, R. and Koirala, M. (2020) Indigenous practice in agro-pastoralism and carbon management from a gender perspective: a case from Nepal. In: Shang, Z., Degen, A., Rafiq, M. and Squires, V. (eds) *Carbon Management for Promoting Local Livelihood in the Hindu Kush Himalayan* (HKH) *Region*. Springer, Cham, Switzerland.

Dong, S.K., Long, R.J. and Kang, M.Y. (2003) Milking and milk processing: traditional technologies in the yak farming system of the Qinghai-Tibetan Plateau, China. *International Journal of Dairy Technology* 56(2), 86–93. DOI: 10.1046/j.1471-0307.2003.00088.x

Elwell, N. (2016) If you care about the environment, you should care about gender. World Resources Institute. Available at: https://www.wri.org/insights/if-you-care-about-environment-you-should-care-about-gender#:~:text=Without%20careful%20planning%2C%20environmental%20projects,achieving%20environment%20and%20development%20objectives (accessed 25 February 2024).

Ergas, C. and York, R. (2012) Women's status and carbon dioxide emissions: a quantitative cross-national analysis. *Social Science Research* 41(4), 965–976. DOI: 10.1016/j.ssresearch.2012.03.008

Goodrich, C.G., Gurung, D.D. and Bastola, A. (2021) *State of Gender Equality and Climate Change in Nepal.* United Nations Entity for Gender Equality and the Empowerment of Women (UN Women) United Nations Environment Programme (UNEP).

Gou, Y.J., Liu, X.Y., Zhang, W.M., Xu, C.L. and Long, R.J. (2015) An investigation of women's social status and their recognition of the grassland policies in the pastoral area of the Qilian Mountains. *Acta Ecologica Sinica* 35(10), 3472–3479. (in Chinese).

Griffin, S. (1978) *Woman and Nature: The Roaring Inside Her*. Harper & Row, New York.

Guangming Daily (2018) Climate change has a greater impact on women. Availabe at: https://epaper.gmw.cn/wzb/html/2018-03/22/nw.D110000wzb_20180322_6-07.htm (accessed 25 February 2024).

Gurung, J., Giri, K., Setyowati, A.B. and Lebow, E. (2011) Getting REDD+ right for women: an analysis of the barriers and opportunities for women's participation in the REDD+ sector in Asia. *United States Agency International Development* 1–113.

Hayes, B.C. (2001) Gender, scientific knowledge, and attitudes toward the environment: a cross-national analysis. *Political Research Quarterly* 3, 657–671. DOI: 10.1177/106591290105400309

Holthaus, E. (2015) Yak dung is making climate change worse and there's no easy solution. Available at: http://www.slate.com/articles/technology/future_tense/2014/12/yak_dung_is_making_climate_change_worse_and_new_cookstoves_don_t_help.html/ (accessed 28 September 2023).

Huang, J. and Kisgen, D.J. (2013) Gender and corporate finance: Are male executives overconfident relative to female executives? *Journal of Financial Economics* 108(3), 822–839.

Hunter, L.M., Hatch, A. and Johnson, A. (2004) Cross-national gender variation in environmental behaviors. *Social Science Quarterly* 3, 677–694. DOI: 10.1016/j.jfineco.2012.12.005

ICIMOD (2020) COVID-19 and gender inequality in Pakistan. Available at: https://www.icimod.org/article/covid-19-and-gender-inequality-in-pakistan/ (accessed 28 September 2023).

ICIMOD (2021) The gendered impacts of indoor air pollution in Nepal's Koshi basin. Available at: https://www.icimod.org/article/the-gendered-impacts-of-indoor-air-pollution-in-nepals-koshi-basin/ (accessed 28 September 2023).

ICIMOD, UNEP and UN Women (2021) State of gender equality and climate change in Nepal. United Nations Entity for Gender Equality and the Empowerment of Women (UN Women) United Nations Environment Programme (UNEP).

IPCC (2007) *Climate Change 2007. The Fourth Assessment Report: The Physical Science Basis.* Cambridge University Press, Cambridge, UK.

IUCN (2008) Azraq Oasis Restoration Project, Jordan.

Kathryn, M. 'Ecofeminism'. *Encyclopedia Britannica*. Available at: https://www.britannica.com/topic/ecofeminism/ (accessed 28 September 2023).

Li, Y.H. and Wang, L. (2004) Gender analysis of the conceptual structure and behavioral styles of Tibetan women: a case study of Gannan Tibetan autonomous prefecture in Gansu province. *Gansu Social Sciences* 3, 125–127.

Liu, C. (2018) Are women greener? Corporate gender diversity and environmental violations. *Journal of Corporate Finance* 52, 118–142. DOI: 10.1016/j.jcorpfin.2018.08.004

Liu, H.O., Chen, H.J. Liu, L., Yan, W. and Xue, D. (2017) Incorporating gender equality into biodiversity conservation. *Biodiversity Science* 25(11), 1176–1181. DOI: 10.17520/biods.2017236

Liu, J.L. (2007) The study on rural women's participation in the community forestry in China. Doctoral thesis, Beijing Forestry University.

Long, K.S. and Liu, L. (2019) Research on ecological protection behavior of rural residents from the perspective of gender differences. *Population and Society* 35(6), 98–110.

Ma, J. (2022) Research on the Impact of women's status on household carbon emission based on empirical analysis of CGSS 2017. MSc thesis, Zhongnan University of Economics and Law, Wu Han, China.

Malika, A.M., Ahmed, A., Mohammed, A., Assi, R., Sbieh, Y. *et al.* (2008) Gender dimension in the conservation and sustainable use of agro-biodiversity in West Asia. *The Journal of Socio-Economics* 37(1), 265–383. DOI: 10.1016/j.socec.2007.06.007

Manase, G., Ndamba, J. and Makoni, F. (2003) Mainstreaming gender in integrated water resources management: the case of Zimbabwe. *Physics and Chemistry of the Earth* 28, 967–971. DOI: 10.1016/j.pce.2003.08.023

Norgaard, K. and York, R. (2005) Gender equality and state environmentalism. *Gender & Society* 19(4), 506–522. DOI: 10.1177/0891243204273612

Pope, C.A. and Dockery, D.W. (2006) Health effects of fine particulate air pollution: lines that connect. *Journal of Air Waste Management* 56, 709–742. DOI: 10.1080/10473289.2006.10464545

Sellmyer, A., Vasily, L. and Thomas, S. (2019) *Annual Report 2018*. International Centre for Integrated Mountain Development, Kathmandu.

Sen, G. and Nayar, A. (2015) Powerful synergies: gender equality, economic development and environmental sustainability. United Nations Development Programme. Available at: https://repository.iimb.ac.in/handle/2074/12707 (accessed 28 September 2023).

Shang, Z., White, A., Degen, A.A. and Long, R. (2016) Role of Tibetan women in carbon balance in the Alpine Grasslands of the Tibetan Plateau: a review. *Nomadic Peoples* 20(1), 108–122. DOI: 10.3197/np.2016.200107

Sharma, B. (2010) Challenges and opportunities for sustainable energy solutions in rangelands – experiences and lessons from the Hindu-Kush Himalayas. ICIMOD, Kathmandu.

Shen, M.Y. and Tang, L. (2023) Research on water resources protection and rural women's livelihood promotion in Western Sichuan Plateau. *Resource Development & Market* 39(5), 556–561+590.

UN (2015) Transforming our world: the 2030 agenda for sustainable development. Available at: https://www.un.org/zh/documents/treaty/A-RES-70-1/(25 Sep 2015)(accessed 26 September 2023).

UN (2022) Explainer: how gender inequality and climate change are interconnected. Available at: https://www.unwomen.org/en/news-stories/explainer/2022/02/explainer-how-gender-inequality-and-climate-change-are-interconnected (accessed 26 September 2023).

UN (2023) Facts and figures: women's leadership and political participation. Available at: https://www.unwomen.org/en/what-we-do/leadership-and-political-participation/facts-and-figures/ (accessed 28 September 2023).

UNDP (2005) *Human Development Report 2005*. Available at: https://hdr.undp.org/content/human-development-report-2005 (accessed 28 September 2023).

Wang, C.X., Zhang, Y., Guo, Z.G. and Liu, H. (2011) Relationship between ecofeminism and grassland management in the northwest pastoral areas of China. *Pratacultural Science* 28(3), 467–471.

WHO (2009) *Country Profiles of Environmental Burden of Disease, Public Health and the Environment.* World Health Organization (WHO), Geneva, Switzerland.

WMO (2010) *Greenhouse Gas Bulletin: The State of the Greenhouse Gases in the Atmosphere until December 2009*. World Meteorological Organization, Geneva, Switzerland.

WMO (2021) *Greenhouse Gas Bulletin: Another Year Another Record.* World Meteorological Organization, Geneva, Switzerland.

Xiao, C.Y. and McCright, A.M. (2014) A test of the biographical availability argument for gender differences in environmental behaviors. *Environment and Behavior* 2, 241–263. DOI: 10.1177/0013916512453991

Yan, Z.L. and Wang, N.X. (2000) Ecological restoration and gender in Western Sichuan. *World Sci-tech R&D* S1, 23–25.

Yang, C. (2019) *Research on the Capacity Building of Rural Women's Participation in Ecological Poverty Alleviation in Tibetan Area Based on the Survey of Four Counties in Sichuan Tibetan Area.* Master's dissertation, Sichuan Academy of Social Sciences.

Yang, M.Y., Dong, S.K., Dong, Q.N. *et al.* (2020) Cooperative grassland management practices promoted by land tenure system transformation benefit social-ecological systems of pastoralism on the Qinghai-Tibetan plateau, China. *Journal of Environmental Management* 261, 110215. DOI: 10.1016/j.jenvman.2020.110215

Zeng, D., Jiang, F., Fan, Z. and Du, X. (1999) Ecosystem health and sustainable development for human. *Chinese Journal of Applied Ecology* 10(6), 751–756.

20 Conservation Strategies: Multi-scale Approaches and Policies in the Hindu Kush Himalayan Region

Binyu Luo[1,*], Allan Degen[2] and Zhanhuan Shang[1]
[1]*College of Ecology, Lanzhou University, Lanzhou, China;* [2]*Institutes for Desert Research, Ben Gurion University of the Negev, Israel*

Abstract

Over the past 70 years, the number of protected areas of the Hindu Kush Himalayan (HKH) region increased substantially, but there is a lack of connectivity among them. Despite the increasing number of initiatives on transboundary landscape management, exchanges and co-operation among member states on environmental and ecological protection are still insufficient. Innovative institutions and arrangements should be adopted to promote inter-governmental dialogue, common policy deliberation and transnational practice. For the development of mountain areas, ensuring the establishment of transportation and infrastructure in the region can strengthen the exchange and trade between mountain areas and other areas. In general, the member states should strengthen co-operation and exchanges and reduce border conflicts to prevent the HKH region from becoming an ecological and cultural island.

Keywords: policy, co-operation, protected area, culture, livelihood, management strategy

20.1 The Zoning Classification of Ecological Reserves in Each Member Country

A protected area is a 'geographically defined area that is designated or managed to achieve a specific conservation objective' (CBD, 1992). Protected areas are also one of the main strategies for conservation and sustainable development, managing critical habitats, ensuring the maintenance of natural processes, and directly supporting the livelihoods of some 1.1 billion people (CBD, 2019). In addition, protected areas have been identified as key to solving the 'biodiversity crisis' (Hoekstra *et al.*, 2005) and preventing species extinctions (IUCN, 2017). The Hindu Kush (HKH) region harbours the world's major mountains and is rich in ecosystems, species and cultural diversity. In a broad sense, extreme changes in elevation, topography, soil and climate gradients (Fig. 20.1) have created complex and diverse ecosystems and habitats at the intersection of two of the world's eight terrestrial domains. However, the region is highly vulnerable to climate change and human activities. With increasing impacts, including land-use change, the HKH region, particularly the rainforests of southern Tibet, India and Myanmar, has been identified as an

*Corresponding author: luoby18@lzu.edu.cn

DOI: 10.1079/9781800622579.0020

Fig. 20.1. Rainforests in southern Tibet, China. (Photo: Binyu Luo)

area of imminent extinction (Ricketts *et al.*, 2005). Jantz *et al.* (2015) predicted that the HKH region might lose 80–86% of its original habitats by 2100. Therefore, there is an urgent need to establish protected areas in the HKH region to ensure the maintenance of biodiversity and sustainable development of the region.

20.1.1 Distribution of Hindu Kush Himalayan protected areas

The number of protected areas has increased substantially in the last century (1900–2020). Until the 1970s, there were only 64 protected areas; but by 1981, 78 more protected areas were established. Over the next ten years, by 1991, the number of protected areas increased by 176 to 318. Over the next two decades, the number of protected areas grew exponentially, reaching 467 by 2000 and 559 by 2010; while 16 protected areas were added between 2010 and 2020 (Fig. 20.2). At present, there are 575 protected areas in the HKH region (Fig. 20.3), covering an area of 1,721,894 km^2 (Chaudhary *et al.*, 2022), which accounts for 40.2% of the area of the HKH region and 8.49% of the global protected areas (Chaudhary *et al.*, 2022).

The number of protected areas varies greatly among the member countries (Fig. 20.4), with China contributing 45% of the total, Pakistan 35%, India 11%, Nepal 3%, Myanmar 2%, Afghanistan 1.83%, Bhutan 1.7% and Bangladesh 0.11% (Chaudhary *et al.*, 2022). A lack of economic opportunities, cross-scientific nature of conservation actions, institutional development, skilled human resources and large-scale conservation methods hinder the establishment and management of protected areas. In Myanmar, for example, anthropogenic pressures and a lack of financial and skilled human resources are hampering the effective management of protected areas. But half of the countries in the HKH region – Bhutan, China, Nepal and Pakistan – have already met, surpassed or are on track to reach the global target of 17% protected area coverage by 2022.

20.1.2 Classification of protected area management in the Hindu Kush Himalayan region

Within HKH's protected areas, there are eight categories of protected area management. Of the protected areas, 45 fall under the IUCN category (I–VI), while 30 fall under World Heri-

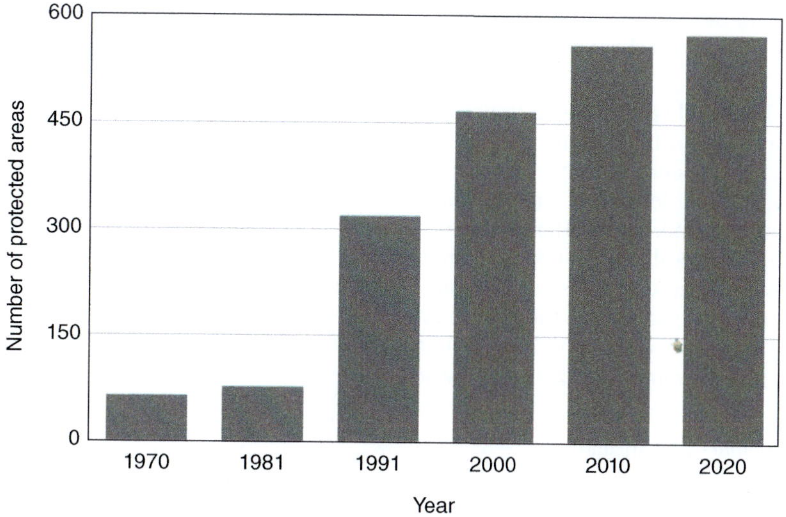

Fig. 20.2. The number of protected areas in the Hindu Kush Himalayan region from 1970 to 2020 (Data from Chaudhary *et al.*, 2022; figure: the author)

Fig. 20.3. Yarlung Zangbo Grand Canyon National Nature Reserve. (Photo: Binyu Luo)

tage Sites (WHS) of international importance. The remaining 91 protected areas are not classified and are listed as unreported in the world database on protected areas (WDPA).

In the IUCN classification, 32% of the protected areas are managed as 'protected landscape', 28% are managed as category IV (habitat/species management area), 10% are classified as category II (national parks) and 8% are managed as category VI (protected areas for sustainable use of natural resources). Only one reserve is in the wilderness area and four are in the strict nature reserve (Fig. 20.5).

20.1.3 Altitude distribution of the Hindu Kush Himalayan region protected areas

The HKH reserves have a large elevation span, with lower elevations in the south of below 2000 m above sea level (masl) and higher elevations in the north, such as the Tibet Autonomous Region of China, with an average elevation of more than 4000 masl. Rudrasager lake, a Ramsar site in Tripura, India, is only 19 masl; whereas China's Qomolangma National Park and Nepal's Sagarmatha National Park are 8849 masl. Abundance of animals and plants is higher at lower than higher elevations, and, as a result, the number of protected areas is higher at lower elevations. In total, about 58% of the protected areas are below 3000 masl and cover an area of 582,524 km^2, while only 5% are above 5000 masl and 0.3% are above 6000 masl (Fig. 20.6). Although species abundance is lower at higher elevations, many endemic species exist at high altitude. Therefore, the establishment of a network of protected areas in high-altitude areas, or the protection of large areas in the form of national parks, can contribute to biodiversity conservation and sustainable development.

20.1.4 Role of religions and cultures in protected areas

Religious and cultural beliefs related to natural resources have played important roles in the creation and maintenance of protected areas in the Himalayas. Some forests of Garhwal and Kumaon in Uttarakhand, such as Hariyali Sacred Landscape and Haat Kali Sacred Grove, cannot be used by anyone because the forests and streams within are considered sacred and, therefore they are dedicated to the gods (Negi, 2005; Sinha and Mishra, 2015). In Chhyangru Village, Byas Village Development Committee, Apinampa Reserve, Nepal, a temple called Shyanchho was built in the forest to prevent further deforestation and degradation and to preserve the forest as sacred. In India, Buddhists influenced the establishment of areas to protect natural landscapes, including the high-altitude lakes and their basins in Ladakh and Sikkim. In Yunnan and Tibet in south-west China, cultural diversity (Fig. 20.7) plays an important role in protected areas, mainly including natural resources and socio-cultural values (Fig. 20.7) (Anderson *et al.*, 2005; Brandt *et al.*, 2013). In China's high-altitude areas, where the killing of wild animals is considered an affront to the gods due to their Buddhist faith, the local residents adhere to conservation practices. In addition, in about 90,000 km^2 of southern Tibet in south-west China, the cultural landscape is protected by various social and religious institutions that use traditional ecological knowledge to manage protected areas sustainably.

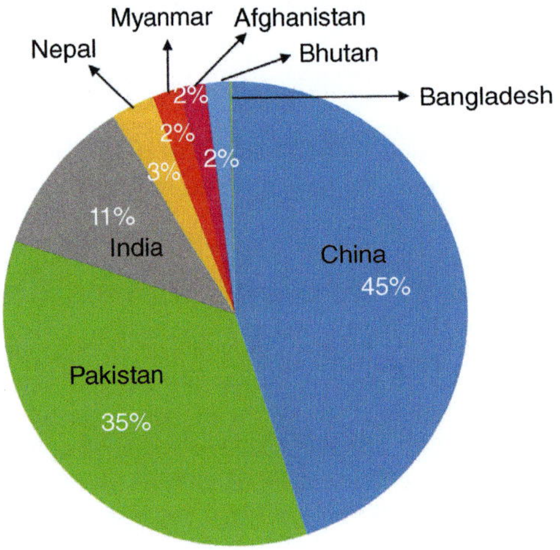

Fig. 20.4. Proportion of protected areas in each member state of the Hindu Kush Himalayan region (Data from Chaudhary *et al.*, 2022; figure: the author)

20.2 Transboundary Ecological Conservation in the Hindu Kush Himalayan Region

As the HKH region covers a large area and spans eight member countries, strengthening communication and exchanges among member countries is important for the region's sustainable development. With the rapid expansion of opportunities for international co-operation and

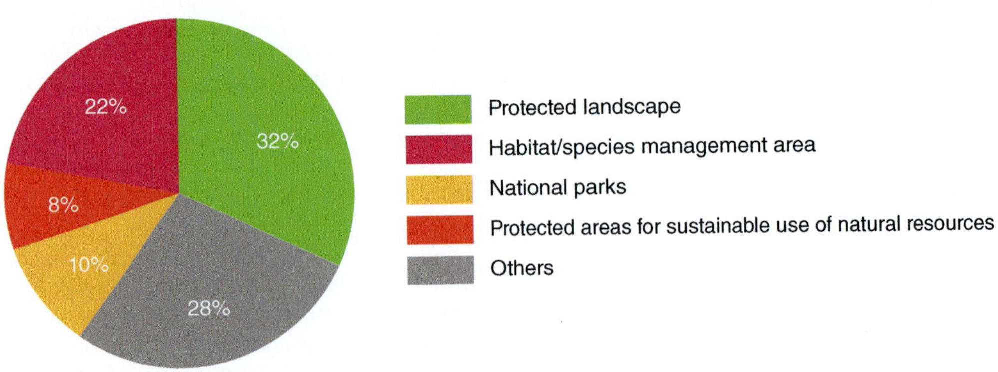

Fig. 20.5. Proportion of all types of protected areas in the Hindu Kush Himalayan region. (Data from Chaudhary *et al.*, 2022; figure: the author)

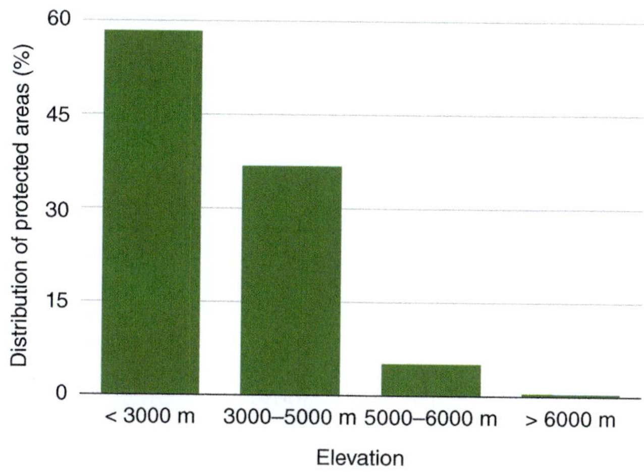

Fig. 20.6. Elevation distribution of protected areas in the Hindu Kush Himalayan region. (Data from Chaudhary *et al*., 2022; figure: the author)

Fig. 20.7. Temples in Tibet. (Photo: Binyu Luo)

cross-environmental management, regional or transnational co-operation has become key for global ecological protection. Although the Himalayas may be one of the most politically complex regions in the world, it also offers a range of examples of ecological conservation and services (Blaikie and Muldavin, 2004; Grumbine and Pandit, 2013). With the economic rise of China and India on either side of the mountain range, the region has also become a geopolitical and political-economic centre of global importance (Wirsing, 2013; Zhang, 2016); but, concomitantly, it also brings ecological and political challenges. The case of water management is the most glaring example of the importance of regional co-operation. For example, the Yarlung Zangbo river originates in Tibet, China, and flows into India through Bashika, and the upstream and downstream are increasingly closely related (Fig. 20.8). Strengthening international co-operation between the two member states is essential for ecological protection near the basin.

Ten river systems flow through the borders of the HKH countries, and, consequently, bilateral and multi-lateral treaties and agreements among member states are indispensable if ecosystems are to be managed on a regional scale. For example, Afghanistan and Pakistan are developing a water-sharing agreement for the Kunar river in the Kabul basin (Vick, 2014). The Mahakali River Agreement between Nepal and India, the Ganges Water Treaty between India and Bangladesh, and the Indus River Water Treaty between India and Pakistan are primarily aimed at ensuring electricity generation and sharing; however, they are often perceived as lacking equitable, comprehensive arrangements for the use and distribution of resources (Mustafa 2010; Uprety and Salman, 2011; Butler, 2016). Formal co-operation between China and India on cross-border resource management remains limited (Rahaman and Varis, 2009).

20.2.1 Cross-border co-operation among member countries

Co-operation among member countries can benefit the ecological conservation and well-being of the HKH region; however, countries with

Fig. 20.8. The Brahmaputra river in the Chinese Himalaya. (Photo: Binyu Luo)

larger economies often find it difficult to co-operate with smaller economies. For example, India has established the Power Trading Corporation to develop and trade hydropower in the region; however, power trading agreements have not been promoted through regional co-operation (Karki and Vaidya, 2009).

Border co-operation at the subnational level is bearing fruit, particularly in addressing issues related to ecological hotspots such as transboundary wildlife movements, grazing management, and soil and water management. Although these practices are often irregular compared to formal co-operation between states, they are, nonetheless, good examples of co-operation between countries. These practices are common in some open and semi-open border control areas such as between Nepal and India, Nepal and China, Bhutan and India, and Bangladesh and India. Border communities between these countries have historically co-operated in rangeland management, promoting forest diversity, water-quality regulation of wells, religious protection of ecological land landscapes, and co-developing livelihood opportunities (Ning *et al.*, 2013; 2016).

In recent years, several new funds have been allotted to facilitate cross-border exchanges and co-operation among member countries in the HKH region. For example, the International Centre for Integrated Mountain Development (ICIMOD), with financial support from Germany's Federal Ministry for Economic Cooperation and Development and the UK's Department for International Development, is promoting the conservation and development of Kailash in a partnership among China, India and Nepal. In addition, ICIMOD is initiating development and conservation projects in the Himalayan region, including Mount Everest and the Junggar region (Sherpa *et al.*, 2003; Chettri *et al.*, 2008). These projects demonstrate that co-operation among member states is both necessary and feasible.

Despite a series of collaborations in border areas between member countries of the HKH region, these projects have produced few ecological benefits. Because the main goal of such co-operation is economic development between countries, environmental and ecological protection are not prioritized.

20.2.2 Lack of co-operation at the country level

Although there are increasing initiatives for transboundary landscape management, they are not supported by official documents of member states, so there is still insufficient exchange and co-operation on environmental and ecological protection. Moreover, even exchanges and co-operation among member states do not include all countries. Examples include Bhutan, Bangladesh, India and Nepal (BBIN) and the Bay of Bengal Initiative for Multi-Sectoral Technical and Economic Cooperation (BIMSTEC), where the only countries included are Bangladesh, India, Myanmar, Thailand, Bhutan and Nepal, while China and Pakistan are excluded. Moreover, such co-operation is driven mainly by political interests of individual countries. HKH spans multiple countries, and to ensure the sustainable development of the region, it is necessary to support close co-operation among all member countries, with HKH as a shared platform. The South Asian Association for Regional Cooperation Comprehensive Framework on Disaster Management and Disaster Prevention is another initiative on collaborative governance at the regional level. Despite the initiative's potential, lack of trust among member states and political tensions in border areas have hampered its implementation.

The lack of horizontal and vertical linkages between different government agencies and actors has hindered efforts to collaborate on HKH resource governance (Kohler *et al.*, 2015). Although sub-national and border co-operation is gaining momentum, policies and actions at the national level still demonstrate that countries are not aware of the need for ecological protection in the HKH region. The efforts of inter-governmental bodies such as the South Asian Association for Regional Cooperation (SAARC) and ICIMOD are not binding and do not have sufficient influence to convince member countries that co-operation is vital. As a result, there is still a lack of decision makers at the national level to manage the HKH region.

Due to the policy sensitivities among the member countries in the HKH region, it is important to provide a platform for co-operation between individual countries. Such platforms not only contribute to the common policy agenda but also to exchanges between countries to achieve common conservation goals. For example, knowledge-sharing platforms such as climate-change adaptation, poverty reduction, watershed management and biodiversity conservation should be established to increase co-operation and exchanges among countries.

20.2.3 The important role of super economies in Hindu Kush Himalayan ecological conservation

Responsibility for water resources and basin management has been a challenge, and future co-operation will depend on how resources are distributed among upstream and downstream member countries. Often, power imbalances among participating countries and the dominance of larger economies in the exploitation of natural resources are obstacles to the establishment of equitable regimes. Smaller economies are often more dependent on the supply of natural resources than larger economies, which can create conflicts between them.

Ecological connectivity across borders and the development framework of eco-regions are key to ensuring sustainable development in the HKH region. Current natural resource management and ecological conservation still use old methods, such as cross-border grazing between China and Nepal. However, the current co-operation methods are still inadequate to meet the sustainable development of the HKH region. Co-operation and exchanges between China and India, two important economies on both sides of the Himalayas, can take the lead in promoting co-operation among member countries to address major ecological security issues in the HKH region. However, due to border conflicts between these two countries, co-operation and exchanges are still greatly hindered.

As more than half the HKH area is located in China, the actions taken by China, the world's second largest economy, play a leading role in the ecological preservation. For example, in the future, innovative institutions and arrangements will be adopted to promote inter-governmental dialogue, common policy deliberations, and transnational practices. Large-scale investment, which can be provided by China, and multi-scale regional co-operation are crucial to achieve prosperity and sustainable development in the region.

20.3 External Forces to Protect the Ecology and Improve Livelihoods

20.3.1 Integration of the Hindu Kush Himalayan region for ecological protection

Ecological protection and sustainable development of under-developed countries are complex, and are mutually promoted and restricted by population, resources, environment, economy, science and technology, and society. Under-developed countries usually prioritize national economic development and neglect ecological issues in the HKH region. This occurs because the under-developed mountain areas lack economic and technological means and the implementation of relevant measures in the protected areas. For example, biodiversity-protected areas in Myanmar have poor infrastructure, inadequate management staff and lack of expertise, which makes it difficult to implement effective management activities in protected areas. In countries with strong economic development, the focus shifts from economic development to ecological security and other issues. Therefore, promoting the integration of HKH plays a vital role in the ecological protection of the HKH region.

Through co-operation between under-developed and economically strong countries, an independent management body can be established to manage the HKH region in a unified manner. This approach would allow resources to be distributed more evenly among protected areas, thereby avoiding the neglect of ecological protection due to the lack of economic strength in under-developed countries. In addition, through the establishment of a unified management body, data sharing among countries could be promoted, including statistics, remote sensing, geographic information, monitoring, model simulation and network data, and waste management in the HKH region.

20.3.2 Livelihoods in under-developed countries

Most of the countries in the HKH region are developing rapidly, creating many economic opportunities both nationally and locally. For example, in the past few decades, with the policy of reform and opening, China has become the fastest-developing country in the world, with an average annual GDP growth rate of nearly 10%. It was expected that in 2021, the entire population of China would be out of poverty. In addition, with the economic rise in India, south Asia has also experienced strong economic growth, becoming the fastest-developing region in the world in recent years.

While these developments are impressive, the region's economic development is not synchronized among and within countries. For example, there is a huge gap between the development of China's western regions and the coastal areas in the east and south, while in India, Himachal Pradesh has developed well above the national average. At the same time, countries with unstable societies and long periods of war, such as Afghanistan, Pakistan and Myanmar, have experienced slower economic growth, while countries with very rugged terrain, such as Bhutan and Nepal, have also lagged behind. As a result, economic development in the HKH region remains fragile.

Livelihoods in the HKH region are based mainly on agriculture and animal husbandry; however, as the population increases, limited pasture resources make the development of the HKH region unsustainable. Under-developed regions in China such as Tibet can rely on the better-developed eastern regions of China and other economically prosperous regions for economic aid. However, in regions such as Pakistan, Bhutan and Nepal, the income is mainly provided by countries such as China and India. Despite large foreign donations, this system is not sustainable. Only by strengthening co-operation and exchanges among countries can the economic growth of these under-developed countries be ensured.

20.3.3 Co-operation between under-developed and strongly developed countries

Regional cross-border trade is crucial to the economic development of mountain countries and the income of the local people. For example, driven by Chinese demand, the harvesting and sale of the fungus *Cordyceps* has become an important income for mountain-dwellers in under-developed countries such as India, Nepal and Bhutan (Winkler, 2009). However, south Asia remains one of the least integrated regions in the world due to the complexity of border procedures (Ahmed and Ghani, 2007).

The South Asian Association for Regional Cooperation and the south-Asian Sub-regional Economic Cooperation Programme aim to improve economic opportunities in the region, promote connectivity between China and south-east Asia, and ensure rapid economic growth in some under-developed countries. Other large regional and international initiatives have also sought to strengthen regional integration:

- India's 'Look East' policy has been successful since the opening of Myanmar in 2014;
- Pakistan and China are building an economic corridor with substantial investments in infrastructure and energy production;
- China announced a modern version of its 'Belt and Road' initiative, which aims to connect much of Asia with countries in Europe and East Africa by sea (one belt) and land (one road);
- Bangladesh, China, India and Myanmar are building an economic corridor linking Mandalay, Dhaka and Kunming in China's Yunnan province with Kolkata in India.

These large-scale initiatives are likely to shape the socio-economic dynamics of the HKH region in the near future. Given the increasingly rapid globalization and the vulnerability and marginality of mountain areas, further economic integration requires more targeted policies to promote the development of mountain towns in the HKH region.

20.3.4 Infrastructure development to facilitate exchanges with developing and strong countries

Transportation and infrastructure development in remote mountain areas can strengthen the exchange and trade between mountain areas and other areas. Improvement in infrastructure and road network are desperately needed in the

HKH region to prevent the region from becoming an 'isolated island'. Although it is generally accepted that the construction of roads, increased tourism and increased trade are accompanied by environmental problems and population displacement, these problems can be reduced by such methods as designating core areas as protected areas.

There are great differences in transportation development and infrastructure among south-Asian countries. In China, for example, there is a comprehensive road network in the HKH region, but this is not the case in countries like Nepal, Pakistan and Myanmar. Due to the complex terrain of the HKH region, the difference in infrastructure among countries has become an obstacle to communication between under-developed and prosperous countries.

One of the keys to promoting regional integration of the HKH countries is the improvement of the infrastructure. Regional cross-border trade plays an important role in food transportation and food security. Because most of the HKH region is mountainous, the cost of transporting food domestically is usually higher than transporting food across the border. China has built many crossings along the border (Zhangmu, Pulan, Gyirong and Riwu), which enable residents of the two countries to pass through easily, and has brought new opportunities for the development of the region. For example, Zhangmu crossing carries more than 80% of China's and more than 90% of Tibet's trade with Nepal. Due to the large difference in altitude between the two sides of the crossing, the resources produced are also different. For China, the main imports are rice, flour, peppers and perfume; whereas the main exports are wool, tea and salt. The establishment of these crossings has solved the problems of the lack of vegetables north of the crossing and lack of meat resources south of the crossing.

Road construction in the HKH region is very costly compared to the plains. In addition to the high cost of engineering and materials due to the remoteness of the HKH region, the climate conditions are another major factor. In August and September, for example, heavy monsoon rains in the HKH region make construction and maintenance very difficult, while the summer rainy season and winter snowfall make roads impassable for many months of the year. China is the only country in the world with a major railway to the mountains. The spectacular railways on the permafrost of the Tibetan Plateau linking the capital Beijing with the Shigatse region in the west, has greatly boosted Tibet's development. Therefore, if under-developed countries in the HKH region could improve travel to the domestic mountain areas, their regional integration in the HKH region would be promoted, and their trade co-operations and exchanges with other countries would be strengthened.

20.4 What Policies Are in Place at the National Level to Protect Ecosystems and Livelihoods?

Establishing ecological and social robustness of ecosystems in the HKH region is critical to achieving Sustainable Development Goal (SDG) 15: to protect, restore and promote the sustainable use of terrestrial ecosystems, manage grasslands and forests sustainably, prevent desertification, halt and reverse land degradation, and halt biodiversity loss. By 2030, ensure that the mountain ecosystems, including biodiversity, are protected in order to enhance their capacity to provide benefits essential for sustainable development. By protecting the ecosystem stability of the HKH region, it also guarantees the livelihoods of local people.

20.4.1 Biodiversity conservation policy

Over the past few decades, the HKH region's ecological conservation has undergone a major transformation. For example, population growth and over-exploitation of ecological resources (Sandhu and Sandhu, 2014; Chettri and Sharma, 2016), as well as the continuous degradation of alpine meadows on the Tibetan Plateau (Fig. 20.9), indicate that ecological conservation in the HKH region faces serious management challenges (Klein *et al.*, 2004; Luo *et al.*, 2010). Therefore, the enactment of policies at the national level is crucial for ecological protection. Through adaptive management and appropriate policies, the concept of ecological conservation changed markedly; for example, the 1992 Convention on Biological Diversity, which was a

Fig. 20.9. The kiang (*Equus kiang*) on the Tibetan Plateau. (Photo: Binyu Luo)

milestone in the history of biodiversity conservation.

The successful establishment of the International Geosphere-Biosphere Programme and the Millennium Development Goals between 1992 and 2000 provided additional support to ecological conservation. With the progress of the millennium development process, the importance of ecosystem services as biodiversity to human well-being has been widely recognized by countries around the world. Before 1992, the HKH region implemented many policies to promote ecological protection. The earliest was the Forest Act of 1927, which is still in force in India and Pakistan. The Wildlife Protection Act of 1972 and the Islamabad Wildlife Protection Act of 1979 in Pakistan were ground-breaking legal instruments. In China, the Wildlife Protection Law of 1988 was the first law. Since 1999, China has protected forests by the successful 'returning farmland to forest' programme. Bhutan's Forest Policy in 1974 was the country's first national protection policy, and the Forest and Nature Conservation Act in 1995 was the second.

The Panchayat (Extension to Scheduled Areas) Act (1996), the Biological Diversity Act (2002), the Conservation Area Rules (1996), the National Parks and Wildlife Conservation Act (1972) and its amendments (1993), the Forest Act (1993), Forest Policy (2015) and Nepal's Buffer Zone Management Regulation (1996) are good examples of promoting ecological conservation in the HKH. Each case represents a devolution of control from the government to local authorities, a shift that represents an emphasis on ecological conservation in the HKH region. These policies also address local competing/conflicting interests. India has also announced that it will implement these changes further, with a goal of reforestation of 6 million ha of degraded forest land and expanding the forest area. Current laws and regulations at the national level have brought new hope for ecological protection in the HKH region.

20.4.2 Adaptation policies to climate change

Sustainable Development Goal 13 (SDG 13) aims to 'take urgent action to address climate change and its impacts'. Strengthening the resilience and adaptation of countries to climate-related natural disasters and integrating climate change measures into national policies, strategies and planning are major objectives of Sustainable Development Goal 13. As the warming rate in the HKH region exceeds the global average, especially on the Tibetan Plateau (Wang *et al.*, 2008), HKH member countries are developing a range of policies to address climate change. For example, the trust fund established by Bangladesh and the Climate Change Budget law established by Nepal were enacted to address climate change. China, Pakistan, Afghanistan and Bhutan developed policies that incorporated climate change into their implementation strategies. Bangladesh discussed national development priorities under future climate change (Saito, 2012), while Bhutan included climate change as one of the key national outcomes of the eleventh Five-Year Plan (2013–2018). How to adapt to climate change is one of the themes discussed by member countries in the HKH region for national development.

Although the policies formulated by the HKH member countries differ, they all move in a unified direction. China, India, Pakistan, Nepal, Bhutan, Bangladesh and Myanmar all incorporated climate change into policy making across all sectors. Bhutan's Integrated National Water Resources Management Plan 2016, Framework for Mainstreaming Environment, Climate Change and Poverty Alleviation in Five-Year Plans, National Action Plan to Combat Land Degradation 2014 and National Biodiversity Strategy and Action Plan 2014 all focused on climate change. Afghanistan integrated climate change NAPA (2009) with the 2009 self-assessment of national capacity needs for global environmental management to include climate change into policies and programmes (e.g. national environmental action plans). Bangladesh's policies, including Bangladesh CCSAP (2009), BCCTF and Bangladesh Climate Change Resilience Fund, and NAPA (2005) aimed to adapt to climate change. India's National Climate Change Action Plan aimed to mainstream climate change issues at the subnational level with the establishment of the National Climate Change Adaptation Fund (NAFCC – 2015) and the Indian Climate Change Assessment Network (INCCA – 2010). In Myanmar, the Myanmar Climate Change Alliance was established (2013) and major projects funded by the Adaptation Fund and the Global Environment Facility (GEF) are being implemented. Nepal's National Action Plan Development Process (2015), Climate Change and Algorithms (2012), Climate Change Policy (2011), National Action Plan Framework (2011) and National Action Plan (2010) were initiated to combat climate change.

20.4.3 Land protection policy

In recent years, with the intensification of human activities, land-use change and land degradation have become important ecological problems in the HKH area, for example, soil desertification caused by over-grazing. In Nepal, the National Land Degradation and Desert Action Plan estimates that about 3.3 million ha are undergoing desertification, with 70% of degraded land being forests and grassland (Fig. 20.10). Desertification is also a major concern for India and Pakistan in the HKH region, and EPASSA (2008) estimates that desertification will become more severe in these countries. In the alpine regions of Pakistan, the forest area decreased by 30% from 1968 to 2007, of which a third was due to agricultural expansion, with agricultural land expanding by 70% and forest area decreasing by 50% in the middle and high altitudes (Qasim *et al.*, 2013). India is losing 0.2% of its forest area per year (Reddy *et al.*, 2013) and Myanmar is losing 0.3% per year (Leimgruber *et al.*, 2015). On the Tibetan Plateau, 90% of grasslands have been degraded to varying degrees due to a combination of climate change and over-grazing (Bardgett *et al.*, 2021). As land degradation intensifies, governments have taken various measures for restoration.

In China, since the 10th Five-Year Plan (2001–2005), the Tibet autonomous region has received a total of 31.6 billion yuan in ecological compensation from the state for forests,

Fig. 20.10. Land degradation on the Tibetan Plateau. (Photo: Binyu Luo)

grasslands, wetlands and key ecological functional areas. During the 12th Five-Year Plan period (2011–2015), the state allocated a total of 10.88 billion yuan in subsidies for ecological protection of Tibetan grasslands. In addition, since 2013, China has continuously increased its investment in ecological construction in Qinghai province. The central government has allocated a total of 16.4 billion yuan in funds, and successively initiated ecological benefit compensation projects such as grasslands, forests and wetlands. With the implementation of a series of grassland ecological protection projects, such as the return of grazing land to grassland, the policy of subsidies and incentives for grassland ecological protection, and the control of grassland rodents and pests, the degradation of the Qinghai-Tibet Plateau has been curbed. In addition, the Chinese government has invested substantial additional funds in the protection and restoration of ecosystems. China launched large ecological service projects including the Natural Forest Conservation Project (NFCP) and the Slope Land Improvement Project (SLCP) (Chen *et al.*, 2015), with cumulative investments of more than $50 billion. The SLCP alone has benefitted tens of millions of families and hundreds of millions of farmers (Ouyang *et al.*, 2016).

20.4.4 High-altitude rangeland management policy

Over the past few decades, under the influence of globalization and socio-economic transformations, resource utilization patterns and development activities in the HKH region have undergone unprecedented changes. These factors, combined with a rapidly changing climate, have severely affected mountain ecosystem services and the well-being of residents. For example, most high-altitude ecosystems are suffering from degradation, desertification and soil erosion, and are further exacerbated by climate and human factors. Due to the general lack of livelihood options for residents in the HKH region, they generally rely on grazing livestock or slash-and-burn farming, and, as a result, their livelihoods are under serious threat.

Management of high-altitude rangelands requires a strong participatory and adaptive approach. Countries and governments should provide policy support for the use of rangeland resources so that landscape elements such as wetlands, biological corridors, biodiversity and watersheds are protected from rapid land-use change and over-development. Much of the HKH region is an important corridor for the seasonal migration of high-altitude animals and is also an important habitat for a large number of species.

To protect natural grassland and restore degraded grassland, the Chinese government has introduced a series of policies, such as the Natural Conservation Project and the Ecological Protection and Construction Project in the Source Area of Three Rivers, to ban grazing on non-degraded grassland and manage and restore degraded grassland. For extremely degraded or desertified grassland, the Chinese government sows seeds to increase the aboveground biomass and the carbon sequestration capacity of the grassland, and thus increases the grazing capacity of the grassland. For grazing prohibition, the Chinese government provides subsidies to herders to reduce grazing intensity.

20.4.5 Livelihoods of Hindu Kush Himalayan residents

The main limiting factors for residents in the HKH region are geographical remoteness (Fig. 20.11), inaccessibility and high dependence on natural resources. In addition, issues such as economic and gender inequalities, and discrimination based on the caste system, also contribute to poverty in the region. Overall, the HKH region is home to hundreds of millions of people who are socially, economically and politically isolated and marginalized.

Due to the different levels of economic strength among member HKH countries, the poverty index differs greatly; Afghanistan's poverty index is 59%, while China's is only 5%. According to the multi-dimensional poverty index, Afghanistan ranks highest in poverty (0.35), followed by India (0.28), whereas China ranks lowest (0.02). Despite the strong economy in India, the gap between rich and poor is the main cause of poverty in the mountain people. In China, Tibet's per capita GDP even exceeds that of some eastern Chinese provinces.

The ecosystems of the HKH region are fragile and it is difficult for the local population to rely wholly on the natural resources provided by the ecosystem services to survive, as these services have undergone intense changes due to human activities and climate change. This has forced the residents in the HKH region to diversify their livelihoods from total dependence on land and natural resources during the past few decades. Government subsidies and tourism development, the improvement of education levels, the expansion of the labour force, changes

Fig. 20.11. Transportation in the Hindu Kush Himalayan region. (Photo: Binyu Luo)

in government policies and the local population's ideology are gradually improving their livelihoods.

In addition, the inhabitants of the HKH region have shifted from a subsistence smallholder farming economy to high-value agricultural products (Singh *et al.*, 2011). In China, for example, high-value cash crops provide farmers with four times the income of rice (Fig. 20.12), while in Nepal, export earnings from high-value crops such as tea, coffee, cardamom, ginger and turmeric exceed the value of cereals and dairy product imports. Local residents diversified their livelihood in accordance with the environmental conditions. These livelihoods, which are a key to poverty alleviation, are generally related to local transportation conditions, and in areas with more developed transportation, residents tend to cultivate cash rather than subsistence crops (Salick *et al.*, 2005).

Many male labourers from the HKH countries move away for months or decades for employment elsewhere and send remittances to their families. In some countries, these remittances account for a substantial share of the GDP: Bangladesh (11.6%), Afghanistan (16.3%) and Nepal (28.5%) (IFAD, 2013). However, these labour migrations are often not sustainable strategies as they are not covered by social protection. This labour migration is usually by men, which increases the women's role in agriculture greatly, besides their responsibility for the household and children.

20.4.6 Livelihood protection policies

Of the total population in Tibet, 80% live in rural areas and farm for their livelihood, and are in poverty. In 2015, China promulgated the Decision of the Central Committee of the Communist Party of China and The State Council on Winning the Battle against Poverty, which aimed to lift the rural poor out of poverty by

Fig. 20.12. Cash crops in the Tibetan Plateau. (Photo: Binyu Luo)

2020 by identification, classification and assistance. At the Sixth Symposium of the CPC Central Committee held in Beijing in 2015, it was pointed out that Tibet, like other parts of China, has entered the decisive stage of building a moderately prosperous society. It was necessary to take the improvement in people's livelihood and their cohesion as a starting point and focus on poverty alleviation by promoting economic and social development of the Tibet Autonomous Region, identifying and solving the causes of poverty, and improving the living conditions of people in extreme poverty. The Poverty Alleviation Office of the Tibet Autonomous Region and the Poverty Alleviation Command Corps have repeatedly carried out assessments of the implementation of the Poverty Alleviation Plan for the 13th Five-Year Plan Period. It is difficult for residents to escape poverty, but these efforts reflect the Chinese government's determination to improve the livelihoods of these people.

Pakistan has developed the National Forest Policy (MoCC, 2015) to expand and promote the sustainable use of forests, protected areas, natural habitats and watersheds to restore their ecological functions and improve livelihoods and human health, in line with national priorities and international agreements. In a forest-rich area of Khybol Pakhtunkhwa province in Pakistan, forestry sector projects institutionalized forest management, thereby improving the natural capital and economic conditions of the local people (Ali, 2007).

References

Ahmad, S. and Ghani, E. (2007) South Asia growth and regional integration: an overview. In: Ahmed, S. and Ghani, E. (eds) *South Asia: Growth and Regional Integration.* The International Bank for Reconstruction and Development, The World Bank, Washington, DC, pp. 3–42.

Ali, J. (2007) Livestock sector development and implications for rural poverty alleviation in India. *Livestock Research for Rural Development* 19(2), (online edition).

Anderson, D.M., Salick, J., Moseley, R.K. and Xiaokun, O. (2005) Conserving the sacred medicine mountains: a vegetation analysis of Tibetan sacred sites in Northwest Yunnan. *Biodiversity and Conservation* 14(13), 3065–3091. DOI: 10.1007/s10531-004-0316-9

Bardgett, R.D., Bullock, J.M., Lavorel, S., Manning, P., Schaffner, U. *et al.* (2021) Combatting global grassland degradation. *Nature Reviews Earth & Environment* 2(10), 720–735. DOI: 10.1038/s43017-021-00207-2

Blaikie, P. and Muldavin, J.S.S. (2004) Upstream, downstream, China, India: the politics of environment in the Himalayan region. *Annals of the Association of American Geographers* 94(3), 520–548. DOI: 10.1111/j.1467-8306.2004.00412.x

Brandt, J.S., Wood, E.M., Pidgeon, A.M., Han, L.X., Fang, Z. *et al.* (2013) Sacred forests are keystone structures for forest bird conservation in southwest China's Himalayan Mountains. *Biological Conservation* 166, 34–42. DOI: 10.1016/j.biocon.2013.06.014

Butler, C.J. (2016) Knowledge, nature, and nationalism: the Upper Karnali dam in Nepal. Doctoral dissertation, University of California, Santa Cruz, California.

CBD (Convention on Biological Diversity) (1992) United Nations Environment Programme (UNEP), Nairobi.

CBD (2019) Protected areas: an overview. Available at: https://www.cbd.int/protected/overview/ (accessed 28 September 2023).

Chaudhary, S., Uddin, K., Chettri, N., Thapa, R. and Sharma, E. (2022) Protected areas in the Hindu Kush Himalaya: A regional assessment of the status, distribution, and gaps. *Conservation Science and Practice* 4, e12793.

Chen, Y., Wang, K., Lin, Y., Shi, W., Song, Y. *et al.* (2015) Balancing green and grain trade. *Nature Geoscience* 8, 739. DOI: 10.1038/ngeo2544

Chettri, N. and Sharma, E. (2016) Reconciling the mountain biodiversity conservation and human wellbeing: drivers of biodiversity loss and new approaches in the Hindu-Kush Himalayas. *Proceedings of the Indian National Science Academy* 82, 53–73. DOI: 10.16943/ptinsa/2016/v82i1/4837

Chettri, N., Shakya, B., Thapa, R. and Sharma, E. (2008) Status of a protected area system in the Hindu Kush-Himalayas: an analysis of PA coverage. *International Journal of Biodiversity Science and Management* 4, 164–178.

EPASSA (2008) Ecosystem services and poverty alleviation study in South Asia: a situation analysis for India and the Hindu Kush Himalayan region (Annexure 3). Available at: https://lib.icimod.org/record/14027 (accessed 28 September 2023).

Grumbine, R.E. and Pandit, M.K. (2013) Threats from India's Himalaya dams. *Science* 339, 36–37. DOI: 10.1126/science.1227211

Hoekstra, J.M., Boucher, T.M., Ricketts, T.H. and Roberts, C. (2005) Confronting a biome crisis: global disparities of habitat loss and protection. *Ecology Letters* 8, 23–29. DOI: 10.1111/j.1461-0248.2004.00686.x

IFAD (2013) *Sending Money Home to Asia*. ISBN: 978-92-9072401-8

IUCN (2017) *Programmatic Strategy for Strengthening the Presence of IUCN in Asia*. Version: May 2017.

Jantz, S.M., Barker, B., Brooks, T.M., Chini, L.P., Huang, Q. *et al.* (2015) Future habitat loss and extinctions driven by land-use change in biodiversity hot-spots under four scenarios of climate-change mitigation. *Conservation Biology* 29, 1122–1131. DOI: 10.1111/cobi.12549

Karki, M.B. and Vaidya, R. (2009) Adaptation to climate change impacts and regional cooperation on water and hazards in the Himalayan Region. *South Asia Water Journal* 1(2), 117–131.

Klein, J.A., Harte, J. and Zhao, X.Q. (2004) Experimental warming causes large and rapid species loss, dampened by simulated grazing, on the Tibetan Plateau. *Ecology Letters* 7, 1170–1179. DOI: 10.1111/j.1461-0248.2004.00677.x

Kohler, T., Balsiger, J., Rudaz, G., Debarbieux, B., Pratt, D.J. *et al.* (2015) *Green Economy and Institutions for Sustainable Mountain Development: From Rio 1992 to Rio 2012 and Beyond.* Centre for Development and Environment (CDE); Swiss Agency for Development and Cooperation (SDC); University of Geneva; Geographica Bernensia.

Leimgruber, P., Kelly, D.S., Steininger, M., Brunner, J., Müller, T. *et al.* (2005) Forest cover change patterns in Myanmar (Burma) 1990–2000. *Environmental Conservation* 32, 356–364. DOI: 10.1017/S0376892905002493

Luo, C., Xu, G., Chao, Z., Wang, S., Lin, X. *et al.* (2010) Effect of warming and grazing on litter mass loss and temperature sensitivity of litter and dung mass loss on the Tibetan plateau. *Global Change Biology* 16, 1606–1617. DOI: 10.1111/j.1365-2486.2009.02026.x

MoCC (2015) National Forest Policy 2015. Ministry of Climate Change, Government of Pakistan, Islamabad.

Mustafa, D. (2010) *Hydropolitics in Pakistan's Indus Basin*. US Institute of Peace, Washington, DC.

Negi, C.S. (2005) Socio-cultural and ethnobotanical value of a sacred forest, Thal Ke Dhar, central Himalaya. *Indian Journal of Traditional Knowledge* 4, 190–198.

Ning, W., Rawat, G.S., Joshi, S., Ismail, M. and Sharma, E. (2013) *High-altitude Rangelands and Their Interfaces in the Hindu Kush Himalayas*. International Centre for Integrated Mountain Development (ICIMOD). Kathmandu.

Ning, W., Shaoliang, Y., Joshi, S. and Bisht, N. (2016) *Yak on the Move: Transboundary Challenges and Opportunities for Yak Raising in a Changing Hindu Kush Himalayan Region*. International Centre for Integrated Mountain Development (ICIMOD), Kathmandu.

Ouyang, Z., Zheng, H., Xiao, Y., Polasky, S., Liu, J. *et al.* (2016) Improvements in ecosystem services from investments in natural capital. *Science* 352, 1455–1459. DOI: 10.1126/science.aaf2295

Qasim, M., Hubacek, K., Termansen, M. and Fleskens, L. (2013) Modelling land use change across elevation gradients in district Swat, Pakistan. *Regional Environmental Change* 13, 567–581. DOI: 10.1007/s10113-012-0395-1

Rahaman, M.M. and Varis, O. (2009) Integrated water management of the Brahmaputra basin: perspectives and hope for regional development. *Natural Resources Forum* 33, 60–75. DOI: 10.1111/j.1477-8947.2009.01209.x

Reddy, C.S., Sreelekshmi, S., Jha, C.S. and Dadhwal, V.K. (2013) National assessment of forest fragmentation in India: landscape indices as measures of the effects of fragmentation and forest cover change. *Ecological Engineering* 60, 453–464. DOI: 10.1016/j.ecoleng.2013.09.064

Ricketts, T.H., Dinerstein, E., Boucher, T., Brooks, T.M., Butchart, S.H. *et al.* (2005) Pinpointing and preventing imminent extinctions. *Proceedings of the National Academy of Sciences* 102, 18497–18501. DOI: 10.1073/pnas.0509060102

Saito, N. (2012) Mainstreaming climate change adaptation in least developed countries in South and Southeast Asia. *Mitigation Adaptation Strategies for Global Change* 18, 825–849. DOI: 10.1007/s11027-012-9392-4

Salick, J., Yongpin, Y. and Amend, A. (2005) Tibetan land use and change near Khawa Karpo, Eastern Himalayas. *Economic Botany* 59, 312–325. DOI: 10.1663/0013-0001(2005)059[0312:TLUACN]2.0.CO;2

Sandhu, H. and Sandhu, S. (2014) Linking ecosystem services with the constituents of human well-being for poverty alleviation in eastern Himalayas. *Ecological Economics* 107, 65–75. DOI: 10.1016/j.ecolecon.2014.08.005

Sherpa, L.N., Peniston, B., Lama, W. and Richard, C. (2003) *Hands around Everest: Transboundary Cooperation for Conservation and Sustainable Livelihoods*. International Centre for Integrated Mountain Development (ICIMOD), Kathmandu.

Singh, S.P., Khadka, I.B., Karky, B.S. and Sharma, E. (2011) *Climate Change in the Hindu Kush-Himalayas: The State of Current Knowledge*. International Centre for Integrated Mountain Development. Kathmandu.

Sinha, B. and Mishra, S. (2015) Ecosystem services valuation for enhancing conservation and livelihoods in a sacred landscape of the Indian Himalayas. *International Journal of Biodiversity Science, Ecosystem Services & Management* 11, 156–167.

Uprety, K. and Salman, S.M. (2011) Legal aspects of sharing and management of transboundary waters in South Asia: preventing conflicts and promoting cooperation. *Hydrological Sciences Journal* 56, 641–661. DOI: 10.1080/02626667.2011.576252

Vick, M.J. (2014) Steps towards an Afghanistan-Pakistan water-sharing agreement. *International Journal of Water Resources Development* 30, 224–229. DOI: 10.1080/07900627.2014.886471

Wang, B., Bao, Q., Hoskins, B., Wu, G. and Liu, Y. (2008) Tibetan Plateau warming and precipitation changes in East Asia. *Geophysical Research Letters* 35, L14702. DOI: 10.1029/2008GL034330

Winkler, D. (2009) Caterpillar fungus (*Ophiocordyceps sinensis*) production and sustainability on the Tibetan Plateau and in the Himalayas. *Asian Medicine* 5, 291–316.

Wirsing, R.G. (2013) Melting the geopolitical ice in south Asia. *Asia Policy* 16, 19–25. DOI: 10.1353/asp.2013.0029

Zhang, H. (2016) Sino-Indian water disputes: the coming water wars? *Wiley Interdisciplinary Reviews: Water* 3, 155–166. DOI: 10.1002/wat2.1123

21 Strategies for Forest and Grassland Restoration in the Himalayan Regions

Ramesh Prasad Sapkota[1,*], Rashila Deshar[1] and Madan Koirala[1]
[1]*Central Department of Environmental Science, Tribhuvan University, Nepal*

Abstract

Forests and grasslands ecosystems provide important ecological resources to the people. However, these ecosystems, particularly in the Hindu Kush Himalayan (HKH) region, are being degraded due to anthropogenic activities. The restoration of degraded, damaged and destroyed forest and grassland ecosystems in the HKH region is essential, and provisions for their restoration should be included in national plans and strategies. This chapter describes: (i) the status of forests and grasslands; (ii) history and present status of forest and rangeland restoration; and (iii) possible future strategies for landscape-level restoration to enhance ecosystem functions and support livelihood options.

Key words: degradation, ecological resources, livelihood

21.1 Introduction

21.1.1 Forest and grassland degradation

Degradation is the deterioration of the productive capacity of lands and affects human livelihood and well-being. One of the major drivers for biodiversity loss in forests and grasslands is land-use change, which has impacted the habitat and richness of biodiversity and ecosystem services. Terrestrial ecosystems have been converted mainly from natural land to cultivated and urban lands. Invasive alien species, over-exploitation of forests, over-grazing of grasslands, climate change, and fires (Fig. 21.1) have further accelerated forest and grassland degradations.

Gibbs and Salmon (2015) reviewed the continental and global estimates of land degradation in different areas. Globally, land degradation was 470 million ha in 2008 (Campbell *et al.*, 2008) and increased to 991 million ha in 2011, while in Asia, 118 million ha were degraded in 2008 and increased to 490 million ha in 2011. Forests and grasslands have the potential to store large quantities of carbon, estimated at up to 200 gigatonnes (IPCC, 2019), and trillions of dollars can be derived annually from their ecosystem services (FAO, 2018). However, 50% of the world's forests have been degraded or destroyed (WWF, 2020) and grasslands are being lost at 2% per year (FAO, 2019).

Restoring forests and rangelands is important in safeguarding the diverse flora and fauna

*Corresponding author: sapkotaramesh@hotmail.com

DOI: 10.1079/9781800622579.0021

Fig. 21.1. Fire can act both as an ecological tool and as a stressor for ecosystem management. In the figures, fire is acting as stressors in forest ecosystems.

in these ecosystems as they provide habitats and other benefits such as enhanced water filtration. Additionally, forests and rangelands act as carbon sinks, aiding in climate-change mitigation. Restoration efforts maintain essential ecosystem services like pollination, erosion control and flood regulation by bolstering carbon storage within these ecosystems (Daily, 1997; MEA, 2005). Therefore, it is necessary to prioritize and implement effective restoration strategies of forests and grasslands to secure a sustainable future.

21.1.2 History of forest and rangeland restoration

The history of forest and rangeland restoration can be traced back to the early 1900s, when Aldo Leopold and other conservationists began to advocate for the restoration of degraded ecosystems (Leopold, 1949). In the 1960s and 1970s, there was a growing awareness of the need for restoration, and the field began to develop as a formal discipline (Forman, 1990). One of the earliest examples of forest restoration was the work of the Civilian Conservation Corps (CCC) during the Great Depression. The CCC planted millions of trees and built thousands of miles of trails in national forests and parks (Brock, 2005). This work helped to restore degraded forests and create new recreation opportunities. The first attempts at rangeland restoration date back to the 19th century in the USA, where the US Army was tasked with restoring rangelands that had been degraded by over-grazing. In the mid-20th century, there was a growing interest in rangeland restoration, which was partly due to the increasing awareness of the environmental impacts of over-grazing and the development of new technologies that made restoration more feasible (NRC, 1994). In the late 20th and early 21st centuries, there was a renewed focus on rangeland restoration due to the damage caused by climate change (Foley *et al.*, 2005). Rangeland restoration is an ongoing process at present and there have been many successes, but much work still needs to be done. In the 1960s, the US Forest Service began experimenting with new restoration techniques, such as controlled burning and seeding (Waldrop and Goodrick, 2018), which were used to restore fire-adapted ecosystems degraded by fire suppression. In the 1970s, the National Park Service began to restore degraded rangelands, which included removing invasive species, planting native grasses and controlling grazing (Westbrooks, 1998). In the 1980s, the restoration of forests and rangelands became a more widespread practice, partly due to the increasing awareness of the importance of these ecosystems for biodiversity, water quality and carbon storage (Jones *et al.* 2018).

In the 1990s, there was a growing emphasis on using adaptive management in restoration projects, which involves using scientific data to monitor the progress of a restoration project and make adjustments as needed (Pickett, 1999). In the 2000s, there was an initiative of restoring degraded forests and rangelands in developing countries by a number of organizations, including the World Wildlife Fund, the United Nations Environment Programme, and the International Union for Conservation of Nature (MEA, 2005).

The history of forest and rangeland restoration is a long and complex one; however, significant progress has been made in recent decades.

21.1.3 Present status of forest and rangeland restoration

There has been significant progress in forest restoration globally in recent years, as evidenced by the FAO and UNEP (2020) report, which noted that over 150 million ha of degraded land has been restored since 2000. However, much more restoration remains to be done, as the IPCC (2019) estimates that an additional 2 billion ha of forest restoration is needed by 2050 to mitigate climate change. The Millennium Ecosystem Assessment (MEA, 2005) also highlighted the importance of forest restoration in providing ecosystem services, conserving biodiversity, improving water quality, and providing food and livelihoods for local communities. Forest restoration has progressed slowly, with only 27 million ha restored between 2000 and 2019, which is well below the 150 million ha pledged by 61 countries under the 2011 Bonn Challenge. The FAO and UNEP reported that Latin America has seen the most progress in forest restoration, with 10.7 million ha, followed by east Asia with 9.5 million ha, south-east Asia with 2.5 million ha, North America with 1.8 million ha, and Europe and Russia with 1.3 million ha (FAO and UNEP, 2020). The report, however, reported that deforestation remains as a major problem, with an estimated 10 million ha lost annually between 2015 and 2020. To achieve the Bonn Challenge, it is essential to both increase the pace of forest restoration and reduce deforestation substantially.

There are a number of key challenges to forest restoration, including land tenure, costs, technical difficulties and social challenges. However, there are also a number of reasons to be optimistic about the future of forest restoration including the growing awareness of the importance of forests and the need for restoration, the work of organizations such as the WRI and IUCN, and the development of new technologies that can improve the efficiency and effectiveness of restoration. The future of forest restoration is bright. With continued investment and effort, it is possible to restore degraded forests and reap the many benefits these ecosystems provide.

Rangelands are being degraded at an alarming rate. Over 50% of the world's rangelands are degraded, and this percentage is increasing (UNEP, 2014). Rangeland degradation can lead to food insecurity for millions of people. Several promising restoration techniques are being implemented including planting of native plants, grazing management, and water harvesting. Rangeland restoration is a growing field and has made important progress in recent years. The Food and Agriculture Organization of the United Nations estimated that over 100 million ha of degraded rangelands have been restored since 2000 (FAO and UNEP, 2020). However, much more work remains to be done, as the Intergovernmental Panel on Climate Change (IPCC) estimated that an additional 1.5 billion ha of rangeland restoration is needed by 2050 to mitigate climate change (IPCC, 2019).

Forest and rangeland restorations in Nepal date back to the 19th century. Different community forest-related projects proliferated during the 1950s, while the master plan for the forestry sector was prepared in 1988. The Department of Forests was established in 1973 and Community Forestry Programmes were launched after 1992. However, it was not until the late 20th century that restoration efforts began to gain momentum. In the early 1990s, the government of Nepal launched the Community Forestry Programme, which aimed to devolve forest management responsibilities to local communities. This programme restored millions of ha of degraded forest and rangeland (Ranjit, 2019). In the late 1990s and early 2000s, Nepal implemented several other restoration programmes, such as the Terai Arc Landscape Programme and the National Adaptation Programme of Action (NAPA), which aimed to restore forests and rangelands, and improve the livelihoods of millions of rural communities. The National Climate Change Policy and The Forest Act were enacted during 2019.

According to Laudari *et al.* (2022), Nepal's forest restoration interventions from 1956 to the early 1980s had mixed results because of the limited integration of socio-economic and ecological concerns in restoration programmes. However, forest restoration projects after the mid-1980s were more successful because of policies favouring decentralized decision making and local institutions; devolution of rights and responsibilities; firm commitment for and adoption of multi-stakeholder partnership for the forest and landscape restoration; recognition of multi-functionality of forest ecosystems; accommodation of socio-economic and ecological concerns in restoration programmes; adoption of multiple restoration approaches at different scales; and capacity development and extension services.

21.2 Why Forest and Grassland Restoration?

Ecosystems across the globe are under pressure with degradation and fragmentation, loss of soil nutrients, spread of invasive species, overexploitation of natural resources, pollution, and climate change. The prevalence of these factors and their intensification are reducing ecosystem services (https://www.iucn.org/), as is occurring globally (Roe, 2019). These losses are affecting livelihoods and the cost of items, which is being felt more by the poor than the rich, especially in developing countries. One example is the cost of water in most developing and developed countries. Fifty years ago, drinking water was free with no restrictions on its use for domestic purposes, but today, most people purchase water.

MEA (2005) estimated that 60% of the world's ecosystem services are being degraded or used unsustainably but the reduction due to the loss and degradation of these ecosystem services are difficult to measure. Outside Antarctica, the proportion of land remaining in natural condition ranges from 17% to 52% (Dudley, 2011), and is being reduced in many countries. Protected areas are being established and extended to conserve sites of importance for biodiversity and ecosystem services (SCBD, 2010). The current rate and scale of deforestation, especially in tropical regions, and large tracts of degraded lands, require urgent interventions to restore biodiversity, ecological functioning, and the supply of goods and ecological services, especially for poor rural communities (Cairns and Heckman, 1996; Lamb and Gilmour, 2003; Aronson and Alexander, 2013). Ecosystem services that have been reduced or degraded include fisheries, water supply, waste treatment and detoxification, water purification, natural hazard protection, regulation of air quality, regional and local climate, and erosion and aesthetic enjoyment (MEA, 2005).

Depending upon temporal and spatial needs, ecosystem restoration can be applied to a single species, community assemblage or to whole ecosystems and landscapes (Dobson *et al.*, 1997). Ecosystems are dynamic and restoration goals cannot be based on static attributes (Hobbs and Harris, 2001), and should consider the site conditions (Aplet *et al.*, 2000). The difficulty and expense in the restoration increases with the severity of the degradation of the ecosystem (Hobbs *et al.*, 2004).

21.3 Forest and Grassland Restoration Strategies

Historic range of variability (HRV) describes the changes in the ecosystem with time and space. It provides information on potential vegetation types and ecological processes that shape the composition of vegetation. Knowledge on HRV enables forest and grassland managers to better define structural and functional diversity of ecosystems.

21.3.1 Concept of the historic range of variability

There was a net decrease in global forest area of 3% (between 1990 and 2015) (Keenan *et al.*, 2015) and, therefore, there is still a need to employ better options and approaches to forest and rangeland management. One of the promising ways of forest management, to improve ecosystem health and integrity, conceived since 1990s, is the approach of HRV. The HRV concept was introduced to understand the past spatial and temporal variability in ecosystems and has proved to be beneficial for landscape ecosystem restoration and management (Cissel *et al.*, 1994; Morgan *et al.*, 1994; Kimmins, 2004) by using historic patterns to guide forest management and harvesting patterns (Perry *et al.*, 2008). HRV is basically a coarse filter approach, which emulates disturbance and aims to maintain the pattern and process of whole landscapes, including geochemical cycling and energy flow. This system uses the historical ecosystem information for determining reference conditions for the forest/ecosystem management (Swetnam *et al.*, 1999).

Historical perspectives enable us to understand the dynamics of the landscape and frame a reference for future patterns and processes (Keane *et al.*, 2009). When considering HRV, scale is paramount because the response variables differ when quantified for different time periods and landscape scales (Turner *et al.*, 2001).

Detailed work on HRV principles and objectives is required before implementing HRV techniques. Practitioners and planners first need to work at a conceptual level to determine the effect of HRV trajectories and reference conditions on the socio-economic spectrum. Wherever the HRV is employed, it should not focus on just a single target from historical evidence but should focus on a wide range of variabilities and possibilities (Keane *et al.*, 2002). There is a need for using spatial analysis tools. In cases of scarce, inconsistent and/or limited vegetation data, it is difficult to predict HRV. However, there are studies in which digital maps and aerial photos are used to determine HRV of forests (Tinker and Knight, 2001; Tinker *et al.*, 2003).

21.3.2 Advantages of historic range of variability

HRV has several advantages in forest and rangeland management. It can be used for single or multiple characteristics that describe the ecosystem. Some other advantages are:

- It is flexible (can be used for any scale), robust and dependable (Keane *et al.*, 2009).
- It is useful in the forests, where there are frequent stand-replacing disturbances and the disturbances match the desired frequency of HRV.
- It is a suitable coarse filter in which there is a high probability of sustaining all species and values across landscapes (especially useful in developing countries, where there is limited understanding of ecology and ecological systems).
- Historical conditions may not be the only references for HRV analysis, other time-series scenarios (disturbances, exotics, grazing etc.) can also be modelled for different sets of reference conditions (Keane *et al.*, 2009).
- It fosters improved understanding of ecosystem dynamics and ecosystem responses to changing conditions. Understanding HRV helps to identify and access ecological complexities in landscape management (Keane *et al.*, 2009).
- With sufficient background data, modelling tools can create good HRV conditions and comprehensive land management applications.

21.3.3 Shortcomings of historic range of variability

HRV also has shortcomings. The following examples are applicable in forest and rangeland restoration and management:

- There can be a problem in finding a historical baseline time, i.e. 100, 200 or 1000 years. The different time periods have different phytosociological characteristics of vegetation and landscapes.
- Some data cannot be used for predicting (or may not have appropriate resolution) spatial and temporal landscape distributions (Keane *et al.*, 2009).
- Some historical and time series data may show autocorrelation, so it becomes very difficult to predict HRV.
- Using a limited temporal and spatial scale to evaluate landscapes across larger regions can lead to errors. This is because there is variation between vegetation responses across larger and smaller domains.
- Though spatial statistics and modelling have progressed in recent years, there are still limited analysis and modelling techniques to evaluate multiple observations in historical time series.
- The models used for assessing HRV are complicated. Complicated models tend to have high variation in results. Models may not represent the full range of landscapes (Keane *et al.*, 2009).
- HRV may not be a good model of forest management in case of larger scale/infrequent or very severe disturbances.

- If different alternative stable states have developed in the landscapes, it is difficult to return the ecosystems to the reference historical conditions.
- There are still conceptual problems in framing a series of historical conditions for landscape pattern and processes (Keane *et al.*, 2009).

21.3.4 Historical range of variability in the context of global change

There are criticisms about HRV that it will be no longer a concept in the changing situations of environment and increasing human pressures (Kimmins, 2004; Millar *et al.*, 2007). These pressures can be global climate change (drought) or other disturbances by humans. If the future climatic conditions, as predicted by studies and large-scale disturbances, occur, then HRV might not be useful for landscape management. This is because extreme new/damaging conditions change the ecological systems and structure of existing vegetation (composition and type), which may be more favourable for new types of plant species. In such conditions, HRV may not be appropriate, as the predictions will fail to manage landscapes, forests and rangelands. For example, highland landscapes are considered to be one of the most vulnerable ecosystems due to changing climate and disturbances (Tsering *et al.*, 2010). Therefore, the HRV predictions will not coincide with the high-altitude geographical changes and vegetation dynamics. The concern is whether the large and infrequent disturbances (LID) will occur or not. The knowledge on climate change impacts still needs to be improved (Stainforth *et al.*, 2005). In addition, our knowledge base on the interaction between climate and plant and animal life (especially on reproduction vigour and mortality) is also very limited. If LIDs due to climate or anthropogenic causes do not occur, then HRV will be a better management strategy for landscapes, forests and rangelands. If we can develop better stochastic and mechanistic models, which integrate disturbances, biological processes, land-use change, hydrology, and climate change phenomena, we can simulate HRV predictions better and provide a good reference for developing strategies for forest and rangelands restoration and management.

21.4 Success Stories of Restoration Interventions and Way Forward

Ecosystem restoration initiatives and practices are receiving increasing attention throughout the globe (Erwin 2009; Schmutz *et al.* 2014; Barral *et al.*, 2015) with efforts ranging from localized interventions to landscape or 'mega' projects (Hobbs and Norton, 1996). The practices have been essential for human health, biodiversity conservation and climate stability.

China is facing the problem of desertification, sandstorms and floods which have been attributed to extensive land degradation (Liu *et al.*, 2008). Restoration projects were initiated for restoring those areas. Logging was prohibited in most natural areas, and cultivated land on slopes of more than 25 degrees was terraced or restored with vegetation that protects against erosion (Feng *et al.*, 2005). Since 2001, ecological conditions of these key project regions have improved. Forests increased by 482,000 km^2 and forest coverage by 23% over a decade (CBD, 2014). Forest reserves reached 13.72 billion m^3, an increase of over 20% in a decade, and these projects enhanced restoration of habitats and contributed to a rise in the population of wild species (CBD, 2014).

The successful restoration projects due to the increasing concerns of environmental degradation in developing and developed countries have improved the capacities to manage intact ecosystems and repair damaged, destroyed or degraded ecosystems (Holl *et al.*, 2003; Lamb, 2011; Suding, 2011). However, there is need of more integrated and transdisciplinary approaches (SER, 2004; Falk *et al.*, 2006; Clewell and Aronson, 2013). The following improvements should be considered in the future:

- Restoration should maintain the heterogeneity in vegetation assemblages to support multi-functionality. Future strategies should consider improving diversity in such stands.
- Monoculture stands should be converted to mixed stands. Some forest stands cannot support multiple species at the initial stages of community development, so replacement with mixed species should be done gradually.
- Maintaining biodiversity at local scale fosters ecological functions and services, so future

management should focus on a variety of native species spectra. Restoring forests based on native species has been recognized widely as an effective strategy of recovering ecosystems and biodiversity in degraded areas.

- Despite the increase in overall forest area, some forest stands have undergone degradation, so these forest stands should be conserved, restored and monitored.
- Disturbances such as climate change have created unprecedented pressure, so newer approaches such as ecological engineering and genetic conservation are required.
- Empirical studies, which focus on larger landscape scales restoration, forest regeneration strategies in different ecological regions, resistance and resilience capacity of ecosystems towards biotic and abiotic perturbations, are needed.

21.5 Conclusions

Restoration of degraded ecosystems increases the services which improve the livelihoods of communities. The improved understanding of the global, regional and local database is essential for ecosystem restoration. Historic range of variability offers a successful option for guiding restoration strategies and should be implemented for restoration of forest and rangeland ecosystems of Himalayan landscapes.

References

Aplet, G., Thomson, J. and Wilbert, M. (2000) Indicators of wildness: using attributes of the land to assess the context of wilderness. In: McCool, S.F., Cole, D.N., Borrie, W.T. and O'Loughlin, J. (eds) *Wilderness Science in a Time of Change*. USDA Forest Service Rocky Mountain Research Station, RMRS-P-15-VOL-2, Missoula, Montana, pp. 89–98.

Aronson, J. and Alexander, S. (2013) Ecosystem restoration is now a global priority: time to roll up our sleeves. *Restoration Ecology* 21(3), 293–296. DIO: 10.1111/rec.12011

Barral, M.P., Benayas, J.M.R., Meli, P. and Maceira, N.O. (2015) Quantifying the impacts of ecological restoration on biodiversity and ecosystem services in agroecosystems: a global meta-analysis. *Agriculture Ecosystems and Environment* 202, 223–231. DOI: 10.1016/j.agee.2015.01.009

Brock, J. (2005) *A History of the CCC in Rocky Mountain National Park*. Rocky Mountain National Park, Colorado.

Cairns, J. and Heckman, J.R. (1996) Restoration ecology: the state of emerging field. *Annual Review of Energy and Environment* 21, 167–189. DOI: 10.1146/annurev.energy.21.1.167

CBD (2014) *Global Biodiversity Outlook 4*. Convention on Biological Diversity Secretariat, Montreal, Canada.

Cissel, J.H., Swanson, F.J., McKee, W.A. and Burditt, A.L. (1994) Using the past to plan the future in the Pacific Northwest. *Journal of Forestry* 92, 46.

Clewell, A.F. and Aronson, J. (2013) *Ecological Restoration: Principles, Values, and Structure of an Emerging Profession*. 2nd edn. Island Press, Washington, DC.

Conniff, R. (2014) Rebuilding the natural world: a shift in ecological restoration. Available at: https://e360.yale.edu/features/rebuilding_the_natural_world_a_shift_in_ecological_restoration#:~:text=From%20forests%20in%20Queens%20to,nature%20and%20what%20does%20not (accessed 27 February 2024).

Daily, G.C. (1997) *Nature's Services: Societal Dependence on Natural Ecosystems*. Island Press, Washington, DC.

Davis, M.A. and Slobodkin, L.B. (2004) The science and values of restoration ecology. *Restoration Ecology* 12, 1–3. DOI: 10.1111/j.1061-2971.2004.0351.x

Dobson, A.P., Bradshaw, A.D. and Baker, A.J.M. (1997) Hopes for the future: restoration ecology and conservation biology. *Science* 277, 515–522. DOI: 10.1126/science.277.5325.515

Dudley, N. (2011) *Authenticity in Nature: Making choices about Naturalness of Ecosystems*. Earthscan Publications, London.

Erwin, K.L. (2009) Wetlands and global climate change: the role of wetland restoration in a changing world. *Wetlands Ecology and Management* 17, 71–84. DOI: 10.1007/s11273-008-9119-1

Falk, D.A., Palmer, M. and Zedler, J. (2006) *Foundations of Restoration Ecology*. Island Press, Washington, DC.

FAO (2018) *The State of the World's Forests*. Food and Agriculture Organization of the United Nations, Rome.

FAO (2019) *Global Assessment of Land Degradation and Restoration*. Food and Agriculture Organization of the United Nations, Rome.

FAO and UNEP (2020) *The State of the World's Forests 2020. Forests, Biodiversity and People*. Food and Agriculture Organization of the United Nations, Rome.

Feng, Z., Yang, Y., Zhang, Y., Zhang, P. and Li, Y. (2005) Grain-for-green policy and its impacts on grain supply in West China. *Land Use Policy* 22, 301–312. DOI: 10.1016/j.landusepol.2004.05.004

Foley, J.A., DeFries, R., Asner, G.P., Barford, C., Bonan, G. *et al.* (2005) Global consequences of land use. *Science* 309(5734), 570–574. DOI: 10.1126/science.1111772

Forman, R.T. (1990) The beginnings of landscape ecology in America. In: Baudry, J., Zonneveld, I.S. and Forman, T.T. (eds) *Changing Landscapes: An Ecological Perspective*. Springer, New York, pp. 35–41.

Gibbs, H.K. and Salmon, J.M. (2015) Mapping the world's degraded lands. *Applied Geography* 57, 12–21. DOI: 10.1016/j.apgeog.2014.11.024

Hobbs, R.J. and Harris, J.A. (2001) Restoration ecology: repairing the Earth's ecosystems in the new millennium. *Restoration Ecology* 9, 239–246. DOI: 10.1046/j.1526-100x.2001.009002239.x

Hobbs, R.J. and Norton, D.A. (1996) Towards a conceptual framework for restoration ecology. *Restoration Ecology* 4, 93–110. DOI: 10.1111/j.1526-100X.1996.tb00112.x

Hobbs, R.J., Davis, M.A., Slobodkin, L.B., Lackey, R.T., Halvorson, W. *et al.* (2004) Restoration ecology: the challenge of social values and expectations. *Frontiers in Ecology and the Environment* 2, 43–48.

Holl, K.D., Crone, E.E. and Schultz, C.B. (2003) Landscape restoration: moving from generalities to methodologies. *BioScience* 53(5), 491–502. DOI: 10.1641/0006-3568(2003)053[0491:LRMFGT]2.0.CO;2

IPCC (2019) *Global Warming of 1.5°C: An IPCC Special Report on the Impacts of Global Warming of 1.5°C Above Pre-industrial Levels and Related Global Greenhouse Gas Emission Pathways, in the Context of Strengthening the Global Response to the Threat of Climate Change, Sustainable Development, and Efforts to Eradicate Poverty*. IPCC, https://www.ipcc.ch/sr15/ (accessed 15 April 2024).

Jones, K.R., Venter, O., Fuller, R.A., Allan, J.R., Maxwell, S.L. *et al* (2018) One-third of global protected land is under intense human pressure. *Science* 360(6390), 788–791. DOI: 10.1126/science.aap9565

Keane, R.E., Parsons, R.A. and Hessburg, P.F. (2002) Estimating historical range and variation of landscape patch dynamics: limitations of the simulation approach. *Ecological Modelling* 151, 29–49. DOI: 10.1016/S0304-3800(01)00470-7

Keane, R.E., Hessburg, P.F., Landres, P.B. and Swanson, F.J. (2009) The use of historical range and variability (HRV) in landscape management. *Forest Ecology and Management* 258, 1025–1037. DOI: 10.1016/j.foreco.2009.05.035

Keenan, R.J., Reams, G.A., Achard, F., de Freitas, J.V., Grainger, A. *et al.* (2015) Dynamics of global forest area: results from the FAO global forest resources assessment 2015. *Forest Ecology and Management* 352, 9–20. DOI: 10.1016/j.foreco.2015.06.014

Kimmins, J.P. (2004) *Forest Ecology: A Foundation for Sustainable Forest Management and Environmental Ethics in Forestry*. Prentice-Hall, Englewood Cliffs, New Jersey.

Koebel, Jr, J.W. and Bousquin, S.G. (2014) The Kissimmee River Restoration Project and Evaluation Program Florida. *Restoration Ecology* 22(3), 345–352. DOI: 10.1111/rec.12063

Lamb, D. (2011) *Regreening the Bare Hills: Tropical Forest Restoration in the Asia-Pacific Region*. Springer, Dordrecht, The Netherlands.

Lamb, D. and Gilmour, D. (2003) *Rehabilitation and Restoration of Degraded Forests*. IUCN, Gland, Switzerland and Cambridge, UK and WWF, Gland, Switzerland.

Laudari, H.K., Aryal, K., Maraseni, T., Pariyar, S., Pant, B. *et al.* (2022) Sixty-five years of forest restoration in Nepal: lessons learned and way forward. *Land Use Policy* 115, 106033. DOI: 10.1016/j.landusepol.2022.106033

Leopold, A. (1949) *A Sand County Almanac and Sketches Here and There*. Oxford University Press, New York.

Liu, J., Li, S., Ouyang, Z., Tam, C. and Chen, X. (2008) Ecological and socioeconomic effects of China's policies for ecosystem services. *Proceedings of the National Academy of Sciences of the United States of America* 105:9477–9482. DOI: 10.1073/pnas.0706436105

MEA (2005) *Ecosystems and Human Well-being: Synthesis, Millennium Ecosystem Assessment*. Island Press, Washington, DC.

Millar, C.I., Stephenson, N.L. and Stephens, S.L. (2007) Climate change and forest of the future: managing in the face of uncertainty. *Ecological Applications* 17, 2145–2151. DOI: 10.1890/06-1715.1

Morgan, P., Aplet, G.H., Haufler, J.B., Humphries, H.C., Moore, M.M. *et al.* (1994) Historical range of variability: a useful tool for evaluating ecosystem change. *Journal of Sustainable Forestry* 2, 87–111. DOI: 10.1300/J091v02n01_04

NRC (1994) *Rangeland Health: New Methods to Classify, Inventory, and Monitor Rangelands*. National Academy Press, Washington, DC.

Perry, D.A., Oren, R. and Hart, S.C. (2008) *Forest Ecosystems*. The Johns Hopkins University Press, Baltimore, Maryland.

Pickett, S.T.A. (1999) Ecosystem management: restoring the natural capital of ecosystems. *Ecological Applications* 9(3), 723–735.

Ranjit, Y. (2019) History of forest management in Nepal: an analysis of political and economic perspective. *Economic Journal of Nepal*, 42(3–4), 12–28. DOI: 10.3126/ejon.v42i3-4.36030

Roe, D. (2019) Biodiversity loss – more than an environmental emergency comment. *Lancet Planetary Health* 3(7), e287–e289. DOI: 10.1016/S2542-5196(19)30113-5

SCBD (2010) *Strategic Plan for Biodiversity 2011–2020 and the Aichi Target: Living in Harmony with Nature*. CBD Secretariat, Montreal, Canada.

Schmutz, S., Kremser, H., Melcher, A., Jungwirth, M., Muhar, S. *et al.* (2014) Ecological effects of rehabilitation measures at the Austrian Danube: a meta-analysis of fish assemblages. *Hydrobiologia* 729, 49–60. DOI: 10.1007/s10750-013-1511-z

SER (2004) *The SER Primer on Ecological Restoration*. Society for Ecological Restoration International, Washington DC. Available at: http://www.ser.org/resources/resources-detail-view/ser-international-primer-on-ecological-restoration (accessed 27 February 2024).

Stainforth, D.A., Aina, T., Christensen, C., Collins, M., Faull, N. *et al.* (2005) Uncertainty in predictions of the climate response to rising levels of greenhouse gases. *Nature* 433(6). DOI: 10.1038/nature03301

Suding, K.N. (2011) Toward an era of restoration in ecology: successes, failures, and opportunities ahead. *Annual Review of Ecology, Evolution, and Systematics* 42, 465–487. DOI: 10.1146/annurev-ecolsys-102710-145115

Swanson, F.J., Jones, J.A., Wallin, D.O. and Cissel, J.H. (1994) Natural variability – implications for ecosystem management. Vol. II: Ecosystem management principles and applications. In: Jensen, M.E. and Bourgeron, P.S. (eds) *Eastside Forest Ecosystem Health Assessment*. USDA Forest Service Pacific Northwest Research Station, pp. 80–94.

Swetnam, T.W., Allen, C.D. and Betancourt, J.L. (1999) Applied historical ecology: using the past to manage for the future. *Ecological Applications* 9, 1189–1206. DOI: 10.1890/1051-0761(1999)009[1189:AHEUTP]2.0.CO;2

Tinker, D. and Knight, D. (2001) Temporal and spatial dynamics of coarse woody debris in harvested and unharvested lodgepole pine forests. *Ecological Modelling* 141, 125–149. DOI: 10.1016/S0304-3800(01)00269-1

Tinker, D.B., Romme, W.H. and Despain, D.G. (2003) Historic range of variability in landscape structure in subalpine forests of the Greater Yellowstone Area, USA. *Landscape Ecology* 18, 427–439.

Tsering, K., Sharma, E., Chettri, N. and Shrestha, A. (2010) *Climate Change Impact and Vulnerability in the Eastern Himalayas – Synthesis Report*. International Centre for Integrated Mountain Development, Kathmandu.

Turner, M., Gardner, R. and O'Neill, R. (2001) *Landscape Ecology: In Theory and Practice*. Springer, New York.

UN Climate Summit New York Declaration on Forests (2014) Available at: https://unfccc.int (accessed 27 February 2024).

UNEP (2014) UNEP/CBD/SBSTTA/18/14 – Report on issues in progress: ecosystem conservation and restoration. Available at: http://www.cbd.int/sp/targets/ (accessed 27 February 2024).

USDA (2014) Success stories – ecosystem and wetland restoration. US Dept. of Agriculture.

Waldrop, T.A. and Goodrick, S.L. (2018) *Introduction to Prescribed Fire in Southern Ecosystems*. Government Printing Office, Washington, DC.

Westbrooks, R. (1998) *Invasive Plants: Changing the Landscape of America*. The Federal Interagency Committee for Management of Noxious and Exotic Weeds. National Wildlife Refuge System, Washington, DC.

WWF (2020) Forests. Available at: https://www.worldwildlife.org/threats/deforestation (accessed 27 February 2024).

Websites

https://www.iucn.org/about/union/commissions/cem/cem_work/cem_restoration/ (accessed 5 February 2023)

https://www.landscapeperformance.org/case-study-briefs/cheonggyecheon-stream-restoration-project#:~:text=By%20demolishing%20an%20elevated%20freeway,in%20the%20center%20of%20Seoul (accessed 17 July 2023)

https://www.wri.org/insights/what-does-it-take-successful-forest-landscape-restoration (accessed 5 February 2023)

22 Sustainable Livestock Production in the Hindu Kush Himalayan Region

Allan Degen*
Institutes for Desert Research, Ben Gurion University of the Negev, Israel

Abstract
Decision makers regard pastoralism as a threat to the Himalayan region, and they insist that over-grazing has caused widespread degradation of the grassland. Consequently, the government policy is to settle the pastoralists, restrict mobility and convert grazing land to cultivated land. Besides the loss of grazing lands, pastoralists must contend with climate warming and forage shortage, especially in the winter and spring. Crop aftermaths, like rice and wheat straw, could provide large quantities of feed for ruminants. Ensiling of straw could improve the nutritive value and reduce enteric methane production by the animal. Alternative feeds like jujube by-products and some herbs could improve productivity and health of livestock. Artificial grasslands could restore degraded grasslands and provide feed all year. The exchange of germplasm, especially for yaks, should be of high priority to prevent inbreeding. This chapter discusses strategies which would enable sustainable livestock production in the HKH region.

Keywords: alternative feeds, artificial grassland, greenhouse gases, germplasm exchange, pastoralism, food insecurity

22.1 Introduction

The Hindu Kush Himalayan (HKH) region spans 3500 km from Afghanistan in the north-west to Myanmar in the south-east, and includes all or part of eight countries, namely Afghanistan, Bangladesh, Bhutan, China, India, Myanmar, Nepal and Pakistan (Fig. 22.1). The region covers approximately 4.3 million km^2, and the parts of the countries within the HKH range from 0.013 million km^2 for Bhutan to 2.4 million km^2 for China (Table 22.1). Elevation varies greatly, ranging from sea level to higher than 8000 m above sea level, with many mountain peaks exceeding 6000 m. Except for the Antarctica and Arctic, the HKH region has the largest cryosphere in the world and is the source of ten major rivers, including the Brahmaputra, Ganges, Indus, Mekong, Yangtze and Yellow, and possesses large stores of water in the form of snow and glaciers (You *et al.*, 2017). Approximately 1.65 billion people live in the downstream basins of these rivers.

Warming is predicted to continue over the entire HKH region in the 21st century. Average air temperature in the mountains is rising at a rate of 0.06°C per year, which is substantially faster than the global average, and is causing the loss of snow cover and the rapid melting and shrinking of glaciers (Shrestha and Aryal, 2011).

*Address for correspondence: degen@bgu.ac.il

DOI: 10.1079/9781800622579.0022

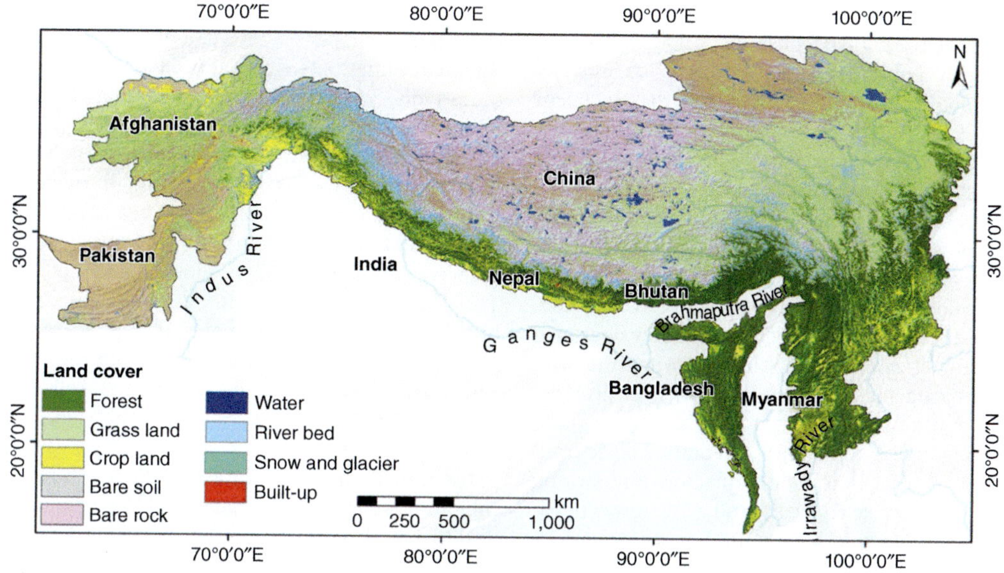

Fig. 22.1. Countries and rivers of the Hindu Kush Himalaya (HKH) region with description of land cover. (From ICIMOD, 2022; https: doi.org10.26066RDS.1972511)

Table 22.1. The area and population of countries and the average annual rainfall in the Hindu Kush Himalayan region. (Adapted from Chettri *et al*., 2008; UNDESA, 2017)

Country	Area (1000 km²)		Population (million)		Rainfall (mm/year)
	Total	In HKH (%)	Total	In HKH (%)	
Afghanistan	652.2	390.5 (59.9)	35.5	28.3 (79.7)	327
Bangladesh	144.0	13.3 (9.2)	164.7	1.33 (0.81)	2320
Bhutan	46.5	46.5 (100.0)	0.81	0.81 (100.0)	2200
China	9597.0	2420.3 (25.2)	1409.5	29.5 (2.1)	645
India	2387.6	461.1 (19.3)	1339.2	72.4 (5.4)	1170
Myanmar	676.6	317.6 (46.9)	53.4	11.0 (20.6)	2341
Nepal	147.2	147.2 (100.0)	29.3	29.3 (100.0)	1500
Pakistan	796.1	490.0 (61.6)	197.0	39.4 (20.0)	494
Total	14,447.1	4286.5 (29.7)	3229.4	211.9 (6.6)	

Air temperature across the mountains of the HKH region is predicted to increase by 1–2°C by 2050, and by even more at higher elevations (Shrestha *et al*., 2015). By the end of the 21st century, the average air temperature is projected to rise by 2.5°C according to the moderate RCP4.5 scenario (Van Vuuren *et al*., 2011), and by 5.5°C according to the more extreme RCP8.5 scenario, with winters becoming warmer at a faster rate than summers. Precipitation projections are less certain, and the spatial variability is expected to be high. Extreme weather events such as high-intense rainfall leading to flash floods, landslides and debris flow (Shrestha *et al*., 2015) will likely increase in frequency and magnitude and result in many deaths. The poor infrastructure, weak development, lack of credit institutions and difficulty in transferring techniques in the HKH hinder the implementation of strategies to mitigate climate change (Adenle *et al*., 2015).

The principal vegetation consists of tropical and subtropical rain forests, temperate broadleaf deciduous or mixed forests, and temperate coniferous forests, and also includes high-altitude steppe and cold desert shrubs (Chettri *et al.*, 2008). The region is known for its unique flora and fauna, high level of endemism and numerous hotspots of global importance (Chettri *et al.*, 2008; Elalem and Pal, 2015). Approximately 50.5% of the HKH region is comprised of grassland, 22.6% of forests, 9.0% of cultivated land, less than 0.1% of urban development, and 17.8% of bare land, water, snow and ice. In 2004, approximately 92% of the total area of usable grassland was classified as degraded grassland, of which 38% was severely and 54% was moderately degraded (Wang *et al.*, 2008). Climate warming and anthropogenic activities, such as over-grazing, reclamation, shrub uprooting and poor land management, were the main causes of degradation. The areas of alpine meadow and swamp meadow decreased by 4% and 25%, respectively, in the 1970s and 1980s, and by another 25% and 35%, respectively, in the 1990s (Zhou *et al.*, 2008). Degradation of grassland threatens the subsistence of local herders and the region's economic sustainability (Cheng and Shen, 2000; Lin *et al.*, 2011), and has resulted in a decline in ecological services and productivity (Mao *et al.*, 2008; Gao *et al.*, 2010; Dong, 2013).

22.2 People in the Hindu Kush Himalayan Region

The HKH region supports approximately 20% of the world's population, either directly or indirectly, and is characterized by a fragile ecosystem, low livelihood security and a vulnerable agricultural ecosystem (Sandhu and Sandhu, 2015; Shukla *et al.*, 2016; Pandey *et al.*, 2017). There are approximately 211.9 million people in the eight countries of the HKH region, ranging between 0.81 million people from Bhutan and 39.4 million people from Pakistan. The inhabitants in the HKH region comprise 6.6% of the total population of the countries and inhabit 29.7% of the total land area (Table 22.1). The region's population is increasing at 1.4% annually – the highest rates in Pakistan (2.1%) and Afghanistan (2.8%) and lowest rates in China (0.5) and Myanmar (0.9%) – while the per capita availability of land and other natural resources is declining (World Bank, 2015). Poverty in the HKH is widespread (Gerlitz *et al.*, 2012), and, in general, is higher in mountain regions, where people are at greater risk than in the plains. This can be explained by the inequalities between highland and lowland, mainly due to inaccessibility, inadequate infrastructure, limited economic opportunities, poor market access and harsh environmental conditions in the highlands. Furthermore, many policies directed towards the remote areas over the years have failed or were insufficient and did not incorporate mountain conditions adequately. The high poverty, both uni-dimensional and multi-dimensional, is one of the major causes of food and nutrition insecurity in the region. The uni-dimensional poverty line is based on only the income per person in a household, whereas the multi-dimensional poverty aggregates household-level data on ten indicators including years of schooling, child enrolment, child mortality, nutrition, access to and type of electricity, drinking water, sanitation, type of flooring, cooking fuel and assets (Alkire and Santos, 2010; UNDP, 2010). Close to 50% of the inhabitants of most countries in the HKH region are deprived of essentials, with Pakistan (53.5%), India (55.3%) and Afghanistan (58.8%) ranked the highest (Table 22.2). The multi-dimensional poverty index (MPI) is calculated as the product of the number of people living in multi-dimensional poverty and the average intensity of deprivation as measured by indicators (Alkire and Santos, 2010). Of the 795 million under-nourished people worldwide, 415 million (52%) are from HKH countries (FAO-IFAD-WFP, 2015). Greater than 30% of the HKH population is faced with food insecurity and approximately 50% suffer from malnutrition. Women and children are the most vulnerable, as 20–50% of children under the age of 5 suffer from stunting, with a high incidence of wasting and being underweight, while a substantial number of women suffer from anaemia. The situation is most severe for the indigenous people who are often the most under-privileged and vulnerable groups in the mountains (Gerlitz *et al.*, 2012).

Most of the people in the HKH region live in rural areas, while 60–85% subsist directly through ecosystem services. However, the urban population increased rapidly in the last three decades due to urbanization, industrialization and migration to urban centres. In addition, in general, the policy of the governments is to settle the people and turn some of the grasslands to

Table 22.2. Percentage of inhabitants below the poverty line, with food insecurity (Food ins), access to sanitary conditions and drinking water, and who are at high mortality risk, and the number of people killed in one event, such as a severe flood, between 1980 and 2015. (From UN-Water, 2013)

	Below poverty line (%)		Multi-dimensional poverty		Food ins (%)	Sanitation access (%)	Drinking water (%)	Population at risk (%)	People killed
Country	HKH	Country	%	Index					
Afghanistan	42	33	58.8	0.353	33	37	50	46.0	1,193
Bangladesh	46	37	49.5	0.237	53	n.a.	80	97.7	1,230
Bhutan	23	23	29.4	0.128	6	65	92	60.8	
China	n.a.	17	5.2	0.023	11	n.a.	89	33.4	
India	34	36	55.3	0.282	18	31	88	27.2	2,357
Myanmar	n.a.	32	41.4	0.197	9	81	71	16.8	
Nepal	40	31	45.6	0.237	51	62	88	97.4	1,048
Pakistan	32	25	53.5	0.210	57	91	48	49.6	2,113
Total	31	26			31				

Note: Both sanitation access and drinking-water percentages are higher in urban than in rural populations. n.a., not available

croplands. In China, the urban population in the HKH region rose from 16% to 56% during the period from 1960 to 2014. All other countries in the region had a similar percentage rise during this period – between 30% and 40% – except for Nepal (19%) and Afghanistan (27%), which were considerably lower (UNDESA, 2017). Outmigration from the region with dependencies on foreign remittances is reported to be the highest in the world and is due largely to the poverty and food insecurity (Karki *et al.*, 2012). Towns attract people by promising better livelihoods, income, medical care and education. Mainly men have migrated, leaving women over-burdened and responsible for the household in mountain villages. Frequent labour shortages during some critical periods of agricultural activities is common. Increased outmigration coupled with reduced interest by the younger generation in farming contribute to low productivity as well as the abandoning of some farmland. In addition, due to a massive demand from China's eastern provinces, the *yarsagumba* or caterpillar fungus (*Cordyceps sinensis*) trade has become very profitable in recent years in the Tibetan Plateau, India, Nepal and Bhutan Himalaya. Income is realized not only from the harvesting of the fungus but also from the sale and leasing of land to others for that purpose. This trade has become one of the most important sources of cash income for many mountain people (Cencetti, 2011).

22.3 Livestock Production

Livestock are an important source of income in the HKH region, more so with increasing elevation. Most farmers practise a mixed crop-livestock farming system (Tulachan, 2000); but even in the

> ...lower valleys where cropping and forestry are possible, livestock is still the main activity and major means of sustaining livelihoods and food security. This means that nearly half of the population of Tibet is deriving some livelihood from farm animals and that nearly 30% depend almost entirely on livestock. Animal products from rangeland areas meet more than 50% of the people's food and agricultural needs.
>
> (Tashi and Partap, 2000)

In some areas of the HKH countries, such as the Indian Himalayan Region (IHR), livestock is the sole source of livelihood and is considered as 'the engine and inspiration' of the mountain economy (Maikhuri *et al.*, 2018). Men are responsible for trade activities and managing the livestock, while women are responsible for taking care of the home and children, cooking, gathering fodder, fuelwood (Fig. 22.2) and dung, supplying water for the household (Fig. 22.3) and doing much of the farming.

In general, yaks, cattle, goats and sheep are raised on pasture and at high densities, at 13–15 head/km^2 (Table 22.3), which is higher than the average in Asia of 11 head/km^2 and the global

Fig. 22.2. A woman transporting firewood from a community forest to her home in the mid-hills of Nepal. (Photo: Allan Degen)

average of about two head/km^2. Buffaloes in the HKH countries are distributed mainly in northern India, south-west China, Pakistan and Nepal, in areas away from the high, cold and arid environment. Monogastric animals are raised mostly in the lowland settlement areas at an average of six pigs/km^2, 66 chickens/km^2 and five ducks/km^2. These densities are generally lower than the average in Asia of 13 pigs/km^2, 250 chickens/km^2 and 26 ducks/km^2 because of large areas in the HKH region that are unsuitable for raising these animals; however, averages in the HKH region are several-fold greater than global averages (Wu *et al.*, 2014). Except for buffaloes, China raises the greatest number of animals of all species in the HKH countries (Table 22.3).

Livestock are important for daily sustenance, providing milk (Fig. 22.4) and meat, and their products, and occasionally blood for food, wool (Fig. 22.5), hair and hides for clothing and weaving, horns and skulls for spiritual reasons and ornaments, and dung for heating and cooking (Fig. 22.6) and for fertilizer (Fig. 22.7). In addition, livestock are used as pack animals (Fig. 22.8), for ploughing (Fig. 22.9) and for transportation (Degen *et al.*, 2022), and livestock products are traded for essentials and luxuries, such as tea, barley and spices (Rhode *et al.*, 2007). Furthermore, livestock, especially yaks, are central to many cultural traditions and religious festivals (Long *et al.*, 2008). Buddhists use yak butter (Fig. 22.10) as fuel for the butter

Fig. 22.3. Tibetan women bringing water to the household on the Qinghai-Tibetan Plateau. (Photo: Allan Degen)

lamp, and they believe it provides wisdom and light of knowledge to eradicate darkness and ignorance (Degen *et al.*, 2019). In Uttarakhand Himalaya, it is believed that every Hindu household should own a cow, irrespective of its economic status, as the cow is worshipped (Sati and Singh, 2010). Muslims throughout the HKH slaughter small ruminants in many festive occasions and in holidays such as *Eid al-adha* (Feast of the Sacrifice) (Degen and el-Meccawi, 2011).

Pastoralism, agropastoralism and agroforestry are dominant land uses of the vast rangelands of the HKH (Dong *et al.*, 2016). The alpine grassland is considered to be one of the best grazing lands for livestock production in Asia (Miller, 1990) even though the climate is extremely harsh (Fig. 22.11). Consequently, livestock grazing plays a crucial role in the livelihoods of the pastoralists in the highlands of the countries in the HKH region, in particular in China, Nepal, India and Bhutan (Rhode *et al.*, 2007; Wang *et al.*, 2016). Agroforestry is often combined with livestock production, as tree leaves are lopped and used as fodder for livestock (Fig. 22.12) (Degen *et al.*, 2010). The depletion of forest resources has also had an adverse effect on local agricultural systems because of the reduced supply of manure from livestock fed on forest fodder. In the high-elevation areas of the HKH, usually above 3000 m, and including

Table 22.3. Main livestock numbers in the countries within the Hindu Kush Himalayan region in 2010. (Adapted from Li *et al.*, 2019)

Country	Species							
	Cattle	Buffaloes	Sheep	Goats	Horses	Pigs	Chickens	Ducks
Afghanistan	4,957,470	3927	10,121,468	4,785,893	180,600	n.a.	10,606,645	320,763
Bangladesh	550,881	58,306	26,184	1,179,778	4632	140,356	7,982,598	556,739
Bhutan	381,611	1537	14,482	56,129	17,655	23,349	455,447	338,697
China	17,134,195	1,378,784	32,655,768	24,897,560	1,457,824	18,191,304	91,954,874	15,464,382
India	11,928,360	2,070,362	4,715,141	6,471,851	220,034	3,152,750	28,725,225	2,417,245
Myanmar	2,971,110	1,359,151	150,808	482,915	71,340	2,839,921	33,999,526	1,411,188
Nepal	7,321,860	4,908,587	799,097	8,947,710	45,929	1,073,489	25,982,787	386,934
Pakistan	8,077,874	1,621,196	13,843,156	18,434,484	137,538	n.a.	78,749,284	1,014,892
Total	**53,323,361**	**11,401,850**	**62,326,104**	**65,256,320**	**2,135,552**	**25,421,169**	**278,456,386**	**21,910,840**

Note: Yaks are classified with cattle. There are approximately 16 million domestic yaks and 15–20,000 wild yaks in the world. Approximately 94% are raised in China, but yaks are also present in Nepal, Bhutan, India, Pakistan, Mongolia, Russia, Afghanistan, Tajikistan and Kyrgyzstan. n.a., not available (Joshi *et al.*, 2020)

Fig. 22.4. Tibetan woman milking a yak after the calf initiated milk letdown on the Qinghai-Tibetan Plateau, China. (Photo: Allan Degen)

Fig. 22.5. Tibetan woman spinning sheep wool on the Tibetan Plateau. A solar collector for heating water can be seen in the background. (Photo: Allan Degen)

Tibet, Qinghai province in China, Bhutan and Nepal, semi-nomadism is the main practice, while in lower-elevation sites, including Afghanistan, India, Nepal and Pakistan, a more sedentary lifestyle is predominant (Wu *et al.*, 2014).

Pastoralism in the Himalayas is based on transhumance practices and involves seasonal migration of livestock to the highlands in the summer and lowlands in the winter (Degen *et al.*, 2007; Bhasin, 2011; Dong *et al.*, 2016). The pastoral production systems have been regarded as primitive and backward and are largely overlooked by policy makers. State policies have become increasingly counter-productive by restricting mobile forms of land use adopted by the pastoralists. Decision makers and politicians are of the view that pastoralism is a threat to the Himalayan region because of over-grazing and over-stocking (Dong *et al.*, 2016). Thus, pastoralists in many HKH countries are facing an uncertain future due to pressure from government authorities, and, as a consequence, sustainability is being threatened (Maikhuri *et al.*, 2018). Transhumance, at least in Nepal, may even disappear (Banjade and Paudel, 2008; Aryal *et al.*, 2014). Namgay *et al.* (2013) concluded that there are other constraints to transhumant pastoralism including modifications of livelihood options due to changes in demography, migration,

Fig. 22.6. Tibetan woman stacking dry yak dung on the Qinghai-Tibetan Plateau. (Photo: Allan Degen)

Fig. 22.7. Women spreading livestock manure to fertilize cultivated land in the mid-hills of Nepal. (Photo: Allan Degen)

shortage of labour, diversification of agriculture, market influence on rural economy, and privatization and nationalization of rangelands. Conflicts between communities, government officials, herders and non-herders in rangeland use have also been identified as a threat to transhumant pastoralism (Banjade and Paudel, 2008). The pastoral system in the HKH is under the process of shifting from traditional extensive ruminant grazing to intensive industrial production (Kreutzmann,

Fig. 22.8. *Jhopkyos* (yak x cow hybrid) near the Kali Gandaki river transporting wood to Jomsom in lower Mustang. *Jhopkyos* are sterile males that are more docile and easier to handle than yaks. (Photo: Allan Degen)

Fig. 22.9. Cattle ploughing rice fields in the mid-hills of Nepal. (Photo: Allan Degen)

Fig. 22.10. Yak butter being sold near the Jokhang Temple Monastery in Llasa. (Photo: Allan Degen)

Fig. 22.11. Tibetan sheep grazing natural grassland on the Qinghai-Tibetan Plateau. (Photo: Allan Degen)

2012; Bai *et al.*, 2021a). Numerous local pastoral communities have increased their livestock production systems in feedlots (Fig. 22.13), while the previous grazing lands have been fenced for restoration. These changes can be termed as 'industrialization', which is limiting livestock grazing while intensifying livestock production.

Generally, traditional nomadic pastoralism can increase income and production only by increasing the number of livestock, that is

Fig. 22.12. Nepalese women carrying lopped leaves from a community forest in the mid-hills of Nepal. The leaves are used as livestock fodder. (Photo: Allan Degen)

Fig. 22.13. Yaks raised in a feedlot on the Qinghai-Tibetan Plateau. (Photo: Allan Degen)

livestock size and carrying capacity of the pasture affect the livelihood of local herders. The increase in the demand for animal products has encouraged an increase in livestock carrying capacity on grassland in the HKH. It is estimated that the global demand for livestock products will double by 2050, and production is expected to meet only 60% of the demand. This creates an imbalance between supply and demand, which leads to grassland degradation and an unsustainable situation. This imbalance is most prominent in winter and spring when there is often insufficient forage, and livestock can perish in large numbers in the face of the harsh conditions. For livestock production to be sustainable in the HKH countries, strategies must be generated to overcome this difficult period. The lack of forage can be mitigated by supplying the animals with supplementary feed and/or establishing artificial grassland. For reducing over-grazing and starvation by livestock in alpine grasslands, livestock could be brought from higher to lower altitudes, fattened and then marketed. This system would reduce the stocking rate and the pressure on the natural grassland. If livestock numbers are to be increased, there must be a decoupling between livestock numbers and GHG production and if livestock numbers are not increased and are even decreased, as is the aim of most governments, there must be a decoupling between the number of livestock and the income from the livestock or to raise livestock that are more productive.

22.4 Alternative Feeds

The use of novel forage plants and herbs, especially as supplementary feed for ruminants, has stimulated much interest in livestock production research. The jujube tree (*Ziziphus jujuba* Mill) is one of these plants. More than 50 countries grow jujube trees, including four from the HKH. China produces most of the jujube products in the world, but jujube trees are also distributed in Afghanistan, India and Pakistan. Approximately 15–25% of harvested jujube fruits are classified as residual (Wen *et al.*, 2019), which consists of defective, low-quality, split or diseased fruits. In China, this amounts to 1.0–1.5 million tons per year. Jujube residuals as a supplement or as a replacement for part of the feed ingredients for ruminants, monogastrics (including poultry) and fish can increase palatability, feed intake, nutrient digestibility, reduce the feed-to-gain conversion rate and improve the intestinal microbial system and quality of animal products. Given its ability to improve immunity, and, thereby, to reduce the use of antibiotics, jujube fruit residual has broad potential use in animal feed as a growth promoter, antioxidant and flavouring agent. In general, the protein and fat contents in jujube are relatively low, but the total sugar content and gross energy yield are high; and, hence, jujube could be considered as a high-energy feed supplement (Xu *et al.*, 2022).

Better use should be made of crop aftermaths as a feed source for ruminants. Rice straw is one of the most abundant crop residues in many HKH countries (Fig. 22.14). In China, nearly 21 million metric tons are produced annually (Chen, 2016), and much of the straw is discarded or burnt directly in the field, generating massive pollution and greenhouse gases (Kuriqi *et al.*, 2016). As a potential source of feed for ruminants, rice straw has the advantages of low cost and abundant availability (Fig. 22.15). However, its high fibre content reduces its utilization by microorganisms, resulting in low degradation in ruminal fermentation (Yu *et al.*, 2016). The addition of urea to straw, at 4% of the intake, increased the DMI, provided nitrogen for rumen microbes to produce protein and increased milk production by lactating buffaloes (Chemjong, 1991). Urea is a relatively cheap non-protein nitrogen that can improve the utilization rate of carbohydrates and the feed conversion rate when added to poor-quality forage (Ding *et al.*, 2015). Adding other nutrient sources such as sugar molasses and ligninolytic enzymes could increase intake, enhance microbial fermentation and improve the use of straw for ruminants (Malik *et al.*, 2015). Ensiling is an effective method for improving feed utilization of rice straw and improving performance of livestock. A combination of lactic acid bacteria, consisting of *Lactobacillus plantarum*, *L. buchneri* and *Pediococcus acidilactici*, was added at a ratio of 2:1:1 for ensiling rice straw. Ensiling ruptured the physical structure of rice straw, reduced the contents of dry matter, neutral detergent fibre, hemicellulose and water-soluble carbohydrates, and increased the content of neutral detergent solutes and concentrations of lactic acid, acetic acid and other organic acids. These changes promoted rumen microbial colonization and fibre digestion during early fermentation, which increased bacterial diversity and reduced the relative abundance of the methanogen *Methanobrevibacter* and CH_4 production, increased *in vitro* degradability of nutrients, and, ultimately, improved

Fig. 22.14. Women harvesting rice in mid-hills of Nepal. (Photo: Allan Degen)

Fig. 22.15. Woman chopping rice straw to feed livestock in mid-hills of Nepal. (Photo: Allan Degen)

average daily gain of growing Hu lambs (Xu *et al.*, 2023). The silage produced could be stored and supplemented to the livestock during the harsh winter, and the reduction in CH_4 emission is important to minimize climate warming.

The root extract of the herb *Astragalus membranaceus*, supplemented at 5–8% dry-matter intake (DMI), improved the average daily gain, antioxidant capacity and immunity response in early-weaned yak calves (Wei *et al.*, 2021) and Tibetan sheep (Wang *et al.*, 2021a, b). *Agriophyllum squarrosum*, fed at 20% DMI to Tan lambs, increased the water-holding capacity of the meat, made the meat more tender and made the colour more appealing to the consumer (Jiang *et al.*, 2022; Liang *et al.*, unpubl. data). Other herbs, such as *Angelica sinensis*, *Codonopsis pilosula* and *Glycyrrhiza uralensis*, have proved to be beneficial for the health of ruminants (Jiang *et al.*, 2021; 2022). Hempseed meal, a by-product of hemp oil production, can be an alternative feeding resource for goats (Abrahamsen *et al.*, 2021).

Lick blocks can improve the performance of livestock. A lick block is a solidified mixture of molasses, urea, minerals, filler, coagulant and binder that is supplemented to livestock mainly in relatively extensive rearing systems. The lick block provides nutrients such as soluble sugars, proteins, minerals and vitamins to balance dietary intake, and can improve rumen fermentation and facilitate digestion and absorption of nutrients. These supplements improve livestock production, reproduction and carcass quality. In addition, lick blocks can partially replace concentrate, serve as a delivery vehicle for additives such as enzymes and drugs, and mediate the distribution of grazing livestock (Zhao *et al.*, 2022).

22.5 Artificial Grassland

The establishment of artificial grassland by planting local forage seeds is the most effective way to restore severely degraded grassland (Wu *et al.*, 2010; Dong *et al.*, 2013), and has been used extensively in rehabilitating severely degraded and even mildly to moderately degraded grasslands. The artificial grassland can increase the above-ground and underground biomass (Li *et al.*, 2014) as well as the productivity of heavily degraded grassland. In addition, although the soil-carbon content of artificial grassland is less than that of natural non-degraded grassland (Wang *et al.*, 2010; Zhu *et al.*, 2015), the establishment of artificial grassland on heavily degraded grassland can increase soil-carbon content (Wang *et al.*, 2005a; Feng *et al.*, 2010; Wu *et al.*, 2010). The high productivity of artificial grassland can provide forage for livestock and mitigate the grass–animal imbalance, which eases the pressure on the natural grassland and enables the restoration of degraded grassland. Of importance, artificial grassland can be harvested and stored for the winter and spring when forage is sparse.

The restoration of natural grassland also increases its carbon sequestration capacity while increasing grassland productivity. With appropriate management, artificial grassland plants are less subject to natural stress, and the productivity can be 3–12 times greater than that of natural grassland (Shen *et al.*, 2016). An area of the artificial grassland can be designated to supplement the needs of excess livestock and, concomitantly, reduce pressure on the natural grassland. The area devoted to artificial grassland is considered a measure of the animal husbandry development in a country (Li *et al.*, 2016a). Of the total grassland area, artificial grassland accounts for 73% in New Zealand, 58% in Australia, 50% in Europe, 22% in Canada and 10% in the USA, while it accounts for only 3.2% in China. In Qinghai province, the main province of the HKH, it covers only 1.04%. While this percentage is well below that of China, it is substantially greater than the rest of the HKH countries (Li *et al.*, 2016b). Artificial grassland should be prioritized by HKH countries to provide forage for grazing livestock and reduce the pressure on natural grassland.

The types of artificial grasslands are diverse, and can be divided into annual and perennial artificial grasslands, and into unicast and mixed artificial grasslands. Unicast artificial grassland is divided mainly into legume grass and gramineous pasture (Bai *et al.*, 2016). At present, the artificial grassland in the HKH region is dominated by oat unicast (Xu, 2005), which yields greater biomass than natural grassland, but has a shorter growing period. As a result, the land of oat unicast grass is exposed for a long period resulting in substantial soil-water loss, which is not conducive to the protection of the ecological environment. Therefore, the establishment of perennial artificial grassland is often the preferred option for the sustainable development of alpine grassland on the HKH (Gu, 2008). Legumes and grasses are the ideal combination of perennial artificial grassland for yield, quality and stability (Wang *et al.*, 2007). Grasses are well adapted to the low air temperature and are the main type of artificial grassland planted in the HKH, as the low air temperature has limited the cultivation of perennial forage legumes.

After the artificial grassland is established, there is often a decline in productivity within a few years due to the loss of nutrients (Shang *et al.*, 2008). With the planting of artificial grass, a large amount of chemical fertilizer is generally applied to provide sufficient nutrients for growth. However, with time after planting, harvesting of grasses and leaching of soil nitrate, there is a gradual reduction of soil nutrients (Li *et al.*, 2017). As there is usually a single plant species in artificial grassland, there is strong competition for common resources, resulting in the decrease of certain nutrients in the soil pool (Wang *et al.*, 2005b), in particular, total soil nitrogen, total phosphorus and total potassium (Dong *et al.*, 2012). The loss of nutrients weakens the ability of the grasses to compete for nutrients, which enables the invasion of miscellaneous grasses, and, with the degradation of artificial grassland, the main species are gradually replaced by poisonous weeds (Zhang *et al.*, 2014). Two approaches can be employed to limit the degradation of artificial grassland: one is to supplement the nutrients and propagules needed by the degraded artificial grassland and the other is to increase the ability of the artificial grassland to resist environmental stress. A community composed of multiple populations can use environmental resources more effectively than a single population and maintain long-term higher productivity and greater stability. Mixed sowing can also extend the pasture grass period, increase soil fertility and contribute to soil carbon stability (Yang *et al.*, 2011). Soil organic carbon plays a key role in maintaining the sustainability of pasture grassland ecosystems (Xu *et al.*, 2018), which affects soil-water conservation and erosion potential and is a key factor affecting soil structure (Tisdall and Oades, 1982). As the richness of species increases, the resistance of artificial grassland communities to weed invasion is enhanced (Li *et al.*, 2009). Increasing diversity, soil nutrients and propagating bodies, and the eradication of weeds, are effective methods to prevent degradation of artificial grassland.

However, there is a risk that the artificial grassland will be degraded within three to five years and become 'bare-land' again (Shang *et al.*, 2018). With proper management, artificial grassland can establish stable soil and good vegetation coverage quickly, and provide much forage for livestock. A study of more than 15 years demonstrated that the artificial grassland with good management can provide sustainable forage production and reclaim soil nutrients and carbon gradually, and improve the ecological and productive functions (Shang *et al.*, 2018). A six-year study on the economic benefits of artificial grassland concluded that good management with investment for seeds, fertilizer, machinery, labour and fencing can earn considerably more income than without investment (Dong *et al.*, 2011b; Table 22.4). Feng *et al.*, (2010) reported that artificial grassland established on bare land in Qinghai province had stable vegetation, the same soil-carbon content as non-degraded meadow, and a 33% higher soil-carbon content than bare land. An artificial grassland at an altitude of 3700–4000 m recovered the soil-carbon level in a moderately degraded meadow in five years; however, the artificial grassland degraded again after four years, and its soil-carbon content declined to the level of bare land (Li *et al.*, 2014).

22.6 Decoupling Livestock Numbers and Earnings

The decoupling of livestock numbers from earnings can be achieved either by raising animals that are more productive or by earning more money from the existing livestock. Local breeds of goats, sheep and bovines are important sources

Table 22.4. Economic statistics of establishing sown grassland on bare soil land at a site 3700 m above sea level. (Adapted from Dong *et al.*, 2011a)

	Years	Hay forage production (kg/hm²)	Stocking rate (sheep/hm²)	Income (sheep/hm²)	Investment cost (sheep/hm²)	Net income (sheep/hm²)	Net income/investment
Sown grassland without management (traditional grazing)	1st	3960	7.59	2277	1538	740	0.48:1
	2nd	4950	9.49	2847	22.5	2825	125:1
	3rd	3555	6.82	2046	22.5	2024	89:1
	4th	2641	5.07	1521	22.5	1499	66:1
	5th	1546	2.96	888	22.5	866	38:1
	6th	792	1.52	456	22.5	434	19:1
Sown grassland with management (planned grazing)	1st	3960	7.59	2277	1538	740	0.48:1
	2nd	7496	14.38	4314	323	3992	12:1
	3rd	6010	11.53	3459	323	3137	10:1
	4th	5742	11.01	3303	323	2981	9:1
	5th	5183	9.94	2982	398	2585	7:1
	6th	5152	9.88	2964	398	2567	6:1

of food security and livelihoods for many mountain people (Rasul and Hussain, 2015). They are generally well adapted to the harsh conditions but are characterized by low production and low reproductive rate. Overall, the population of local breeds of livestock has decreased over time in high-elevation areas, and, in particular, the population of yaks is in decline in India, Nepal and Bhutan (Wu *et al.*, 2016). In China, it is government policy to reduce the number of livestock on the Qinghai-Tibetan Plateau and, since the 1980s, there have been substantial and rapid changes in pastoralism. Large tracts of grasslands were lost to grazing under government programmes, and many households abandoned animal herding (Levine, 2015). Outside of China, yak numbers are declining because of lack of feed, global warming and inbreeding, mainly because the Tibetan borders have been closed to yak transboundary movement (Bhattacharya, 2020; Dorji *et al.*, 2020).

Policies have encouraged the introduction of improved breeds in both mountains and plains, but in mountain areas these breeds are vulnerable to extreme weather conditions, disease and seasons with sparse pasture and fodder. Selective breeding has improved production in some livestock. China has 13 breeds of yaks, some of which produce more or less milk and more or less meat. Yaks can mate with cattle, generally a male yak with a cow, and produce hybrids that are more productive than the pure parents. The hybrid female, called *chauri*, is fertile and produces more milk than yak cows (Fig. 22.16), while the male hybrid, called *jhopkyos*, is sterile and is more docile than yaks (Degen *et al.*, 2007). These crosses are common in the mountain areas of upper Chitral and Gilgit-Baltistan in Pakistan and upper Rasuwa district in Nepal, and Arunachal Pradesh in India. Both *chauris* and *jhopkyos* are maintained at a lower altitude than yaks and, at about 2500 m, are the primary source of earning a livelihood in many areas of the Himalayas (Degen *et al.*, 2007).

There was a shift in priority from high livestock numbers to better-quality and higher-priced animal products in order to stabilize or increase household income. Reducing the number of animals improved their nutrition, survival and fecundity, while also improving the long-term survival and prosperity of the grasslands. Furthermore, a reduction in the number of livestock reduces the risk of large losses of livestock due to climatic disasters, while maintaining commercial profitability by increasing individual animal performance, and, most importantly in the long term, curbing the degradation of grasslands (Shang *et al.*, 2012). The reduction in livestock decreased

Fig. 22.16. A woman in Langtang, Nepal, milking a *chauri* (yak x cow hybrid), whose calf had died. *Chauris* produce more milk than yaks and are common in the mountain areas of upper Chitral and Gilgit-Baltistan in Pakistan, upper Rasuwa district in Nepal, and Arunachal Pradesh in India. The skin of the calf on the ground was used to initiate milk letdown. (Photo: Allan Degen)

the red meat available in the markets while the demand remained high, which, ultimately, increased the price of meat. The percentage of livestock output in agriculture decreased, which reduced GHG emissions; however, in terms of livelihood, earnings from livestock increased. Consequently, the livelihood of the herders did not suffer as they were compensated for their policy-induced reduction in livestock by higher meat prices. However, the large inter-annual fluctuations demonstrated that livestock production was still vulnerable (Bai *et al.*, 2021a).

22.7 Decoupling Livestock Numbers and GHG Emissions

Industry is considered the main source of GHG emissions, with agriculture accounting for 17% of the total (Li *et al.*, 2011). Livestock account for 14.5% of global anthropogenic GHG emissions, mostly from developing countries (Gerber *et al.*, 2013), and produce 44% and 53% of the total CH_4 and N_2O emissions, respectively (Caro *et al.*, 2014). Sources of carbon emissions in the livestock production system include soil, enteric fermentation, manure and energy use (Table 22.5). With the greater demand for meat products and possible increase in livestock numbers, there will be a potential increase in GHG emissions and, consequently, the global low-carbon development of animal husbandry faces severe challenges (Herrero *et al.*, 2016; Gil *et al.*, 2019). In the extensive rangeland system, where the local feed resource-based production is dominant, CH_4 emission is the main component of total GHG emission in the livestock sector. In addition, global warming has increased soil-carbon emission. For example, it has been predicted that increases in air temperature of 1.88°C and 3.19°C on the Tibetan Plateau, would increase CO_2 release from the frozen soil of alpine grassland by approximately 18% and 29%, respectively (Feng *et al.*, 2016).

Decoupling greenhouse gas emissions from economic growth of the livestock industry has become a key component of global climate governance (Herrero *et al.*, 2016; Röös *et al.*, 2017). The government's programme is to reduce GHG emissions with minimal effect on economic growth in the livestock industry. It has been reported that the global GHG emission per unit product in crop and animal production has decreased by 39% and 44%, respectively, since 1970 (Bennetzen *et al.*, 2016), which demonstrated that the decoupling of GHG emissions from farming and animal husbandry occurred. As CH_4 produced by enteric fermentation is the largest source of GHG emissions from animal husbandry (Bennetzen *et al.*, 2016), many efforts are being made to reduce these emissions worldwide (Beauchemin *et al.*, 2020). The CH_4 emitted by ruminants is related strongly to the dietary intake. Studies reported that productivity of intensive systems – that is grain-based and dry-lot feeding – were greater and CH_4 emission intensities were correspondingly lesser when compared to extensive systems, that is grazing based on low inputs of feed and labour (Opio *et al.*, 2013). This occurred mainly because of the decrease in emission intensity (emissions per unit product)

Table 22.5. Methane (CH_4) emission factors of livestock in the Hindu Kush Himalayan (HKH) region in 2010. (Adapted from Li *et al.*, 2019)

	Enteric fermentation	Manure	Enteric fermentation	Manure	Total
Livestock	CH_4 factors (kg/year)	CH_4 factors (kg/year)	CH_4 (million tons/year)	CH_4 (million tons/year)	CH_4 (million tons/year)
Yaks	40	1	0.54	0.01	0.56
Cattle and buffaloes	47	1	2.39	0.05	2.44
Sheep and goats	5	0.1	0.64	0.01	0.65
Horses	18	1	0.04	0.00	0.04
Pigs	1	2	0.03	0.05	0.08
Poultry	0	0.01	0.00	0.00	0.00
Total			3.63	0.13	3.77

Note: Emission factors from Intergovernmental Panel on Climate Change (IPCC, 2006) and Xue *et al.* (2014).

(Beauchemin *et al.*, 2020), but also because of the reduction in enteric CH_4 emissions through feeding higher-quality diets (Zhuang *et al.*, 2017). Consequently, the transition from extensive to intensive livestock production has become the mainstream recommendation, globally, and China has been promoting the transition since 2007 (Zhuang *et al.*, 2017). However, Zhuang *et al.* argued that the conversion of traditional pastoralism into a more intensive system on the Tibetan Plateau has actually increased carbon emissions substantially in the animal husbandry system. It is crucial that GHG emission-intensity assessments are made available, so that development of low-carbon livestock husbandry in the grazing livestock system could be applied (Salvador *et al.*, 2017). Grasslands accounts for 25% of the global land area, so mitigating GHG emissions from grassland animal husbandry would be important in controlling global warming (Hörtnagl *et al.*, 2017).

The combustion of faeces, at least on the Tibetan Plateau, was reported to be the second largest source of GHG emissions from local animal husbandry (Zhuang *et al.*, 2017). In the pastoral areas, this source could be decreased substantially by increasing the proportion of localized clean-energy infrastructure, especially solar energy utilization for rural households. Solar energy is abundant in the HKH, and, at least in the Qinghai province, is becoming an important area for solar energy application. Solar units have been installed in some Tibetan homes to heat water, mainly for tea (Fig. 22.17), and a more extensive use of these units could reduce the amount of dung used. Dung could be used in the production of biochar, which has many uses in agriculture (Bai *et al.*, 2021b).

Intensification of animal production, if managed correctly, can reduce carbon emission; however, the essential inputs including fossil energy, mechanization, fertilizer and imported feed can increase GHG emission intensity (Ogino *et al.*, 2016; Zhuang *et al.*, 2017). Consequently, the balance between free grazing and penned livestock should be examined carefully, as proper management would not only reduce GHG emissions, but also improve grassland and increase livestock production. The ratio of carbon sequestration to emissions in the grass-based livestock system has decreased gradually, mainly due to: (i) decreasing livestock numbers following the policy of the local government (Shang *et al.*, 2014; 2016); (ii) increasing livestock numbers being raised in feedlots (Bai *et al.*, 2021a); and (iii) improving technical efficiency of livestock production with the promotion of clean and low-carbon technologies, such as solar power panels

Fig. 22.17. A solar unit heating water in a kettle on the Qinghai-Tibetan Plateau. (Photo: Allan Degen)

and animal waste recycling. From the trend of the efficiency factor, there are indications that GHG emissions would continue to decrease. Another possibility to reduce emissions from the ruminant livestock sector could be to exploit feed additives with anti-methanogenic properties (Khanal *et al.*, 2022). Enteric CH_4 emission is lesser in yaks than in cattle and in Tibetan sheep than in lowland sheep, due, at least in part, to differences in the composition of the rumen microbiota between the high- and low-altitude ruminants. Transferring rumen contents from the low emitters of enteric CH_4 to the high emitters could reduce the CH_4 emission in the high-emitters. In addition, importation of more livestock resources from outside the system would reduce the need to produce these resources for animal husbandry. For example, by importing commercial forage, the grazing pressure on the grassland would be reduced, and the net carbon fixation would be enhanced. This is rarely practised today in the Tibetan highlands; however, should it become part of animal husbandry management, the net GHG would be reduced.

The intensification of land use in the alpine grassland resulted in increased GHG emissions (Zhuang *et al.*, 2017), namely increases in N_2O emission from soils associated with N fertilizer, in CH_4 from cultivated grassland, and from burning of pasture and agricultural residues. This was coupled with the increase in pesticides, operation of farm machinery and extra energy consumption by industry (Burney *et al.*, 2010). Through expanding the scope of the system, not only structural adjustments within the system are required, but also more integration and optimization of industrial bodies from outside the system are needed, as well as strict controls on materials used. For example, commercial N fertilizer is prohibited when revegetating degraded grassland on the Qinghai-Tibetan Plateau, as it increases N_2O emissions from soils (Wang *et al.*, 2011).

22.8 Germplasm and the Case of Yaks

The yak (*Poephagus grunniens*) is an iconic symbol of high altitude and the animal most identified with the Himalayas, but, in reality, it is in danger of extinction. This is occurring even though, for many HKH inhabitants, 'yaks are a cornerstone of their livelihoods as much as of their social and religious traditions' (Bhattacharya, 2020). The number of yaks has been declining due to climate warming, forage shortage, irregular snowfall and inbreeding. For example, the number of Brokpa families practising transhumant yak rearing in Arunachal Pradesh, India, has plummeted from approximately 800 to 200 households. Inbreeding in yaks has become a major problem, as the number of yaks in each household is relatively small, with access to only a few bulls. Thus, it is urgent to make germplasm available to yak herders and to promote the exchange of yak germplasm in different regions. However, the geopolitics in the area have restricted movements of yaks across borders beginning in the 1950s, and, therefore, yaks in Nepal, India, Bhutan and Pakistan have been isolated for more than 60 years. The prolonged use of the same old bulls is having detrimental effects on the yak population and, according to the Bropka yak herders, the animals are weaker, of smaller size and more disease-prone. There are between 10,000 and 15,000 wild 'pure breed yaks', all within China's Tibetan Autonomous Region. Except for China, no other country has access to these yaks and this gene pool. On the Qinghai-Tibetan Plateau, the genetic improvement of domestic yaks with frozen semen from wild yaks is used, which has played a vital role in the improvement of yak germplasm. In addition, co-operatives on the plateau with large yak herds are increasing rapidly in pastoral regions. Bulls from different small herds are managed together, and artificial insemination with frozen semen is employed. These have facilitated the gene exchange of yaks and helped prevent further germplasm degradation (Jing *et al.*, 2022). However, this is not the case in other yak-raising countries in the HKH region. The borders between countries used to be open, allowing yak movement and the exchange of germplasm (Wu *et al.*, 2016), which was critical in acquiring good-quality breeding stock for sustaining yak production. ICIMOD recognized this problem of inbreeding and prioritized yak germplasm exchange among HKH countries. Transboundary yak festivals were organized in 2017 and 2018 for groups from Nepal, Bhutan and India to strengthen the regional network and co-operation among them. In 2020, under the sponsorship of ICIMOD, Bhutan presented two high-quality breeding bulls to Nepal and one to Sikkim, India, which initiated an important yak germplasm exchange programme. ICIMOD is also exploring the use of artificial insemination and semen exchange to ensure good-quality breeding stock for sustaining yak production (Manandhar, 2020).

22.9 Conclusions

Decision makers regard the pastoral production system as primitive and backward, and their policy is to settle the pastoralists and convert the rangeland into cultivated land. Consequently, pastoralists are faced with a loss of grazing land, as well as climate warming, widespread grassland degradation, a serious shortage of forage in the harsh winter and spring, and outmigration of many of the men herders in search of non-pastoral employment. Increasing livestock numbers is difficult because of the concomitant increase in greenhouse gases (GHG), mainly CH_4. These problems must be mitigated if livestock production is to remain sustainable on the HKH, and improvements must be implemented. Better use of crop aftermaths, such as rice and wheat straw, the establishment of artificial grasslands, and the introduction of novel feeds and herbs could ease the forage shortage. An uncoupling of livestock number and earnings by raising more productive animals and by earning more for animal products would increase the income of the herders without increasing livestock numbers. An uncoupling of livestock number and CH_4 production would allow the herders to raise more animals without increasing GHG. Fattening livestock in the lowlands would ease the pressure on the grassland and enable restoration of the degraded lands. An exchange of germplasm, particularly in yaks, must be prioritized to minimize serious inbreeding.

References

Abrahamsen, F.W., Gurung, N.K., Abebe, W., Reddy, G.P., Mullenix, K. *et al.* (2021) Effects of feeding varying levels of hempseed meal on dry matter intake, rumen fermentation, in vitro digestibility, blood metabolites, and growth performance of growing meat goats. *Applied Animal Science* 37, 681–688. DOI: 10.15232/aas.2021-02153

Adenle, A.A., Azadi, H. and Arbiol, J. (2015) Global assessment of technological innovation for climate change adaptation and mitigation in developing world. *Journal of Environmental Management* 161, 261–275. DOI: 10.1016/j.jenvman.2015.05.040

Alkire, S. and Santos, M.E. (2010) *Acute Multidimensional Poverty: A New Index for Developing Countries.* OPHI Working Papers 38, University of Oxford, UK.

Aryal, S., Maraseni, T.N. and Cockfield, G. (2014) Sustainability of transhumance grazing systems under socio-economic threats in Langtang, Nepal. *Journal of Mountain Science* 11(4), 1023–1034. DOI: 10.1007/s11629-013-2684-7

Bai, Y., Pan, Q.M. and Xing, Q. (2016) Fundamental theories and technologies for optimizing the production functions and ecological functions in grassland ecosystems. *Chinese Science Bulletin* 61(2), 201–212.

Bai, Y., Guo, C., Li, S., Degen, A.A., Ahmad, A.A. *et al.* (2021a) Instability of decoupling greenhouse gas emissions from economic growth in livestock products in the Tibetan highland. *Journal of Environmental Management* 287, 112334. DOI: 10.1016/j.jenvman.2021.112334

Bai, Y., Rafiq, M.K., Li, S., Degen, A.A., Mašek, O. *et al.* (2021b) Biochar from pyrolyzed Tibetan yak dung as a novel additive in ensiling sweet sorghum: an alternate to the hazardous use of yak dung as a fuel in the home. *Journal of Hazardous Materials* 403, 123647. DOI: 10.1016/j.jhazmat.2020.123647

Banjade, M.R. and Paudel, N.S. (2008) Mobile pastoralism in crisis: challenges, conflicts and status of pasture tenure in Nepal Mountains. *Journal of Forest and Livelihood* 7(1), 49–57.

Beauchemin, K.A., Ungerfeld, E.M., Eckard, R.J. and Wang, M. (2020) Review: fifty years of research on rumen methanogenesis: lessons learned and future challenges for mitigation. *Animal* 14–S1, S2–S16. DOI: 10.1017/S1751731119003100

Bennetzen, E., Smith, P. and Porter, J.R. (2016) Decoupling of GHG emissions from global agricultural production: 1970–2050. *Global Change Biology* 22, 763–781. DOI: 10.1111/gcb.13120

Bhasin, V. (2011) Status of women in transhumant societies. *Journal of Sociology and Social Anthropology* 2(1), 1–22. DOI: 10.1080/09766634.2011.11885543

Bhattacharya, B.K. (2020) India's yak herders face the end of ancient tradition in warming Himalayas. Available at: https://news.mongabay.com/2020/01/indias-yak-herders-face-the-end-of-ancient-tradition-in-warming-himalayas/ (accessed 28 September 2023).

Burney, J.A., Davis, S.J. and Lobell, D.B. (2010) Greenhouse gas mitigation by agricultural intensification. *Proceedings of the National Academy of Sciences of the United States of America* 107, 12052–12057. DOI: 10.1073/pnas.0914216107

Caro, D., Davis, S.J., Bastianoni, S. and Caldeira, K. (2014) Global and regional trends in GHG emissions from livestock. *Climatic Change* 126(1–2), 203–216. DOI: 10.1007/s10584-014-1197-x

Cencetti, E. (2011) Tibetan plateau grassland protection: Tibetan herders' ecological conception versus state policies. *Himalaya* 30(1), 39–50.

Chemjong, P.B. (1991) Economic value of urea-treated straw fed to lactating buffaloes during the dry season in Nepal. *Tropical Animal Health and Production* 23, 147–154. DOI: 10.1007/BF02356993

Chen, X.G. (2016) Economic potential of biomass supply from crop residues in China. *Applied Energy* 166, 141–149. DOI: 10.1016/j.apenergy.2016.01.034

Cheng, S.K. and Shen, L. (2000) Approach to dynamic relationship between population, resources, environment and development of the Qinghai-Tibet Plateau. *Journal of Natural Resources* 15(4), 297–304.

Chettri, N., Shakya, B., Thapa, R. and Sharma, E. (2008) Status of a protected area system in the Hindu Kush Himalayas: an analysis of PA coverage. *The International Journal of Biodiversity Science and Management* 4, 164–178.

Degen, A.A. and El-Meccawi, S. (2011) Bedouin chefs catering Bedouin weddings in the Negev Desert of Southern Israel. *International Journal of Business and Globalization* 3, 303–318. DOI: 10.1504/IJBG.2011.042060

Degen, A.A., Kam, M., Pandey, S.B., Upreti, C.R., Pandey, S. *et al.* (2007) Transhumant pastoralism in yak herding in the Lower Mustang district of Nepal. *Nomadic Peoples* 11(2), 57–85. DOI: 10.3167/np.2007.110204

Degen, A.A., Pandey, L.N., Kam, M., Pandey, S.B., Upreti, C.R. *et al.* (2010) Goat production and fodder leaves offered by local villagers in the mid-hills of Nepal. *Human Ecology* 38, 625–637. DOI: 10.1007/s10745-010-9342-2

Degen, A.A., El-Meccawi, S. and Kam, M. (2019) Milk and dung production by yaks (*Poephagus grunniens*): important products for the livelihood of the herders and for carbon recycling on the Qinghai-Tibetan Plateau. In: Shang, Z., Degen, A.A., Rafiq, M.K. and Squires, V.R. (eds) *Carbon Management for Promoting Local Livelihood in the Hindu Kush Himalayan (HKH) Region*. Springer Nature, Cham, Switzerland, pp. 145–162.

Degen, A.A., El-Meccawi, S., Sivan, J. and Kam, M. (2022) The dromedary camel and the yak: very different animals with many similar traits. *Nomadic Peoples* 26, 268–291. DOI: 10.3197/np.2022.260206

Ding, L.M., Lascano, G.J. and Heinrichs, A.J. (2015) Effect of precision feeding high- and low-quality forage with different rumen protein degradability levels on nutrient utilization by dairy heifers. *Journal of Animal Science* 93, 3066–3075. DOI: 10.2527/jas.2014-8260

Dong, Q.M., Shi, J.J., Ma, Y.S., Wang, Y.L. *et al.* (2011a) Analysis on economic and ecological benefit of blacksoil-beach sown grassland. *Acta Agrestia Sinica* 19(2), 195–201. DOI: 10.11733/j.issn.1007-0435.2011.02.003

Dong, Q.M., Zhao, X.Q., Wu, G.L., Shi, J.J. and Ren, G.H. (2013) A review of formation mechanism and restoration measures of 'black soil land' degraded grassland in the Qinghai-Tibetan Plateau. *Environmental Earth Sciences* 70(5), 2359–2370.

Dong, S.K. (2013) *Productive and Ecological Paradigm of Alpine Cultivated Grasslands in the Region of Qinghai-Tibet Plateau*. Science Press, Beijing.

Dong, S.K., Wen, L., Liu, S., Zhang, X., Lassoie, J.P. *et al.* (2011b) Vulnerability of worldwide pastoralism to global changes and interdisciplinary strategies for sustainable pastoralism. *Ecology and Society* 16(2), 10.

Dong, S.K., Wen, L., Li, Y.Y., Wang, X.X., Zhu, L. *et al.* (2012) Soil-quality effects of grassland degradation and restoration on the Qinghai-Tibetan Plateau. *Soil Science Society of America Journal* 76(6), 2256–2264. DOI: 10.2136/sssaj2012.0092

Dong, S.K., Kassam, K.-A.S., Tourrand, J.F. and Boone, R.B. (2016) *Building Resilience of Human-natural Systems of Pastoralism in the Developing World*. Springer, Berlin.

Dorji, N., Derks, M., Groot Koerkamp, P.W.G. and Bokkers, E.A.M. (2020) The future of yak farming from the perspective of yak herders and livestock professionals. *Sustainability* 12, 4217. DOI: 10.3390/su12104217

Elalem, S. and Pal, I. (2015) Mapping the vulnerability hotspots over Hindu-Kush Himalaya region to flooding disasters. *Weather and Climate Extremes* 8, 46–58. DOI: 10.1016/j.wace.2014.12.001

FAO-IFAD-WFP (2015) *The State of Food Insecurity in the World. Meeting the 2015 International Hunger Targets: Taking Stock of Uneven Progress*. Food and Agriculture Organization of the United Nations, International Fund for Agriculture Development and World Food Programme, Rome.

Feng, J.L., Hu, H.P. and Chen, F. (2016) An eolian deposit–buried soil sequence in an alpine soil on the northern Tibetan Plateau: implications for climate change and carbon sequestration. *Geoderma* 266, 14–24. DOI: 10.1016/j.geoderma.2015.12.005

Feng, R., Long, R., Shang, Z., Ma, Y., Dong, S. *et al.* (2010) Establishment of *Elymus natans* improves soil quality of a heavily degraded alpine meadow in Qinghai-Tibetan Plateau, China. *Plant and Soil* 327(1–2), 403–411. DOI: 10.1007/s11104-009-0065-3

Gao, Q., Duan, M., Wan, Y., Li, Y., Guo, Y. *et al.* (2010) Comprehensive evaluation of eco-environmental sensitivity in Northern Tibet. *Acta Ecologica Sinica* 30(15), 4129–4136.

Gerber, P.J., Steinfeld, H., Henderson, B., Mottet, A., Opio, C., *et al.* (2013) *Tackling climate change through livestock: a global assessment of emissions and mitigation opportunities*. Food and Agriculture Organization of the United Nations, Rome.

Gerlitz, J.Y., Hunzai, K. and Hoermann, B. (2012) Mountain poverty in the Hindu-Kush Himalayas. *Canadian Journal of Development Studies* 33, 250–265. DOI: 10.1080/02255189.2012.689613

Gil, J.D.B., Daioglou, V., van Ittersum, M., Reidsma, P., Doelman, J.C. *et al.* (2019) Reconciling global sustainability targets and local action for food production and climate change mitigation. *Global Environmental Change* 59, 101983. DOI: 10.1016/j.gloenvcha.2019.101983

Gu, M. (2008) Study on the productivity and stability of perennial cultivated grassland in Qinghai-Tibetan alpine meadow. PhD thesis, Lanzhou University, China.

Herrero, M., Henderson, B., Havlik, P., Thornton, P.K., Conant, R.T. *et al.* (2016) Greenhouse gas mitigation potentials in the livestock sector. *Nature Climate Change* 6, 452–461. DOI: 10.1038/NCLIMATE2925

Hörtnagl, I., Barthel, M., Buchmann, N., Eugster, W., Butterbach-Bahl, K. *et al.* (2017) GHG fluxes over managed grasslands in central Europe. *Global Change Biology* 24(5), 1843–1872. DOI: 10.1111/gcb.14079

Intergovernmental Panel on Climate Change (IPCC) (2006) *2006 IPCC Guidelines for National Greenhouse Gas Inventories. Agriculture, Forestry and Other Land Use (AFOLU)*. In: Eggleston, S., Buendia, L., Miwa, K., Ngara, T. and Tanabe, K. (eds). Prepared by the National Greenhouse Gas Inventories Programme, Institute for Global Environmental Strategies, Hayama, Japan.

Jiang, C., Ding, L., Dong, Q., Wang, X., Wei, H. *et al.* (2021) Effects of root extracts of three traditional Chinese herbs as dietary supplements on dry matter intake, average daily gain, rumen fermentation and ruminal microbiota in early weaned yak calves. *Animal Feed Science and Technology* 278, 115002. DOI: 10.1016/j.anifeedsci.2021.115002

Jiang, C., Dong, Q., Xin, X., Degen, A.A. and Ding, L. (2022) Effect of Chinese herbs on serum biochemical parameters, immunity indices, antioxidant capacity and metabolomics in early weaned yak calves. *Animals* 12, 2228. DOI: 10.3390/ani12172228

Jiao, D., Liang, Y., Zhou, S., Wu, X., Degen, A.A. *et al.* (2022) Supplementing diets with *Agriophyllum squarrosum* reduced blood lipids, enhanced immunity and anti-inflammatory capacities, and mediated lipid metabolism in Tan lambs. *Animals* 12, 3486. DOI: 10.3390/ani12243486

Jing, X., Ding, L., Zhou, J., Huang, X., Degen, A.A. *et al.* (2022) The adaptive strategies of yaks to live in the Asian highlands. *Animal Nutrition* 9, 249–258. DOI: 10.1016/j.aninu.2022.02.002

Joshi, S., Shrestha, L., Bisht, N., Wu, N., Ismail, M. *et al.* (2020) Ethnic and cultural diversity amongst yak herding communities in the Asian highlands. *Sustainability* 12, 957. DOI: 10.3390/su12030957

Karki, M., Sharma, S., Mahat, T.J., Tuladhar, A. and Aksha, S. (2012) Sustainable mountain development 1992, 2012 and beyond – Rio +20 assessment report for the Hindu Kush Himalayan Region. International Centre for Integrated Mountain Development, Kathmandu.

Khanal, P., Dhakal, R., Khanal, T., Pandey, D., Devkota, N.R. *et al.* (2022) Sustainable livestock production in Nepal: a focus on animal nutrition strategies. *Agriculture* 12, 679. DOI: 10.3390/agriculture12050679

Kreutzmann, H. (2012) Pastoral practices in transition: animal husbandry in high Asian contexts. In: Kreutzmann, H. (ed.) *Pastoral Practices in High Asia. Agency of 'Development' Effected by Modernisation, Resettlement and Transformation*. Springer, Dordrecht, The Netherlands, pp. 1–29.

Kuriqi, A., Kuriqi, I. and Poci, E. (2016) Simulink programing for dynamic modelling of activated sludge process: aerator and settler tank case. *Fresenius Environmental Bulletin* 25(8), 2891–2899.

Levine, N.E. (2015) Transforming inequality: eastern Tibetan pastoralists from 1955 to the present. *Nomadic Peoples* 19(2), 164–188. DOI: 10.3197/np.2015.190202

Li, A., Gu, M.H., Zhang, S.T., Zhou, X.H., Wang, Y. *et al.* (2009) Effects of species richness on weed invasion in an artificial grassland ecosystem in eastern Tibetan Plateau. *Chinese Journal of Ecology* 28(2), 177–181.

Li, B., Zhang, J.B. and Li, H.P. (2011) Research on spatial-temporal characteristics and affecting factors decomposition of agricultural carbon emission in China. *China Population Resource Environment* 21, 80–86. DOI: 10.4028/www.scientific.net/AMM.291-294.1385

Li, L.H., Lu, P., Gu, X.Y., Mao, X.T., Gao, S.Q. *et al.* (2016a) Principles and paradigms for developing artificial pastures. *Chinese Science Bulletin* 61(2), 193–200.

Li, Y., Degen, A.A., Shang, Z. (2019) Carbon management of the livestock industry in the HKH region. In: Shang, Z., Degen, A.A., Rafiq, M.K. and Squires, V.R. (eds) *Carbon Management for Promoting Local Livelihood in the Hindu Kush Himalayan (HKH) Region*. Springer Nature, Cham, Switzerland, pp. 109–124.

Li, Y.J., Sun, C.C. and Cao, G.M. (2016b) Study on plant biomass and soil nutrients under different land use patterns in Three-River headwater region. *Grassland and Turf* 36(4), 48–53.

Li, Y.J., Wang, Y.Y., Cao, G.M., Long, R. and Yao, T. (2017) Effect of land use patterns on soil nitrogen characteristics in Three-River headwater region. *Agricultural Research in the Arid Areas* 35(3), 272–277.

Li, Y.Y., Dong, S.K., Wen, L., Wang, X.X. and Wu, Y. (2014) Soil carbon and nitrogen pools and their relationship to plant and soil dynamics of degraded and artificially restored grasslands of the Qinghai-Tibetan Plateau. *Geoderma* 213, 178–184. DOI: 10.1016/j.geoderma.2013.08.022

Lin, X., Zhang, Z., Wang, S., Hu, Y., Xu, G. *et al.* (2011) Response of ecosystem respiration to warming and grazing during the growing seasons in the alpine meadow on the Tibetan Plateau. *Agricultural and Forest Meteorology* 151(7), 792–802. DOI: 10.1016/j.agrformet.2011.01.009

Long, R.J., Ding, L.M., Shang, Z.H. and Guo, X.S. (2008) The yak grazing system on the Qinghai-Tibetan Plateau and its status. *The Rangeland Journal* 30(2), 241–246. DOI: 10.1071/RJ08012

Maikhuri, R.K., Rawat, L.S., Phondani, P.C., Maletha, A. and Bahuguna, Y.M. (2018) Livestock: the engine and inspiration of mountain economy. LEISA-India, Bangalore, India. Available at: https://leisaindia.org/ livestock-the-engine-and-inspiration-of-mountain-economy/ (accessed 28 September 2023).

Malik, K., Tokkas, J., Anand, R.C. and Kumari, N. (2015) Pretreated rice straw as an improved fodder for ruminants: an overview. *Journal of Applied and Natural Science* 7, 514–520.

Manandhar, A. (2020) Yak across borders: Bhutan gifts breeding bulls to Nepal and India for gene pool improvement. ICIMOD News. Available at: https://www.icimod.org/yak-across-borders-bhutan-gifts-breeding-bulls-to-nepal-and-india-for-gene-pool-improvement/ (accessed 28 September 2023).

Mao, F., Zhang, Y.H., Hou, Y.Y., Tang, S.H., Zhi, T. *et al.* (2008) Dynamic assessment of grassland degradation in Naqu of northern Tibet. *Chinese Journal of Applied Ecology* 19(2), 278–284.

Miller, D.J. (1990) Grasslands of the Tibetan Plateau. *Rangelands* 12(3), 159–163.

Namgay, K., Millar, J., Black, R. and Samdup, T. (2013) Transhumant agro-pastoralism in Bhutan: exploring contemporary practices and socio-cultural traditions. *Pastoralism* 3(1), 1–26. DOI: 10.1186/2041-7136-3-13

Ogino, A., Sommart, K., Subepang, S., Mitsumori, M., Hayashi, K. *et al.* (2016) Environmental impacts of extensive and intensive beef production systems in Thailand evaluated by life cycle assessment. *Journal of Cleaner Production* 112, 22–31. DOI: 10.1016/j.jclepro.2015.08.110

Opio, C., Gerber, P., Mottet, A., Falcucci, A., Tempio, G. *et al.* (2013) *Greenhouse Gas Emissions from Ruminant Supply Chains: A Global Life Cycle Assessment*. Food and Agriculture Organization of the United Nations, Rome.

Pandey, R., Jha, S.K., Alatalo, J.M., Archie, K.M. and Gupta, A.K. (2017) Sustainable livelihood framework-based indicator for assessing climate change vulnerability and adaptation for Himalayan communities. *Ecological Indicators* 79, 338–346. DOI: 10.1016/j.ecolind.2017.03.047

Rasul, G. and Hussain, A. (2015) Sustainable food security in the mountains of Pakistan: towards a policy framework. *Ecology of Food and Nutrition* 54(6), 625–643. DOI: 10.1080/03670244.2015.1052426

Rhode, D., Madsen, D.B., Brantingham, P.J. and Tsultrim, D. (2007) Yaks, yak dung, and prehistoric human habitation of the Tibetan Plateau. *Developments in Quaternary Science* 9(07), 205–224. DOI: 10.1016/S1571-0866(07)09013-6

Röös, E., Bajželj, B., Smith, P., Patel, M., Little, D. *et al.* (2017) Greedy or needy? Land use and climate impacts of food in 2050 under the different livestock futures. *Global Environmental Change* 47, 1–12. DOI: 10.1016/j.gloenvcha.2017.09.001

Salvador, S., Corazzin, M., Romanzin, A. and Bovolenta, S. (2017) Greenhouse gas balance of mountain dairy farms as affected by grassland carbon sequestration. *Journal of Environmental Management* 196, 644–650. DOI: 10.1016/j.jenvman.2017.03.052

Sandhu, H. and Sandhu, S. (2015) Poverty, development, and Himalayan ecosystems. *AMBIO* 44, 297–307. DOI: 10.1007/s13280-014-0569-9

Sati, V.P. and Singh, R.B. (2010) Prospects of sustainable livestock farming in the Uttarakhand Himalaya, India. *Journal of Livestock Science* 1(1), 9–16.

Shang, Z.H., Ma, Y.S., Long, R.J. and Ding, L.M. (2008) Effect of fencing, artificial seeding and abandonment on vegetation composition and dynamics of 'black soil land' in the headwaters of the Yangtze and the Yellow Rivers of the Qinghai-Tibetan Plateau. *Land Degradation and Development* 19(5), 554–563. DOI: 10.1002/ldr.861

Shang, Z.H., Gibb, M.J. and Long, R.J. (2012) Effect of snow disasters on livestock farming in some rangeland regions of China and mitigation strategies: a review. *The Rangeland Journal* 34, 89–101. DOI: 10.1071/RJ11052

Shang, Z., Gibb, M., Leiber, F., Ismail, M., Ding, L. *et al* (2014) The sustainable development of grassland-livestock systems on the Tibetan plateau: problems, strategies and prospects. *The Rangeland Journal* 36, 267–296.

Shang, Z., Dong, Q., Degen, A.A. and Long, R. (2016) Ecological restoration on Qinghai-Tibetan Plateau: problems, strategies and prospects. In: Squires, V.R. (ed.) *Ecological Restoration: Global Challenges, Social Aspects and Environmental Benefits*. Nova Science Publishers, Hauppauge, New York, pp. 151–176.

Shang, Z.H., Dong, Q.M., Shi, J.J., Zhou, H.K., Dong, S.K. *et al.* (2018) Research progress in recent ten years of ecological restoration for 'black soil land' degraded grassland on Tibetan plateau – concurrently discuss of ecological restoration in Sanjiangyuan region. *Acta Agrestia Sinica* 26(1), 1–21.

Shen, M.H., Zhu, Y.K., Zhao, X., Geng, X.Q., Gao, S.Q. *et al.* (2016) Analysis of current grassland resources in China. *Chinese Science Bulletin* 61(2), 139–154.

Shrestha, A.B. and Aryal, R. (2011) Climate change in Nepal and its impact on Himalayan glaciers. *Regional Environmental Change* 11, S65–S77. DOI: 10.1007/s10113-010-0174-9

Shrestha, A.B., Agrawal, N.K., Alfthan, B., Bajracharya, S.R., Maréchal, J. *et al.* (2015) *The Himalayan Climate and Water Atlas: Impact of Climate Change on Water Resources in Five of Asia's Major River Basins*. GRID-Arendal, CICERO and ICIMOD, Kathmandu.

Shukla, R., Sachdeva, K. and Joshi, P.K. (2016) Inherent vulnerability of agricultural communities in Himalaya: a village-level hotspot analysis in the Uttarakhand state of India. *Applied Geography* 74, 182–198. DOI: 10.1016/j.apgeog.2016.07.013

Tashi, N. and Partap, T. (2000) Livestock based livelihoods in Tibet, China, and sustainability concerns. In: Tulachan, P.M., Saleem, M.A.M., Maki-Hokkonen, J. and Partap, T. (eds) *Contribution of Livestock to Mountain Livelihoods*. ICIMOD, Kathmandu, pp. 171–182.

Tisdall, J.M. and Oades, J.M. (1982) Organic matter and water-stable aggregates in soils. *European Journal of Soil Science* 33(2), 141–163. DOI: 10.1111/j.1365-2389.1982.tb01755.x

Tulachan, P.M. (2000) Livestock production and management strategies in the mixed farming areas of the Hindu-Kush Himalayas. In: Tulachan, P.M., Saleem, M.A.M., Maki-Hokkonen, J. and Partap, T. (eds) *Contribution of Livestock to Mountain Livelihoods*. ICIMOD, Kathmandu, pp. 123–134.

UNDP (2010) The real wealth of nations: pathways to human development. *Human Development Report 2010*. UNDP, New York.

UNDESA (United Nations Department of Economic and Social Affairs) (2017) World population prospects: the 2017 revision, key findings and advance tables. Working paper ESA/P/WP/248.

UN-Water (2013) Water security and the global water agenda: a UN-water analytical brief. United Nations University, Hamilton, Ontario.

Van Vuuren, D.P., Edmonds, J., Kainuma, M., Riahi, K., Thomson, A. *et al.* (2011) The representative concentration pathways: an overview. *Climatic Change* 109(1), 5–31. DOI: 10.1007/s10584-011-0148-z

Wang, H., Zhou, X., Wan, C., Fu, H., Zhang, F. *et al.* (2008) Eco-environmental degradation in the north-eastern margin of the Qinghai-Tibetan Plateau and comprehensive ecological protection planning. *Environmental Geology* 55(5), 1135–1147. DOI: 10.1007/s00254-007-1061-7

Wang, J., Xiong, Z. and Yan, X. (2011) Fertilizer-induced emission factors and background emissions of N_2O from vegetable fields in China. *Atmospheric Environment* 45(38), 6923–6929. DOI: 10.1016/j.atmosenv.2011.09.045

Wang, J., Wang, Y., Li, S.C. and Qin, D. (2016) Climate adaptation, institutional change, and sustainable livelihoods of herder communities in northern Tibet. *Ecology and Society* 21(1), 5. DOI: 10.5751/ES-08170-210105

Wang, W.Y., Wang, Q.J., Wang, C.Y., Shi, H.L., Li, Y. *et al.* (2005a) The effect of land management on carbon and nitrogen status in plants and soils of alpine meadows on the Tibetan plateau. *Land Degradation and Development* 16(5), 405–415. DOI: 10.1002/ldr.661

Wang, W.Y., Wang, Q.J. and Wang, H.C. (2005b) The effect of land management on plant community composition, species diversity, and productivity of alpine Kobersia steppe meadow. *Ecological Research* 21(2), 181–187. DOI: 10.1007/s11284-005-0108-z

Wang, X., Zeng, Z.H., Hu, Y.G. and Zhu, B. (2007) Progress and prospect on mixture of gramineae herbage and leguminosae herbage. *Chinese Journal of Grassland* 29(4), 92–98.

Wang, X., Willms, W.D., Hao, X.Y., Zhao, M. and Han, G. (2010) Cultivation and reseeding effects on soil organic matter in the mixed prairie. *Soil Science Society of America Journal* 74(4), 1348–1355. DOI: 10.2136/sssaj2009.0366

Wang, X., Ding, L.M., Wei, H.Y., Jiang, C.X., Yan, Q. *et al.* (2021a) Astragalus membranaceus root supplementation improves average daily gain, rumen fermentation, serum immunity and antioxidant indices of Tibetan sheep. *Animal* 15, 100061. DOI: 10.1016/j.animal.2020.100061

Wang, X., Hu, C., Ding, L., Tang, Y., Wei, H. *et al.* (2021b) Astragalus membranaceus alters rumen bacteria to enhance fiber digestion, improves antioxidant capacity and immunity indices of small intestinal mucosa, and enhances liver metabolites for energy synthesis in Tibetan sheep. *Animals* 11, 3236. DOI: 10.3390/ani11113236

Wei, H., Ding, L., Wang, X., Yan, Q., Jiang, C. *et al.* (2021) Astragalus root extract improved average daily gain, immunity, antioxidant status and ruminal microbiota of early weaned yak calves. *Journal of the Science of Food and Agriculture* 101, 82–90. DOI: 10.1002/jsfa.10617

Wen, X., Tang, X., Zhang, S., Wang, Y., Song, Z. *et al.* (2019) Effect of jujube residue on apparent nutrient digestibility, blood biochemical indexes and fattening performance of beef cattle. *Heilongjiang Animal Science and Veterinary Medicine* 10, 124–127.

World Bank (2015) Population data in the world. Available at: http://data. worldbank.org/indicator/SP.POP.TOTL (accessed 28 September 2023).

Wu, G.L., Liu, Z.H., Zhang, L., Hu, T.M. and Chen, J.M. (2010) Effects of artificial grassland establishment on soil nutrients and carbon properties in 'black soil land' degraded grassland. *Plant and Soil* 333 (1–2), 469–479. DOI: 10.1007/s11104-010-0363-9

Wu, N., Ismail, M., Joshi, S., Liang, Y.S., Shrestha, R. *et al.* (2014) Livelihood diversification as an adaptation approach to change in the pastoral Hindu-Kush Himalayan region. *Journal of Mountain Science* 11(5), 1342–1355. DOI: 10.1007/s11629-014-3038-9

Wu, N., Yi, S., Joshi, S. and Bisht, N. (eds) (2016) *Yak on the Move: Transboundary Challenges and Opportunities for Yak Raising in a Changing Hindu Kush Himalayan Region*. ICIMOD, Kathmandu.

Xu, C.L. (2005) The study on production potential of nutritive agriculture of oat artificial grassland in Qinghai-Tibetan plateau. *Grassland of China* 27(6), 64–66.

Xu, S., Silveira, M.L., Sollenberger, L.E., Viegas, P.R.A., Lacerda, J. *et al.* (2018) Conversion of native rangelands into cultivated pasturelands in subtropical ecosystems: impacts on aggregate-associated carbon and nitrogen. *Journal of Soil and Water Conservation* 73(2), 156–163. DOI: 10.2489/jswc.73.2.156

Xu, T., Zhou, X., Degen, A.A., Yin, J., Zhang, S. *et al.* (2022) The inclusion of jujube by-products in animal feed: a review. *Sustainability* 14, 7882. DOI: 10.3390/su14137882

Xu, Y., Aung, M., Sun, Z., Zhou, Y., Xue, T. *et al.* (2023) Ensiling of rice straw enhances the nutritive quality, improves average daily gain, reduces in vitro methane production and increases ruminal bacterial diversity in growing Hu lambs. *Animal Feed Science and Technology* 295, 115513. DOI: 10.1016/j.anifeedsci.2022.115513

Xue, B., Wang, L.Z. and Yan, T. (2014) Methane emission inventories for enteric fermentation and manure management of yak, buffalo and dairy and beef cattle in China from 1988 to 2009. *Agriculture Ecosystem and Environment* 195, 202–210. DOI: 10.1016/j.agee.2014.06.002

Yang, H.S., Tai, J.C. and Fan, F. (2011) Effects of sowing methods on the oxidation stability and the chemical bound forms of soil organic carbon in artificial grassland. *Acta Prataculturae Sinica* 20(3), 36–42.

You, Q., Ren, G., Zhang, Y., Ren, Y.Y., Sun, X.B. *et al.* (2017) An overview of studies of observed climate change in the Hindu Kush Himalayan (HKH) region. *Advances in Climate Change Research* 8, 141–147. DOI: 10.1016/j.accre.2017.04.001

Yu, Q., Zhuang, X.S., Wang, W., Qi, W., Wang, Q. *et al.* (2016) Hemicellulose and lignin removal to improve the enzymatic digestibility and ethanol production. *Biomass and Bioenergy* 94, 105–109. DOI: 10.1016/j.biombioe.2016.08.005

Zhang, R., Wang, Y., Ma, L.N., Sang, C., Wang, L. *et al.* (2014) Species diversities of plant communities of degraded artificial grassland, 'black soil land' and natural grassland in the Three-River headwaters region. *Acta Agrestia Sinica* 22(6), 1171–1178. DOI: 10.11733/j.issn.1007-0435.2014.06.004

Zhao, X., Degen, A.A., Liu, S. and Hao, L. (2022) Ruminant lick blocks, particularly in China: a review. *Sustainability* 14, 7620. DOI: 10.3390/su14137620

Zhou, H.K., Zhao, X.Q., Zhao, L., Li, Y.N., Wang, S.P. *et al.* (2008) Restoration capability of alpine meadow ecosystem on Qinghai-Tibetan Plateau. *Chinese Journal of Ecology* 27, 697–704.

Zhu, P., Chen, R.S., Song, Y.X., Liu, G.G., Chen, T. *et al.* (2015) Effects of land cover conversion on soil properties and soil microbial activity in an alpine meadow on the Tibetan Plateau. *Environmental Earth Sciences* 74(5), 4523–4533. DOI: 10.1007/s12665-015-4509-1

Zhuang, M., Gongbuzeren and Li, W. (2017) Greenhouse gas emission of pastoralism is lower than combined extensive/intensive livestock husbandry: a case study on the Qinghai-Tibet plateau of China. *Journal of Cleaner Production* 147, 514–522. DOI: 10.1016/j.jclepro.2017.01.126

23 Ecosystem Services and Compensation in the Hindu Kush Himalayan Region: Strategies and Suggestions

Dawei Zhang[1], Jiayi He[1] and Hui Xu[2,*]
[1]*State Key Laboratory of Herbage Improvement and Grassland Agro-ecosystems, College of Ecology, Lanzhou University, Lanzhou, China*; [2]*School of Economics, Lanzhou University, Lanzhou, China*

Abstract
This chapter discusses the status, opportunities and challenges of ecosystem services in the Hindu Kush Himalayan (HKH) region. The fragile ecosystem of the HKH region faces the impact of multiple factors, including climate change, the gradual loss of ecosystem services, and the difficulty of ecological protection and restoration, which can lead to social vulnerability and threaten human well-being. The HKH region urgently needs to develop ecological compensation strategies based on ecosystem services to promote ecosystem protection and restoration. This chapter puts forward a comprehensive strategic concept for the future development of the HKH region from the dimensions of ecology, socio-culture and economy to promote sustainable development goals.

Keywords: human well-being, ecological compensation, protection and restoration, Sustainable Development Goals

23.1 Ecosystem Services in the Hindu Kush Himalayan Region: The Source of Human Well-being

23.1.1 Overview of natural resources

The Hindu Kush Himalayan (HKH) region is often referred to as the 'Third Pole' and the 'Water Tower of Asia' (Bajracharya *et al.*, 2015). From the Tibetan Plateau in the north to the Ganges basin in the south, it extends over 3500 km from east to west and covers an area of 4.3 million km^2, including eight countries, namely Afghanistan, Bangladesh, Bhutan, China, India, Myanmar, Nepal and Pakistan. It is one of the largest high-altitude regions, the largest high-latitude permafrost region and the largest mountain system on the planet. It is also the origin of ten major river basins in Asia (Sharma *et al.*, 2019).

As the highest ecosystem in the world, the HKH region is home to more than 35,000 plant and 200 animal species, including more than 15,000 endemic species, making it a unique global repository of biodiversity and ecosystem services. According to the International Union

*Corresponding author: xhhui@lzu.edu.cn

DOI: 10.1079/9781800622579.0023

for Conservation of Nature (IUCN), the region hosts all or parts of four global biodiversity hotspots supporting diverse flora and fauna – the Himalaya, the Indo-Burma, the mountains of Central Asia, and the mountains of south-west China (Jaeger, 2021). Its multi-national, contiguous and diverse ecosystems are important sources of ecosystem services in the region. According to the latest data from the International Centre for Integrated Mountain Development (ICIMOD), the lives and livelihoods of approximately 240 million people are directly dependent on the region's ecosystems, and nearly 1.9 billion people living in ten river basins also benefit directly or indirectly from the resources. Moreover, more than 3 billion people rely on the food produced in the region's river basins (Sharma *et al.*, 2019). Therefore, ecosystem services play an essential role in regional and global socio-economic development, the preservation of vital environments, and the enhancement of human well-being throughout the HKH region.

However, human activities and climate change are seriously threatening the sustainable development of the region's biodiversity, hydrologic systems and people's livelihoods. The HKH region faces multiple challenges, including land-use and land-cover changes, climate change, and invasive plant species, resulting in species loss and ecosystem degradation (Sharma *et al.*, 2010). In particular, changes in regional glaciers, snow and permafrost caused by global warming are also 'unprecedented and mostly irreversible', and 55–80% of the total volume of glaciers is predicted to disappear by the end of the century (Wester *et al.*, 2023). Thus, it is particularly important and urgent to conduct scientific assessments to protect ecosystem services in the HKH region.

23.1.2 Ecosystem services and their evaluation

Human well-being enhancement is highly dependent on the health of ecosystems and biodiversity (Díaz *et al.*, 2015). Therefore, ecosystem services have gradually evolved into a distinct hot research topic (Costanza and Kubiszewski, 2012), and have become an important part of the global discourse on environmental management (Chaudhary *et al.*, 2015). With the implementation of the Millennium Ecosystem Assessment (MEA) of the United Nations, the concept and application of ecosystem services have been greatly developed. In 2010, ecosystem services became an important indicator for measuring human well-being and were incorporated into the Convention on Biological Diversity (CBD) as part of the Aichi Targets. Subsequently, the concept gained global recognition through the establishment of the Intergovernmental Science-Policy Platform on Biodiversity and Ecosystem Services (IPBES) and inclusion in the 2030 Sustainable Development Goals (SDGs) of the United Nations (Kandel *et al.*, 2021). In the context of the SDGs, a sound understanding of ecosystem services will help define a path on how to achieve targets under the UN holistic approach (Wood *et al.*, 2018).

What is ecosystem services?

Ecosystem services are the benefits people derive, directly or indirectly, from functioning ecosystems (Guerry *et al.*, 2015). According to the MEA of the United Nations, ecosystem services are generally subdivided into the categories of provisioning, regulating, cultural and supporting services (Elias, 2005), which are also widely recognized and adopted function definitions (Box 23.1).

Ecosystem services in the Hindu Kush Himalayan region

It is estimated that 70–80% of the population in the HKH region lives in rural areas, while

Box 23.1. Specific definitions of four types of ecosystem services.

- Provisioning service: refers to the products obtained from the ecosystem, such as food, wood, medicine, fibre, fresh water and fuel.
- Regulating service: refers to the benefits obtained from the process of ecosystem regulating, such as water conservation, soil conservation, wind prevention and sand fixation, pollution purification, disaster mitigation, carbon sequestration and oxygen release.
- Cultural service: refers to the non-material benefits that people obtain from the ecosystem through spiritual enrichment, perceptual development, entertainment, aesthetics and other ways.
- Supporting services: refers to the services that support the above three types of services.

60–85% of the population depends directly on ecosystem services (Xu *et al.*, 2019). All types of ecosystem services are prominent in the region, and the flows of resources and services provided by the rich ecosystems are critical in underpinning the life-support systems and human well-being across the HKH region (Jaeger, 2021). The main ecosystem services in the HKH region are presented in Fig. 23.1.

PROVISIONING SERVICES. The HKH region is highly dependent on ecosystems, including grasslands, wetlands and forests, for subsistence livelihoods and local economies (Jaeger, 2021). Direct services and benefits of diverse ecosystems are important sources of food, fuelwood, pasture, fodder, medicinal plants and non-timber forest products (NTFPs). Furthermore, the HKH region is rich in freshwater resources, encompassing the largest area of permanent ice cover in the world outside the polar region, which increases groundwater recharge through watershed functions such as snow, glaciers, permafrost, wetlands and rivers (Xu et al., 2019). It not only meets the freshwater demands of regional agriculture, industry and households, but also functions as an important economic source for tourism and hydropower (Mishra *et al.*, 2020).

REGULATING SERVICES. The HKH mountain ecosystems are unique in terms of their role in regulating climate, nutrients, water storage and disaster-risk reduction. Especially in south Asia, the regulating services are essential to agriculture through water supply, climate and wind regulation, groundwater recharge and maintenance of wetland ecosystems (Aryal *et al.*, 2019). In recent years, the contribution of forest, rangeland and wetland ecosystems to carbon sequestration (Yin *et al.*, 2023) and soil conservation (Zhao *et al.*, 2023) has been widely acknowledged.

CULTURAL SERVICES. Cultural ecosystem services are consistently recognized, but not yet adequately defined or integrated within the ecosystem services framework (Daniel *et al.*, 2012). They are often considered subjective, intangible and difficult to be quantified in monetary terms, and, thus, are often neglected or completely excluded from evaluation (Chan *et al.*, 2012). Multicultural ecosystem services of the HKH region, characterized by diverse ethnic and religious

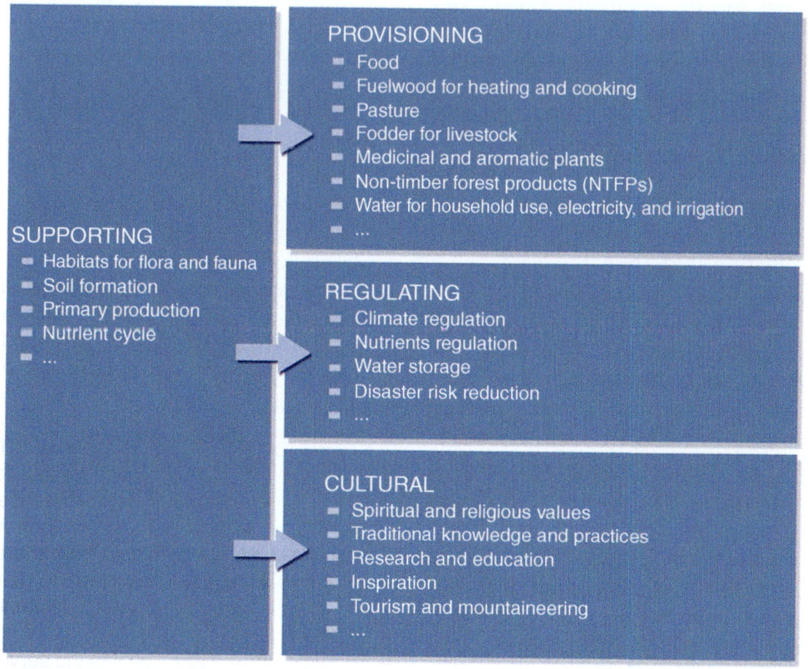

Fig. 23.1. Major services provided by ecosystems in the HKH region. (From ICIMOD, 2023a)

cultures, impact the biophysical landscapes accessible to the ecosystem, such as sacred valleys, trees, groves, rivers, lakes and peaks (Jaeger, 2021). They are part of the co-evolution of society and ecosystems, which are inextricably linked to rural livelihoods, land use, human health and climate change (Xu *et al.*, 2008). The rich natural resources and multi-ethnic cultural characteristics of the region also have created a good foundation for the development of ecotourism (Wei, 2023). In recent years, tourism and entertainment based on ecosystem services have grown in importance as a source of income for the HKH region.

SUPPORTING SERVICES. The structure, composition, and functions of ecosystems provide essential habitats for a range of flora, fauna and micro-organisms (O'Connor *et al.*, 2020). There are numerous organisms which forage, reproduce and develop in cryosphere habitats and form unique food chains (Su *et al.*, 2019). All these supporting services are important for the sustainability of biodiversity.

There is no doubt that the diversified service flows provided by the ecosystems in the HKH region not only have obvious impacts on food security, energy security, economic growth and mountain livelihoods in the region, but also are indispensable to the overall human well-being of the HKH downstream region and in the world.

Ecosystem services evaluation

Nearly all decision-making processes related to ecosystem services should consider ecosystem services evaluation, although the assessment methodology is still controversial (Xie *et al.*, 2017). It is urgent to quantify the value of ecosystem services through optimization methods and to understand the difference between ecosystem protection and utilization, so that policy-makers and the general public can recognize the enormous benefits provided by a healthy ecosystem and the huge losses brought on by a damaged ecosystem (Zhang *et al.*, 2022a).

In recent years, the importance of ecosystem services in terms of economic value has been rationalized as a result of the development of ecosystem services research. However, the amount of research on the economic value is much less than in other areas due to the methodology's constraints and the geographic complexity prevalent in the HKH region (Xu *et al.*, 2019). However, there are indications that the economic value of both marketed and non-marketed goods is substantial for rural communities which depend largely on ecosystem services (Franco and Luiselli, 2015). In the case of the assessments on Bhutan, the total estimated value of ecosystem services is approximately US$ 15.5 billion per year, substantially greater than the gross domestic product of US$ 3.5 billion per year. While GDP is the final goods and services provided by the economic system, the large gap between the estimated numbers indicates that the final goods and services provided by the ecosystem are far greater than what is currently provided by the economic production (Kubiszewski *et al.*, 2013).

The evaluation of ecosystem services is still very important and has a wide range of possible applications on various spatial and temporal scales. The value of ecosystem services should be measured in monetary terms, since people are accustomed to value measurements expressed in such units. Doing so will help raise social awareness and the general public's attention, which would benefit relevant policies, land-use planning, and the formulation of ecological compensation plans (Jiang *et al.*, 2021).

Trade-offs and synergies of ecosystem services

Ecosystem services and human activities influence and restrict each other. Under a specific spatial and temporal scale, it is difficult for different ecosystem services to maximize their benefits concomitantly, which means that there is a competitive relationship of varying degrees among the services (Hein *et al.*, 2006). In addition, synergies of mutual promotion or inhibition may also be formed (Bennett *et al.*, 2009). Biological interdependence and economic interdependence are two unique modes of action that produce complicated interaction linkages between various ecosystem services (Abler, 2004). Table 23.1 lists some cases of trade-offs and synergies between ecosystem services.

This is particularly prevalent in the highly fragile HKH region. Studies have shown that there are complicated interactions and dynamic linkages among mountain ecosystem services

Table 23.1. A few cases of trade-offs and synergies between ecological services.

	Ecosystem services A	Ecosystem services B	Trade-off or synergy
Among provisioning services	Agricultural irrigation	Fishery culture	Trade-offs
Between provisioning services and regulating services	Water quantity	Carbon sequestration	Trade-offs
Between provisioning services and cultural services	Agricultural production	Cultural tourism	None
Between provisioning services and supporting services	Agricultural production	Sediment deposition	Trade-offs and synergies
Among regulating services	Soil and water conservation	Water conservation	Synergies
Between regulating services and cultural services	Water quality regulation	Landscape entertainment	Synergies

(Adrienne *et al.*, 2008). Therefore, when considering the production of one or more ecosystem services, it is important to also consider the potential, even large-scale, impacts on others. Taking six ecosystem services, including crop production, timber production, carbon sequestration, soil conservation, water yield, and habitat quality (food, fibre, climate regulation, water regulation, erosion control, and biodiversity conservation) as examples, the trade-offs and synergies were analyzed. The results showed that not only did the distribution of ecosystem services display different proportions in the HKH region, but also the nature of the inter-relationships among these ecosystem services changed on different spatial scales. The southern lowlands were rich but the northern uplands were poor in crop and timber production. All ecosystem services were correlated positively to each other, except crop production, which was correlated negatively with soil conservation. Crop production and timber production were mutually exclusive at the pixel level, but were correlated positively at the watershed level (Aryal *et al.*, 2023).

From the perspective of ecosystem services and human well-being, synergy (win-win) is the ultimate goal of human activities to maximize the benefits of ecosystem services (Howe *et al.*, 2014). Investigation into the dynamics of the trade-offs and synergistic relationships in regional ecosystem services can shed light on interactions and feedbacks between ecosystem services and the stakeholders at different spatial and temporal scales, as well as avoid double-counting issues in ecosystem services assessment. In addition, information and progress on trade-offs and synergies in the delivery of ecosystem services can provide a scientific basis for designing and implementing ecological compensation schemes and optimize the management of ecosystem services, and, ultimately, improve human well-being associated with trade-offs and synergies among different ecosystem services (Li *et al.*, 2012). Therefore, sustainable ecosystem management in the HKH region should focus on minimizing trade-offs and maximizing synergies among ecosystem services holistically (Aryal *et al.*, 2023). It is necessary to weigh the spatial distribution and the temporal dynamic evolution of ecosystem services, as well as the structure changes of the interactions in different ecosystem services, so as to make solid scientific decisions to support the sustainable development of ecosystem services in the HKH region.

23.1.3 Research status of ecosystem services

So far, researches about ecosystem services very limited focus on the Asian region (Zhang *et al.*, 2019b). Summarizing research of ecosystem services in the HKH region can provide knowledge on what is known, identify research priorities, and provide a strong theoretical foundation for integrating ecosystem services into policies and practices in the region (Kandel *et al.*, 2021).

Spatial and temporal distributions

In terms of a time-series viewpoint, the study of ecosystem services in the HKH region has progressed rapidly in the last 20 years, which is commensurate with the global trend of growing research on ecosystems. Most studies on ecosystem services in the HKH region have been conducted in China, followed by Nepal and India, with almost no relevant study from Myanmar and Afghanistan. In addition, there are obvious differences between the results of studies from different countries. For instance, most studies from India concluded that the Indian Himalayas' ecosystem services have been degraded due to population growth and rapid development activities such as urban expansion, over-exploitation of natural resources and so on. But active human intervention can also restore ecosystem services. Studies from China reported an increase in the flow of ecosystem services after the implementation of several major ecological compensation programmes, such as the Natural Forest Conservation Programme, the Tilted Land Conversion Programme, and the Food for Green Programme (Song *et al.*, 2014). The number of publications in the HKH region for the eight countries during 2002–2020 is presented in Fig. 23.2.

Theme distribution

Most studies have assessed the benefits for humans derived directly from natural ecosystems in terms of fresh water, fuelwood, timber, NTFPs and hydropower. Among them, both NTFPs and water conservation displayed a consistent decreasing trend, as well as the biodiversity values and overall ecosystem services. However, cultural services and development of tourism displayed increasing trends in the HKH region. Of all research topics, water conservation is the most studied ecosystem service (accounting for about 15.9%), but few publications have focused on the governance and management of services in the region, while

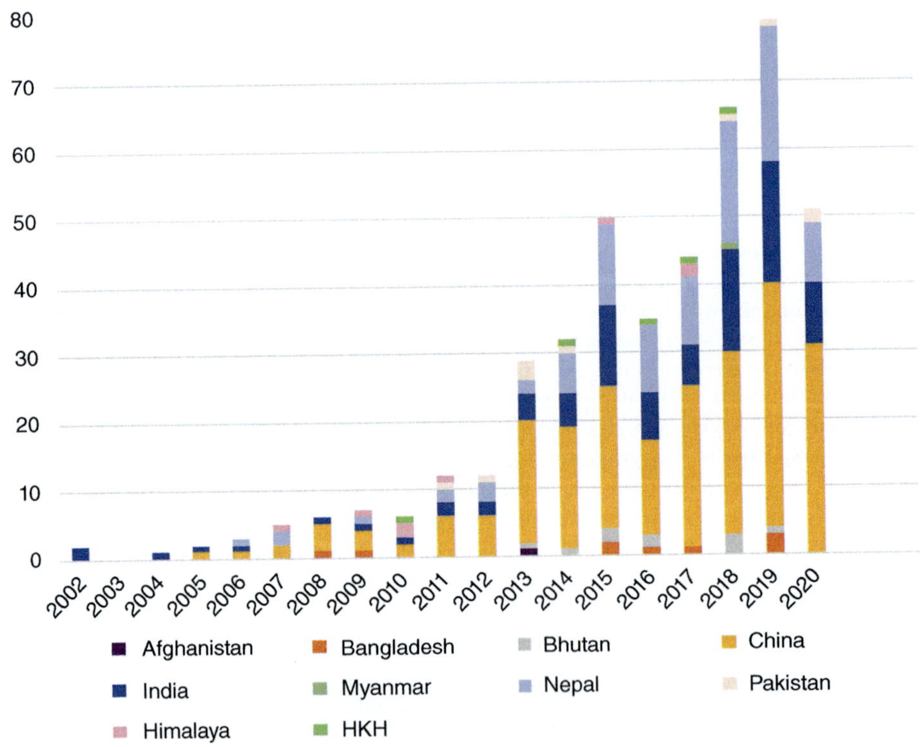

Fig. 23.2. Number of relevant publications in the eight Hindu Kush Himalayan countries from 2002 to 2020. (From Kandel *et al.*, 2021)

the other relatively understudied subject is the role of Himalayan ecosystems in health security. The research topics of ecosystem services in the HKH region are presented in Fig. 23.3 (Xu *et al.*, 2019).

Shortcomings

A holistic approach that can solve problems is based on natural resources endowments and human well-being enhancement. While countries in the HKH region have conducted extensive research on issues such as hydropower, water and carbon storage, there has been little qualitative or quantitative evaluation of ecosystem services, and most studies have focused on a limited topic without a holistic view that would help to better understand the upstream-downstream linkages, resulting in their relevance for macro-policy decisions.

The lack of data in the HKH region has severely limited the understanding of regional socio-ecological processes. Relevant organizations should work together to fill knowledge gaps through a more centralized and co-ordinated collection of data, particularly in the alpine regions. To improve scientific understanding of ecosystem structure and function, as well as the drivers of change, regional data of human well-being and the connections to the natural resources should be improved continuously, serving as a basis for developing integrated ecosystem management approaches and strategies linked to human well-being and poverty reduction (ICIMOD, 2023).

23.2 Ecological Compensation Based on the Value of Ecosystem Services

Humans must preserve the ecosystem, but ecosystem services should be developed rationally and utilized to produce high-quality ecological products and provide services for people and development (Guerry *et al.*, 2015). However, with climate change, growing population and economic pressures in the region, damage to the ecosystem in the HKH region has increased and intensified, which, ultimately, affected livelihoods and the socio-economy in the local communities (Mishra *et al.*, 2017). Therefore, it is urgent to thoroughly study and evaluate the ecosystem services provided by HKH, and to develop ecological compensation that is appropriate for the area.

23.2.1 Ecosystem service value and ecological compensation

Although there is not a fully agreed-upon definition of ecological compensation, the understanding among groups is basically the same. In China, ecological compensation is defined as a policy instrument aiming to maintain or improve ecological conditions through various incentives (Yu *et al.*, 2020), with the aim of preserving and using ecosystem services sustainably. Stakeholder relationships are adjusted based on financial resources through various cash or non-cash ways, referring to the cost of ecological protection, missed opportunities for development,

Fig. 23.3. Topics of ecosystem services in the Hindu Kush Himalayan region. (Data from Xu *et al.*, 2019)

and the value of ecosystem services as the basis (Fei *et al.*, 2021).

Payment for ecosystem services (PES) is a 'conservation scheme that uses economic means to encourage service providers to ensure the sustainable supply of ecosystem services' (Wunder, 2015). It is a progressively widely used policy tool since 2005 (Yu *et al.*, 2020), and the concept of PES is basically similar to that of ecological compensation (Pan *et al.*, 2017). Since 2007, ecological compensation is always expressed as a PES-like policy (Wang *et al.*, 2016). It is regarded as an important political tool, which can help people understand the value and scarcity of resources, integrate environmental concerns in enacting economic policies (Zhang *et al.*, 2023), while also contributing to the local livelihoods of resource-dependent communities, especially in the HKH region (Mishra *et al.*, 2017).

Relationship between ecosystem services and ecological compensation

Ecological compensation is an economic means to promote ecological environmental protection (Yu *et al.*, 2020). Ecosystem service assessment is the value basis for determining the amount of ecological compensation and putting ecological compensation into practice. It conducts health diagnosis and evaluation of the current situation of the ecosystem, so that the value of ecosystem services can enter the overall accounting system in the form of specific currency (Pagiola, 2008). Ecological compensation can be regarded as the specific practice of ecosystem service value assessment (Guo *et al.*, 2016). To promote the preservation and restoration of ecosystem functions in ecologically damaged areas, it is important to implement ecological compensation based on the value of ecosystem services.

Environmental taxes should be levied on the beneficiaries of the downstream river basin, and the green bonus of ecosystem services should be paid to the native people of the upstream regions, so that people can enjoy rights equally. For example, ecotourism services, which are playing an increasingly important role, can be used by the local people by providing logistic services. Besides, with abundant hydropower resources, employment in the infrastructure development projects, including hydropower projects, should be offered to the local people. A suitable model for the sustainable distribution of drinking water could be developed through people's participation. These could all enhance the livelihoods in the rural mountain areas of the HKH region and help in reducing outmigration (Sati, 2023).

Application of ecosystem service value accounting in the field of ecological compensation

The HKH region is not only the highest mountain region in the world, but also one of the most ecologically vulnerable, densely populated and impoverished regions (Gerlitz *et al.*, 2012). In recent years, countries have gradually begun to implement some ecological compensation projects in the HKH region. Although overall funding for the preservation and restoration of the HKH ecosystem is rising, it is still constrained and biased in favour of particular nations and ecosystems due to a lack of knowledge and research (Kandel *et al.*, 2021). In particular, PES has grown to be a crucial tool for balancing the development of various HKH regions and the interplay between nature and people.

ICIMOD mode

To achieve the desired objectives, the ecological compensation mechanism must make the necessary institutional arrangements among all parties involved. The research and practice of the International Centre for Integrated Mountain Development (ICIMOD) concluded that shifting from solely market-based mechanisms to incentive-based mechanisms can ensure that mountain communities reap the benefits of their efforts in ecosystem conservation and management. If the incentive mechanism is properly embedded in a policy, it can effectively assure the long-term supply of ecosystem services in the HKH region (Xu *et al.*, 2019). To this end, ICIMOD is collaborating actively with member countries to continuously explore ways to effectively create incentives for ecosystem services, including studying the economic value of ecosystem services, bringing stakeholders together to reach effective agreements and institutional arrangements, and providing policy support based on current situations.

ICIMOD evaluated a productive attempt at a tripartite institutional arrangement and found that it is effective for implementing incentive mechanisms. Where active participation, mediation and facilitation from local governments are required for the successful implementation of the mechanism, and consumers of an ecosystem service, such as water resources, agree to provide incentives to upstream communities (or producers); then these actions would not only encourage upstream communities to protect upstream ecosystems but would also provide support for alternative livelihood options (Mishra *et al.*, 2017).

ERCCP model – an example of payment for ecosystems (PES) project in Tianzhu County, China

China has been implementing a series of ecological compensation policies. For example, China launched the Natural Forest Conservation Programme (NFCP) and the Sloping Land Conversion Programme (SLCP), which are two of the largest payments for ecosystem services (Chen *et al.*, 2015), and also provide a good solution to the difficult problem of promoting economic growth and environmental protection. The total cumulative investment through NFCP and SLCP exceeds 50 billion US dollars, with SLCP alone benefitting over 120 million farmers in 32 million households (Ouyang *et al.*, 2016).

Taking the PES, Ecological Relocation and Capital Compensation Programme (ERCCP) in Tianzhu county, Qinghai-Tibetan Plateau, as an example, the extensive production and development mode of agriculture and animal husbandry in Tianzhu county has had a significant impact on the ecosystem. Consequently, from 2010 to 2020, the ERCCP, as a payment mechanism for ecosystem services, invested a total of US$ 330 million in the ecological relocation and ecosystem restoration of livestock herders at an altitude of ~2800 m. Incentives such as direct financial assistance, urban housing subsidies and subsidies for restricting grazing were provided, and herdsmen were encouraged to migrate to urban areas in search of better social public services. Concurrently, ecological restoration initiatives of degraded rangelands and forests are implemented to restore and enhance ecosystem services. Implementation of the ERCCP is not only essential in achieving rural revitalization and enhancing people's well-being, but also is an efficient means of preventing environmental degradation.

ERCCP not only resolves the conflict between local ecological protection and human well-being, but also builds a generalized model that illustrates the win-win path to promote ecosystem services and the co-ordinated development of regional health (Fig. 23.4). Particularly, by ensuring the interests of the participants, the project has changed the perception and attitude of

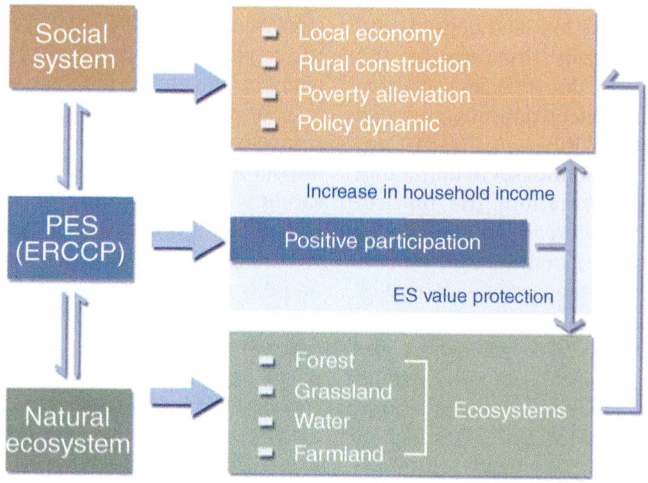

Fig. 23.4. Ecological Relocation and Capital Compensation Programme (ERCCP) model. (Adapted from Yang *et al.*, 2022)

farmers and herders towards the project, from passive participation to acceptance and then to active participation. The active participation of farmers and herders could improve the efficacy of ecological restoration strategies, thereby enhancing the value of regional ecosystem services, resulting in the formation of a stable ecosystem. The ERCCP model plays a crucial role in co-ordinating the conflict between ecological protection and people's livelihoods, which not only results in a win-win situation but could also be applied to other regions of the globe (Yang *et al.*, 2022).

23.2.2 Influence of internal and external compensation mechanisms

Ecological compensation protects the ecological environment on the basis of ensuring economic development (Kangas and Ollikainen, 2019). In the practice of ecological compensation, at present, most PES projects are small-scale and mainly economic means, and play an important role in ecological environment protection and regional sustainable development. However, at the regional or national level, a systematic, comprehensive and universal ecological environment system needs to be established in the HKH region.

Ecological benefits are the most direct impact of implementing ecological compensation in the HKH region. The overall trend of land-use and land-cover change (LULCC) in the HKH region, including 'slow down the loss and degradation of rangelands and wetlands', and reversing the reduction of forest cover due to excessive deforestation, are being implemented through a range of compensatory means (Rai *et al.*, 2018). Concomitantly, along with technological upgrading and industrial transformation, the natural resources in the HKH region are promoted to be more rationally developed and utilized, so as to reduce the loss of ecosystem services caused by over-exploitation (Peh *et al.*, 2016).

From the perspective of social welfare, the implementation of ecological compensation not only improves livelihoods, encourages household income growth, reduces the number of poor, and attracts people to migrate to access better social public services, but it also modifies the traditional perception of local residents and motivates them to engage in ecological compensation projects with a more positive mindset (Meli *et al.*, 2023). In the long run, environmental awareness and the development of green industries could be truly sustainable and need to be widely applied to large-scale ecosystem conservation (Benham, 2017). The ecological benefits from the HKH region through ecological compensation will have a huge positive impact on the sustainable development of many downstream countries and the well-being growth of hundreds of millions of people.

23.2.3 Performance evaluation of ecological compensation

Whether ecological compensation can achieve the anticipated policy goals to achieve a win-win situation between economic and social development and environmental protection is a main topic of discussion (Ren *et al.*, 2022). The ecological impact, economic effectiveness and social welfare of ecological compensation implementation can be inferred from performance evaluation. The core of ecological compensation benefit assessment is the accounting of ecological benefit, which assesses the change trend of ecological assets, ecological products and ecological services in the compensated area after the implementation of ecological compensation. At present, there is a variety of qualitative and quantitative methods for ecological compensation. The qualitative methods employ mainly comparative analysis and questionnaire surveys to compare and analyze the ecological compensation policies before and after implementation (Zhang *et al.*, 2010).

To optimize and enhance the ecological compensation system, accurate evaluation of the impact and effectiveness of ecological compensation is required (Tang *et al.*, 2023). The policy evaluation of ecological compensation efficiency should be based on the 'cost-benefit' principle, and it is necessary to dynamically monitor how ecosystem services are changing in the compensated areas and consider the positive social impacts. Research should concentrate on ways to increase the supply capacity of ecological goods, lower the transaction costs associated with ecological compensation, improve the efficiency of ecological compensation, and boost participant enthusiasm in order to increase the effectiveness of ecological compensation (Fei *et al.*, 2021).

The effectiveness of ecological compensation evaluation is helpful to summarize the experience and lessons learned from pilot projects, and provides theoretical reference and practical experience for the transformation and upgrading of ecological compensation schemes (Tang *et al.*, 2023). However, on the whole, the HKH region still lacks a long-term guarantee for compensation funds; there is no long-term systematic comprehensive data-support system for ecological compensation, and there is no investigation, monitoring or evaluation system for ecological compensation, which results in an inadequate scientific foundation for ecological compensation and irrational compensation standards. In addition, there is a large gap in the development of ecological compensation practices in various countries and regions in the HKH region, and there are insufficient impact analyses on the effectiveness and impact of ecological compensation. Moreover, the evaluation techniques and indicators used are very dissimilar, and the performance effectiveness of ecological compensation evaluation is poor. There are still many challenges in this area, and a reasonable performance assessment system needs to be developed in the future (ICIMOD, 2023b).

23.3 Ecological Compensation and Ecosystem Protection and Restoration

23.3.1 Ecosystem protection and restoration

On 1 March 2019, the UN General Assembly declared 2021–2030 the UN Decade on Ecosystem Restoration. It called for actions to accelerate the global restoration of degraded ecosystems (Waltham *et al.*, 2020), to protect ecosystems and provide society with a more sustainable future. Ecological restoration improves ecological health through the repair of damaged ecosystems, and the reconnection of society with nature (Breed *et al.*, 2021). The protection of the fragile ecological environment and the management of rich ecological resources of the HKH region require complicated policies, institutions, practices and implementation procedures. There are many examples of good practices in the HKH region for protecting and restoring degraded habitats that are closely related to community development, such as protected areas (PAs), and these need to be increased and scaled up (Box 23.2) (Ojha *et al.*, 2019). Overall, the current HKH region is unable to cope with future ecological risks.

In general, to implement ecosystem conservation and restoration in the HKH region, the following points need attention:

Close-to-nature restoration

Natural restoration, artificial intervention, and ecological engineering are examples of traditional restoration methods for degraded ecosystems (Ren and Peng, 2019). However, in recent years, a ‘close-to-nature restoration’ concept has gained popularity and has been put into practice. This concept is based on ecological theory and relies on natural ecological processes. Through scientific and effective artificial assistance and management measures, it can transform degraded ecosystems into ones with species composition, diversity and community structure close to

Box 23.2. Protected areas.

For centuries, protected areas (PAs) have played a fundamental role in the conservation of biodiversity and ecosystems (Palomo *et al.*, 2014). In 2022, in the Kunming-Montreal Global Biodiversity Framework (GBF), adopted during the fifteenth meeting of the Conference of the Parties (COP 15), it was clearly stated that by 2030, at least 30% of biodiversity and areas with important ecosystem functions and services should be effectively protected and managed through ecologically representative, well-connected and equitably governed systems of PAs and other effective area-based conservation measures. As signatories to the Convention on Biological Diversity (CBD), all eight HKH member countries are committed to conservation and, as a measure towards the immediate protection of globally significant landscapes, have set aside a considerable proportion of their most biologically rich terrain as PAs (Chettri *et al.*, 2008). The HKH has made significant progress in the establishment of PAs in recent decades, but the actual implementation of conservation measures within PAs still varies greatly across the region (Xu *et al.*, 2019), and the average coverage in the HKH region is lower than the global average (UNDP, 2021).

those of zonal communities, which will improve diversity and stability and achieve sustainability of ecosystem structure and function (He *et al.*, 2020).

By way of illustration, the Qinghai-Tibetan Plateau's alpine rangeland ecosystem has experienced serious degradation of its vegetation and soil, a decline in biodiversity and productivity, and a threat to the local ecology and economy (Zhang *et al.*, 2019a). Currently, restoration measures include mainly artificial reseeding, enclosure, rotational grazing, fertilization, and sod cutting (Jiang *et al.*, 2020). Although these restoration efforts have, in the short term, reduced rangeland degradation and increased productivity, there are still issues with single ecosystem functions and services, poor stability after restoration, and a lack of focus on the special characteristics of alpine grassland. It is challenging to achieve sustainable recovery (Qian *et al.*, 2019). Relevant studies on alpine rangeland show that close-to-nature restoration should consider not only the effects of biodiversity, but also how native plant species respond to changes in their environment and the make-up of their communities in order to achieve ecosystem stability after restoration. It should also strive to increase species diversity of the community through beneficial interactions among species. As a result, communities may be more productive and interference-resistant (He *et al.*, 2020).

As a novel concept in ecological restoration, the close-to-nature restoration can better maintain the ecosystem's multiple functions and services since it can restore the biodiversity and community composition more effectively, and promote greater productivity. In addition, it can enhance the ability to sequester carbon, conserve water resources, maintain landscape diversity, and regulate the circulation of elements within the ecosystem, so as to achieve the co-ordinated development of production, ecology and economy.

The systematic social system

Human and natural systems are integrated in which people interact with natural components. Thus, each ecological component cannot be isolated when protecting and restoring an ecosystem, and must be considered with a holistic perspective (Liu *et al.*, 2007). In a word, the implementation of ecosystem restoration and protection should be in accordance with the inherent laws of ecosystem, and management planning should consider spatial and temporal patterns.

In specific ecological practices, the needs of various stakeholders in PES projects should be met to achieve sustainable management. In addition to residents, government authorities at all levels, as important stakeholders of ecological and environmental policies (Zhang *et al.*, 2021), should improve the effectiveness of policy management and co-ordinate the activities of various stakeholders through multi-level systematic management and cross-regional co-operation (Palomo *et al.*, 2014).

Incentive programmes for all parties involved

As the main players in the operation of environmental economic practices (i.e. PES or ecological compensation), local residents' attitudes affect PES and ecological compensation performance (Ren *et al.*, 2020). Box 23.3 uses Nepal as an example to introduce the different roles of collective and individual willingness to pay (WTP) in PES projects. Active participation promotes the effectiveness of ecological restoration strategies, which is crucial to the health and sustainability of natural ecosystems. Therefore, the WTP for maintaining and improving ecological services should be fully respected in related projects.

Box 23.3. Collective and individual willingness to pay in the Hindu Kush Himalayan region.

Measuring the value of particular ecological services in terms of consumer willingness to pay (WTP) can aid in negotiations and agreement formation. However, collective WTP may differ from individual WTP, which determines incentives for proximal communities. In Shardukhola watershed in Nepal, the total WTP for water services is US$ 118,000 per annum to support watershed management activities, while individual household WTP is US$ 3.48 per month, which is 18% more than the present water tariff. In this case, water consumers agreed to provide incentives to upstream communities under an institutional framework and agreement (Bhatta *et al.*, 2014). This greatly promoted the steady advancement of the PES.

The adoption of effective incentive measures and programmes can maximize the enthusiasm, accessibility and social participation, which then ensures the sustainability of the project (Ureta *et al.*, 2022).

In conclusion, to promote the protection and restoration of ecosystems in the HKH region and effectively maintain the ecological balance and sustainable development, a realistic and feasible approach is to integrate the overall ecological protection resources of the region, promote the parallel and diversified ecological compensation mechanisms, and form an ecological compensation system led by governments of countries and participation by social entities. Moreover, it is also necessary to adopt methods based on close-to-nature restoration, build a social structure with systematic planning, and design incentive programmes that involve all parties, to form a composite ecosystem protection and restoration programme that integrates 'ecology-socio-culture-economy'.

23.3.2 Ecological compensation financing model and its community role in ecological protection and restoration

Ecological compensation, as a tool for environmental policy, is a key strategy for achieving ecosystem protection and restoration. The practice of ecological compensation can further offer ideas for exploring and improving governance concepts, models, and paths for ecological protection and restoration (Zou *et al.*, 2018), which is crucial for the preservation and recovery of ecosystems in the HKH region.

Role and mode selection of ecological compensation

Ecological compensation is generally divided into government- and market-based ecological compensation. Co-operation between government and market ecological compensation, a hybrid type, was derived from these two. Ecological compensation calls for a sizeable capital investment and is a key method of financing ecological protection and restoration in the HKH region (Jin *et al.*, 2019). The comparison of the three ecological compensation methods is presented in Table 23.2.

The government-based ecological compensation is the primary source of funding for ecological protection and restoration, especially when market players cannot generate profits from ecological protection and restoration, in which case they rely primarily on government-funded ecological compensation to achieve ecological protection and restoration. For instance, the Grain for Green Programme (GGP) in China played an important role in the conservation of rangeland ecology, vegetation cover and reduction of related disasters. The market-based ecological compensation is applicable mostly to

Table 23.2. Comparison of three types of ecological compensation.

Types of ecological compensation	Function	Disadvantages	Advantages
Government-based	The government is the funder and provides the basic funding	Single source of funding, low efficiency, high regulatory costs	Promote regional co-ordinated development, social equity, and improve regional governance capacity
Market-based	The suppliers and consumers of ecological resources conduct direct transactions through market-oriented or quasi-market-oriented channels	The market mechanism is not sound, the protection and restoration areas are difficult to cover, and the transaction costs are large	Alleviate the pressure of fiscal expenditure, improve the effective use of resources, and broaden the investment path of the capital market
Co-operation between government and market	The mixed mode of the above two basic forms	It needs government guidance and supervision	Incentive the role of the government and market, and stimulate the public's enthusiasm for environmental protection

ecological protection and restoration projects that generate economic benefits, including land-preferential policy attraction, trading of water rights, forestry carbon sinks, green labelling and green finance. It plays an important role in reducing the pressure on government financial expenditure, expanding the sources of funds for ecological restoration, and increasing investment opportunities in the capital market. It is a crucial development direction in ecological protection and restoration financing. The co-operation of government and market ecological compensation includes public-private partnership (PPP) and government service procurement in the field of ecological protection and restoration. The government can take advantage of the market's role in resource allocation and fully utilize its coordination role by attracting social capital and creating mutually beneficial and win-win business models. By doing so, it can play to each other's strengths, co-operate with each other, and commit jointly to ecological preservation and restoration. According to the specific conservation and restoration projects and regional natural, social, economic and cultural traits, the HKH region can choose the appropriate models for ecological compensation and ecological protection and restoration.

The role of communities

The community, in addition to the roles of the government and social capital, play an extremely important role in many PES projects in the HKH region. In recent years, the objectives of PES have gradually expanded from ensuring the sustainable supply of ecosystem services to alleviating poverty (Wunder *et al.*, 2008) and contributing to the well-being of service providers by providing financial payments (Wunder *et al.*, 2018). As the natural commodification of purely market-based PES is not necessarily beneficial for poor groups, incentives for ecosystem services are one of the solutions that can effectively engage communities in managing their resources to ensure that HKH communities benefit from their efforts in conservation and ecosystem management (Mishra *et al.*, 2017).

Taking PES practices and programmes in Nepal as an example, the current PES projects in Nepal include mainly the government–community, community–local government/municipality, and community–community (Table 23.3) (Bhatta *et al.*, 2014). The participation of the community can strengthen the knowledge, innovation and practices of local groups to ensure equitable access to and sustainable management of natural resources. The design of corresponding payment systems for ecosystem services by community members can be an incentive that is more suited to social interactions and be friendly to the environment.

23.4 Driving Force of Changes and Coping Strategies

23.4.1 Driving force of change

Broad distinctions have been made among environmental, socio-cultural and economic drivers, while highlighting the complex interrelationships among these dimensions (Wang *et al.*, 2019). In the HKH region, the rapid changes in these drivers have impacts on the environmental conditions and sustainability, as they are interconnected, inextricably linked and increasingly influenced by regional and global developments.

Ecological variables, biodiversity and livelihoods are influenced by such driving forces as LULCC, over-exploitation of natural resources, environmental pollution, invasive alien plant species, mountain hazards and climate change (Rai *et al.*, 2018). However, the impact of indirect socio-cultural and economic driving factors on the HKH region, such as demographic change, socio-cultural change, governance system and institutions, technological implementation, economic growth and differentiation, infrastructure development, and urban expansion, still requires further research (Gerlitz *et al.*, 2012). Among these drivers, demographic change, governance systems and institutions, and climate change are likely to have the most devastating or challenging impact on sustainable development for the HKH region (Wang *et al.*, 2019). Drivers of sustainability in the HKH region are presented in Fig. 23.5.

Within the overlapping influence of the drivers, their evaluation and management are complex, as changes in one driver have interactive or feedback impacts on others, with no clear simple causal relationship among them. The origins and impacts of the major drivers of change

Table 23.3. Comparison of payment for ecoservices in Nepal.

PES-type	Example	Participation	Effect
Government–community	Kulekani hydropower	Makwanpur DDC invests part of the revenues from the Kulekhani hydropower facility in upstream watershed conservation as payment for community efforts in conservation	Maintain water flow to the reservoir
Community–local government/ municipality	Dhulikhel drinking-water supply	Dhulikhel municipality and the village development committee (VDC) agreed on payments both in cash and as in-kind compensations to reward upstream communities for their contributions to the conservation of the water source	Ensure sustainable supply of upstream water
Government–community	Incentive buffer-zone communities	Government provides 30–50% of total revenue generated by a PA to the buffer-zone communities	Conserve biodiversity and landscape integrity in and surrounding protected areas (PAs)
Community–community	Haldehar irrigation scheme	Upstream and downstream communities signed a formal contract to provide irrigation water, and downstream users pay NRS 30 (US$ 0.30) per hour for water use to the upstream community forests user groups (CFUGs)	Maintain sustainable management and conservation of water supplies and surrounding forests

in the HKH region can be examined on three main scales:

- the direct local impacts of drivers on land, natural resources and their management within a coupled socio-ecological framework (Liu *et al.*, 2007);
- the regional effects of local drivers are primarily mediated by the provision of ecosystem services in the context of upstream and downstream connections (Sati, 2023);
- regional and global impacts are also mediated by remote coupled systems, where decisions and actions taken outside the region have a significant impact on the HKH region (Jing *et al.*, 2020).

23.4.2 Strategy

The changes mentioned above may provide new opportunities for the sustainable development of the HKH's ecosystem services, but may also pose challenges. However, it has become clear that the HKH ecosystem services are declining, and the deterioration of ecosystem conditions could result in severe and irreversible losses, heightening the risk of extreme weather events, the loss of watershed functions, the decline in the stability of agro-ecological systems, and the aggravation of environmental pollution. In addition, the deterioration could cause problems such as food and water shortages, which have an impact on billions of people downstream and, ultimately, imperil human well-being.

To reverse the current dilemma and improve the HKH's ecosystem protection and development, a comprehensive strategy should be formulated encompassing three dimensions: environment, socio-culture and economy. Box 23.4 provides a brief overview of the interconnections between ecosystem services and SDGs in the HKH region. The following broad strategic directions are likely to contribute to the long-term provision of HKH's ecosystem services and, ultimately, promote progress towards the SDGs.

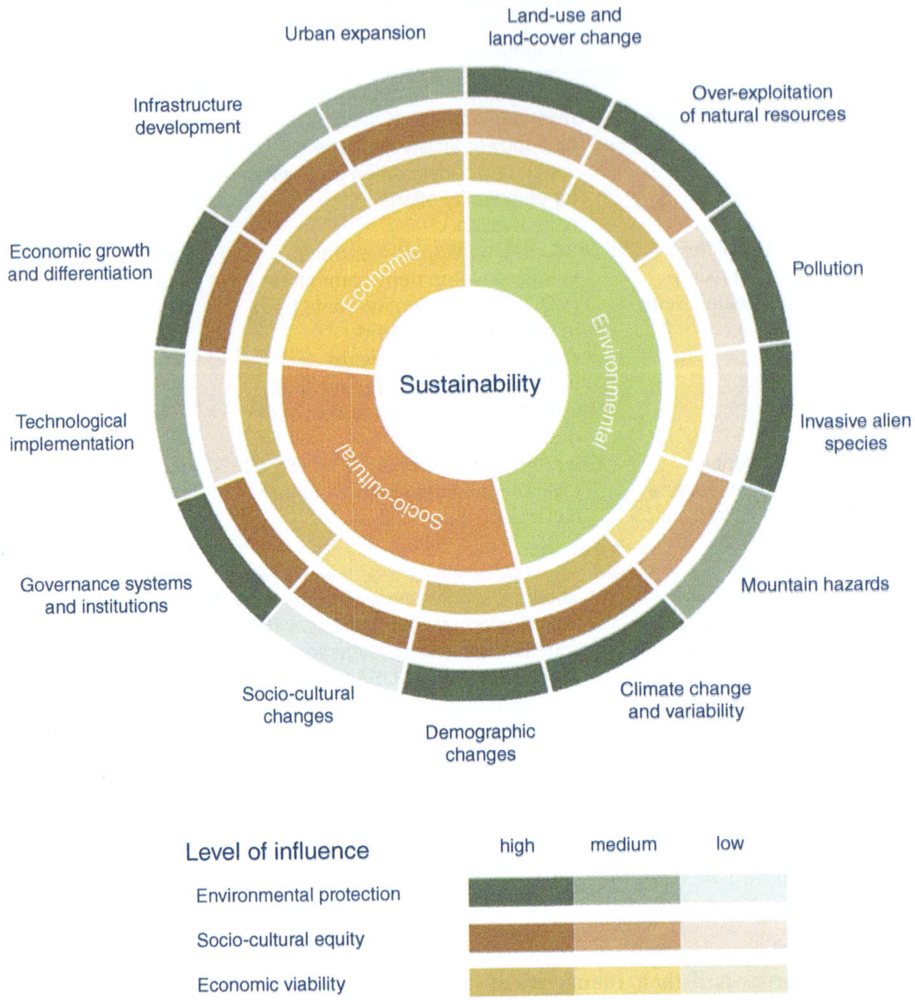

Fig. 23.5. Key drivers affecting sustainability in the Hindu Kush Himalayan region. (Adapted from Jaeger, 2021)

Green resilient ecological management method

The HKH region should analyze environmental changes scientifically by employing a better method of ecosystem services valuation. To better respond to the challenges and opportunities brought on by the rapid changes in the socio-economy, climate and environment, the HKH region should also form a green, inclusive and resilient ecosystem management system and improve ecological compensation to take into account development policies, planning and investment.

The strategies should include, but are not limited to, the following:

1. The high-altitude ecosystems need landscape-based restoration and governance approaches such as PAs and close-to-nature restoration to achieve resilience and multiple socio-ecological benefits.
2. Administrative boundaries for ecosystem conservation should be removed to reduce bias in assessment results and policy formulation due to scope limitations.

Box 23.4. The interconnections between ecosystem services and Sustainable Development Goals in the Hindu Kush Himalayan region.

Ecosystem services in the HKH region contribute directly or indirectly to each of the 16 Sustainable Development Goals (SDGs), with SDG 1 (no poverty), SDG 2 (zero hunger), SDG 3 (health and well-being), SDG 6 (clean water and sanitation) and SDG 15 (living on land) being most prominent. The provisioning services contribute directly to the 16 goals. Regulating and supporting services are critical to the multiple goals of promoting agricultural development (SDG 2), eradication of extreme poverty (SDG 1), conservation of terrestrial biodiversity (SDG 15), economic growth (SDG 8), access to clean drinking water (SDG 6), and renewable energy (SDG 7). Cultural services contribute indirectly to sustainable tourism and income generation (SDG 1), improving food security (SDG 2) and improving health and well-being (SDG 3), primarily through recreational and religious values (Zhang *et al.*, 2022b).

3. Livelihoods should be modified to build a green and resilient mountain economy, and to incorporate ecological industrialization.
4. Incentives should be introduced to strengthen ecosystem management and determine the PES in the HKH region to identify and solve the problems such as ecosystem services beneficiary areas, compensation methods, compensation standards and recipients.
5. For relevant stakeholders, the willingness to pay and the willingness to be paid should be clarified, and third-party evaluations should be strengthened to improve the objectivity, fairness and acceptability of ecological compensation standards.
6. The ecological compensation should be combined with major ecological development strategies of various countries.

Framework for comprehensive social governance

The complex relationship between ecosystems and human well-being is shaping ecosystem services and supporting diversity (Meli *et al.*, 2023), as humans have become a global force in transforming the ecology of the planet. To better understand, predict and successfully manage these ecological processes or changes, it is necessary to have a better understanding of why and how humans are reshaping these ecological patterns, processes and changes. Human socio-cultural dynamics should be considered in ecological theory and practice, as are biological and geophysical variables. In addition to growing synergies between man-made and natural factors, the state of biodiversity and ecosystem services in the HKH region is changing rapidly as a result of weak institutional capacity and inadequate governance.

The response strategies should include, but are not limited to, the following:

1. Gender-inclusive and socially supportive policies should be developed at national and regional levels.
2. Reform-governance policies and institutional and legal frameworks should be improved to address the complexity of ecosystems and people.
3. The sustainable use of resources should support agricultural output, tourism development, and downstream clean energy supply (Rasul, 2014).
4. Research should be integrated and inclusive in light of local realities, so as to avoid being dominated by global perspectives and marginalized by local concerns.
5. The progress of diversified participants in achieving scenario goals should be monitored.

Diversified scientific knowledge

The HKH region is witnessing the emergence of new methods for conserving biodiversity and ecosystems that also mitigate and adjust to environmental change, making it possible to combine the protection of the HKH's rich natural and cultural assets. The development of new technologies over the last few decades, including information technology, geospatial tools, and participatory methodologies, has opened up new avenues for the collection of fundamental data in the HKH region. The involvement of new research institutes and academics, global development partners and residents is essential to the implementation of ecological policies, such as the National Council for Himalayan Research (NMHS) of India and funding programmes of the National Natural Science Foundation of China. Beyond that, the SDGs and progressive outlook offer a fresh perspective on the region's conservation and development pathways.

The response strategies should include, but are not limited, to the following:

1. The long-term data-monitoring ecosystems should be improved, and the study on the temporal and spatial coverage of ecosystem climate response should be strengthened.
2. Evaluation methods should be standardized and models should be established for purposes of reproducibility, comparability and generalization of the findings.
3. The knowledge system in line with the cultural reality of HKH should be integrated to provide a new scientific basis for ecological governance practices.
4. Studies on the cascading effects of ecosystem services on society and well-being should be initiated.

Future-oriented regional co-operation investment

Investments at the national and international levels are necessary to maintain the sustainability of ecosystem services and strengthen ecosystem resilience to support human well-being. Eight neighbouring nations share the HKH ecosystem, which not only creates upstream and downstream connections at the level of ecosystem services, but also enables these nations to forge strong ties in areas like trade, tourism and culture. As a result of the current over-exploitation of resources in the HKH region, ecosystems are deteriorating, necessitating the transition to a more sustainable, low-carbon, and circular-economy model. It is urgently necessary to boost regional co-operation among the eight nations, to increase investments for a greener, more inclusive and climate-resilient environment, and to create an environment that is more favourable to sustainable development in the HKH region. Decision-makers, including individuals, companies and policy-makers should combine actions at various stages of the HKH's development, weigh the benefits and risks, and then create an environment for sustainable and inclusive production and trade of high-value ecological goods and services to better meet challenges, while seizing unique and emerging opportunities.

The response strategies should include, but are not limited to, the following:

1. Large-scale investments in sustainable development should be attracted by mobilizing human resources and focusing on projects to develop natural resources.
2. Bottom-up investments with co-operation across multiple levels should be established to mobilize the participation of non-state development institutions and social entities.
3. Co-operation among countries, enterprises and organizations at the national and international levels should be strengthened to increase investment intensity and narrow the investment gap among countries.
4. Countries should share resources, encourage communication and improve infrastructure for greater trade mobility.

These four strategies, with the interweaving of ecological, socio-cultural and economic dimensions, and actions at individual, local, regional and global levels, would create a more comprehensive ecosystem management system for the HKH region, which can better protect and restore the ecosystem, maintain the flow of ecosystem services, improve relationships between humans and natural systems, fill the knowledge gap and increase diversified investment projects.

References

Abler, D. (2004) Multifunctionality, agricultural policy, and environmental policy. *Agricultural and Resource Economics Review* 33, 8–17.

Adrienne, G.-R., Ariane, W. and Peter, B. (2008) Valuing ecosystem services for sustainable landscape planning in alpine regions. *Mountain Research and Development* 28, 156–165. DOI: 10.1659/mrd.0951

Aryal, J.P., Sapkota, T.B., Khurana, R., Khatri-Chhetri, A., Rahut, D.B. *et al.* (2019) Climate change and agriculture in South Asia: adaptation options in smallholder production systems. *Environment, Development and Sustainability* 22(3), 5045–5075. DOI: 10.1007/s10668-019-00414-4

Aryal, K., Maraseni, T. and Apan, A. (2023) Spatial dynamics of biophysical trade-offs and synergies among ecosystem services in the Himalayas. *Ecosystem Services* 59, 101503. DOI: 10.1016/j.ecoser.2022.101503

Bajracharya, S.R., Maharjan, S.B., Shrestha, F., Guo, W., Liu, S. *et al.* (2015) The glaciers of the Hindu Kush Himalayas: current status and observed changes from the 1980s to 2010. *International Journal of Water Resources Development* 31(2), 161–173. DOI: 10.1080/07900627.2015.1005731

Benham, C.F. (2017) Aligning public participation with local environmental knowledge in complex marine social-ecological systems. *Marine Policy* 82, 16–24. DOI: 10.1016/j.marpol.2017.04.003

Bennett, E.M., Peterson, G.D. and Gordon, L.J. (2009) Understanding relationships among multiple ecosystem services. *Ecology Letters* 12(12), 1394–1404. DOI: 10.1111/j.1461-0248.2009.01387.x

Bhatta, L.D., Vanoort, B.E.H., Rucevska, I. and Baral, H. (2014) Payment for ecosystem services: possible instrument for managing ecosystem services in Nepal. *International Journal of Biodiversity Science, Ecosystem Services & Management* 10(4), 289–299. DOI: 10.1080/21513732.2014.973908

Breed, M.F., Cross, A.T., Wallace, K., Bradby, K., Flies, E. *et al.* (2021) Ecosystem restoration: a public health intervention. *EcoHealth* 18(3), 269–271. DOI: 10.1007/s10393-020-01480-1

Chan, K.M.A., Guerry, A.D., Balvanera, P., Klain, S., Satterfield, T. *et al.* (2012) Where are cultural and social in ecosystem services? A framework for constructive engagement. *BioScience* 62, 744–756. DOI: 10.1525/bio.2012.62.8.7

Chaudhary, S., Mcgregor, A., Houston, D. and Chettri, N. (2015) The evolution of ecosystem services: a time series and discourse-centered analysis. *Environmental Science & Policy* 54, 25–34. DOI: 10.1016/j.envsci.2015.04.025

Chen, Y., Wang, K., Lin, Y., Shi, W., Song, Y. *et al.* (2015) Balancing green and grain trade. *Nature Geoscience* 8, 739–741. DOI: https://doi.org/10.1038/ngeo2544

Chettri, N., Shakya, B., Thapa, R. and Sharma, E. (2008) Status of a protected area system in the Hindu Kush-Himalayas: an analysis of PA coverage. *International Journal of Biodiversity Science & Management* 4, 164–178. DOI: 10.3843/Biodiv.4.3:4

Costanza, R. and Kubiszewski, I. (2012) The authorship structure of 'ecosystem services' as a transdisciplinary field of scholarship. *Ecosystem Services* 1(1), 16–25. DOI: 10.1016/j.ecoser.2012.06.002

Daniel, T.C., Muhar, A., Arnberger, A., Aznar, O., Boyd, J.W. *et al.* (2012) Contributions of cultural services to the ecosystem services agenda. *Proceedings of the National Academy of Sciences* 109(23), 8812–8819. DOI: 10.1073/pnas.1114773109

Díaz, S., Demissew, S., Carabias, J., Joly, C., Lonsdale, M. *et al.* (2015) The IPBES Conceptual Framework – connecting nature and people. *Current Opinion in Environmental Sustainability* 14, 1–16. DOI: 10.1016/j.cosust.2014.11.002

Elias, S.A. (2005) Reference module in earth systems and environmental sciences. In: Douglas, I. (ed.) *Ecosystems and Human Well-being*. Island Press, Washington, DC.

Fei, D., Guiyang, Z. and Shouxian, Z. (2021) Policy demand and development direction of ecological compensation mechanism in China during the 14th Five-Year Plan period. *Jiangxi Social Sciences* 41, 59–69+255 (in Chinese).

Franco, D. and Luiselli, L. (2015) The shared knowledge behind payment for rural ecosystem services: a case study. *International Journal of Environmental Studies* 72, 191–206. DOI: 10.1080/00207233.2014.947727

Gerlitz, J.Y., Hunzai, K. and Hoermann, B. (2012) Mountain poverty in the Hindu-Kush Himalayas. *Canadian Journal of Development Studies/Revue canadienne d'études du développement* 33, 250–265.

Guerry, A.D., Polasky, S., Lubchenco, J., Chaplin-Kramer, R., Daily, G.C. *et al.* (2015) Natural capital and ecosystem services informing decisions: from promise to practice. *Proceedings of the National Academy of Sciences* 112(24), 7348–7355. DOI: 10.1073/pnas.1503751112

Guo, R., Shen, H. and Yang, M. (2016) Studies on ecosystem service value and ecological compensation strategy in Lishui river basin. *Research of Environmental Sciences* 29, 774–782.

He, J., Bu, H., Hu, X., Feng, Y., Li, S. *et al.* (2020) Close-to-nature restoration of degraded alpine grasslands: theoretical basis and technical approach. *Chinese Science Bulletin* 65, 3898–3908 (in Chinese).

Hein, L., Van Koppen, K., De Groot, R.S. and Van Ierland, E.C. (2006) Spatial scales, stakeholders and the valuation of ecosystem services. *Ecological Economics* 57(2), 209–228. DOI: 10.1016/j.ecolecon.2005.04.005

Howe, C., Suich, H., Vira, B. and Mace, G.M. (2014) Creating win-wins from trade-offs? Ecosystem services for human well-being: a meta-analysis of ecosystem service trade-offs and synergies in the real world. *Global Environmental Change* 28, 263–275.

ICIMOD (2023a) *Water, Ice, Society and Ecosystems in the Hindu Kush Himalaya: An Outlook* (Eds P. Wester, S. Chaudhary, N. Chettri, M. Jackson, A. Maharjan, S. Nepal and J. F. Steiner.) International Centre for Integrated Mountain Development, Kathmandu.

ICIMOD (2023b) *ICIMOD Strategy 2030: Moving Mountains*. International Centre for Integrated Mountain Development, Kathmandu.

Jaeger, T. (2021) Leveraging the World Heritage Convention for conservation in the Hindu Kush Himalaya: an independent assessment of natural World Heritage potential. Available at: https://portals.iucn.org/library/node/49840 (accessed 29 August 2023).

Jiang, S., Feng, T., Liu, G. and He, J. (2020) A bibliometric analysis of the application of grassland ecological restoration technology. *Pratacultural Science* 37, 685–702 (in Chinese).

Jiang, W., Wu, T. and Fu, B. (2021) The value of ecosystem services in China: a systematic review for twenty years. *Ecosystem Services* 52(4), 101365. DOI: 10.1016/j.ecoser.2021.101365

Jin, L., Chu, Z. and Zou, C. (2019) Role of various types of eco-compensation in ecological protection and restoration of mountains-rivers-forests-farmlands-lakes-grasslands. *Acta Ecologica Sinica* 39, 8709–8716 (in Chinese).

Jing, S., Liu, J., Yang, X., Zhao, F. Qin, Y. *et al.* (2020) Sustainability in the Anthropocene: Telecoupling framework and its applications. *Acta Geographica Sinca* 11(75), 2408–2416 (in Chinese).

Kandel, P., Chettri, N., Chaudhary, S., Sharma, P. and Uddin, K. (2021) Ecosystem services research trends in the water tower of Asia: a bibliometric analysis from the Hindu Kush Himalaya. *Ecological Indicators*, 121.

Kangas, J. and Ollikainen, M. (2019) Economic insights in ecological compensations: market analysis with an empirical application to the Finnish economy. *Ecological Economics* 159, 54–67.

Kubiszewski, I., Costanza, R., Dorji, L., Thoennes, P. and Tshering, K. (2013) An initial estimate of the value of ecosystem services in Bhutan. *Ecosystem Services* 3(3), e11–e21. DOI: 10.1016/j.ecoser.2012.11.004

Li, P., Jiang, L., Feng, Z. and Yu, X. (2012) Research progress on trade-offs and synergies of ecosystem services: an overview. *Acta Ecologica Sinica* 32, 5219–5229. DOI: 10.5846/stxb201109161360 (in Chinese).

Liu, J., Dietz, T., Carpenter, S.R., Alberti, M. and Folke, C. (2007) Complexity of coupled human and natural systems. *Science* 317(5844), 1513–1516. DOI: 10.1126/science.1144004

Meli, P., Vieli, L., Spirito, F., Reyes-Riveros, R. *et al.* (2023) The importance of considering human well-being to understand social preferences of ecosystem services. *Journal for Nature Conservation* 72(3), 126344. DOI: 10.1016/j.jnc.2023.126344

Mishra, A., Agrawal, N.K. and Gupta, N. (2017) *Building Mountain Resilience: Solutions from the Hindu Kush Himalaya*. International Centre for Integrated Mountain Development (ICIMOD), Kathmandu.

Mishra, S.K., Veselka, T.D., Prusevich, A.A., Grogan, D.S., Lammers, R.B. *et al.* (2020) Differential impact of climate change on the hydropower economics of two river basins in High Mountain Asia. *Frontiers in Environmental Science* 8. DOI: 10.3389/fenvs.2020.00026

O'Connor, B., Bojinski, S., Röösli, C. and Schaepman, M.E. (2020) Monitoring global changes in biodiversity and climate essential as ecological crisis intensifies. *Ecological Informatics* 55, 101033.

Ojha, H.R., Ghate, R., Dorji, L., Shrestha, A., Paudel, D. *et al.* (2019) Governance: key for environmental sustainability in the Hindu Kush Himalaya. In: Wester, P., Mishra, A., Mukherji, A. and Shrestha, A.B. (eds) *The Hindu Kush Himalaya Assessment: Mountains, Climate Change, Sustainability and People*. Springer International Publishing, Kathmandu.

Ouyang, Z., Zheng, H., Xiao, Y., Polasky, S., Liu, J. *et al.* (2016) Improvements in ecosystem services from investments in natural capital. *Science* 352(6292), 1455–1459. DOI: 10.1126/science.aaf2295

Pagiola, S. (2008) Payments for environmental services in Costa Rica. *Ecological Economics* 65(4), 712–724. DOI: 10.1016/j.ecolecon.2007.07.033

Palomo, I., Montes, C., Martín-López, B., González, J. A., García-Llorente, M. *et al.* (2014) Incorporating the social-ecological approach in protected areas in the anthropocene. *BioScience* 64(3), 181–191. DOI: 10.1093/biosci/bit033

Pan, X., Xu, L., Yang, Z. and Yu, B. (2017) Payments for ecosystem services in China: policy, practice, and progress. *Journal of Cleaner Production* 158, 200–208. DOI: 10.1016/j.jclepro.2017.04.127

Peh, K.S.H., Thapa, I., Basnyat, M., Balmford, A., Bhattarai, G. *et al.* (2016) Synergies between biodiversity conservation and ecosystem service provision: lessons on integrated ecosystem service valuation from a Himalayan protected area, Nepal. *Ecosystem Services* 22 (Part B), 359–369. DOI: 10.1016/j.ecoser.2016.05.003

Qian, Z., Li, M., Zhonghua, Z., Wenhua, X., Bingrong, Z. *et al.* (2019) Ecological restoration of degraded grassland in Qinghai-Tibet alpine region: degradation status, restoration measures, effects and prospects (in Chinese). *Acta Ecologica Sinica* 39, 7441–7451.

Rai, R., Zhang, Y., Paudel, B., Acharya, B.K. and Basnet, L. (2018) Land use and land cover dynamics and assessing the ecosystem service values in the trans-boundary Gandaki River Basin, Central Himalayas. *Sustainability* 10(3052), 22. DOI: 10.3390/su10093052

Rasul, G. (2014) Food, water, and energy security in South Asia: a nexus perspective from the Hindu Kush Himalayan region. *Environmental Science & Policy* 39, 35–48. DOI: 10.1016/j.envsci.2014.01.010

Ren, D., Liu, L., Gong, X., Jiang, P., Liu, S. *et al.* (2022) Effect evaluation of ecological compensation for strategic mineral resources exploitation based on VIKOR-AISM model. *Sustainability* 14(23), 15969. DOI: 10.3390/su142315969

Ren, H. and Peng, S. (2019) *Introduction to Restoration Ecology*, 3rd edn. Science Press, Beijing (in Chinese).

Ren, Y., Lu, L., Zhang, H., Chen, H. and Zhu, D. (2020) Residents' willingness to pay for ecosystem services and its influencing factors: a study of the Xin'an river basin. *Journal of Cleaner Production* 268(10), 122301. DOI: 10.1016/j.jclepro.2020.122301

Sati, V.P. (2023) Ecosystem services valuation and payment for livelihood sustainability in the Indian Central Himalayan region. *Journal of Resources and Ecology* 14(3), 468–478. DOI: 10.5814/j.issn.1674-764x.2023.03.004

Sharma, E., Chettri, N. and Oli, K.P. (2010) Mountain biodiversity conservation and management: a paradigm shift in policies and practices in the Hindu Kush-Himalayas. *Ecological Research* 25(5), 909–923. DOI: 10.1007/s11284-010-0747-6

Sharma, E., Molden, D., Rahman, A., Khatiwada, Y.R., Zhang, L. *et al.* (2019) Introduction to the Hindu Kush Himalaya assessment. In: Wester, P., Mishra, A., Mukherji, A. and Shrestha, A.B. (eds) *The Hindu Kush Himalaya Assessment: Mountains, Climate Change, Sustainability and People*. Springer International Publishing, Kathmandu.

Song, X., Peng, C., Zhou, G., Jiang, H. and Wang, W. (2014) Chinese grain for green program led to highly increased soil organic carbon levels: a meta-analysis. *Scientific Reports* 4(1), 4460. DOI: 10.1038/srep04460

Su, B., Xiao, C., Chen, D., Qin, D. and Ding, Y. (2019) Cryosphere services and human well-being. *Sustainability* 11(11), 4365. DOI: 10.3390/su11164365

Tang, W., Wang, Q., Cheng, H. and Zhu, Z. (2023) Performance evaluation of ecological compensation at the county level: a case study of Anyuan County in Dongjiangyuan Watershed, China. *Journal of Resources and Ecology* 14(2). DOI: 10.5814/j.issn.1674-764x.2023.02.004

UNDP (2021) *Creating a Nature-Positive Future: The Contribution of Protected Areas and Other Effective Area-based Conservation Measures*. UNDP, New York.

Ureta, J.C., Motallebi, M., Vassalos, M., Seagle, S. and Baldwin, R. (2022) Estimating residents' WTP for ecosystem services improvement in a payments for ecosystem services (PES) program: a choice experiment approach. *Ecological Economics* 201(2), 107561. DOI: 10.1016/j.ecolecon.2022.107561

Waltham, N.J., Elliott, M., Lee, S.Y., Lovelock, C., Duarte, C.M. *et al.* (2020) UN Decade on Ecosystem Restoration 2021– 2030 – What chance for success in restoring coastal Eeosystems? *Frontiers in Marine Science* 7. DOI: 10.3389/fmars.2020.00071

Wang, H., Dong, Z., Xu, Y. and Ge, C. (2016) Eco-compensation for watershed services in China. *Water International* 41(2), 271–289. DOI: 10.1080/02508060.2015.1138023

Wang, Y., Wu, N., Kunze, C., Long, R. and Perlik, M. (2019) Drivers of change to mountain sustainability in the Hindu Kush Himalaya. In: Wester, P., Mishra, A., Mukherji, A. and Shrestha, A.B. (eds) *The Hindu Kush Himalaya Assessment: Mountains, Climate Change, Sustainability and People*. Springer Nature Switzerland, Cham, Switzerland.

Wei, S. (2023) Sustainable development strategies of eco-tourism in the Third Pole based on environmental carrying capacity assessment study. *Journal of Sustainable Development* 16(2), 85. DOI: 10.5539/jsd.v16n2p85

Wester, S., Chaudhary, N., Chettri, M., Jackson, A., Maharjan, S. *et al.* (2023) Water, ice, society, and ecosystems in the Hindu Kush Himalaya: an outlook. Available at: https://lib.icimod.org/record/36322 (accessed 15 April 2024).

Wood, S.L.R., Jones, S.K., Johnson, J.A., Brauman, K.A., Chaplin-Kramer, R. *et al.* (2018) Distilling the role of ecosystem services in the Sustainable Development Goals. *Ecosystem Services* 29(A), 70–82. DOI: 10.1016/j.ecoser.2017.10.010

Wunder, S. (2015) Revisiting the concept of payments for environmental services. *Ecological Economics* 117, 234–243. DOI: 10.1016/j.ecolecon.2014.08.016

Wunder, S., Engel, S. and Pagiola, S. (2008) Taking stock: a comparative analysis of payments for environmental services programs in developed and developing countries. *Ecological Economics* 65(4), 834–852. DOI: 10.1016/j.ecolecon.2008.03.010

Wunder, S., Brouwer, R., Engel, S., Ezzine-de-blas, D., Muradian, R. *et al.* (2018) From principles to practice in paying for nature's services. *Nature Sustainability* 1, 145–150. DOI: 10.1038/s41893-018-0036-x

Xie, G., Zhang, C., Zhen, L. and Zhang, L. (2017) Dynamic changes in the value of China's ecosystem services. *Ecosystem Services* 26, 146–154. DOI: 10.1016/j.ecoser.2017.06.010

Xu, J., Sharma, R., Fang, J. and Xu, Y. (2008) Critical linkages between land-use transition and human health in the Himalayan region. *Environment International* 34(2), 239–247. DOI: 10.1016/j.envint.2007.08.004

Xu, J., Badola, R., Chettri, N., Chaudhary, R.P., Zomer, R.J. *et al.* (2019) Sustaining biodiversity and ecosystem services in the Hindu Kush Himalaya: mountains, climate change, sustainability and people. In: Wester, P., Mishra, A., Mukherji, A. and Shrestha, A.B. (eds) *The Hindu Kush Himalaya Assessment: Mountains, Climate Change, Sustainability and People*. Springer Nature Switzerland, Cham, Switzerland.

Yang, H.J., Gou, X.H., Yin, D.C., Du, M.M., Liu, L.Y. *et al.* (2022) Research on the coordinated development of ecosystem services and well-being in agricultural and pastoral areas. *Journal of Environmental Management* 304(19), 114300. DOI: 10.1016/j.jenvman.2021.114300

Yin, C., Zhao, W., Ye, J., Muroki, M. and Pereira, P. (2023) Ecosystem carbon sequestration service supports the Sustainable Development Goals progress. *Journal of Environmental Management* 330(8), 117155. DOI: 10.1016/j.jenvman.2022.117155

Yu, H., Xie, W., Yang, L., Du, A., Almeida, C.M.V.B. *et al.* (2020) From payments for ecosystem services to eco-compensation: conceptual change or paradigm shift? *Science of the Total Environment* 700, 134627. DOI: 10.1016/j.scitotenv.2019.134627

Zhang, C., Li, J. and Zhou, Z. (2022a) Ecosystem service cascade: concept, review, application and prospect. *Ecological Indicators* 137(2), 108766. DOI: 10.1016/j.ecolind.2022.108766

Zhang, G., Zhang, Q., Yang, X., Fang, R., Wu, H. *et al.* (2023) Living environment shaped residents' willingness to pay for ecosystem services in Yangtze River middle reaches megalopolis, China. *Geography and Sustainability* 4(2), 213–221. DOI: 10.1016/j.geosus.2023.03.007

Zhang, J., Zhao, X.-Z., Zhou, R., Tian, T., Cui, J.-Y. *et al.* (2021) Labor force transfer, vegetation restoration and ecosystem service in the Qilian Mountains. *Journal of Environmental Management* 288(80), 112387. DOI: 10.1016/j.jenvman.2021.112387

Zhang, J., Zhang, W., Liu, S., Kong, W. and Zhang, W. (2022b) Cryosphere services to advance the national SDG priorities in Himalaya-Karakoram region. *Sustainability* 14(5). DOI: 10.3390/su14052532

Zhang, L., Dang, W., Zheng, H. and Wang, T. (2010) Implimention effect evaluation and mechanism of soil and water conservation eco-compensation in Yellow River basin. *Bulletin of Soil and Water Conservation* 30, 176–181 (in Chinese).

Zhang, W., Xue, X., Peng, F., You, Q. and Hao, A. (2019a) Meta-analysis of the effects of grassland degradation on plant and soil properties in the alpine meadows of the Qinghai-Tibetan Plateau. *Global Ecology and Conservation* 20, e00774. DOI: 10.1016/j.gecco.2019.e00774

Zhang, X., Estoque, R.C., Xie, H., Murayama, Y. and Ranagalage, M. (2019b) Bibliometric analysis of highly cited articles on ecosystem services. *PLOS ONE* 14(2), e0210707. DOI: 10.1371/journal.pone.0210707

Zhao, J., Yang, W., Ji-Shi, A., Ma, Y., Tian, L. *et al.* (2023) Shrub encroachment increases soil carbon and nitrogen stocks in alpine grassland ecosystems of the central Tibetan Plateau. *Geoderma* 433, 116468. DOI: 10.1016/j.geoderma.2023.116468

Zou, C., Wang, Y., Wang, W., Xu, D., Lin, N. *et al.* (2018) Theory of mountain-river-forest-farmland-lake-grass system and ecological protection and restoration research. *Journal of Ecology and Rural Environment* 34, 961–967 (in Chinese).

Index

Note: Page numbers in **bold** type refer to **figures** and Page numbers in *italic* type refer to *tables.*

Printed and bound by CPI Group (UK) Ltd, Croydon, CR0 4YY

13/09/2026

14961109-0003